Probability and Mathematical Statistics (Continued)

LEHMANN • Testing Statistical Hypotheses, *Second Edition*
LEHMANN • Theory of Point Estimation
MATTHES, KERSTAN, and MECKE • Infinitely Divisible Point Processes
MUIRHEAD • Aspects of Multivariate Statistical Theory
OLIVER and SMITH • Influence Diagrams, Belief Nets and Decision Analysis
PRESS • Bayesian Statistics: Principles, Models, and Applications
PURI and SEN • Nonparametric Methods in General Linear Models
PURI and SEN • Nonparametric Methods in Multivariate Analysis
PURI, VILAPLANA, and WERTZ • New Perspectives in Theoretical and Applied Statistics
RANDLES and WOLFE • Introduction to the Theory of Nonparametric Statistics
RAO • Asymptotic Theory of Statistical Inference
RAO • Linear Statistical Inference and Its Applications, *Second Edition*
RAO • Real and Stochastic Analysis
RAO and SEDRANSK • W.G. Cochran's Impact on Statistics
ROBERTSON, WRIGHT and DYKSTRA • Order Restricted Statistical Inference
ROGERS and WILLIAMS • Diffusions, Markov Processes, and Martingales, Volume II: Îto Calculus
ROHATGI • An Introduction to Probability Theory and Mathematical Statistics
ROHATGI • Statistical Inference
ROSS • Stochastic Processes
RUBINSTEIN • Simulation and The Monte Carlo Method
RUZSA and SZEKELY • Algebraic Probability Theory
SCHEFFE • The Analysis of Variance
SEBER • Linear Regression Analysis
SEBER • Multivariate Observations
SEBER and WILD • Nonlinear Regression
SEN • Sequential Nonparametrics: Invariance Principles and Statistical Inference
SERFLING • Approximation Theorems of Mathematical Statistics
SHORACK and WELLNER • Empirical Processes with Applications to Statistics
STOYANOV • Counterexamples in Probability
WHITTAKER • Graphical Models in Applied Multivariate Statistics

Applied Probability and Statistics

ABRAHAM and LEDOLTER • Statistical Methods for Forecasting
AGRESTI • Analysis of Ordinal Categorical Data
AGRESTI • Categorical Data Analysis
AICKIN • Linear Statistical Analysis of Discrete Data
ANDERSON and LOYNES • The Teaching of Practical Statistics
ANDERSON, AUQUIER, HAUCK, OAKES, VANDAELE, and WEISBERG • Statistical Methods for Comparative Studies
ARTHANARI and DODGE • Mathematical Programming in Statistics
ASMUSSEN • Applied Probability and Queues
BAILEY • The Elements of Stochastic Processes with Applications to the Natural Sciences
BARNETT • Interpreting Multivariate Data
BARNETT and LEWIS • Outliers in Statistical Data, *Second Edition*
BARTHOLOMEW • Stochastic Models for Social Processes, *Third Edition*
BARTHOLOMEW and FORBES • Statistical Techniques for Manpower Planning
BATES and WATTS • Nonlinear Regression Analysis and Its Applications
BECK and ARNOLD • Parameter Estimation in Engineering and Science
BELSLEY, KUH, and WELSCH • Regression Diagnostics: Identifying Influential Data

Applied Probability and Statistics (Continued)

BHAT • Elements of Applied Stochastic Processes, *Second Edition*
BHATTACHARYA and WAYMIRE • Stochastic Processes with Applications
BLOOMFIELD • Fourier Analysis of Time Series: An Introduction
BOLLEN • Structural Equations with Latent Variables
BOX • R. A. Fisher, The Life of a Scientist
BOX and DRAPER • Empirical Model-Building and Response Surfaces
BOX and DRAPER • Evolutionary Operation: A Statistical Method for Process Improvement
BOX, HUNTER, and HUNTER • Statistics for Experimenters: An Introduction to Design, Data Analysis, and Model Building
BROWN and HOLLANDER • Statistics: A Biomedical Introduction
BUNKE and BUNKE • Nonlinear Regression, Functional Relations and Robust Methods: Statistical Methods of Model Building
BUNKE and BUNKE • Statistical Inference in Linear Models, Volume I
CHAMBERS • Computational Methods for Data Analysis
CHATTERJEE and HADI • Sensitivity Analysis in Linear Regression
CHATTERJEE and PRICE • Regression Analysis by Example
CHOW • Econometric Analysis by Control Methods
CLARKE and DISNEY • Probability and Random Processes: A First Course with Applications, *Second Edition*
COCHRAN • Sampling Techniques, *Third Edition*
COCHRAN and COX • Experimental Designs, *Second Edition*
CONOVER • Practical Nonparametric Statistics, *Second Edition*
CONOVER and IMAN • Introduction to Modern Business Statistics
CORNELL • Experiments with Mixtures: Designs, Models and The Analysis of Mixture Data, Second Edition
COX • A Handbook of Introductory Statistical Methods
COX • Planning of Experiments
DANIEL • Applications of Statistics to Industrial Experimentation
DANIEL • Biostatistics: A Foundation for Analysis in the Health Sciences, *Fourth Edition*
DANIEL and WOOD • Fitting Equations to Data: Computer Analysis of Multifactor Data, *Second Edition*
DAVID • Order Statistics, *Second Edition*
DAVISON • Multidimensional Scaling
DEGROOT, FIENBERG and KADANE • Statistics and the Law
DEMING • Sample Design in Business Research
DILLON and GOLDSTEIN • Multivariate Analysis: Methods and Applications
DODGE • Analysis of Experiments with Missing Data
DODGE and ROMIG • Sampling Inspection Tables, *Second Edition*
DOWDY and WEARDEN • Statistics for Research
DRAPER and SMITH • Applied Regression Analysis, *Second Edition*
DUNN • Basic Statistics: A Primer for the Biomedical Sciences, *Second Edition*
DUNN and CLARK • Applied Statistics: Analysis of Variance and Regression, *Second Edition*
ELANDT-JOHNSON and JOHNSON • Survival Models and Data Analysis
FLEISS • The Design and Analysis of Clinical Experiments
FLEISS • Statistical Methods for Rates and Proportions, *Second Edition*
FLURY • Common Principal Components and Related Multivariate Models
FOX • Linear Statistical Models and Related Methods
FRANKEN, KÖNIG, ARNDT, and SCHMIDT • Queues and Point Processes
GALLANT • Nonlinear Statistical Models
GIBBONS, OLKIN, and SOBEL • Selecting and Ordering Populations: A New Statistical Methodology
GNANADESIKAN • Methods for Statistical Data Analysis of Multivariate Observations

(cardplate continues in back of book)

Graphical Models in Applied Multivariate Statistics

Graphical Models in Applied Multivariate Statistics

Joe Whittaker
Department of Mathematics
University of Lancaster
UK

JOHN WILEY & SONS
Chichester • New York • Brisbane • Toronto • Singapore

Other Wiley Editorial Offices

John Wiley & Sons, Inc., 605 Third Avenue,
New York, NY 10158-0012, USA

Jacaranda Wiley Ltd, G.P.O. Box 859, Brisbane,
Queensland 4001, Australia

John Wiley & Sons (Canada) Ltd, 22 Worcester Road,
Rexdale, Ontario M9W 1L1, Canada

John Wiley & Sons (SEA) Pte Ltd, 37 Jalan Pemimpin 05-04,
Block B, Union Industrial Building, Singapore 2057

Library of Congress Cataloging-in-Publication Data:

Whittaker, J. (Joe)
Graphical models in applied multivariate statistics / Joe Whittaker.
p. cm.
Includes bibliographical references.
ISBN 0 471 91750 8
1. Multivariate analysis. 2. Graph theory. I. Title
QA278.W55 1989
519.5′35—dc20 89-22679

British Library Cataloguing in Publication Data:

Whittaker, Joe
Graphical models in applied multivariate statistics.
1. Multivariate analysis
I. Title
519.5′35

ISBN 0 471 91750 8

Printed in Great Britain by Courier International, Tiptree, Essex

Contents

Preface

Embedded in the theories of probability and statistics is the fundamental notion of conditional independence; one only needs point to Markov chains and to the statistical concept of sufficiency to make the case. More recently there has developed an increasing recognition of its importance in applied statistics, specifically for describing and modelling multivariate systems. Much of the impetus has come from developments in the analysis of categorical data, where standard statistical practice altered radically during the 1970's and the development of log-linear models made it possible to formulate complex models for the dependences between the variables cross-classifying a contingency table. The elucidation of the family of graphical models as a subset of log-linear models unravelled the connection between these models and the fundamental notion of conditional independence.

Graphical modelling is that body of statistical techniques based on fitting graphical models to data. Its claim for recognition are the benefits of:

- *Interpretation:* central to applied multivariate statistical analysis are the manifest inter-relationships between several variables. To describe these and explain them by conditioning on or controlling for some other factor is the primary objective of graphical modelling. Conditioning is the key theoretical concept underpinning graphical models, and it leads to the global Markov property that entails an explicit set of rules for interpreting the independence graph.

- *Simplification:* any systematic procedure for the analysis of multivariate observations has to condense the data set without eliminating or obscuring any interesting associations. The standard statistical modelling approach, for data summarised in a contingency table, is to select and fit a parsimonious log-linear interaction model. The fact that graphical models span the set of all log-linear models and are easy to fit and interpret, simplifies the whole process of model search and selection.

- *Unification:* graphical models provide a unified framework for the statistical analysis of continuous data summarised by a correlation matrix and discrete data summarised by a contingency table; and this unification suggests generalisation to mixed variable systems.

Even if no simple graphical model furnishes an adequate description of the data set, the exploratory graphical analysis still provides useful information about the relative importance of the interactions, and may well suggest alternative avenues of analysis.

Readership: The book is directed at two classes of readers: the student who requires a course on applied multivariate statistics unified by the concept of conditional independence; and the research worker essentially concerned to apply the techniques of graphical modelling. Writing for the student implied the need for some exercises and solutions, and in part, determined the theorem-proof style. It encourages a more precise statement of the proposition; it makes clear what constitutes a proof; and additionally, as proofs are separate in the text, they can be deferred on a first reading. Writing for the applied worker meant including many case studies, of which some have a substantially practical nature.

We expect the reader to be familiar with concepts from elementary probablity theory such as the notions of random variables, probability density functions, expectation, and correlation; and to have some idea of the elementary theory of statistical inference with its concepts of parameters, repeated sampling, likelihood, estimation and hypothesis testing. Mathematics to the level of a first course on differential calculus and matrix algebra is assumed, but the material on graphs is self-contained.

Acknowledgements: My interest in log-linear models stemmed from a year spent at the University of Chicago and research into graphical models was stimulated by a visit of Terry Speed to Lancaster, and further developed through many contacts with Steffen Lauritzen and Nanny Wermuth.

The book developed from a course on multivariate statistics to final-year undergraduates at the University, and I should like to thank these students for all their helpful comments and remarks. Thanks also to Roger Penn and the ESRC for allowing me to make use of part of their data collected under the project 'Social Change and Economic Life in Rochdale' (G13250011), and to the ESRC again, for a grant (H00232071) enabling graphical modelling workshops which stimulated my interest in applications. I gratefully acknowledge the assistance of the following publishers who have granted permission for the reproduction of extracted material quoted in the text, sources being given in the References at the end of the book: Biometric Society; Biometrika Trust; Royal Statistical Society; Academic Press: London; Butterworths: London; Free Press of Glencoe: London; Griffin: London; Iowa Univ. Press: Iowa; M.I.T. Press: Cambridge, Mass; McGraw-Hill: New York; and Sage Publications: London.

I would like to acknowledge the encouragement and interest of Vic Barnett, Phil Dawid, Anders Ekholm, Antoine de Falguerolles, Antonio Forcina, Tomas Havranek, Peter van der Heijden, Bent Jorgensen, Svend Kreiner, Steffen Lauritzen, Jean Pierre Pages and Andy Scott who have commented on various parts of the manuscript, and the energies of Peter Diggle and Peter Smith, who read it all. Finally, thanks to David Edwards, Brian Francis and Svend Kreiner for use and development of graphical modelling software without which this book would not exist.

JOE WHITTAKER

High Bentham, October 1989

Glossary

MATHEMATICS

$G = (K, E)$ graph, $K = \{1, 2, \ldots, k\}$ vertex set, E edge set. $G^{\prec} = (K, E^{\prec})$ directed graph, G^m, G^u associated moral and undirected graphs.

a, b subsets of K, $\cup$ union, $\cap$ intersection, $a \backslash b$ remainder, a^c complement; $\mathrm{bd}(a)$ boundary of a, $\mathrm{pa}(a)$ parents of a.

g arbitrary real valued function, g_a coordinate projection notation.

Brackets: () for vectors, {} for sets, [] for inner products.

R^p p-dimensional Euclidean space.

RANDOM VECTORS AND DENSITY FUNCTIONS

$X = (X_1, X_2, \ldots, X_k)$ $(= X_K)$ vector of full set of random variables, X_i i-th coordinate of X_K.

X_a sub-vector of X_K containing X_i for $i \in a$, (X_a, X_b) partition of $X_{a \cup b} = X_{ab}$.

f_X density function of X, $f_X(x)$ value at x, f_a marginal density function of X_a, $f_K(x_K) = f_{12\ldots k}(x_1, x_2, \ldots, x_k)$ joint density function, $f_{a\cup b}$ $(= f_{ab})$ marginal density of (X_a, X_b), $f_{b|a}$ conditional density of X_b given X_a, $f_{b|a}(x_b; x_a)$ evaluated at x_b for given x_a.

DISCRETE MODELS

p $(= p_K)$ k-way table of probabilities classified by X, parameter of Multinomial distribution, $r_1, r_2, \ldots, r_k$ number of levels of variables $X_1, X_2, \ldots, X_k$. p_a, p_{ab} marginal tables corresponding to classifying variables X_a and (X_a, X_b), $(p_{a\cup b} = p_{ab})$, $p_a(x_a)$ element of marginal table.

u-terms parameters of log-linear expansion of p_K, u_a, u_{12} coordinate projection notation (more explicitly u_a^K).

n $(= n_K)$ observed table of counts, n_a marginal table, n_ϕ the total count $(=$ sample size $N)$, $n_a(x_a)$ element of marginal table, $n_{b|a}(x_b; x_a) = n_{a\cup b}(x_a, x_b)$ slice of full table, N_K random table of counts.

EXPECTATION

$E(X)$ expectation of X, $\mathrm{var}(X)$ variance, $E_{b|a}(X_b)$ conditional expectation of X_b given X_a (also $E_{Y|X}(Y)$), $\mathrm{var}_{b|a}(X_b)$ conditional variance.

$\mathrm{var}(X_a, X_b)$ variance of (X_a, X_b), $\mathrm{cov}(X_a, X_b)$ covariance between X_a and X_b.

$\mathrm{var}(X_b|X_a)$ partial variance of X_b given X_a, $\hat{X}_b(X_a)$ linear least squares predictor of X_b from X_a.

CONTINUOUS MODELS

$\sim N(\mu, V)$ Normally distributed with mean μ and variance V, $n(x;\mu,V)$ multivariate Normal density function.

V $(= V_{KK})$ variance matrix of X_K of order $k \times k$, V_{aa} $(= V_{a,a})$ variance matrix of X_a, (also V_{XX} for variance matrix of X; $V(X)$, $V(X,X)$ are not defined), V_{ab} covariance matrix between X_a and X_b.

$V_{aa|b}$ partial variance matrix of X_a partialled on X_b.

$D = V^{-1}$ inverse variance matrix, D_{aa} block with coordinates in a of inverse variance matrix of V_{KK},and is *not* the inverse variance matrix of X_a.

v_{ij} covariance of X_i and X_j equals (i,j)-th element of V, d_{ij} (i,j)-th element of D.

$\bar{x}$ sample mean, S sample variance matrix, S_{aa} sample variance matrix of X_a, $\mathrm{var}_N(.)$ variance operator determined by sample variance.

INTERACTION AND INFORMATION

$\mathrm{cpr}(X_1, X_2)$ cross-product ratio, $\mathrm{corr}(X_1, X_2)$ correlation, $R^2(Y;X)$ multiple squared correlation coefficient.

i_{12} mixed derivative measure for X_1 and X_2, $i_{12|a}$ mixed derivative measure for X_1 and X_2 conditioned on X_a.

$I(f,g)$ information divergence between f and g, $I(f,g||X)$ divergence based on X, $I(f^p, f^q)$ (or $I(p,q)$) divergence between two tables of probabilities, $I(f^V, f^W)$ (or $I(V,W)$) divergence between two variance matrices.

INFERENCE

α, β, θ parameters, N sample size $(= n_\phi)$.

M model formula, p^M parameters constrained to M, f^S, f^V density functions with parameters given by S and V.

$\hat{p}$ mle of p, $\hat{V}$ mle of V.

Chapter 1

Introduction

We begin with two examples: an analysis of student marks achieved in some University mathematics examinations followed by an analysis of data on factors relating to infant mortality.

1.1 Mathematics Marks

This data set comes from Mardia, Kent and Bibby (1979) and consists of the examination marks of 88 students in the five subjects mechanics, vectors, algebra, analysis and statistics. Mechanics and vectors were closed book examinations and the remainder were open book. The full set of marks is given in Table 1.1.1. Stem and leaf plots of the marks give a quick assessment of normality and a qualitative overview of the data; for the algebra and statistics marks:

	algebra	statistics
0 -		9
10 -		45778889
20 -	1	012455699
30 -	1266677889	0011122333344556677799
40 -	011333334556666777788899999	000000011234445555556679
50 -	000000111223333344455666778899	0113346
60 -	000111123455578	11233447888
70 -	12	033
80 -	0	1111
90 -		

and suggest that the marginal distributions of marks are fairly well approximated by the Normal distribution. On average, the statistics marks are lower than the algebra marks by about 10 points and also have a wider scatter about their mean.

Table 1.1.1: Marks in five mathematics exams for 88 students. From Mardia, Kent and Bibby (1979).

me	ve	al	an	st	me	ve	al	an	st	me	ve	al	an	st
77	82	67	67	81	30	69	50	52	45	62	44	36	22	42
63	78	80	70	81	46	49	53	59	37	48	38	41	44	33
75	73	71	66	81	40	27	54	61	61	34	42	50	47	29
55	72	63	70	68	31	42	48	54	68	18	51	40	56	30
63	63	65	70	63	36	59	51	45	51	35	36	46	48	29
53	61	72	64	73	56	40	56	54	35	59	53	37	22	19
51	67	65	65	68	46	56	57	49	32	41	41	43	30	33
59	70	68	62	56	45	42	55	56	40	31	52	37	27	40
62	60	58	62	70	42	60	54	49	33	17	51	52	35	31
64	72	60	62	45	40	63	53	54	25	34	30	50	47	36
52	64	60	63	54	23	55	59	53	44	46	40	47	29	17
55	67	59	62	44	48	48	49	51	37	10	46	36	47	39
50	50	64	55	63	41	63	49	46	34	46	37	45	15	30
65	63	58	56	37	46	52	53	41	40	30	34	43	46	18
31	55	60	57	73	46	61	46	38	41	13	51	50	25	31
60	64	56	54	40	40	57	51	52	31	49	50	38	23	9
44	69	53	53	53	49	49	45	48	39	18	32	31	45	40
42	69	61	55	45	22	58	53	56	41	8	42	48	26	40
62	46	61	57	45	35	60	47	54	33	23	38	36	48	15
31	49	62	63	62	48	56	49	42	32	30	24	43	33	25
44	61	52	62	46	31	57	50	54	34	3	9	51	47	40
49	41	61	49	64	17	53	57	43	51	7	51	43	17	22
12	58	61	63	67	49	57	47	39	26	15	40	43	23	18
49	53	49	62	47	59	50	47	15	46	15	38	39	28	17
54	49	56	47	53	37	56	49	28	45	5	30	44	36	18
54	53	46	59	44	40	43	48	21	61	12	30	32	35	21
44	56	55	61	36	35	35	41	51	50	5	26	15	20	20
18	44	50	57	81	38	44	54	47	24	0	40	21	9	14
46	52	65	50	35	43	43	38	34	49					
32	45	49	57	64	39	46	46	32	43					

It is a particularly pleasant homogeneous data set: the array of marks is rectangular, all the variables are measured on the same scale, and there are no missing values.

Scatter diagrams give some idea of the interrelationships between the subjects: they display the substantial correlations between these marks and suggest approximate Normality for the bivariate distribution of marks.

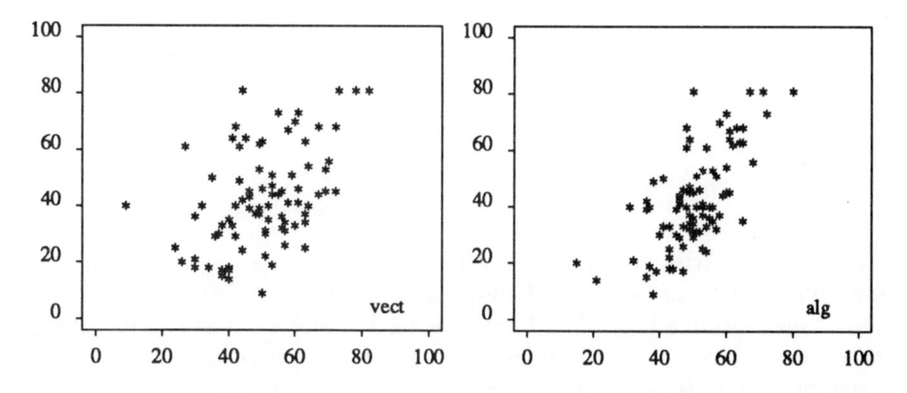

Figure 1.1.1: Scatter diagrams of the statistics marks against vectors and algebra.

The correlation matrix

The scatter diagram is a two dimensional projection of an essentially five dimensional structure. If we can assume that the distribution of the vector of mathematics marks is multivariate Normal then there is no loss of information by condensing these data into the sample mean vector and the sample variance-covariance matrix. Algebra and vectors obtained the highest averages, followed by analysis, statistics and mechanics. The variance matrix is composed of the sample variances along the diagonal and sample covariances in the off diagonal entries. Only the lower triangle is printed as the matrix is symmetric.

Table 1.1.2: The sample variance matrix of the mathematics marks.

	mech	vect	alg	anal	stat
mech	302.29				
vect	125.78	170.88			
alg	100.43	84.19	111.60		
anal	105.07	93.60	110.84	217.88	
stat	116.07	97.89	120.49	153.77	294.37
means	38.95	50.59	50.60	46.68	42.31

This matrix shows that the variance of the marks differs considerably between subjects (in fact the subjects with the lowest average scores have the highest variances). The correlation matrix is obtained from the variance matrix by scaling rows and columns so that the diagonal entries are all unity:

mech	1.0				
vect	0.55	1.0			
alg	0.55	0.61	1.0		
anal	0.41	0.49	0.71	1.0	
stat	0.39	0.44	0.66	0.61	1.0
	mech	vect	alg	anal	stat

That all the entries are positive can be interpreted as a reflection that good students tend to do well on all subjects and bad ones badly. The range of values varies from a high of 0.71 between algebra and analysis to a low of 0.39 between statistics and mechanics. Apart from this it is difficult to discern much pattern in these correlation coefficients as they stand.

The inverse correlation matrix

All the above is well understood in the statistical fraternity. What is less well known is how to interpret the elements of the inverse variance and correlation matrices. First, and most importantly, if an element in the inverse variance matrix is zero then the corresponding variables are conditionally independent given the remaining variables. This immediately motivates the calculation of the inverse variance. Experience dictates that it is easier to interpret a correlation rather than a covariance; so instead the inverse of the correlation matrix is computed:

mech	1.60				
vect	-0.56	1.80			
alg	-0.51	-0.66	3.04		
anal	0.00	-0.15	-1.11	2.18	
stat	-0.04	-0.04	-0.86	-0.52	1.92
	mech	vect	alg	anal	stat

Each diagonal element of the inverse correlation matrix is related to the proportion of variation in the corresponding variable explained by regressing on the remaining variables. More explicitly each diagonal element equals $1/(1-R^2)$ where R is the multiple correlation coefficient between that variable and the rest. For example, the proportion of explained variation for mechanics is $R^2(\text{mech; rest}) = (1.60 - 1)/1.60 = 37.5\%$ while it is $R^2(\text{alg; rest}) = (3.04 - 1)/3.04 = 66.7\%$ for algebra. In fact algebra is the most predictable mathematics mark and mechanics is the least.

The inverse correlation matrix is now scaled to have unit entries on the diagonal, in the same way as a variance matrix is scaled to give a correlation matrix:

mech	1.0				
vect	-0.33	1.0			
alg	-0.23	-0.28	1.0		
anal	0.00	-0.08	-0.43	1.0	
stat	-0.02	-0.02	-0.36	-0.25	1.0
	mech	vect	alg	anal	stat

The off diagonal elements of this scaled inverse correlation matrix are the negatives of the partial correlation coefficients between the corresponding pair of variables given the remaining variables. The correlation between algebra and analysis persists after adjusting for the other variables but the correlation between mechanics and analysis does not.

A graphical model

Unlike the original correlation matrix of the mathematics marks, there is a discernible structure to this matrix: the elements in the lower lefthand block are all near zero, which suggests that the matrix can be approximated by one with the following structure:

mech	*				
vect	*	*			
alg	*	*	*		
anal	0	0	*	*	
stat	0	0	*	*	*
	mech	vect	alg	anal	stat

where * denotes a non-zero entry.

A powerful way to convey this information is to construct a *graph*, in which the vertices of the graph correspond to the variables, and in which the absence of an edge between two vertices corresponds to a zero term in the inverse variance (or scaled inverse correlation). It will transpire that the graph constructed in this manner is nothing but the conditional independence graph for jointly distributed multivariate Normal random variables.

In Figure 1.1.2 we give the independence graph for the mathematics marks. Of the ten possible edges four are missing: mechanics and analysis are independent conditional on the other remaining variables, as are mechanics and statistics, vectors and statistics and finally vectors and analysis.

The most important tool for interpreting independence graphs is the global Markov property: variables are independent conditional on the separating set

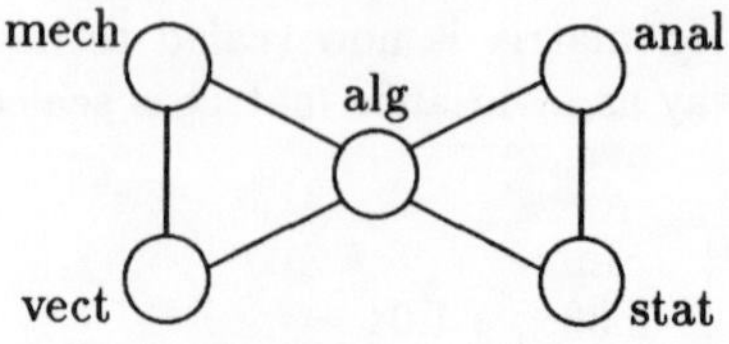

Figure 1.1.2: Independence graph of the mathematics marks.

alone. For example, as the partial correlation between mechanics and analysis is zero then these variables are conditionally independent given the variables, vectors, algebra and statistics. But since algebra alone separates mechanics and analysis it follows that these two variables are independent conditionally on algebra alone.

The graph resembles a butterfly composes of two wings separated by a single vertex for the body. Application of the global Markov property summarises this structure in the form: (mech,vect) is independent of (anal,stat) conditionally on algebra alone. This is an important conclusion for several reasons:

- it reduces the complex five dimensional object into two simpler three dimensional objects;
- it groups the variables into two sets (mechanics, vectors, algebra) and (algebra, analysis, statistics);
- it highlights algebra as the one crucial examination in analysing the interrelationship between different subjects in exam performance;
- it asserts that algebra and analysis alone will be sufficient to predict a statistics mark and that algebra and vectors will be sufficient to predict mechanics; but that all four marks are needed to predict algebra.

Much of this text is concerned with deriving the theoretical properties of the graph constructed in this way and with improved procedures for suggesting, fitting and selecting the underlying graphical model.

1.2 Infant Survival

The following data set has been made famous by the book of Bishop, Fienberg and Holland (1975). The data gives information on the survival rate of 715 infants attending two clinics and the amount of care received by the mother, where the amount of care is classified as either 'more' or as 'less':

	survival		
care	no	yes	(%)
less	20	373	5.1
more	6	316	1.9

From this table one would conclude that the more maternal care received the lower the infant mortality rate, with the rate dropping by more than half. However this is not the whole story: the two-way table relating care and survival has been obtained from a three-way table, Table 1.2.1, classifying infants by care, survival and clinic and then collapsing this table over clinic. This paints an entirely different picture. Within the first clinic the mortality

Table 1.2.1: Counts of infant survival according to maternal care and clinic. From Bishop, Fienberg and Holland (1975).

		survival		
clinic	care	no	yes	(%)
clinic 1	less	3	176	1.7
	more	4	293	1.4
clinic 2	less	17	197	7.9
	more	2	23	8.0

rate for the less care group is practically the same as for the more care group; the same is true for the second clinic. In neither clinic is there a relationship between care and survival, or in other words, given clinic, care and survival are independent. A graph that describes this structure is

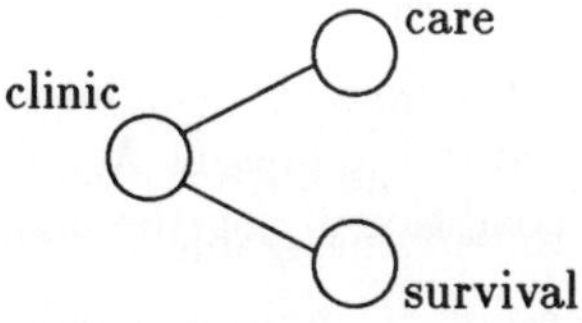

care
clinic
survival

where the vertices correspond to the factors classifying the table and lack of an edge between care and survival indicates that these factors are conditionally independent given clinic.

The two clinics have markedly different survival rates and have very different numbers of patients receiving more care and less care. This could be due to a number of different reasons, one being variation in the social class composition of the patients. A consequence is that when the table is collapsed over clinic a spurious association between care and survival is induced. The lack of independence suggests the graph

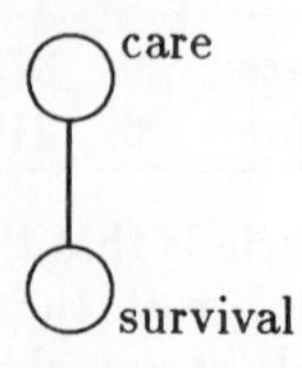

A lesson to learn is that it is dangerous to analyse a three-way table solely by inspecting its two-way margins.

1.3 Graphical Models and Modelling

The basic concept in probability that underpins the theory of graphical models is conditional independence.

Conditional independence

These analyses, of the mathematics marks and of the care-clinic-survival data, are underpinned by the common theoretical concept of conditional independence. To illustrate the extensive nature of conditional independence in the theories of probability and statistics we give some examples.

EXAMPLE 1.3.1 Markov chains. An early, if not the first, explicit use of the notion of conditional independence is Markov (1906) in his paper on Markov chains. Consider a sequence of random variables observed at discrete points in time $\ldots, X_{-1}, X_0, X_1, X_2, \ldots$ and suppose the index t represents the present instance. The Markov assumption is that the conditional probability density function of a future observation at $t+1$ given the past and the present is just dependent on the present, so that, whatever t,

$$f_{t+1}(x_{t+1}|x_t, x_{t-1}, x_{t-2}, \ldots) = f_{t+1}(x_{t+1}|x_t).$$

The past $X_{t-1}, X_{t-2}, \ldots$, has no additional information in it about the future X_{t+1} above that contained in the present X_t. Equivalently, the past and the future are independent conditionally on the present. □

EXAMPLE 1.3.2 Regression. Conditional independence has a role to play in a corner stone of applied statistics: regression analysis. Though not immediately evident in conventional presentations, the choice of variables in a regression equation is an invocation of conditional independence. Suppose the dependent variable, Y, has a Normal distribution with a possible linear regression function

$$E(Y|x_1, , x_2) = \mu + \beta_1 x_1 + \beta_2 x_2,$$

on the explanatory variables x_1 and x_2, and a constant variance σ^2. The test of the hypothesis that $\beta_2 = 0$ is a test that x_2 can be excluded from the equation, and means that after adjusting for x_1, the second variable x_2 has no power to predict Y. A direct reformulation of the hypothesis is that, conditionally on x_1, the distribution of Y is independent of x_2. □

The vast class of essentially multivariate statistical models, which includes topics such as factor analysis, latent structure and latent trait analysis, item analysis, measurement error models, random effect models, variance component models, linear structural and simultaneous equation models, has important applications in education, psychology, agriculture, economics, and sociology. Though apparently disparate these models have a common core: each is composed of latent or unobservable random variables together with independent measurement errors. Here is a simple example:

EXAMPLE 1.3.3 The random effects model. Suppose records are kept over a period of time of the milk yield, Y, of a herd of dairy cows, recording yield at each milking . To analyse such data one might construct a model in which the yield in the j-th milking of the i-th cow is expressed as

$$Y_{ij} = \mu + \alpha_i + \beta_{ij}$$

where the α parameters represent the cow effect and the β parameters the milking effect on each cow. But, on reflection, we would not want to fit this model because there are more parameters than observations, no simplification is taking place and the model is not interesting. So we turn the β's into random variables:

$$Y_{ij} = \mu + \alpha_i + B_{ij},$$

with mean, $E(B_{ij}) = 0$, and variance, $\text{var}\,(B_{ij}) = \sigma_B^2$. The model now states that yield = const + cow effect + random error, which with the assumption of Normal errors is an example of the one-way analysis of variance model with fixed effects, and a perfectly proper model to fit.

While in days gone by, cows had names, Cherry, Daisy, ..., and estimates of specific cow effects made sense, in today's era of agri-business with its larger herds and anonymous cattle, the farmer is no longer interested in individual effects but in the overall magnitude of variation between cows. For the statistician the transition is easy, the specific cow effect α_i becomes a Normally distributed random variable A_i and the model becomes

$$Y_{ij} = \mu + A_i + B_{ij}$$

where $E(A_i) = 0$ and $\text{var}\,(A_i) = \sigma_A^2$. There are now only 3 parameters, μ, σ_A^2 and σ_B^2. There are two sets of random variables, the cow effects, A_i, and the noise, B_{ij}; it is assumed that the variables within each set are mutually independent, and that the sets are independent.

The model has the important corollary: observations on the same cow are correlated. The covariance between the observations Y_{ij} and Y_{ik} is

$$\text{cov}\,(\mu + A_i + B_{ij}, \mu + A_i + B_{ik}) = \text{var}\,(A_i) = \sigma_A^2,$$

as the B's are mutually independent and independent of the A's. Consequently

$$\text{corr}\,(Y_{ij}, Y_{ik}) = \sigma_A^2/(\sigma_A^2 + \sigma_B^2);$$

the correlation between observations on different cows is zero. It is this feature of the model that makes it useful: even today's farmer knows there are high milkers and low milkers, and one beast does not become the other overnight.

But where, one might ask, is the conditional independence structure? Let us look at this from another point of view. We observe a rectangular array of observations $\{Y_{ij}\}$ in which observations on the same cow are correlated, while observations on different cows are not. Our model is to suppose there exist unobserved random variables A_i, one for each cow, that explain both the difference between cows and the similarity within cows. More specifically it is assumed that (i) the A_i's are independent, each coming from the Normal $N(0, \sigma_A^2)$ distribution; (ii) conditionally on given A_i's, the measurements Y_{ij} are identically distributed according to a Normal distribution, $Y_{ij}|A_i \sim N(\mu, \sigma_B^2)$; (iii) conditionally on given A_i's, the measurements Y_{ij} are independent.

The two models outlined here are mathematically identical. Without the last requirement of conditional independence of the observations given the cow effects the model is not identifiable. This second formulation makes explicit the structural independence statements and suppresses the explicit additivity of the Normal linear models. By choosing alternative distributions to the Normal while retaining the conditional independence distribution, the structural features of the model generalise to many of the techniques, such as latent structure analysis, listed above. □

EXAMPLE 1.3.4 Sufficient statistics. A classical example of the application of conditional independence at the foundations of the theory of statistics is the standard definition of a sufficient statistic, see Cox and Hinkley (1974): given a random sample $X = (X^1, X^2, \ldots, X^N)$ of identically distributed observations with a density function indexed by a parameter θ, the statistic T, a function of X, $T = t(X)$, is sufficient for θ if and only if for any other statistic $H = h(X)$ the conditional density $f_{H|T,\theta} = f_{H|T}$. That is, conditionally on T, the statistic H is independent of θ. In this case, all information about the unknown parameter is conveyed in T, and all other statistics are redundant. □

Many students see applied multivariate statistics as a collection or list of statistical methods, assembled together because of their utility as tools, but with little underlying reason, and, as with tools in a tool bag, arranged in no apparent order. To give one example, texts often treat non-metric multidimensional scaling at the same level as they treat factor analysis; but this cannot be right. The former is purely a computational algorithm while the latter concerns the statistical analysis of a fully fledged probability model. Similarly, the meaning of the word 'analysis' in principal components analysis has an entirely different status to its meaning in factor analysis. This lack of unity and rigour is intellectually unsatisfying, and the little unity there is, is restricted to methods based on the variance-covariance matrix assuming Normality. Multivariate statistics needs a wider unifying theoretical framework, in which practical applications can grow, and it is the contention of this text that conditional independence provides such a framework; the theory of independence graphs and their associated graphical modelling applications are just one manifestation.

Graphical models: the ingredients

Graphical models have emerged from a mix of log-linear and covariance selection models, with the constructs of path analysis and the concepts of independence and conditional independence. We give a thumbnail sketch of their genesis.

The past two decades witnessed a rapid growth in the literature on log-linear models and their applications. Beginning in the early papers of Birch (1963, 1964) much of this has been inspired by the pioneering work of statisticians such as Bishop (1969), Goodman (1969, 1970), Haberman (1974) and Bishop, Fienberg and Holland (1975). They have been extensively used by social scientists for the analysis of contingency table data and their development has made it possible to formulate and fit complex patterns of association between the factors cross-classifying a multidimensional table. However it is not always obvious how such a fitted model may be interpreted.

On the other hand, standard analyses of continuous data summarised by a variance or correlation matrix have traditionally depended on techniques that examine linear transformations, for example, principal components or canonical variates. Dempster (1972) suggested an alternative approach that takes the multivariate Normal distribution as a model for the data and specifies that certain elements in the inverse of the variance matrix are zero: the covariance selection models. However, they have not attracted much attention by practising statisticians, though the close analogy between these covariance selection models and the class of log-linear models that exhibit conditional independence was demonstrated by Wermuth (1976a).

An entirely different input stems from the work of Wright (1923, 1934) in the years before the second world war. Essentially a geneticist, he discovered path analysis, path diagrams and established the importance of partial correlation coefficients. The graphic display of the relationship between variables was extremely popular and led to extensive volumes of work in the areas of econometrics and psychological testing. We should not digress to describe these here, many references can be found in Kiiveri and Speed (1982); we only note that another of Goodman's many contributions was to formulate a version of path analysis appropriate to log-linear models and contingency tables, Goodman (1973).

There is substantial scientific importance in developing techniques intended to determine if an observed interaction between two variables can be explained by conditioning or controlling for a third factor. The final necessary ingredient for the development of graphical models was to make the assumption of conditional independence explicit. This already existed in applications to factor analysis and latent structure analysis where the conditioning takes place on an unobserved variable; what was slow in arriving was the recognition that conditional independence has a similar importance for observed variables.

Graphical models

Graphical models are those probability models for multivariate random observations whose independence structure is characterised by a graph, the (conditional) independence graph. The idea of graphs is not new, nor is the idea of conditioning, and nor are log-linear models; but the seminal paper of Darroch, Lauritzen and Speed (1980) put them together in a most illuminating and exciting way. They showed how a subset of log-linear models, the graphical models, can be easily interpreted by application of the Markov properties of their associated independence graphs. The paper summarised and developed collaborative work by Speed and Lauritzen, some of which had been published in Speed (1979) and Lauritzen (1979, 1982). It demonstrated the remarkable Markov properties held by such models, and conclusively demarcated the tractable class of log-linear models with direct estimates through the triangulation theorem: models with direct estimates are those graphical models with triangulated graphs.

Since that time, this area has generated a good deal of activity. An intelligent summary rather presupposes a working understanding of the elements of the theory of independence graphs and graphical models so I refer the reader to a survey paper by Lauritzen (1988) that describes many extensions. Here I shall be content just to list the headings of the research topics and make no attempt to explain the issues: for example: notions of decomposability, reducibility, and collapsibility; extensions to recursive (or directed) and block recursive (or chain) graphs models with a mixture of continuous and discrete random variables; the distinction between graphical and hierarchical models;

relationships to path analysis, causal modelling and latent variable techniques, the development of algorithms for handling graphs and for fitting graphical models; research into the statistical matters of fitting graphical models, model selection techniques, the relative power of procedures, Bayesian and sequential updating methods, methods for handling large numbers of variables; connections to probabilistic expert systems, relational data bases, and spatial analysis; applications in many diverse areas such as sociology, medicine, and psychometrics.

Parametric statistical modelling

A graphical model is a family of probability density functions that incorporates a specific set of conditional independence constraints listed in an independence graph and graphical modelling is the statistical activity of fitting graphical models to data.

In this sense, graphical modelling is just another statistical technique, such as regression analysis, the analysis of variance, log-linear modelling, survival analysis, and time series analysis, to name but a few, whose general motivation and philosophy can be described in the framework of *parametric statistical modelling*, see Cox and Snell (1974) for a fuller exposition. Some of the essential features are described here.

Inference from data proceeds in the following way: the data is represented by a putative *probability model* that includes terms to model the underlying probability mechanism and to model the sampling design that generated the data. This provides a smooth representation of the data which is determined up to the values of some unknown *parameters*. Certain values of these parameters correspond to substantive hypotheses about the subject matter, for example, the omission of a particular variable from an explanatory set. The data and the probability model combine together in the *likelihood function* which gives a measure of relative support to different values of the parameters. There are several schools of thought as to how the likelihood function should be employed, with inference based on either the repeated sampling properties of the function, or on a Bayesian interpretation of the parameters, or even on pure likelihood alone; though there are few differences in practise. In this text we adopt the repeated sampling framework and compare likelihood ratio tests against their theoretical sampling distributions. A *model selection* procedure addresses the problem of choosing a model from a wide class of possible contending models, and, *diagnostic* checks on the model assumptions may suggest improvements to the model first thought of, leading to an iteration of the whole procedure.

In principle we should like to draw *distribution-free* conclusions; for example, we should like to assert that the independence graph of the variables in the mathematics mark set has a butterfly structure, irrespective of the assumption of multivariate Normality. However, to allow the possibility of making inference from small data sets, the position adopted here is to make the distributional assumptions and then use diagnostic procedures to check their credibility. While more research needs to be done in this area for continuous measurements, there is some truth in the claim that our techniques are distribution-free for discrete data. However, one must admit this is more by luck than judgement, because even the most arbitrary distribution for categorical variables requires only a finite number of parameters.

The techniques and principles of graphical modelling fall entirely in this framework: it is nothing more or less than the statistical modelling approach to data analysis based on choosing an independence graph to determine the particular probability model.

1.4 Notational Preliminaries

Every new research area generates its own notational conventions, and the theory of graphical models is no exception, though wherever possible the notation employed in this text has been chosen to conform with the more general statistical literature. The necessary exceptions made to lighten the load are discussed here, and a list is given in the glossary at the end of the preface.

Random variables and density functions

We use upper case for a random variable and lower case for a value that it takes or its realisation. We are only interested in random variables which are either entirely discrete or entirely continuous and use the phrase 'probability density function' and the symbol f, which always denotes a density function, in both cases. It is a function, from either the real line or the non-negative integers to the non-negative reals, and is specified by a formula: $x \rightarrow f(x)$. Thus $f_X(x)$ is the value of the density function f_X, of the random variable X at the value x. The expectation of the random variable is written as $E(X)$ and defined as $\int xf(x)dx$, if X is continuous, and as $\sum xf(x)$, if discrete. We suppose it, and other moments such as the variance, are always well defined. Similarly f_{XY} is the joint density function of the random variables X and Y, and f_X denotes the marginal density function of X, given by integrating or summing out Y from the joint density, and f_Y denotes the marginal density of Y. The conditional density of Y given X is $f_{Y|X} = f_{XY}/f_X$ and its formula is $f_{Y|X}(y) = f_{XY}(x,y)/f_X(x)$. If we wish to make x explicit in the formula we can write $f_{Y|X}(y;x)$. We shall always assume that the density is sufficiently smooth, regular and positive to enable our proofs to go through, and we make no claim to supply the most general statement of results.

Random vectors

A *random vector* is an ordered finite list of random variables, so that if X and Y are two random variables then (X, Y) is a random vector. The vectors (Y, X) and (X, Y) are distinct though the sets $\{Y, X\}$ and $\{X, Y\}$ are identical. This is a somewhat loose usage of the term vector, because vectors consisting of discrete random variables do not, in general, form elements of a vector space, as closure under the operation of scalar multiplication usually fails. Furthermore, we refer to X and Y as random vectors even when they are of different dimensions; and similarly the partitioned vector (X, Y) is itself a random vector. If X is a vector in k-dimensional Euclidean space then we think of it as a column vector, even though for typographical reasons it may be written as a row.

Coordinate projection functions

Let X denote the random vector containing the full complement of random variables under study so that $X = (X_1, X_2, \ldots, X_k)$ where X_i is the i-th coordinate of X. The density function f_X of X is defined as the joint density function $f_{X_1, X_2, \ldots, X_k}$, and is more compactly denoted by $f_{12\ldots k}$.

We shall have occasion to refer to entities such as the partial covariance of two variables given the remaining ones and need to write expressions such as

$$\text{cov}\,(X_i, X_j | X_1, \ldots, X_{i-1}, X_{i+1}, \ldots, X_{j-1}, X_{j+1}, \ldots, X_k).$$

Having wrtten out a few of these, the reader will concur that something better is needed. A now standard device, employed by Darroch, Lauritzen and Speed (1980), is to use set theory at the suffix level to extract the appropriate sub-vector of X. Put $K = \{1, 2, \ldots, k\}$, the index set consisting of the full set of suffices, and let $a = \{i_1, i_2, \ldots, i_p\}$ denote an arbitrary subset of K. Define the random vector X_a as the ordered tuple

$$X_a = (X_{i_1}, X_{i_2}, \ldots, X_{i_p}) = (X_i;\ i \in a).$$

Then $X_{a \cup b}$ and $X_{a \cap b}$ are well-defined, and so are other set operations, in particular, $X_{K \setminus \{i\}}$ denotes the the sub-vector of X obtained by excluding X_i. The full vector is $X_K = X$, and X_ϕ is empty. For example, let $X = (X_1, X_2, X_3)$, so that $K = \{1, 2, 3\}$ and the sub-vector (X_1, X_3) can be denoted by $X_{\{1,3\}}$ or as $X_{K \setminus \{2\}}$. For convenience, we drop the commas and the brackets in the sets and write X_{13} for $X_{\{1,3\}}$. One further remark about partitioning is necessary. In this example, $k = 3$, $a = \{1, 3\}$ and $b = \{2\}$, a and b partition K, so we should like (X_a, X_b) to be identical to (X_1, X_2, X_3). For this to be so a reordering has to take place when the lists of random variables are expanded: thus

$$(X_a, X_b) = ((X_1, X_3), X_2) = (X_1, X_2, X_3) = X.$$

With these conventions, the notation $X_{a\cup b} = (X_a, X_b)$ is consistent and the partial covariance above can be expressed as

$$\operatorname{cov}(X_b, X_c | X_a),$$

where the sets $b = \{i\}$, $c = \{j\}$ and $a = K \backslash \{i, j\}$ partition $K = \{1, 2, \ldots, k\}$. The density function of $X = X_a$ is f_{X_a} but to avoid the double subscript, it is written as f_a. In the example, the density function of $X_{\{1,3\}} = (X_1, X_3)$ is $f_{\{1,3\}}$ which avoids the double subscript, and as above, we drop the commas and the brackets and write f_{13} for $f_{\{1,3\}}$. More generally the density function of $X_{a\cup b} = (X_a, X_b)$ is denoted by $f_{a\cup b}$ or f_{ab}. The partition may vary in the course of an argument or derivation. For example, at one time we may let $a = \{1, 3\}$ while at another $a = \{2, 3\}$.

We shall make the same notational conventions for finite multi-way tables of probabilities and table of counts. The variables in the vector X classify the dimensions of the table and observing $X = x$ determines to which cell of the table the observation belongs. If X is k-dimensional and classifies a k-way table of probabilities, p_K, and X_a is a marginal sub-vector of X, then p_a is the marginal table of probabilities. Similarly if n_K is a table of counts then n_a is the marginal table of counts with dimension given by the number of elements in the subset a. The probability attached to a cell of a k-dimensional table is $p_K(x)$, where x is the value of the k-dimensional vector X. Thus in a two-way table classified by X_1 and X_2, the reader familiar with the notation p_{ij} to denote the entry in the (i, j)-th cell, has to adapt to $p_{12}(i, j)$, or more precisely, $p_{12}(x_1, x_2)$.

Now because f_a is the marginal density of X_a obtained by integrating out the other variables f_K, there is a close relationship between f_K and f_a :

$$f_a(x_a) = \sum f_K(x_a, x_b),$$

where the sum is taken over all variables with coordinates in $b = a^c$. We suppose that exactly the same relationship holds between p_a and p_K, for tables of probabilities, and between n_a and n_K, for tables of counts.

A similar notational device is useful in a more general context, especially when discussing the factorisation criterion. Suppose g is an arbitrary function of x, where x is k-dimensional, with formula $g(x)$, but with the special *coordinate projection* property, that the function only depends on a particular subset of coordinates, x_a of x. Then we write g as g_a and $g(x)$ as $g_a(x)$. This function then has the property that if $x = (x_a, x_b)$ then

$$g_a(x) = g_a(x_a, x_b) = g_a(x_a),$$

and is invariant to the value of x_b.

The specific functions for probability densities, f, tables of probabilities, p, and counts, n, behave slightly differently. Firstly note that $f_a(x)$ is not defined, while $f_a(x_a)$ and $f_K(x_a, x_b)$ are defined and related; and secondly, that the existence of g_a has no implication for the existence of g_K let alone a relationship between the two.

This notation has the one disadvantage that all the action takes place at the suffix level, so to get over this we sometimes adopt the the following shorthand. Rather than partition $X_K = (X_a, X_b)$, we sometimes write $X_K = (Y, Z)$ where $Y = X_a$ and $Z = X_b$ and suppress reference to a and b. The coordinate projection property can be expressed by: if $X_K = (Y, Z)$ then $g_Y(y, z) = g_Y(y)$.

The covariance

Let X_a and X_b denote two random vectors, not necessarily of the same dimension. We shall define the covariance in the standard way: $\text{cov}(X_a, X_b) = E(X_a - EX_a)(X_b - EX_b)^T$, in terms of the expectation operator. This in turn defines the variance of X_a from $\text{var}(X_a) = \text{cov}(X_a, X_a)$ and the variance of (X_a, X_b) as the partitioned matrix

$$\text{var}(X_a, X_b) = \begin{pmatrix} \text{var}(X_a) & \text{cov}(X_a, X_b) \\ \text{cov}(X_b, X_a) & \text{var}(X_b) \end{pmatrix}.$$

Note that $\text{var}(X_a, X_b)$ and $\text{cov}(X_a, X_b)$ are entirely different objects, in fact

$$\text{var}(X_a, X_b) = \text{cov}((X_a, X_b), (X_a, X_b)).$$

Many of our results will be phrased in terms of operators such as these. For example, the linear least squares predictor of X_b from X_a is

$$EX_b + \text{cov}(X_b, X_a)\text{var}(X_a)^{-1}(X_a - EX_a),$$

where E denotes the expectation operator.

An alternative expression for such a formula is

$$\mu_b + V_{ba}V_{aa}^{-1}(X_a - \mu_a),$$

where μ_a and μ_b are vectors of mean values of X_a and X_b, and V is the variance covariance matrix of X_a and X_b, appropriately partitioned as

$$V = \begin{pmatrix} V_{aa} & V_{ab} \\ V_{ba} & V_{bb} \end{pmatrix}.$$

But there are subtle differences between these two expressions. For example, it is not easily possible to analyse the behaviour of the latter as X_b varies. We shall use V for the variance matrix regarded as a parameter of the density function, and var (.) as an operator on a random vector.

1.5 Overview

The skeleton of this book is approximately described by the directed graph of its chapter headings.

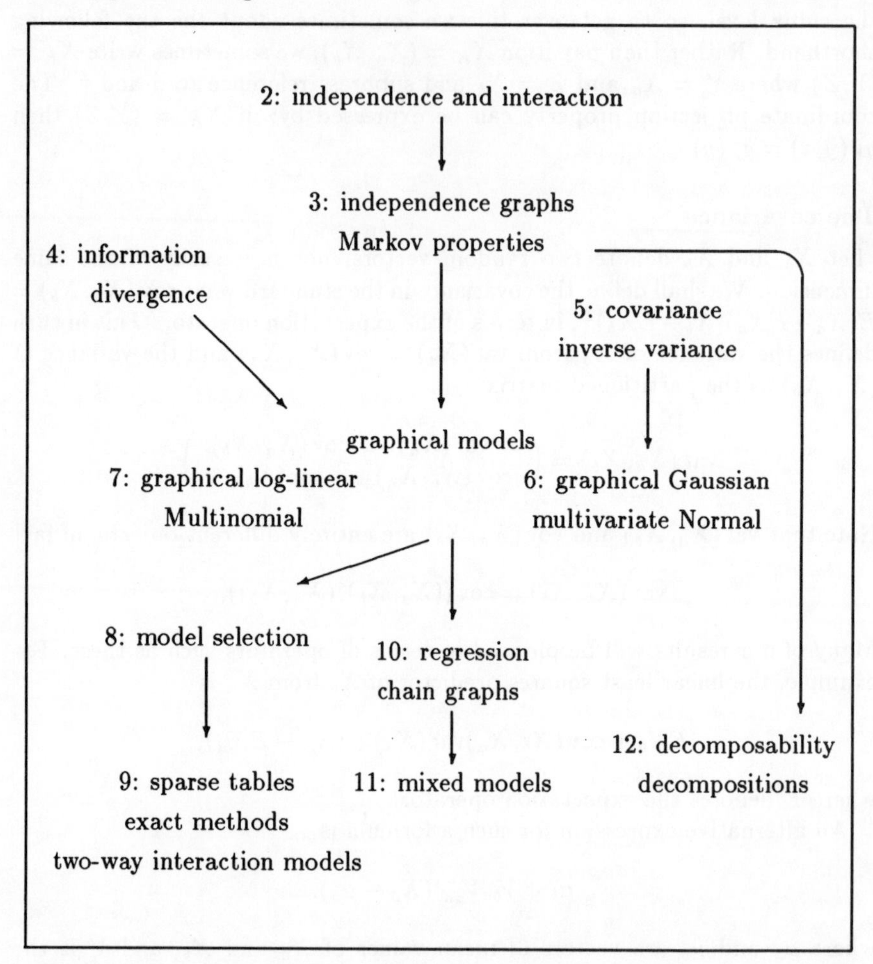

Chapter 2: Independence and Interaction. The chapter begins with the calculus of independence and conditional independence, first of events and then of random variables and random vectors. Important preliminary spadework needed to establish the essential characteristics of independence graphs is done here, and is summarised in three lemmas: the factorisation criterion, the reduction lemma and the block independence lemma.

The factorisation criterion is used to motivate a method of constructing measures of interaction and conditional interaction between random variables: the mixed derivative measure.

The Normal and Bernoulli probability models serve as paradigms for observations on continuous and discrete random variables respectively. Application of the factorisation criterion for independence to two dimensional random variables leads to the correlation coefficient and the cross-product ratio as measures of dependence. In three dimensions the criterion for conditional independence leads to partial correlations and conditional cross-product ratios. The relationship between interactions in two and in three dimensions raises the issue of when a table or correlation matrix is collapsible onto its margins. The infant survival data is an example of Simpson's paradox which arises when the table is not collapsible.

Chapter 3: Independence Graphs. The conditional independence graph is a powerful way of summarising the interactions manifest within a set of variables and the chapter shows how such graphs are constructed from pairwise conditional independences. The global Markov property is the tool needed to interpret the graph and the separation theorem shows that this follows from the pairwise independence property. Proofs are based on the factorisation criterion for conditional independence which allows a unified treatment of the probability theory for graphs composed of discrete and of continuous variables. Directed graphs provide models for studying causal mechanisms, and the final sections of the chapter extend this theory to independence graphs with a mixture of directed and undirected edges.

Chapter 4: Information Divergence. Not only do we need to measure the distance between two random variables but also how close two probability distributions can be to each other. The natural measure is the Kullback-Leibler information divergence. It has two basic roles to play in graphical modelling, one probabilistic and one statistical. The first application of the divergence is as a means to measure the strength of an edge connecting two variables in an independence graph. The second, entirely different application, is to represent the likelihood function as a divergence, which neatly solves some maximum likelihood problems. The chapter reviews the derivation of the divergence from first principles and then discusses its application as a measure of interaction strength.

Chapter 5: The Inverse Variance. This chapter provides the necessary machinery for an adequate mathematical treatment of the multivariate Normal distribution. Linear least squares theory is of interest in its own right and is closely related to the regression techniques of applied statistics. Partial covariance is the theoretical concept that leads to the inverse variance lemma and shows how the elements of the inverse matrix are to be interpreted. The simple criterion for conditional independence in the multivariate Normal distribution is phrased in terms of the inverse variance matrix, and a block version of the Cholesky decomposition shows why. This derivation of the inverse of a partitioned matrix demonstrates the connection between the inverse variance and partial correlation.

Chapter 6: Graphical Gaussian Models. The famous elementary result that uncorrelated Normal random variables are independent suggests that the independence structure of the full set is determined by the variance matrix; this, indeed, is the case: the independence graph of a jointly Normal set of random variables is determined by zeros in the inverse variance. For this reason, they are sometimes known as covariance selection models, after Dempster (1972). The first few sections of this chapter review the basic theory associated with the multivariate Normal distribution, especially the fact that the distribution is closed under marginalisation and conditioning, and evaluation of the information divergence. The standard machinery of maximum likelihood estimation is used to fit a graphical model to data: surprisingly it turns out that the likelihood function can be represented as an information divergence and this gives us a cheap way to extract some of the answers. The edge exclusion deviances are important tools for model selection. The chapter ends with a brief discussion of diagnostics and improved likelihood ratio tests.

Chapter 7: Graphical Log-linear Models. There is a remarkable unity between graphical log-linear and Gaussian models, both in terms of the underlying independence structure and in terms of the mathematics of inference, and a virtually parallel treatment of the two models is possible. The sampling model presumes that each observation is an independent realisation from the cross-classified Multinomial distribution. Two parameterisations of the distribution are discussed: the table of probabilities and the u-terms in the log-linear expansion of the table. The additional richness of models for categorical data is demonstrated by embedding the set of all graphical models within the set of all hierarchical log-linear models. The likelihood function is maximised by routine differentiation as well as using properties of the divergence.

Chapter 8: Model Selection. Even in a moderate number of dimensions there are thousands of possible Gaussian and log-linear models to choose from, many of them very similar. The problem of finding a well fitting parsimonious model is discussed in the context of case studies. It is seen that simple variants of stepwise edge inclusion and deletion procedures reach sensible conclusions relatively quickly. While there is a theoretical preference for backwards elimination, it is surprising how well forwards inclusion does in practice.

Chapter 9: Methods for Sparse Tables. As the number of observed categorical variables increases the number of cells with no entries rapidly grows and the procedure of fitting graphical log-linear models runs into the problems of infinite parameter estimates. We consider two techniques for dealing with the problems posed by sparse tables: one is to restrict attention to two-way interaction models, the other is to condition on the margins of the table and to perform exact tests. The sufficient statistic for the all two-way interaction model is the set of all two-way marginal tables, but, unlike the full table, these are not sparse. Each two-way interaction model has an associated independence graph, which sets up a direct correspondence to a graphical log-linear model. An illustration of the techniques is provided by the analysis of an original data set, the Rochdale survey. The chapter then turns to a discussion of exact tests for conditional independence based on the Hypergeometric distribution, as an alternative method of analysing sparse tables.

Chapter 10: Regression and Graphical Chain Models. New models are required when the variables may be partitioned into response and explanatory variables. The simplest instance of this situation is the classical Normal linear regression model with one dependent variable and p covariates; in which the Normal distribution models the conditional distribution of the response. The relevant property of the independence graph is the local Markov property. More generally, for both discrete and continuous observations, and for multivariate responses, regression models are models for conditional distributions. The techniques generalise to chain graphical models, appropriate for a partition of the variables into a chain of blocks, and formed by combining the regression models for each block on covariates chosen from the preceding blocks. The chain independence graph has a mix of directed and undirected edges, and is summarised by the Markov properties of the associated moral graph developed in Chapter 3.

Chapter 11: Models for Mixed Variables. Studies in which some variables are discrete and some are continuous, arise in many observational and experimental set-ups and require mixed graphical models for their analysis. Several well known statistical techniques are related to this division, for instance, linear models for the analysis of variance, which have continuous response and discrete treatment variables; logistic regression models, which have discrete response and continuous treatment variables; and discriminant analysis. These are unified by the notion of hierarchical mixed interaction models.

Independence graphs for mixed variables are enhanced by marking vertices in a manner corresponding to the two types of variables. While the basic Markov properties of graphs goes through without alteration, the statistical analysis poses new problems. The essential difficulty lies in specifying a flexible family of parametric models, rich enough to encompass the two type of vertices and simple enough to deal with tractably. Substantial progress has been made by Lauritzen and Wermuth (1984, 1989) who introduce the *conditional Gaussian* family of distributions, which has the necessary property, of containing the special cases of the Normal and Multinomial distributions. A problem that arises with these models is closure under marginalisation. One is either faced with the analysis of mixture distributions, which are notoriously problematic as they fall outside the exponential family of distributions; or one must place restrictions on the types of graphs allowed to ensure they do not arise.

We do not attempt a full exposition of the mixed case as a general treatment would overwhelm the essential simplicity of the special cases corresponding to the multivariate Normal and Multinomial distributions.

Chapter 12: Decompositions and Decomposability. The last chapter of the book gives a discussion of decomposability, a notion of some practical importance. It turns out that those log-linear models which possess analytic, or direct, maximum likelihood estimates are decomposable, and that there is a simple characterisation of the independence graph of a decomposable model: it is triangulated. Our discussion is based on the existence of factorisations of the density function into products of marginal density functions. The treatment given here is an introduction to more advanced material available in the research literature, and in particular, we only consider the theory for unmarked and undirected graphs.

Computing. A major ingredient of the text is the analysis of practical examples, and it is hoped that the reader will duplicate and extend these applications. The appendix discusses computing requirements. No one software package unifies all the available methodology to fit, test and diagnose all graphical models, and so a brief introduction to using GLIM, MIM, and EXA, to fit and test a graphical model is given. Further examples can be obtained on disc by writing to the author.

Exercises. Exercises are included at the end of each chapter and outline solutions to some of these are given in an appendix. The questions involving data presume that the student has ready access to computing packages.

Readership. The reader whose central interest is in the applications of graphical models can limit the underlying theory to a brief perusal. The first few sections of Chapter 3 give a good idea of an independence graph, the first few of Chapters 7 and 8, give a quick snapshot of graphical models for discrete and continuous data. Many applications and applied techniques are discussed in Chapters 8 on model selection, 9 on sparse tables, 10 on regression and chain structures, and 11 on mixed models. In contrast, a one term course for final year undergraduates at Lancaster consists of the material in the first seven chapters.

Chapter 2

Independence and Interaction

Independence is cemented into the very foundations of probability theory: it is a theme that recurs in the far frontiers of research, and one that permeates all applications of probability to scientific investigation. Early ideas of independence can be discerned in the calculations of the early probabilists of the 16th, 17th and early 18th centuries; in particular, ideas associated with the names of Bernoulli, Montmort, Pascal and De Moivre, who formulated the calculus of probability and first enunciated the notions of independent events and of conditional probability. But surprisingly, it is not until the late 19th century and early 20th with Markov's first paper and Galton's work on genetics, that the concept of a a probability model for dependent observations occurs. The latter discovered the correlation coefficient, and the symbol 'r' was taken from the initial of 'reversion'. The partial correlation coefficient was discovered at the beginning of this century by Karl Pearson, but the corresponding measures for categorical data had to wait well into the 20th for discovery, see Goodman and Kruskal (1979) for a review.

This chapter first considers independence and conditional independence, of events and then of random variables and vectors. Three important lemmas are proved: the factorisation criterion, which provides a simple check for independence, the reduction lemma which enables conclusions to be drawn about marginal distributions, and a technical result, here called the block independence lemma. These results are used in the following chapter to prove the most characteristic property of independence graphs, the separation theorem. The factorisation criterion is used to motivate a method of constructing measures of interaction and conditional interaction between random variables: the mixed derivative measure.

The standard probability model for the analysis of multivariate categorical data presented in the form of contingency tables is the cross-classified Multinomial distribution. Here we look at a special case appropriate for binary data, the Bernoulli distribution, and its analogue for the analysis of continuous data, the Normal distribution. This chapter introduces the Bernoulli and the Normal random variable, their density functions and extensions to two and three dimensions. It is shown that the mixed derivative measure of interaction applied to these distributions generates the cross-product ratio and a measure closely related to the correlation coefficient. Applied to three dimensional versions it elicits the partial correlation coefficient and the conditional cross-product ratios as the corresponding measures of conditional dependence, and the notion of collapsibility is related to the equality of these interaction measures in the joint and marginal distributions. Simpson's paradox illustrates that they can substantially differ.

2.1 Independent Events

We begin with a definition of independent events. Let P be a probability on the sample space of all possible outcomes to a probability experiment and let A and B denote two events defined on the space. The conditional probability of A given B is $P(A|B) = P(A \cap B)/P(B)$ and is defined only if $P(B) > 0$.

Definition. The events A and B are ***independent*** if and only if $P(A \cap B) = P(A)P(B)$. This is denoted by $A \perp\!\!\!\perp B$. □

Independence is a relation between events. It is not reflexive because A is not independent of A, nor transitive because $A \perp\!\!\!\perp B$ and $B \perp\!\!\!\perp C$ do not imply that $A \perp\!\!\!\perp C$. On the other hand the relation is symmetric, so that

$$A \perp\!\!\!\perp B \text{ implies } B \perp\!\!\!\perp A;$$

even though an equivalent, apparently non-symmetric, formulation of the definition is

$$A \perp\!\!\!\perp B \text{ if and only if } P(A|B) = P(A).$$

Independence also implies $A \perp\!\!\!\perp B^c$, where B^c is the complement of B, so that the independence of A and B is equivalent to the independence of A^c and B^c. Another formulation is that

$$A \perp\!\!\!\perp B \text{ if and only if } P(A|B) = P(A|B^c).$$

These equivalences suggest that independence is a stronger assumption than first meets the eye.

The odds and cross-product ratios

If A is an event with probability $P(A)$ then the *odds* of A is defined as the ratio $P(A)/P(A^c)$. It gives a basic assessment of whether A or A^c is the more probable and it is unity if and only if they are equi-probable. The conditional odds ratio of A given an event B,

$$\text{odds}\,(A|B) = \frac{P(A|B)}{P(A^c|B)},$$

just conditions the probability of A on B. The *cross-product ratio* between the events A and B is defined as

$$\text{cpr}\,(A, B) = \frac{P(A \cap B)P(A^c \cap B^c)}{P(A^c \cap B)P(A \cap B^c)},$$

and is one of many measures of association between the events A and B. The cross-product ratio is unity if and only if the events A and B are independent, see Exercise 2, so that the value $\text{cpr} = 1$, or equivalently $\log \text{cpr} = 0$, is a natural base to assess departures from independence. It can be computed from the conditional probabilities of A and A^c given B and B^c because

$$\text{cpr}\,(A, B) = \frac{P(A|B)P(A^c|B^c)}{P(A^c|B)P(A|B^c)},$$

or from those of B and B^c given A and A^c, and so it can be interpreted as a measure of dependence, as well as association. It compares the odds of A given B with the odds of A given B^c because

$$\text{cpr}\,(A, B) = \text{odds}\,(A|B)/\text{odds}\,(A|B^c),$$

and as it is symmetric it also contrasts the odds of B given A with the odds of B given A^c.

EXAMPLE 2.1.1 As illustration of these ideas consider a fictitious example involving electoral support of British political parties. The joint and conditional probabilities of selecting a voter for one of two parties classified by class are

(a) joint probabilities

	party	
class	Labour	Tory
working	0.42	0.28
bourgeoisie	0.06	0.24

(b) conditional probabilities

	party		
class	Labour	Tory	
working	0.60	0.40	1.00
bourgeoisie	0.20	0.80	1.00

so the cross-product ratio between A=Labour and B=working class voters can be calculated from either (0.41)(0.24)/(0.06)(0.28) or (0.60)(0.80)/(0.20)(0.40), and equals 6, rather far from 1. As $\text{cpr}(A^c, B^c) = \text{cpr}(A, B)$ this ratio is identical to the corresponding ratio between the Tory party and the bourgeoisie. Thus it is not so much a measure of the interaction between Labour and working class voters as a measure of the interaction between social class and party affiliation. The cross-product ratio is just as well calculated from the conditional probabilities of class given party as from the conditional probabilities of party given class. Can we conclude that 'class support of the parties' is identical to 'party support of the classes'? □

There are several directions in which to generalise the cross-product ratio. One extension is to non-binary classifying factors, where the events A and A^c are replaced by disjoint events, $A_1, A_2, \ldots, A_k$, say. Another is to three dimensional tables and to consider cross-product ratios between two events conditioned on a third event, $\text{cpr}(A, B|C)$ say. We take this up in the course of this chapter.

Interpreting a cross-product ratio in terms of counts

One way to get a feeling for the value of a cross-product ratio is to construct a table of counts that has an interaction with that value. There are an infinite number of tables with the same value of the interaction, but only one with given margins. Hence a sensible comparison to make is between tables with the same margins but with different interactions. Finding such tables is not numerically trivial, but can be easily acomplished by the iterative proportional fitting procedure, discussed in Section 4.7.

EXAMPLE 2.1.2 The following four 2×2 tables, classified by variables taking the values 0 and 1, have equal row and column margins and log-cpr's of 0.5, 1.0, 1.5 and 2.0 respectively:

0.5	0	1
0	11.244	8.756
1	8.756	11.244

1.0	0	1
0	12.449	7.551
1	7.551	12.449

1.5	0	1
0	13.586	6.419
1	6.414	13.581

2.0	0	1
0	14.625	5.382
1	5.375	14.618

The convergence is not quite complete in the tables with the larger cprs. Tables exhibiting negative values of the log-cpr are obtained by interchanging rows, which preserves the equality of marginal totals. □

Mutual independence

Three events A, B and C are ***mutually independent*** if and only if each pair of events is independent and $P(A \cap B \cap C) = P(A)P(B)P(C)$. Even if all pairs are independent it may be that the three events are not mutually independent; consider the following counter-example.

EXAMPLE 2.1.3 Marginal does not imply mutual independence. A probability experiment consists of throwing two fair dice, and all the probabilities are determined from the premiss that each of the 36 elementary outcomes is equally probable. Suppose A denotes the event that the 'first die is even', B the event that the 'second die is odd' and C the event that 'either both dice are even or both are odd'. Each pair of events is independent, for example,

$$P(A) = 18/36 = P(B) = P(C) \text{ and}$$
$$P(A \cap C) = P(\text{both are even}) = 9/36 = (1/2)(1/2) = P(A)P(C).$$

But $P(A \cap B \cap C) = 0$ and not the 1/8 needed if the events were mutually independent. □

Underlying this example is the fact that the relations $A \perp\!\!\!\perp B$ and $A \perp\!\!\!\perp C$ do not imply that $A \perp\!\!\!\perp B \cap C$. As we want the statement that A is independent of both B and C to carry this implication we *define* $A \perp\!\!\!\perp [B, C]$ to mean that A is independent of each event in the partition of the sample space generated by B and C. Consequently, as a matter of definition

$$A \perp\!\!\!\perp [B,C] \iff A \perp\!\!\!\perp B \cap C,\ A \perp\!\!\!\perp B \cap C^c,\ A \perp\!\!\!\perp B^c \cap C,\ A \perp\!\!\!\perp B^c \cap C^c.$$

Now taking unions of sets in this partition generates a class containing $4^2 = 16$ sets; one set, for example, is $B = (B \cap C) \cup (B \cap C^c)$; equivalently, this class of sets can be generated by taking finite complements and intersections of the sets B, and C. It is left as an exercise to the reader, Exercise 4, to show that $A \perp\!\!\!\perp [B, C]$ is a powerful enough supposition to deduce that A must be independent of every event in this class. A similar proposition, but one going the other way, and also left as an exercise to the reader, is to show that

$$A \perp\!\!\!\perp B,\ A \perp\!\!\!\perp C,\ A \perp\!\!\!\perp B \cap C \text{ together imply } A \perp\!\!\!\perp [B,C].$$

Mutual independence implies pairwise independence for all pairs of events. In general, the mutual independence of a set of events is defined in terms of the independence properties of its subsets. We break off the discussion here because mutual independence permits no dependences whatsoever and so is uninteresting for the applications we have in mind.

Conditional independence

Conditional independence is a more fruitful way to generalise the relation between two events to the inter-relationships within a class of events. There is a weak and a strong version of the definition, first the weak: the events A and B are *conditionally independent* given the event C (weakly) if and only if $P(A \cap B|C) = P(A|C)P(B|C)$. This is written as $A \perp\!\!\!\perp B|C$. It is assumed that $P(C) > 0$.

It is nothing more than a rewrite of the definition of independence with the unconditional probabilities replaced by conditional probabilities; and if C has probability one then it is just a restatement of the definition of independence. It is a symmetric relation between A and B because $A \perp\!\!\!\perp B|C$ implies $B \perp\!\!\!\perp A|C$. It also implies that $A \perp\!\!\!\perp B^c|C$; however it neither implies nor is implied by $A \perp\!\!\!\perp B|C^c$. To assume that $A \perp\!\!\!\perp B|C$ means that A and B are independent if C occurs and does not say anything about the relation between A and B if C does not occur.

The relationship between marginal and conditional independence is an ever recurrent theme in graphical modelling.

EXAMPLE 2.1.4 A counter-example to the assertion that

$$A \perp\!\!\!\perp B|C,\ A \perp\!\!\!\perp B|C^c \ \Rightarrow\ A \perp\!\!\!\perp B.$$

Take $P(C) = 0.5 = P(C^c)$ and the following specification of the conditional probabilities of the events $A \cap B$, $A^c \cap B$, $A \cap B^c$, $A^c \cap B^c$ given C and given C^c :

conditional probabilities						marginal		
given C	B	B^c	given C^c	B	B^c		B	B^c
A	0.1	0.1	A	0.1	0.4	A	0.10	0.25
A^c	0.4	0.4	A^c	0.1	0.4	A^c	0.25	0.40

Given either C or C^c, A and B are independent. The marginal probability of A and B is 0.1, of A is 0.35 and of B is 0.35. But $0.1 \neq (0.35)(0.35)$ and the events are not independent.

In application the content of this example is often applied in reverse, so that an observed dependence between A and B is 'explained' by controlling for the factor C; recall the infant mortality data discussed in the introduction. □

The *strong* definition of the conditional independence of events comes by defining $A \perp\!\!\!\perp B|[C, D]$ to mean that A and B are independent given any event in the partition of the sample space generated by the events C and D : $C \cap D$, $C^c \cap D$, $C \cap D^c$, $C^c \cap D^c$. That is

$$A \perp\!\!\!\perp B|[C, D] \iff A \perp\!\!\!\perp B|C \cap D,\ A \perp\!\!\!\perp B|C^c \cap D,\ A \perp\!\!\!\perp B|C \cap D^c,\ A \perp\!\!\!\perp B|C^c \cap D^c.$$

The definition naturally extends to a partition generated by an arbitrary number of events, so that $A \perp\!\!\!\perp B|[C]$ for instance, is also meaningful; note that it is a stronger assertion than $A \perp\!\!\!\perp B|C$. Note that $A \perp\!\!\!\perp B||[C, D]$ does not imply that $A \perp\!\!\!\perp B|C$ nor that $A \perp\!\!\!\perp B|C \cup D$ and in fact, Example 2.1.4 above is a counter-example to the assertion that $A \perp\!\!\!\perp B||[C]$ implies $A \perp\!\!\!\perp B$.

The answer to the general question of when statements such as $A \perp\!\!\!\perp B||[C, D]$ imply those such as $A \perp\!\!\!\perp B||[C]$ or $A \perp\!\!\!\perp B$ is not immediately obvious. We shall see later that the notion of an independence graph substantially clarifies the issues involved. In the meantime some insight is attained from the next proposition.

Proposition 2.1.1 Block independence for events. *If P is positive on the partition generated by the events A, B and C then the assertions*

(i) $A \perp\!\!\!\perp [B, C]$, *and*

(ii) $A \perp\!\!\!\perp [B]||[C]$, $A \perp\!\!\!\perp [C]||[B]$

are equivalent.

Proof: To show (ii) implies (i): from the equivalence between $A \perp\!\!\!\perp [B, C]$ and $A \perp\!\!\!\perp B$, $A \perp\!\!\!\perp C$ and $A \perp\!\!\!\perp B \cap C$, established above, it is sufficient to show that (ii) implies that $A \perp\!\!\!\perp B$, $A \perp\!\!\!\perp C$ and $A \perp\!\!\!\perp B \cap C$. Now

$$P(A \cap B \cap C) = P(A \cap B|C)P(C) = P(A|C)P(B|C)P(C),$$

as $A \perp\!\!\!\perp B|C$, so that

$$P(A \cap B \cap C) = P(A|C)P(B \cap C).$$

And, by symmetry, $P(A \cap B \cap C) = P(A|B)P(B \cap C)$, as well, so that, by equating these two expressions for $P(A \cap B \cap C)$ we obtain $P(A|B) = P(A|C)$. Now (ii) also entails $A \perp\!\!\!\perp B|C^c$ so that a similar argument starting with $P(A \cap B \cap C^c)$ establishes that $P(A|B) = P(A|C^c)$. Consequently $P(A|C) = P(A|C^c)$ and so $A \perp\!\!\!\perp C$. Symmetry implies $A \perp\!\!\!\perp B$, as well, and as $P(A \cap B \cap C) = P(A|C)P(B \cap C) = P(A)P(B \cap C)$ we have $A \perp\!\!\!\perp B \cap C$.

To go the other way: from (i), $A \perp\!\!\!\perp [B, C]$ we have that A is independent of $B \cap C$ and of C so that

$$\begin{aligned} P(A \cap B|C) &= P(A \cap B \cap C)/P(C) \\ &= P(A)P(B \cap C)/P(C) \quad \text{as } A \perp\!\!\!\perp B \cap C \\ &= P(A)P(B|C) \\ &= P(A|C)P(B|C) \quad \text{as } A \perp\!\!\!\perp C. \end{aligned}$$

Thus $A \perp\!\!\!\perp B|C$ and hence $A \perp\!\!\!\perp [B]||C$. The same argument with C^c replacing C gives $A \perp\!\!\!\perp [B]||[C]$. The rest follows by symmetry. □

If P is not strictly positive on the partition, for instance, if $P(A \cap B^c \cap C) = 0$, then the proposition may be false. A counter-example is to take $A = B = C$ and note that the proposition then asserts that (i) $A \perp\!\!\!\perp [A, A]$ and (ii) $A \perp\!\!\!\perp A | A$ are equivalent. The former statement implies $A \perp\!\!\!\perp A$, which is generally false. The latter statement, which can be interpreted to mean that knowledge of A imparts no further information about A when A is already known, is generally true. The positivity condition on P rules this counter-example out, because for instance, $P(A \cap A^c \cap A) = 0$.

2.2 Independent Random Vectors

The concepts of independence and conditional independence of events directly extend to random variables and vectors. More results are available because of the extra numerical structure inherent in the notion of a random variable. A natural way to proceed would be to take

$$P((X,Y); X \in A, Y \in B) = P(X; X \in A) P(Y; Y \in B)$$

for all sets A and B, as the definition of independence for two random variables X and Y. But this is rather too general for our purposes. It is the probability density function that plays the key role in statistical applications and so it is more than just convenient to take a definition of independence specified in its terms.

Definition. The random vectors X and Y are ***independent*** if and only if the joint probability density function, f_{XY}, satisfies

$$f_{XY}(x, y) = f_X(x) f_Y(y)$$

for all values of x and y. The relationship is denoted by $X \perp\!\!\!\perp Y$. □

So X and Y are independent if and only if the joint density function factorises into the product of the marginal density functions. The independence relation is symmetric in X and Y. But as with the independence of events the ***conditional formulation***

$$X \perp\!\!\!\perp Y \iff f_{X|Y}(x; y) = f_X(x) \text{ for all } x,$$

is equivalent; that is, $X \perp\!\!\!\perp Y$ if and only if the conditional and the marginal density functions are identical. To see this just note that the conditional density function of X given Y is defined as f_{XY}/f_Y, where we assume that f_Y is non-zero on the relevant set of values. This result rephrases the definition of independence in terms of the conditional and marginal density functions of X. It is used when it is required to apply the concept of independence in the case where Y is not random and so does not have a density function; for example, if Y is a parameter in a probability model. A far reaching discussion of conditional independence specified in these terms is given by Dawid (1979) and a rigorous treatment by Dawid (1980), to whom the notation $\perp\!\!\!\perp$ for independence is due.

The content of the factorisation criterion is that to establish independence it is sufficient to show that the joint density function factorises rather than that it factorises into the product of the marginals.

Proposition 2.2.1 Factorisation criterion for independent random vectors. *The random vectors X and Y are independent if and only if there exist two functions g and h such that*

$$f_{XY}(x,y) = g(x)h(y) \quad \textit{for all } x \textit{ and } y.$$

Proof: If $X \perp\!\!\!\perp Y$ then putting $g = f_X$ and $h = f_Y$ does the trick. To establish the converse, suppose that f_{XY} factorises, as asserted. Integrating this over y gives the marginal density of X as $f_X(x) = \text{ const } g(x)$ and so $g(x)$ is directly proportional to $f_X(x)$. Similarly $h(y)$ is proportional to $f_Y(y)$, and thus $f_{XY}(x,y) = \text{ const} f_X(x)f_Y(y)$. Integrating over both x and y establishes that the constant is unity and so $X \perp\!\!\!\perp Y$. □

Proposition 2.2.2 Reduction: joint independence implies marginal independence. *If (X,Y,Z) is a partitioned random vector then $X \perp\!\!\!\perp (Y,Z)$ implies $X \perp\!\!\!\perp Y$.*

Proof: By definition $X \perp\!\!\!\perp (Y,Z)$ implies $f_{XYZ} = f_X f_{YZ}$. Now integrate out z from both sides to conclude that $f_{XY} = f_X f_Y$. □

This may be viewed as the corollary to the rather stronger result that $X \perp\!\!\!\perp Y \Rightarrow X \perp\!\!\!\perp g(Y)$ for all (measurable) functions g, but a proof requires starting from the more general definition of independence alluded to above, and the proposition is stronger than we need.

Thus $X \perp\!\!\!\perp (Y,Z)$ implies $X \perp\!\!\!\perp Y$ and $X \perp\!\!\!\perp Z$. However Example 2.1.3 rephrased in terms of random variables is a counter-example to the converse assertion that $X \perp\!\!\!\perp Y$ and $X \perp\!\!\!\perp Z$ imply $X \perp\!\!\!\perp (Y,Z)$. Here is another.

EXAMPLE 2.2.1 If the density function over the unit cube $\{x = (x_1, x_2, x_3);$ and $0 < x_i < 1$ for $i = 1,2,3\}$ is

$$f_{123}(x) = \chi_1(x) + \chi_2(x) + \chi_3(x) + \chi_4(x)$$

where the indicator functions

$$\begin{array}{ll}
\chi_1(x) = 1 & \text{when } x_1 < 0.5,\ x_2 < 0.5,\ x_3 < 0.5 \text{ and } = 0 \text{ otherwise,} \\
\chi_2(x) = 1 & \text{when } x_1 < 0.5,\ x_2 > 0.5,\ x_3 > 0.5 \text{ and } = 0 \text{ otherwise,} \\
\chi_3(x) = 1 & \text{when } x_1 > 0.5,\ x_2 < 0.5,\ x_3 > 0.5 \text{ and } = 0 \text{ otherwise,} \\
\chi_4(x) = 1 & \text{when } x_1 > 0.5,\ x_2 > 0.5,\ x_3 < 0.5 \text{ and } = 0 \text{ otherwise,}
\end{array}$$

then the marginal density of (X_1, X_2) is $f_{12}(x_1, x_2) = 1$ for $0 < x_1 < 1$, and $0 < x_2 < 1$. This is the product of the marginal densities of X_1 and X_2, so that $X_1 \perp\!\!\!\perp X_2$. Similarly $X_1 \perp\!\!\!\perp X_3$. But while $f_{123}(x_1, x_2, x_3)$ is given above, $f_1(x_1)f_{23}(x_2, x_3) = 1$ and these functions are not equal. Thus X_1 is not independent of (X_2, X_3). □

To reiterate: it is not true that if X is marginally independent of Y and of Z then X is jointly independent of Y and Z.

Conditional independence

The definition is essentially a rewrite of the definition of independence with the unconditional density functions replaced by conditional density functions.

Definition. The random vectors Y and Z are ***conditionally independent*** on X if and only if

$$f_{YZ|X}(y, z; x) = f_{Y|X}(y; x) f_{Z|X}(z; x)$$

for all values of y and z and for all x for which $f_X(x) > 0$. This is written as $Y \perp\!\!\!\perp Z | X$. □

As we need it to hold for all values of the conditioning variable X, this definition corresponds to the strong conditional independence statement for events, $A \perp\!\!\!\perp B \| [C]$ rather than to the weaker $A \perp\!\!\!\perp B | C$. We write $X \perp\!\!\!\perp Y | (Z, W)$ to denote that X and Y are independent conditional on the partitioned vector (Z, W).

Two other equivalent formulations of the definition are

$$\begin{aligned} f_{Y|XZ}(y; x, z) &= f_{Y|X}(y; x) \quad \text{and} \\ f_{XYZ}(x, y, z) &= f_{XY}(x, y) f_{XZ}(x, z) / f_X(x). \end{aligned}$$

The first expresses the fact that the conditional independence of Y from Z means that Z can be discarded from the conditioning set. The second shows that conditional independence can be rephrased entirely in terms of marginal densities. The factorisation and reduction propositions go through directly, and involve nothing new in their derivation. Together with the block independence lemma, they are used extensively in the proof of the separation theorem, and are set out here for convenience.

Proposition 2.2.3 The factorisation criterion for conditional independence.
The random vectors Y and Z are conditionally independent given X, $Y \perp\!\!\!\perp Z | X$, if and only if there exist functions g and h such that

$$f_{XYZ}(x, y, z) = g(x, y) h(x, z) \quad \textit{for all } y \textit{ and } z$$

and all x with $f_X(x) > 0$.

There is no requirement for the factors $g(x, y)$ and $h(x, z)$ to be unique, only a requirement for the existence of a factorisation into coordinate functions.

Proposition 2.2.4 The reduction lemma. *If (X, Y, Z_1, Z_2) is a partitioned random vector then $Y \perp\!\!\!\perp (Z_1, Z_2)|X$ implies $Y \perp\!\!\!\perp Z_1|X$.*

The next lemma is another technical result needed to prove the separation theorem for conditional independence graphs, and requires proving. It is useful to hold the following diagram in mind:

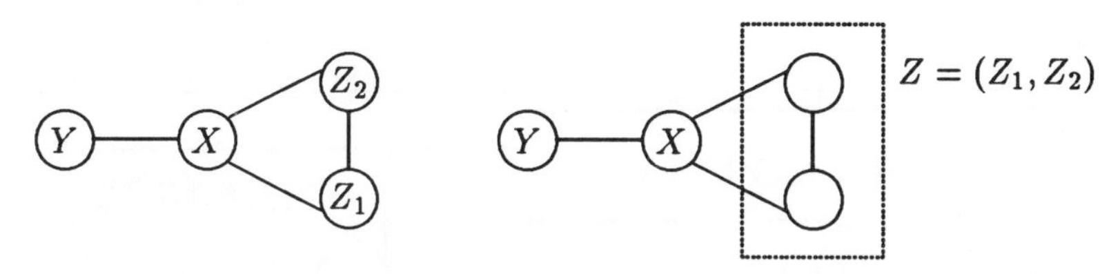

Proposition 2.2.5 The block independence lemma. *If (X, Y, Z_1, Z_2) is a partitioned random vector and f is positive then the following assertions are equivalent:*

(a) $Y \perp\!\!\!\perp (Z_1, Z_2)|X$, and

(b) $Y \perp\!\!\!\perp Z_1|(X, Z_2)$ and $Y \perp\!\!\!\perp Z_2|(X, Z_1)$.

Proof: To prove that (a) implies (b): the factorisation criterion for conditional independence states that

$$Y \perp\!\!\!\perp (Z_1, Z_2)|X \iff \log f(x, y, z_1, z_2) = g(x, y) + h(x, z_1, z_2);$$

that is, there is no function in the sum on the right hand side whose arguments include both y and (z_1, z_2). Consequently no function on the right hand side includes both y and z_1. So by the factorisation criterion, $Y \perp\!\!\!\perp Z_1|(X, Z_2)$. Similarly $Y \perp\!\!\!\perp Z_2|(X, Z_1)$.

To go the other way: if $Y \perp\!\!\!\perp Z_1|(X, Z_2)$ then there exist functions g' and h' such that

$$\log f(x, y, z_1, z_2) = g'(x, y, z_2) + h'(x, z_1, z_2),$$

and if $Y \perp\!\!\!\perp Z_2|(X, Z_1)$ as well, the term involving y and z_2 simplifies, so that there exist functions g'' and g''' such that

$$\begin{aligned} \log f(x, y, z_1, z_2) &= g''(x, y) + g'''(x, z_2) + h'(x, z_1, z_2) \\ &= g''(x, y) + h''(x, z_1, z_2). \end{aligned}$$

A further application of the factorisation criterion implies $Y \perp\!\!\!\perp (Z_1, Z_2)|X$. □

This argument should be compared with the corresponding proof of the block independence of events; this is rather more straightforward.

Putting reduction and block independence together we can conclude that if we start with $Y \perp\!\!\!\perp Z_1 | (X, Z_2)$ and $Y \perp\!\!\!\perp Z_2 | (X, Z_1)$ we may use block independence to deduce that $Y \perp\!\!\!\perp (Z_1, Z_2) | X$, and then apply reduction to conclude $Y \perp\!\!\!\perp Z_1 | X$. We shall see in the next chapter that this is the guts of our proof of the separation theorem.

These independence relationships can be portrayed in a Venn diagram, as in Figure 2.2.1. By the block independence lemma $X \perp\!\!\!\perp Y | Z$ and $X \perp\!\!\!\perp Z | Y$ imply

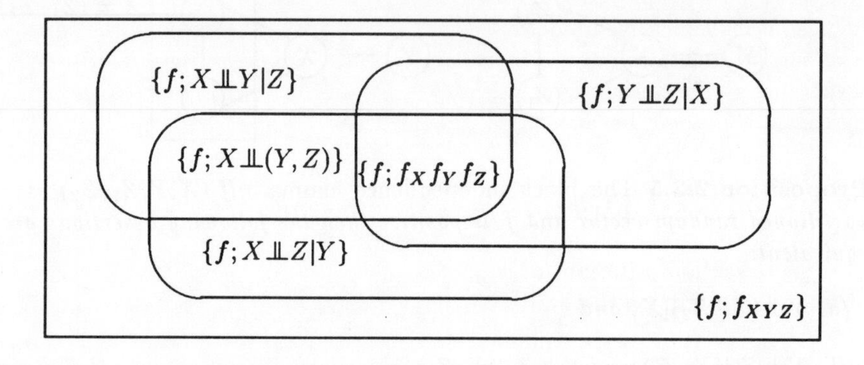

Figure 2.2.1: A Venn diagram of conditional independence statements.

$X \perp\!\!\!\perp (Y, Z)$. And if all three relationships $X \perp\!\!\!\perp Y | Z$, $X \perp\!\!\!\perp Z | Y$ and $Y \perp\!\!\!\perp Z | X$ hold then X, Y and Z are mutually independent. It is not obvious where the set $\{f; X \perp\!\!\!\perp Y\}$ should be placed in the picture.

The next lemma is proved here by reformulating the block independence lemma, but it is certainly as fundamental and perhaps an even deeper result than block independence because of the lack of symmetry in the conditioning set. Firstly, this assymetry provides what is needed for specifying conditional independences in directed independence graphs, and secondly, it removes the necessity for the positivity condition, so enlarging the scope of applications, in particular to include random variables satisfying logical conditions such as mathematical identities.

Proposition 2.2.6 *If (X, Y, Z_1, Z_2) is a partitioned random vector then the following assertions are equivalent:*

(a) $Y \perp\!\!\!\perp (Z_1, Z_2) | X$, and

(b) $Y \perp\!\!\!\perp Z_2 | (X, Z_1)$ and $Y \perp\!\!\!\perp Z_1 | X$.

Proof: To show (a) implies (b): the block independence lemma gives the first assertion of (b), and the reduction lemma gives the second. To show (b) implies (a): consider

$$\begin{aligned} f_{XYZ_1Z_2} &= f_{XYZ_1} f_{XZ_1Z_2} / f_{XZ_1} && \text{as } Y \perp\!\!\!\perp Z_2 | (X, Z_1) \\ &= f_{XY} f_{XZ_1} f_{XZ_1Z_2} / f_X f_{XZ_1} && \text{as } Y \perp\!\!\!\perp Z_1 | X \\ &= f_{XY} f_{XZ_1Z_2} / f_X \end{aligned}$$

and consequently $Y \perp\!\!\!\perp (Z_1, Z_2) | X$. □

Here, we have employed the factorisation criterion for conditional independence, in multipicative mode, rather than additive mode on the logarithms of the density function, only to instill variety. But note that in any attempt to generalise the material to include the possibility of zero probabilities, multiplication is needed, because the logarithm of zero is not defined.

2.3 Mixed Derivative Measures of Interaction

The factorisation criterion for independence naturally delivers a measure of dependence Partition the k-dimensional random vector $X = (X_1, X_2, \ldots, X_k)$ into (X_1, X_2, X_a) where $a = \{3, 4, \ldots, k\}$ so that X_a is the vector $(X_3, \ldots, X_k)$ and X_1 and X_2 are one dimensional random variables. Suppose that their joint density function $f_{12a}(x_1, x_2, x_a)$ is (sufficiently) differentiable. The criterion states that a necessary and sufficient condition for the conditional independence of X_1 and X_2 given X_a, $X_1 \perp\!\!\!\perp X_2 | X_a$, is the existence of functions g and h, such that the density function factorises:

$$\log f_{12a}(x_1, x_2, x_a) = g(x_1, x_a) + h(x_2, x_a),$$

which, if a is empty, is just a condition for marginal independence. Consequently one way to measure the interaction between X_1 and X_2 is to evaluate the first order mixed or cross derivative of the logarithm of the density function.

When the variable X_1 is continuous we take the ordinary partial derivative $D_1 = \frac{\partial}{\partial x_1}$, and when X_1 is discrete or categorical we suppose it takes values in the non-negative integers, and the differential is replaced by the forward difference, ∇_1, given by

$$\nabla_1 g(x_1) = g(x_1 + 1) - g(x_1) \text{ for } x_1 = 0, 1, 2 \ldots.$$

A second derivative or difference, for instance, the second partial derivative with respect to x_3 is written as D_3^2, and a mixed partial derivative, for instance, with respect to x_1 and x_3 is D_{13}^2.

Definition. When the variables X_1 and X_2 are continuous the *mixed derivative measures of marginal and conditional interaction* between X_1 and X_2 are

$$i_{12}(x_1, x_2) = D^2_{12}\log f_{12}(x_1, x_2)$$

and

$$i_{12|a}(x_1, x_2; x_a) = D^2_{12}\log f_{12|a}(x_1, x_2; x_a).$$

If a variable is discrete the partial derivative D is replaced by the forward difference, ∇. □

Generalising to higher dimensions is straightforward by taking successively higher order mixed partial derivatives. Usually both i_{12} and $i_{12|a}$ will be functions of x_1 and x_2, but in several important instances they are constant and so give rise to *global* measures of interaction. These measures have an important property.

Proposition 2.3.1 *A necessary and sufficient condition for conditional independence,* $X_1 \perp\!\!\!\perp X_2 | X_a$, *is that*

$$i_{12|a}(x_1, x_2; x_a) = 0 \text{ for all values of } x_1 \text{ and } x_2.$$

The proof is straightforward and just uses the fact that if the derivative of a (differentiable) function is everywhere zero, then the function is constant. By taking the set a to be empty this is just a statement about marginal independence.

Consider now, i_{12}, the mixed derivative marginal measure of interaction between X_1 and X_2. It is easy to see that the mixed derivatives of the joint and of the two conditional density functions are the same; that is

$$i_{12} = D^2_{12}\log f_{12} = D^2_{12}\log f_{1|2} = D^2_{12}\log f_{2|1}.$$

It is an interesting result because

Proposition 2.3.2 *The mixed derivative is a measure of interaction that is invariant to the sample design, be it sampling using a bivariate response, or using a retrospective scheme where* X_2 *is fixed and* X_1 *is observed, or using a prospective scheme where* X_1 *is fixed and* X_2 *is observed.*

EXAMPLE 2.3.1 In medical trials, the association between two factors such as fat intake with two levels (high/low) and coronary heart disease with two levels (yes/no) can be measured by sampling from the population at risk and then classifying each subject by these two factors. Alternatively a retrospective study might choose equal numbers of subjects with and without the disease and then classify these by their fat intake. It is not immediately obvious that inferences about the association between the factors will be the same for both sampling procedures. However, if association is measured by the mixed derivative then it has the same value under either sampling scheme. □

In general the conditional mixed derivative measure of interaction $i_{12|a}$ will be a function of x_a as well as of x_1 and x_2. In the special case when it does not depend on x_a, it is a *partial* measure of interaction, so called in honour of the partial correlation coefficient. When $X_a = X_3$ is one dimensional as well, then a necessary and sufficient condition for the conditional measure to be a partial measure is that

$$D_3 i_{12|3} = D^3_{123}\log f_{12|3} = D^3_{123}\log f_{123} = 0;$$

that is, there should *no three-way interaction* between X_1, X_2 and X_3. Note that it is a symmetric requirement between the variables X_1, X_2 and X_3.

Collapsibility

If the variables classifying a data set are partitioned into three vectors X_a, X_b and X_c, the vector (X_a, X_b, X_c) is said to be *parametrically collapsible over* X_a with respect to the measure of interaction, if the interaction between X_b and X_c may be estimated equally well from the full data set or from the vector marginalised over X_a. The issue is of some importance to practical data analysis because in many studies it is impossible to measure all potential covariates. It is also important in those theoretical studies which explore the relationship between the joint and marginal distributions.

There are many notions of collapsibility: parametric collapsibility is concerned with the equality of parameters in the joint and the marginal distributions. It has been investigated by Simpson (1951) and Whittemore (1978). The detailed relation between i_{12} and $i_{12|a}$ is explored in Whittaker (1986), see also Ekholm (1985), Wermuth (1986ab). Here we show that conditional independence is a sufficient condition for parametric collapsibility defined in terms of the mixed derivative measures of interaction.

Proposition 2.3.3 *The family of densities f is collapsible over X_a (with respect to the mixed derivative measure of interaction) if either $X_a \perp\!\!\!\perp X_1 | X_2$ or $X_a \perp\!\!\!\perp X_2 | X_1$.*

Proof: First suppose that $X_a \perp\!\!\!\perp X_1 | X_2$, so by the factorisation criterion in Proposition 2.2.1 above we can write $\log f(x_1, x_2, x_a)$ as $g(x_a, x_2) + h(x_1, x_2)$. Now

$$\begin{aligned} i_{12|a}(x_1, x_2; x_a) &= D^2_{12}\log f_{12a}(x_1, x_2, x_a) \\ &= D^2_{12}\{g(x_a, x_2) + h(x_1, x_2)\} \\ &= D^2_{12} h(x_1, x_2), \end{aligned}$$

and so is constant with respect to x_a. Hence $i_{12|a}$ is a partial measure of interaction. Now

$$\begin{aligned} f_{12}(x_1, x_2) &= \int f_{12a}(x_1, x_2, x_a) dx_a \\ &= \int \exp\{g(x_a, x_2) + h(x_1, x_2)\} dx_a \\ &= \exp\{h(x_1, x_2)\} \int \exp\{g(x_a, x_2)\} dx_a. \end{aligned}$$

Consequently the marginal density is of the form

$$\log f_{12}(x_1, x_2) = h(x_1, x_2) + g'(x_2)$$

and so

$$\begin{aligned} i_{12}(x_1, x_2) &= D_{12}^2 \log f_{12}(x_1, x_2) \\ &= D_{12}^2 \{h(x_1, x_2) + g'(x_2)\} \\ &= D_{12}^2 h(x_1, x_2). \end{aligned}$$

But this is identical to $i_{12|a}$ above and so the the family is collapsible. Repeat this argument starting with $X_a \perp\!\!\!\perp X_2 | X_1$. The same argument goes through in the discrete and in the mixed cases. □

This shows that the phenomenon called Simpson's paradox cannot occur when marginalising over X_a if X_a is conditionally independent of at least one of the pair of interacting variables. Conversely, it can and does occur, in examples with a more complex interaction structure, especially when marginalising over potential conditioning variables.

Another, weaker form of collapsibility is concerned with identiticity of independence graphs constructed from the joint and marginal distributions. We discuss it in the final chapter.

2.4 The Bernoulli Distribution

A Bernoulli random variable is the simplest one imaginable. It records the result of an experiment that has only two possible outcomes indicated by the random variable X that takes the value 1 with probability p and 0 with probability $1-p$. It forms the basic building block in a derivation of the Binomial distribution and, using a continuity argument, of the Poisson distribution. The name honours James Bernoulli, a member of the famous Bernoulli family from Basle, who was mathematically active at the turn of the 17th century.

A *Bernoulli random variable*, X, has the probability density function

$$f_X(x) = p^x (1-p)^{1-x} \quad \text{for } x = 0, 1 \text{ and } 0 \leq p \leq 1.$$

One shorthand is to write $X \sim B(1, p)$ where $B(n, p)$ denotes the Binomial density function with parameters n and p. We are interested in a multivariate generalisation of this distribution and accordingly define the bivariate Bernoulli density function.

The *bivariate Bernoulli random vector*, (X_1, X_2), takes the values $(0,0)$, $(0,1)$, $(1,0)$ and $(1,1)$ in the Cartesian product $\{0,1\}^2 = \{0,1\} \times \{0,1\}$ of the two point set $\{0,1\}$ with itself. Its density function f_{12} is determined by

$$f_{12}(x_1, x_2) = p(x_1, x_2)$$

for $x_1 = 0,1$ and $x_2 = 0,1$, where $p(x_1, x_2)$ is the table of parameters (in fact probabilities):

$p(x_1, x_2)$		$x_2 = 0$	$x_2 = 1$	total
	$x_1 = 0$	$p(0,0)$	$p(0,1)$	$p_1(0)$
	$x_1 = 1$	$p(1,0)$	$p(1,1)$	$p_1(1)$
	total	$p_2(0)$	$p_2(1)$	1

and where each $p(x_1, x_2) \geq 0$ and their sum is unity. Each $p(x_1, x_2)$ is a probability and we can refer to the table as a table of probabilities. To ensure we have a consistent notation for marginal densities, we write $p(x_1, x_2)$ as $p_{12}(x_1, x_2)$, then the marginal table of probabilities are

$$\begin{aligned} p_1(x_1) &= p(x_1, 0) + p(x_1, 1), \quad x_1 = 0,1 \\ p_2(x_2) &= p(0, x_2) + p(1, x_2), \quad x_2 = 0,1. \end{aligned}$$

For those who write p_{ij} as the probability for the (i,j)-th cell of a 2-way table, the translation is $p_{12}(i,j)$.

A seemingly trivial remark is to note that the marginal density function of X_1 is univariate Bernoulli with

$$f_1(x_1) = p_1(x_1) \quad \text{for} \quad x_1 = 0,1;$$

and the conditional density function of X_1 given X_2 is again univariate Bernoulli with

$$f_{1|2}(x_1; x_2) = p_{12}(x_1, x_2)/p_2(x_2) \quad \text{for} \quad x_1 = 0,1$$

and for each fixed x_2. Hence X_1 and X_2 are independent if and only if the table of probabilities p satisfies

$$p_{12}(x_1, x_2) = p_1(x_1)p_2(x_2) \text{ for all } (x_1, x_2) \in \{0,1\}^2.$$

The *cross-product ratio between x_1 and x_2* is defined as the corresponding ratio of the events $\{X_1 = 0\}$ and $\{X_2 = 0\}$:

$$\operatorname{cpr}(X_1, X_2) = \frac{p(1,1)p(0,0)}{p(0,1)p(1,0)}.$$

It is the standard measure of the dependence between binary random variables.

Log-linear expansion and the u-terms

We generalise the treatment of the one dimensional Bernoulli density function. Consider the identity

$$p(x_1, x_2) = p(0,0)^{(1-x_1)(1-x_2)} p(0,1)^{(1-x_1)x_2} p(1,0)^{x_1(1-x_2)} p(1,1)^{x_1 x_2}$$

valid for $x_1 = 0, 1$ and $x_2 = 0, 1$. Now the Bernoulli bivariate probability density is determined by $f_{12}(x_1, x_2) = p(x_1, x_2)$, so that taking logarithms of this identity for p, and collecting terms in x_1 and x_2 gives

$$\log f_{12}(x_1, x_2) =$$
$$\log p(0,0) + x_1 \log \frac{p(1,0)}{p(0,0)} + x_2 \log \frac{p(0,1)}{p(0,0)} + x_1 x_2 \log \frac{p(1,1)p(0,0)}{p(0,1)p(1,0)}$$

for (x_1, x_2) in $\{0,1\}^2$. This representation of f lies at the heart of the theory of log-linear models for categorical data, discussed, for example, in the texts by Bishop *et al.* (1975) or Andersen (1980).

It is a bilinear form, linear in both x_1 and in x_2. Reparameterising the right hand side leads to the *log-linear expansion*

$$\log f_{12}(x_1, x_2) = u_\phi + x_1 u_1 + x_2 u_2 + x_1 x_2 u_{12} \quad \text{for } (x_1, x_2) \text{ in } \{0,1\}^2.$$

The coefficients, u, u_1, u_2, u_{12}, are known as the *u-terms.* The term $u_1 = \log p(1,0)/p(0,0)$ and is just the log of the odds of the event $X_1 = 1$ to the event $X_1 = 0$ conditioned on $X_2 = 0$. The coefficient of the product $x_1 x_2$ is the logarithm of the cross product ratio:

$$u_{12} = \log \operatorname{cpr}(X_1, X_2).$$

The density function f_{12} is determined either by its table of probabilities, p, or by its u-terms, $\{u\}$. To calculate systematically the u's from given p's, substitute in the log-linear expansion for $(x_1, x_2) = (0,0), \ldots, (1,1)$ to get

$$\begin{aligned}
\log p(0,0) &= u_\phi \\
\log p(1,0) &= u_\phi + u_1 \\
\log p(0,1) &= u_\phi + u_2 \\
\log p(1,1) &= u_\phi + u_1 + u_2 + u_{12}.
\end{aligned}$$

This is a simple set of linear equations to solve. Note that scaling all the p's will only affect the first u-term, u_ϕ.

EXAMPLE 2.4.1 If p is given by

$p(x_1, x_2)$	x_2	0	1
x_1	0	2/58	8/58
	1	16/58	32/58

the u-terms are

$$u_\phi = \log(2/58),\ u_1 = 3\log(2),\ u_2 = 2\log(2),\ u_{12} = -\log(2),$$

and the log-linear expansion of the density function is

$$\log f_{12}(x_1, x_2) = \log(2/58) + 3\log(2)\,x_1 + 2\log(2)\,x_2 - \log(2)\,x_1 x_2.$$

The log cross-product ratio between x_1 and x_2 is $u_{12} = -\log(2)$. If p is replaced by $58p$ so that it sums to 58 then the only coefficient to change is u, from $\log(2/58)$ to $\log(2)$. □

We must note that the notation $\text{cpr}(X_1, X_2)$ for the cross-product ratio, is not to be thought of as $\text{cpr}(x_1, x_2)$. It is not a function, and for example, $\text{cpr}(0, 1)$ makes no sense. We could have written it, like the u-term u_{12}, as cpr_{12}, but we wish to use this measure in the same way as the covariance, $\text{cov}(X_1, X_2)$, is used. Later, when we generalise to tables with arbitrary numbers of categories, the u-terms become functions and, for instance, the term u_{12} becomes $u_{12}(x_1, x_2)$. Equivalently note that the u-terms are the mixed derivative measures of interaction and do not turn out to be globally constant.

Interaction

The mixed derivative measure of interaction, introduced above, is extracted by evaluating the cross partial forward difference, $\nabla_1\nabla_2 \log f_{12}(x_1, x_2)$, from the log-linear representation of the bivariate Bernoulli density. We have

$$\begin{aligned} i(x_1, x_2) &= \nabla_1\nabla_2 \log f_{12}(x_1, x_2) \\ &= u_{12}, \quad \text{bilinearity of the log-linear expansion} \\ &= \log \text{cpr}(X_1, X_2), \end{aligned}$$

the log cross-product ratio. As x_1 and x_2 only take two values, this interaction is a constant that does not depend on value of x_1 and x_2 and so is a global measure of interaction or dependence.

Proposition 2.4.1 *The random variables X_1 and X_2 are independent if and only if $u_{12} = 0$.*

Proof: One can either appeal to the general result about $i(x_1, x_2)$ in Proposition 2.3.1 or apply the factorisation criterion in Proposition 2.2.1 directly to the log-linear expansion. Inspection shows this is only possible if the coefficient of the product term $x_1 x_2$ is zero. □

The concept of independence has delivered a measure of dependence: the famous log cross-product ratio.

The relationship between the u-terms in the log-linear expansion of the joint density function and the ones for the corresponding expansion of the conditional density function needs to be made clear. The discussion here amplifies the observation that the mixed derivative of interaction, i_{12}, is invariant to a retrospective, prospective or joint sample design, by directly finding the expansion of the conditional density function.

Suppose that (X_1, X_2) is a bivariate Bernoulli random vector. The marginal density function of X_1, is univariate Bernoulli, and has the log-linear expansion

$$\log f_1(x_1) = v_\phi + x_1 v_1, \quad \text{for } x_1 = 0, 1.$$

The conditional probability density of X_2 given X_1 is $f_{2|1} = f_{12}/f_1$ so that $\log f_{2|1} = \log f_{12} - \log f_1$ and hence

$$\log f_{2|1}(x_2; x_1) = u_\phi - v_\phi + u_2 x_2 + (u_1 - v_1)x_1 + u_{12} x_1 x_2.$$

The coefficients of x_2 and of $x_1 x_2$ are identical in the log-linear expansion of the joint and of the conditional density functions; but those of the constant and x_1 are different. The importance of this result is that we can sample from either f_{12} or $f_{1|2}$ or $f_{2|1}$ to obtain estimates of the same measure of association between X_1 and X_2, the cross-product ratio.

2.5 Three Dimensional Bernoulli Distribution

A three dimensional Bernoulli random vector (X_1, X_2, X_3) takes values on the vertices of the cube, $\{0,1\}^3 = \{0,1\} \times \{0,1\} \times \{0,1\}$. Its density function is determined from a three dimensional table of probabilities, p, or better p_{123}, by

$$f_{123}(x_1, x_2, x_3) = p_{123}(x_1, x_2, x_3)$$

for (x_1, x_2, x_3) in $\{0,1\}^3$. The marginal distribution of a pair of variables, say (X_2, X_3), is bivariate Bernoulli with density

$$f_{23}(x_2, x_3) = \sum_{x_1=0}^{1} p_{123}(x_1, x_2, x_3) = p_{23}(x_2, x_3)$$

given by summing out x_1 from the three dimensional table of probabilities. The marginal cross-product ratio between X_2 and X_3, $\text{cpr}(X_2, X_3)$, is defined on this density and is

$$\text{cpr}(X_2, X_3) = \frac{p_{23}(1,1)p_{23}(0,0)}{p_{23}(0,1)p_{23}(1,0)}.$$

The conditional distribution of (X_2, X_3) given X_1 is also bivariate Bernoulli with density

$$f_{23|1}(x_2, x_3; x_1) = p(x_1, x_2, x_3)/p_1(x_1),$$

for (x_2, x_3) in $\{0,1\}^2$, for each value of $x_1 = 0, 1$. The ***conditional cross-product ratio*** between X_2 and X_3 given $X_1 = x_1$ is defined in this conditional distribution and simplifies to

$$\text{cpr}\,(X_2, X_3|X_1 = x_1) = \frac{p(x_1,1,1)p(x_1,0,0)}{p(x_1,0,1)p(x_1,1,0)}, \quad \text{for } x_1 = 0, 1,$$

when expressed in terms of the joint density, because the $p_1(x_1)$ terms cancel.

The log-linear expansion of the density function is derived by extending the formula for the bivariate density function to three dimensions. Writing the first and last of the $2^3 = 8$ terms gives

$$f_{123}(x_1, x_2, x_3) = p(0,0,0)^{(1-x_1)(1-x_2)(1-x_3)} \ldots p(1,1,1)^{x_1x_2x_3}$$

for (x_1, x_2, x_3) in $\{0,1\}^3$. Picking out the coefficients of the constant, of x_2x_3 and of $x_1x_2x_3$ in the expansion for $\log f_{123}$ gives

$$\text{constant}: \quad \log p(0,0,0)$$

$$x_2x_3: \quad \log \frac{p(0,1,1)p(0,0,0)}{p(0,0,1)p(0,1,0)}$$

$$x_1x_2x_3: \quad \log \frac{p(1,1,1)p(1,0,0)}{p(1,0,1)p(1,1,0)} \frac{p(0,1,1)p(0,0,0)}{p(0,0,1)p(0,1,0)}.$$

Collecting terms gives the log-linear expansion

$$\log f(x_1, x_2, x_3) = u_\phi$$
$$+ u_1x_1 + u_2x_2 + u_3x_3 + u_{12}x_1x_2 + u_{13}x_1x_3 + u_{23}x_2x_3 + u_{123}x_1x_2x_3.$$

Identifying the u-terms: the coefficient of the products x_2x_3 and $x_1x_2x_3$

$$\begin{aligned} u_{23} &= \log \text{cpr}\,(X_2, X_3|X_1 = 0) \quad \text{and} \\ u_{123} &= \log \{\text{cpr}\,(X_2, X_3|X_1 = 1)/\text{cpr}\,(X_2, X_3|X_1 = 0)\}. \end{aligned}$$

Hence the highest order u-term u_{123} measures the difference in conditional log cross-product ratios between X_2 and X_3 as X_1 changes from 0 to 1. It is symmetric in X_1, X_2 and X_3 and is a measure of three-way interaction between them.

EXAMPLE 2.5.1 Reconsider the Bishop *et al.* (1975) data about the infant survival rate of patients attending two clinics recorded in Table 1.2.1. For the sake of example, we suppose the counts are not subject to sampling error, and exactly determine the table of probabilities. The u-terms are calculated from the set of 8 linear equations relating the u's to the $\log p$'s. To emphasise the relation of the u's to the variable we write u_1 as u(clin) and so on. We obtain

$$\begin{aligned} u_\phi &= 1.10, & u(\text{clin}) &= 1.74, \\ u(\text{care}) &= 0.29, & u(\text{clin, care}) &= \text{-}2.43, \\ u(\text{surv}) &= 4.07, & u(\text{clin, surv}) &= \text{-}1.62, \\ u(\text{care, surv}) &= 0.22, & u(\text{clin, care, surv}) &= \text{-}0.23. \end{aligned}$$

The log-linear expansion of this table is

$$\begin{aligned} \log f = {} & 1.10 + 1.74 \text{ clin} + 0.29 \text{ care} + 4.07 \text{ surv} \\ & - 2.42 \text{ clin.care} - 1.62 \text{ clin.surv} + 0.22 \text{ care.surv} - 0.23 \text{ clin.care.surv}. \end{aligned}$$

Note the large positive u-term for survival which reflects the large difference between the overall numbers who survive and those who do not. In comparison to the other terms u(care,surv) = 0.22 and u(clin,care,surv)=-0.23 are both small. If these were zero then we could conclude that care and survival are independent given clinic; as it is we can only state that the density is near to this. Later, we regard the counts as subject to random error, and build a statistical model to estimate the u-terms and test if any are zero. □

Conditional independence

The mixed derivative measure of interaction between X_2 and X_3, conditioned on X_1, isolates the coefficients of the terms in x_2x_3 in the log-linear expansion. Application of the definition $i(x_2, x_3; x_1) = \nabla_2\nabla_3 \log f(x_2, x_3; x_1)$ to the log-linear expansion gives

$$i_{23|1}(x_2, x_3; x_1) = u_{23} + x_1 u_{123} = \log \operatorname{cpr}(X_2, X_3 | X_1 = x_1)$$

for $x_1 = 0$ and for $x_1 = 1$.

As with the marginal mixed derivative measure it does not depend on the values of x_2 and x_3, and so is a global measure of the interaction between X_2 and X_3 at each level of X_1. When the three-way interaction is zero then this conditional interaction is the same for all x_1 and is a *partial* interaction. Together with the factorisation theorem for conditional independence this is then a proof of

Proposition 2.5.1 *If (X_1, X_2, X_3) has the trivariate Bernoulli distribution, the three following conditions are equivalent*

(a) $X_2 \perp\!\!\!\perp X_3 | X_1$;

(b) $u_{23} = 0$ *and* $u_{123} = 0$; *and*

(c) $\text{cpr}(X_2, X_3 | X_1 = x_1) = 1$ *for* $x_1 = 0, 1$.

EXAMPLE 2.5.2 The table of probabilities

p_{123}		$x_3 = 0$	$x_3 = 1$
$x_1 = 0$	$x_2 = 0$	1/21	2/21
	$x_2 = 1$	2/21	2/21
$x_1 = 1$	$x_2 = 0$	2/21	4/21
	$x_2 = 1$	4/21	4/21

has the log-linear expansion

$$\log f(x_1, x_2, x_3) = -\log(21) + \log(2)x_1 + \log(2)x_2 + \log(2)x_3 - \log(2)x_2x_3.$$

The coefficients of the three terms x_1x_2, x_1x_3 and $x_1x_2x_3$ are all zero. We may deduce that $X_1 \perp\!\!\!\perp X_2 | X_3$ because $u_{123} = 0$ and $u_{12} = 0$ and $X_1 \perp\!\!\!\perp X_3 | X_2$ because $u_{123} = 0$ and $u_{13} = 0$. We can conclude that $X_1 \perp\!\!\!\perp (X_2, X_3)$ either because of the block independence lemma, or directly, by noting that $\log f(x_1, x_2, x_3)$ has the structure $g(x_1) + h(x_2, x_3)$. □

Collapsibility and Simpson's phenomenon

There are some surprises when a three-way table is collapsed over one or more dimensions, in the terminology, marginalised. The phenomenon goes under the name of Simpson's paradox, Simpson (1951), and has been discussed by several authors, for example, see Cox (1970), Whittemore (1978) and Dawid (1979). We have seen in the analysis of the infant mortality data how misleading the two-way margins can be about the full three-way table. The next example extends this by showing that such behaviour is not confined to examples exhibiting some form of independence or near independence.

EXAMPLE 2.5.3 An illustrative example of Simpson's phenomenon. Consider the following three-way table, scaled by 1/422 to give probabilities:

the table				derived measures			
$p(x_1, x_2, x_3)$		$x_3 = 0$	$x_3 = 1$	*(a)*	*(b)*		*(c)*
$x_1 = 0$	$x_2 = 0$	100	100	10(> 1)	100/110	<	200/211
	$x_2 = 1$	1	10				
$x_1 = 0$	$x_2 = 0$	100	1	10(> 1)	1/11	<	101/211
	$x_2 = 1$	100	10				
margin	$x_2 = 0$	200	101	0.399(< 1)	101/121	>	201/422
	$x_2 = 1$	101	20				

where (a) denotes cpr $(X_2, X_3|X_1 = x_1)$, (b) denotes $P(X_2 = 0|X_3 = 1, X_1 = x_1)$, and (c) denotes $P(X_2 = 0|X_1 = x_1)$; and the last two rows contain the corresponding marginal quantities summed over x_1.

There is a surprising reversal of the inequalities relating to certain derived measures when the table is summed over x_1. In this table the cross-product ratio between X_2 and X_3 is the same whether $X_1 = 0$ or $X_1 = 1$, and so it is a partial measure of interaction, and consequently the three-way interaction term, u_{123}, is zero. Thus cpr $(X_2, X_3|X_1 = x_1) = 10$ for $x_1 = 0, 1$ and is greater than one. But after the table is marginalised over X_1 the resulting ratio is less than one, and in fact cpr $(X_2, X_3) = 0.399$. We have thus constructed an example where

$$i_{23|X_1=x_1} > 0 \text{ for } \underline{\text{every}} \text{ value of } x_1, \text{ but } i_{23} < 0.$$

Recall from Section 2.3 that a sufficient condition for the equality of i_{23} and $i_{23|1}$ is that either $X_1 \perp\!\!\!\perp X_2|X_3$ or that $X_1 \perp\!\!\!\perp X_3|X_2$, and neither of these conditions are satisfied in this example. This seemingly bizarre behaviour is not a peculiarity of the measure of interaction, as it is manifest in terms of probabilities: the relationship between the probability that $X_2 = 0$ and the conditional probability that $X_2 = 0$ given X_3 is reversed in the marginal table. These reversals are surprising because summing over x_1 appears to be a form of averaging, but the attributes of the components are not preserved in the average. □

The next example indicates this is a feature of real life (and death).

EXAMPLE 2.5.4 Florida murderers. The following data concerning sentencing policy, based on 4764 murder cases tried in the American state of Florida over the six year period 1973-79:

	sentence		
murderer	death	other	%
black	59	2448	2.4
white	72	2185	3.2

The table by itself is fairly unremarkable. The proportion of black murderers sentenced to death is about the same as the proportion of whites, and, if anything, the white percentage is higher. However the full table, Table 2.5.1, as published in the *New York Times Magazine* on 11 March 1979, see Krippendorf (1986), breaks this down by colour of victim, and this conclusion is reversed. For both black victims and for white victims, the proportion of black murderers sentenced to death is the higher, and with white victims significantly so.

There is one other point of interest: Krippendorf, when commenting on this table, notes the 'conspicuous absence of death sentences for white convicts

Table 2.5.1: Death sentencing and race in Florida, 1973-79. From Krippendorf (1986).

		sentence		
victim	murderer	death	other	%
black	black	11	2209	0.5
	white	0	111	0.0
white	black	48	239	16.7
	white	72	2074	3.4

when the victim is black'. One might therefore analyse the table either on the basis that this zero is a random value, or on the basis that this is a structural zero, reflecting an institutional racism of the American judiciary. □

2.6 The Normal Distribution

The scalar random variable X has the *standard Normal distribution*, $X \sim N(0,1)$, if its probability density function is

$$f_X(x) = (2\pi)^{-1/2}\exp(-x^2/2) \text{ for } x \text{ in } R^1,$$

the standard Normal density function. This function is denoted by $n(x;0,1)$ and some elementary facts about it are summarised in the next proposition.

Proposition 2.6.1 Properties of the Normal distribution.

(a) The function $n(x;0,1)$ integrates over $-\infty < x < \infty$ to one.

(b) The expected value of X is zero and the variance of X is one, that is, $E(X) = 0$ and $\text{var}(X) = 1$.

(c) The density function of the linear transformation $Y = \mu + \sigma X$ is

$$f_Y(y) = \frac{1}{\sigma(2\pi)^{1/2}}\exp\{-\frac{1}{2\sigma^2}(y-\mu)^2\}$$

and $E(Y) = \mu$, $\text{var}(Y) = \sigma^2$. We write $Y \sim N(\mu,\sigma^2)$ and $f_Y(y) = n(y;\mu,\sigma^2)$.

(d) If $X \sim N(0,1)$ then $Y = X^2$ has a chi-squared distribution with 1 degree of freedom.

(e) If X_1 and X_2 are independently and identically distributed according to $N(0,1)$ then $X_1 + X_2 \sim N(0,2)$ and $X_1^2 + X_2^2$ has a chi-squared distribution with 2 degrees of freedom.

Proofs can be found in any standard probability text.

The bivariate Normal

The essential simplicity of the standard Normal distribution is exemplified by noting that the logarithm of its density function is just a quadratic:

$$\log f_X(x) = -x^2/2 + \text{const}.$$

In two dimensions this generalises to a quadratic form in the vector (x_1, x_2). Let Q be

$$Q(x_1, x_2) = \begin{pmatrix} x_1 & x_2 \end{pmatrix} \begin{pmatrix} 1 & \rho \\ \rho & 1 \end{pmatrix}^{-1} \begin{pmatrix} x_1 \\ x_2 \end{pmatrix}$$

or, on inversion of the 2×2 matrix,

$$Q(x_1, x_2) = \frac{1}{(1-\rho^2)}(x_1^2 - 2\rho x_1 x_2 + x_2^2)$$

with the condition $-1 < \rho < 1$ to ensure that Q is positive definite. The random vector (X_1, X_2) in R^2 has the *standard bivariate Normal distribution* if its joint density function is given by

$$\log f_{12}(x_1, x_2) = -\log 2\pi - \frac{1}{2}\log(1-\rho^2) - \frac{1}{2}Q(x_1, x_2).$$

In a moment we shall see that this expression does define a proper joint density function but before doing so note that the mixed derivative measure of interaction between X_1 and X_2 comes from differentiating the quadratic form in this expression. The coefficient of $x_1 x_2$ in $Q(x_1, x_2)$ is $-2\rho/(1-\rho^2)$; and as there is no other term containing both x_1 and x_2 then

$$i_{12}(x_1, x_2) = D_1 D_2 \log f_{12}(x_1, x_2) = \rho/(1-\rho^2).$$

In fact it is conventional to take ρ itself as the measure of interaction.

Proposition 2.6.2 *If (X_1, X_2) is standard bivariate Normal then the condition $\rho = 0$ is necessary and sufficient for X_1 and X_2 to be independent.*

Proof: Either use the factorisation criterion directly on the bivariate density function, or note that $i(x_1, x_2) = \rho/(1-\rho^2)$ is zero if and only if $\rho = 0$. □

Thus the parameter ρ is a measure of interaction between X_1 and X_2. It turns out that ρ is just the linear *correlation coefficient* between X_1 and X_2, that is $\operatorname{corr}(X_1, X_2) = \rho$, where

$$\operatorname{corr}(X_1, X_2) = \operatorname{cov}(X_1, X_2)/\{\operatorname{var}(X_1)\operatorname{var}(X_2)\}^{1/2}$$

and the matrix

$$\begin{pmatrix} 1 & \rho \\ \rho & 1 \end{pmatrix}$$

determining the quadratic form is the *correlation matrix* of (X_1, X_2). Rather than sweat out the relevant integrals in order to evaluate $\operatorname{cov}(X_1, X_2)$ a more revealing proof of this assertion comes by deriving the conditional probability density function of X_2 given X_1. Though standard, this argument is included here because it serves as a good introduction to the k-dimensional case.

Proposition 2.6.3 *If X_1 and X_2 have the standard bivariate Normal probability density function then the marginal probability density function of X_1 is $n(x_1; 0, 1)$ and the conditional probability density function of X_2 given $X_1 = x_1$ is $n(x_2; \rho x_1, 1 - \rho^2)$.*

Proof: First complete the square in x_2: from the first expression for the quadratic form above

$$Q(x_1, x_2) = \frac{(x_1^2 - 2\rho x_1 x_2 + x_2^2)}{(1 - \rho^2)} = \frac{(x_2 - \rho x_1)^2}{(1 - \rho^2)} + x_1^2.$$

Substituting for Q in the definition of f_{12} leads to

$$f_{12}(x_1, x_2) = \\ (2\pi)^{-1/2} \exp\{-x_1^2/2\} \, (2\pi(1 - \rho^2))^{-1/2} \exp\{-(x_2 - \rho x_1)^2/2(1 - \rho^2)\}.$$

The marginal distribution of x_1 is obtained by integrating out x_2, according to $f_1(x_1) = \int f_{12}(x_1, x_2) dx_2$, so that

$$f_1(x_1) = \\ (2\pi)^{-1/2} \exp\{-x_1^2/2\} \int (2\pi(1 - \rho^2))^{-1/2} \exp\{-(x_2 - \rho x_1)^2/2(1 - \rho^2)\} dx_2.$$

But the integral is unity because the integrand is a univariate probability density function of the form $n(x_2; \rho x_1, 1 - \rho^2)$. And that leaves $f_1(x_1) = n(x_1; 0, 1)$.

The conditional probability density function of X_2 given X_1 is then $f_{2|1} = f_{12}/f_1$; substituting for f_{12} and f_1 gives

$$f_{2|1}(x_2; x_1) = (2\pi(1 - \rho^2))^{-1/2} \exp\{-(x_2 - \rho x_1)^2/2(1 - \rho^2)\};$$

that is, conditionally on $X_1 = x_1$, $X_2 \sim N(\rho x_1, 1 - \rho^2)$, as was to be proved. □

Corollary 2.6.4 *The correlation coefficient in the standard bivariate Normal distribution is ρ, that is* $\operatorname{corr}(X_1, X_2) = \rho$.

Proof: As the margins of the standard bivariate Normal are standard univariate Normal, $EX = 0 = EY$ and $\operatorname{var}(X_1) = 1 = \operatorname{var}(X_2)$. Hence $\operatorname{corr}(X_1, X_2) = E(X_1X_2)$. As $X_2|X_1 \sim N(\rho x_1, 1-\rho^2)$, $E_{2|1}(X_2) = \rho x_1$ so that $E(X_1X_2) = E_1(X_1E_{2|1}(X_2)) = E(\rho X_1^2) = \rho$. □

The trivariate Normal distribution

Our interest in this probability density function is to illustrate the nature of conditional independence, but because it involves matrix inversion, the tractability of the expressions quickly starts to disappear. In fact, manipulation in the general case with an arbitrary number of dimensions proves to be an easier task. This section then, is nothing more than a promise of things to come.

The required extension to the standard bivariate density is to replace the quadratic form in the definition by

$$Q(x_1, x_2, x_3) = \begin{pmatrix} x_1 & x_2 & x_3 \end{pmatrix} \begin{pmatrix} 1 & \rho_{12} & \rho_{13} \\ \rho_{21} & 1 & \rho_{23} \\ \rho_{31} & \rho_{32} & 1 \end{pmatrix}^{-1} \begin{pmatrix} x_1 \\ x_2 \\ x_3 \end{pmatrix}$$

The matrix is symmetric with $\rho_{21} = \rho_{12}$, $\rho_{31} = \rho_{13}$ and $\rho_{32} = \rho_{23}$, and positive definite. It turns out that all the ρ's are correlation coefficients, for instance $\operatorname{corr}(X_2, X_3) = \rho_{23}$, and the matrix in Q is the inverse correlation matrix of (X_1, X_2, X_3). Denoting this by D, the density is

$$\begin{aligned} \log f(x_1, x_2, x_3) &= \text{const} - \frac{1}{2} Q(x_1, x_2, x_3) \\ &= \text{const} - \tfrac{1}{2} \begin{pmatrix} x_1 & x_2 & x_3 \end{pmatrix} D \begin{pmatrix} x_1 \\ x_2 \\ x_3 \end{pmatrix}. \end{aligned}$$

This expansion contains quadratic terms in x_1, x_2 and x_3, bilinear terms in x_1 and x_2, x_1 and x_3, and x_2 and x_3 but, unlike the trivariate Bernoulli, there are no three-way interaction terms such as $x_1x_2x_3$. Taking the mixed partial derivative of the expression with respect to X_2 and X_3 isolates the off diagonal element of D, d_{23} : the mixed derivative measure of conditional interaction, $i_{23|1}$. Because there are no three-way interaction terms this is constant with respect to X_1 and is therefore a *partial* measure of conditional interaction. This is a striking feature of all multivariate Normal distributions. Using the formula that expresses the inverse of a matrix in terms of its minors we have

$$\begin{aligned} i_{23|1} &= D^2_{23}\log f_{123}(x_1, x_2, x_3) \\ &= d_{23} \\ &= \frac{1}{\det(D)} \det \begin{pmatrix} 1 & \rho_{13} \\ \rho_{21} & \rho_{23} \end{pmatrix} \\ &= \frac{1}{\det(D)}(\rho_{23} - \rho_{21}\rho_{13}). \end{aligned}$$

The determinant of D is strictly positive. Application of Proposition 2.3.1 then provides a proof of

Proposition 2.6.5 *If (X_1, X_2, X_3) has the (standard) trivariate Normal distribution, the three following conditions are equivalent:*

(a) $X_2 \perp\!\!\!\perp X_3 | X_1$*;*

(b) the off diagonal element in the inverse variance matrix, $d_{23} = 0$*; and*

(c) the pairwise marginal correlations satisfy $\rho_{23} - \rho_{21}\rho_{13} = 0$.

Though a general argument needs a substantial amount of filling out, we now have some evidence for the assertion that X_2 and X_3 are independent conditional on X_1 if and only if the appropriate element in the inverse correlation matrix of (X_1, X_2, X_3) is zero.

This proposition gives some credence to our analysis of the mathematics marks data in the introduction. Later we see that $\rho_{32} - \rho_{31}\rho_{12}$ is a fundamental component of the partial correlation between X_2 and X_3 after adjustment for X_1. Note that only a single condition has to be satisfied to ensure the conditional independence $X_2 \perp\!\!\!\perp X_3 | X_1$. This is in stark contrast to the trivariate Bernoulli where two conditions have to be simultaneously satisfied.

EXAMPLE 2.6.1 An example of a correlation matrix for which $X_2 \perp\!\!\!\perp X_3 | X_1$ but $X_2 \perp\!\!\!\perp X_3$ is false, and another for which $X_2 \perp\!\!\!\perp X_3$ but $X_2 \perp\!\!\!\perp X_3 | X_1$ is false, are

$$\begin{pmatrix} 1.0 & 0.2 & 0.5 \\ 0.2 & 1.0 & 0.1 \\ 0.5 & 0.1 & 1.0 \end{pmatrix} \quad \text{and} \quad \begin{pmatrix} 1.0 & 0.2 & 0.5 \\ 0.2 & 1.0 & 0.0 \\ 0.5 & 0.0 & 1.0 \end{pmatrix}$$

respectively. To prove this, check that $\rho_{23} - \rho_{21}\rho_{13} = 0.1 - (0.2)(0.5) = 0$ and $\rho_{23} \neq 0$ in the first, while $\rho_{23} - \rho_{21}\rho_{13} = 0.2 - (0.0)(0.5) \neq 0$ and $\rho_{23} = 0$ in the second. Neither condition is satisfied for arbitrary matrices, while both are satisfied if the correlation matrix is proportional to the identity matrix. □

2.7 Exercises

1: Show that $A \perp\!\!\!\perp B$ implies $A \perp\!\!\!\perp B^c$ (and thus that $A \perp\!\!\!\perp B$ is equivalent to $A \perp\!\!\!\perp [B]$). Show that $A \perp\!\!\!\perp \phi$ and $A \perp\!\!\!\perp \Omega$ whatever A, where ϕ denotes the empty set and Ω the whole sample space.

2: Show that the events A and B are independent (a) if and only if $P(A|B) = P(A|B^c)$; and (b) if and only if $\text{cpr}\,(A, B) = 1$.

3: Show that if B and C are exclusive events then $A \perp\!\!\!\perp B$ and $A \perp\!\!\!\perp C$ together imply that $A \perp\!\!\!\perp B \cup C$. Give a counter-example to show that B and C need to be exclusive.

4: Show that
(a) $A \perp\!\!\!\perp [B, C]$ implies $A \perp\!\!\!\perp E$ where E is any event generated by taking complements, intersections and unions of the events B and C; and
(b) $A \perp\!\!\!\perp [B, C]$ if and only if $A \perp\!\!\!\perp B \cap C$, $A \perp\!\!\!\perp B$ and $A \perp\!\!\!\perp C$.

5: Show that '$A \perp\!\!\!\perp [B, C]$ and $B \perp\!\!\!\perp C$' is invariant to permuting the labels A, B and C. Conclude that this is equivalent to mutual independence.

6: Give an example for which events A, B and C
(a) satisfy $P(A \cap B \cap C) = P(A)P(B)P(C)$ but are not mutually independent;
(b) are pairwise independent but not mutually independent.
(c) Generalise the definition of mutual independence appropriate for k events $A_1, A_2, \ldots, A_k$.

7: Suppose A, B and C are events.
(a) Give an explicit counter-example to the assertion that $A \perp\!\!\!\perp B | C$ implies $A \perp\!\!\!\perp B | C^c$.
(b) Show that events A and B are independent conditionally on C if and only if A^c and B^c are conditionally independent on C.

8: A test for cancer gives a false positive with probability 0.1 and a false negative with probability 0.2. The proportion of cancer cases in the population is 0.01. A patient is given two tests. Use Bayes theorem to evaluate the probability of cancer if both tests are positive. Make the assumption that the tests are conditionally independent given cancer and given no cancer.

9: Given that X and Y are discrete variables show that $X \perp\!\!\!\perp Y | X$ is always true and that $X \perp\!\!\!\perp Y$ implies $X \perp\!\!\!\perp Y | g(X)$ for any function g. More interestingly, show that $X \perp\!\!\!\perp Y | Z$ implies $X \perp\!\!\!\perp g(Y) | Z$ for any function g.

10: Write out the proofs that
(a) $X_2 \perp\!\!\!\perp X_3 | X_1$ if and only if $f_{123} = f_{12} f_{13} / f_1$;
(b) $X_2 \perp\!\!\!\perp X_3 | X_1$ if and only if $X_3 \perp\!\!\!\perp X_2 | X_1$;
(c) $X_2 \perp\!\!\!\perp X_3 | X_1$ if and only if $f_{2|13} = f_{2|1}$.

11: The continuous random variables have conditional and marginal probability density functions: $f_{Y|X}(y;x) = x\exp(-xy)$ and $f_X(x) = \exp(-x)$ on $x, y > 0$. Evaluate $f_{X|Y}(x;y)$. Find the mixed derivative measure of interaction between X and Y.

12: Suppose that the vector of random variables $X = (X_1, X_2, X_3, X_4)$ has a joint density function of the form

$$f(x) = \text{const}\exp(x_1x_2 + x_1x_3 + x_2x_4 + x_3x_4),$$

for $0 < x_i < 1$, $i = 1, 2, 3, 4$. Use the factorisation criterion to elicit as many conditional independence statements as possible.

13: Verify that if X is a Bernoulli random variable then $E(X) = p$ and $\text{var}(X) = p(1-p)$.

14: The bivariate Bernoulli random variable (X, Y) has a density determined by the table of probabilities

$p(x,y)$		y	0	1
x	0		0.1	0.2
	1		0.3	0.4

find the marginal density function of X and the density function of X given $Y = 0$ and $Y = 1$. Are X and Y independent?

15: If (X_1, X_2) has the bivariate Bernoulli distribution find $E(X_1)$, $E_{1|2}(X_1)$, $E(X_1X_2)$. Show that $\text{cov}(X_1, X_2) = p_{12}(1,1) - p_1(1)p_2(1)$ and hence deduce that $\text{corr}(X_1, X_2) = 0$ implies $X_1 \perp\!\!\!\perp X_2$ for this density function.

16: (a) Find the cross-product ratios for the following tables of probabilities:

p		y	0	1
x	0		0.1	0.2
	1		0.3	0.4

p		y	0	1
x	0		0.1	0.2
	1		0.4	0.3

p		y	0	1
x	0		0.1	0.3
	1		0.4	0.2

(b) Show by example that the log-cpr can take all values between $-\infty$ and $+\infty$.
(c) Find a table for which the cpr $= 2$. How many such tables are there?

17: The random vector (X, Y) has the bivariate Bernoulli distribution, express $\text{cpr}(X, Y)$ in terms of expectations of X and Y. Show that $\text{cpr}(X, Y) = \text{cpr}(Y, X) = 1/\text{cpr}(1 - X, Y)$.

18: The log-linear expansion for the two-way table of probabilities $p(x, y)$ with $x, y = 0, 1$ is given:

(a) $\log p(x,y) = u + x + y + xy$: find the constant u such that p sums to 1, and find the cross-product ratio $\text{cpr}\,(X,Y)$;
(b) $\log p(x,y) = u + x$; show that Y is equi-probable at each level of X;
(c) $\log p(x,y) = u + x + 2y$; show that $X \perp\!\!\!\perp Y$.

19: Show that the log-linear expansion for the bivariate Bernoulli is expressible as

$$\log f_{XY}(x,y) = \log P(X=0, Y=0) \\ + x \log \text{odds}\,(X|Y=0) + y \log \text{odds}\,(Y|X=0) + xy \log \text{cpr}\,(X,Y).$$

How does this generalise to 3 dimensions?

20: Consider the three-way table classified by gender, treatment and outcome. This is fictitious data discussed by Cox (1970, p109).

	gender	male		female	
	treatment	treated	untreated	treated	untreated
outcome	success	4	8	2	12
	fail	3	5	3	15

Divide by 52 to get probabilities.
(a) Show that outcome and treatment are independent but that they are (positively) dependent at each level of gender.
(b) Modify the entries in the previous table to give an example where outcome and treatment are independent at each level of gender but where they are dependent in the marginal table.

21: Recall the infant mortality data discussed in Section 1.2. The cpr of care and survival is $(20)(316)/(6)(373) = 2.82$. Find the conditional cpr of care and survival given clinic for each of the two clinics, by direct calculation from the table and then from the log-linear expansion given in Example 2.5.1 together with the relation $\log \text{cpr}\,(X_2, X_3|X_1 = x_1) = u_{23} + x_1 u_{123}$.

22: Show that the conditional cross-product ratios satisfy

$$\frac{\text{cpr}\,(X_1, X_2|X_3 = 1)}{\text{cpr}\,(X_1, X_2|X_3 = 0)} = \frac{\text{cpr}\,(X_1, X_3|X_2 = 1)}{\text{cpr}\,(X_1, X_3|X_2 = 0)} = \frac{\text{cpr}\,(X_2, X_3|X_1 = 1)}{\text{cpr}\,(X_2, X_3|X_1 = 0)}.$$

or equivalently that $u_{123} = \log \text{cpr}\,(X_2, X_3|X_1 = 0) - \log \text{cpr}\,(X_2, X_3|X_1 = 1)$ is invariant to permuting the symbols X, X_2 and X_3.

23: The vector (X_1, X_2, X_3) has the three dimensional Bernoulli distribution, show that

$$X_2 \perp\!\!\!\perp X_3 | X_1 \iff E_{23|1}(X_2 X_3) = E_{2|1}(X_2) E_{3|1}(X_3).$$

24: Find the u-terms in the log-linear expansion when $p(x_1, x_2, x_3)$ is proportional to

	x_2	0		1	
	x_3	0	1	0	1
x_1	0	2	8	4	16
	1	16	128	32	256

and comment.

25: Consider the table

	x_2	0		1	
	x_3	0	1	0	1
x_1	0	1	1	1	e^{θ}
	1	e^{θ}	1	1	1

Find the log-linear representation for the full table and for the margin collapsed over X_1.

26: The vector (X, Y) is a standard bivariate Normal random variable, show that $E_{Y|X}(Y) = \rho X$ and $\text{var}_{Y|X}(Y) = 1 - \rho^2$. Use this to show $\text{corr}(X, Y) = \rho$. What are the corresponding results for $X|Y$?

27: Suppose that $Y = \beta X + E$ where β is fixed and $E \sim N(0, 1)$ independently of $X \sim N(0, 1)$. Find (a) the conditional probability density function of Y given X, (b) the joint probability density function of Y and X, (c) the marginal probability density function of Y, and finally, (d) the mixed derivative i_{XY}.

28: In the computation of the conditional probability density function of $Y|X$ from the standard bivariate Normal density in Proposition 2.6.3 where is the condition $-1 < \rho < 1$ used?

29: Given that the correlation matrix of X_1, X_2 and X_3 is

$$\begin{pmatrix} 1 & 0.2 & 0.5 \\ 0.2 & 1 & 0.1 \\ 0.5 & 0.1 & 1 \end{pmatrix}$$

determine if any two variables are conditionally independent given the other.

30: Suppose the correlation matrix of (X_1, X_2, X_3) is such that $X_2 \perp\!\!\!\perp X_3 | X_1$ and $X_1 \perp\!\!\!\perp X_2 | X_3$, show that X_2 is uncorrelated with both X_1 and X_3.

Chapter 3

Independence Graphs

After a summary of the requisite concepts from graph theory, the independence graph of a k-dimensional vector of random variables is introduced and defined. Constructed from selected independences between pairs of variables conditioned on all the remaining variables in the vector, the graph is more correctly called the conditional independence graph. Following some examples, the most characteristic feature of the graph, the global Markov property, is established by proving the separation theorem. It is this property, above all others, that permits such a concise summary of the interaction pattern between the random variables.

The material has been heavily motivated by the the lecture notes of Speed (1978a) to the Institute of Mathematical Statistics in Copenhagen, some of which later appears in Speed and Kiiveri (1986); the seminal Annals paper of Darroch, Lauritzen and Speed (1980); and the elegant exposition by Lauritzen (1982) of the ideas for contingency tables.

The proof of the separation theorem given here has the merit that its basic inductive idea is a relatively straightforward exercise in the application of the factorisation criterion for conditional independence. It is restricted to random variables with a positive joint probability density function so as to apply the lemma of block independence, discussed in the previous chapter. In this sense the proof is distribution free.

There are more elegant proofs for particular distributions: the proof given for the multivariate Normal distribution by Speed and Kiiveri (1986); for Multinomial distributions appropriate for cross-classified categorical data, Lauritzen (1982), and for the family of conditional Gaussian distributions for mixed variables, Lauritzen and Wermuth (1984, 1989). These expositions start by proving that, with respect to a given graph on a finite number of vertices, the following properties are equivalent:

- the pairwise Markov property: that non-adjacent pairs of variables are independent conditional on the remaining variables;

- the local Markov property: that conditional only on the adjacent variables, any variable is independent of all the remaining variables; and

- the global Markov property: that any two subsets of variables separated by a third is independent conditionally only on variables in the third subset.

The separation theorem states that the pairwise property implies the global property.

In this text an independence graph is defined by the pairwise Markov property. There are two essential reasons for this choice: firstly, the pairwise definition of the independence relationship naturally corresponds to the pairwise graph-theoretic definition of the edge set in a graph. Secondly, from the applied point of view, the list of requirements to be verified in constructing the independence graph to model a given data set is the least stringent, while the interpretations that follow are the strongest.

These three properties are indeed equivalent if the density is positive. However, if some of the random variables are logically or mathematically dependent, for example if $X_1 = Y$, $X_2 = Z$ and $X_3 = Y + Z$, the joint density of (X_1, X_2, X_3) is zero whenever $X_1 + X_2 \neq X_3$. In this case, the local and global properties remain equivalent, but cannot be derived from the pairwise property, see the discussion at the end of Section 2.2. One is then obliged to adopt the local Markov property for the definition of an independence graph. However, as the densities in our applications are always positive, we resist the temptation to generalise and stay with the simpler pairwise definition.

When the variables are naturally ordered, for instance, according to time or to some *a priori* causal notion, the definition of the independence graph above is inappropriate. The conditioning set for a pairwise independence containing all remaining variables is too large, and in temporal studies will include the 'future'. A theory for directed independence graphs, based on the ideas of Wermuth and Lauritzen (1983), and Kiiveri, Speed and Carlin (1984), is developed by limiting the conditioning set to the 'past'. This material is very much related to the original path analysis ideas of Wright (1921, 1923). A recent proof of the Markov properties is given in Lauritzen *et al.* (1988).

Finally the extension is made to variables that only satisfy a weaker ordering, one in which the variables can be partitioned into disjoint sets, termed blocks, which are completely ordered and so form a chain. The conditioning sets for each pairwise independence statement consist of all variables in the 'past' blocks and all the remaining variables in the 'present' block. The resultant graph is known as a chain independence graph and provides the conditional independence framework necessary for discussion of multivariate regression and simultaneous equation models.

A different and more general axiomatic approach has been developed by Pearl (1986, 1988) and his co-workers. The idea is to write down a set of rules that encapsulate the relationship of irrelevance within a set of objects. For example, the rule, $A \perp\!\!\!\perp B|C \Rightarrow B \perp\!\!\!\perp A|C$, of symmetry can be interpreted as 'if A is irrelevant to B, given knowledge of C' then 'B is irrelevant to A, given knowledge of C', and is taken as an axiom of the system. Probability with its notion of conditional independence is just one of several application areas that satisfy these axioms of irrelevance. In this context, see Oliver and Smith (1990) and Smith (1989).

Finally we do no more than point out that the material has a deep connection to the probabilistic theory of Markov fields, see for instance Spitzer (1971), Kemeny *et al.* (1976), Speed (1979) and Isham (1981).

3.1 Graph Theory

There are several good books on graph theory, in particular we mention Harary (1969), Berge (1973), and Golumbic (1980). Fortunately its basic notions are easily understandable and a brief summary of the principal concepts and objects of graph theory will suffice for our purposes.

A *graph* G is a mathematical object that consists of two sets, a set of vertices, K, and a set of edges, E, consisting of pairs of elements taken from K. We usually take K to be the set of natural numbers $\{1, 2, \ldots, k\}$. There is a *directed edge* or *arrow* between vertices i and j in K if the set E contains the ordered pair (i, j); vertex i is a *parent* of vertex j, and vertex j is a *child* of vertex i. There is an *undirected edge* or *line* between these vertices if E contains both pairs, (i, j) and (j, i). The graph is *undirected* if all edges are undirected.

We only consider simple graphs, those without multiple edges or loops. The diagram of the graph is a picture, in which circles represent vertices, a *line* represents an undirected edge, and a *arrow* represents a directed edge. The graph with $K = \{1, 2, 3, 4\}$ and $E = \{(1, 2), (2, 1), (1, 3), (4, 3)\}$ has the diagram

Vertices i and j are *adjacent* if the undirected edge between i and j is in E, and a line connects them in the diagram of the graph. Thus 1 and 2 are adjacent, but neither pair 1 and 4, nor pair 1 and 3 are adjacent. The undirected graph G^u associated with G is the graph obtained by replacing all arrows in G by lines.

A *path* is a sequence of distinct vertices, $i_1, i_2, \ldots, i_m$ for which (i_l, i_{l+1}) is in E for each $l = 1, 2, \ldots, m-1$; there is an arrow between each successive pair. It is a *short path* if no subsequence of the sequence is also a path. The path is a *cycle* if the end points are allowed to be the same, $i_1 = i_m$. In an undirected graph each successive pair of vertices in a path are adjacent; and the cycle is *chordless* if no other than successive pairs of vertices in the cycle are adjacent. Two vertices, i and j, are *connected* if there is a path from i to j and a path from j to i, and a graph is *connected* if all pairs of vertices are connected.

A subset of vertices *separates* two vertices, i and j, if every path joining the two vertices contains at least one vertex from the separating subset. A subset separates two subsets a and b of vertices in K if it separates every pair of vertices $i \in a$ and $j \in b$.

Let $a \subseteq K$ denote a subset of vertices of the graph. The *neighbours* of a are all those vertices in K, but not in a, that are adjacent to a vertex in a. The set of *parents* of a is $\text{pa}(a)$, the set of all those vertices in K, but not in a, that have a child in a. The *boundary* of a is $\text{bd}(a)$, the union of the neighbours and the parents of a. In an undirected graph the boundary and the set of neighbours are one and the same.

The induced *subgraph* of a, G_a, is the graph obtained by deleting all the vertices not in a from the graph on K, together with all edges that do not join two elements of a. A graph or subgraph is *complete* if all vertices are joined with either directed or undirected edges.

A *clique* is a subset of vertices which induce a complete subgraph but for which the addition of a further vertex renders the induced subgraph incomplete; that is, a clique is *maximally* complete.

EXAMPLE 3.1.1 The diagram of the undirected graph (K, E) with $K = \{1,2,3,4,5,6,7\}$ and edge set $E = \{(1,2),(1,4),(2,3), (2,5),(3,5),(4,5), (5,6)\} \cup \{(2,1), (4,1),(3,2),(5,2), (5,3),(5,4),(6,5)\}$ is

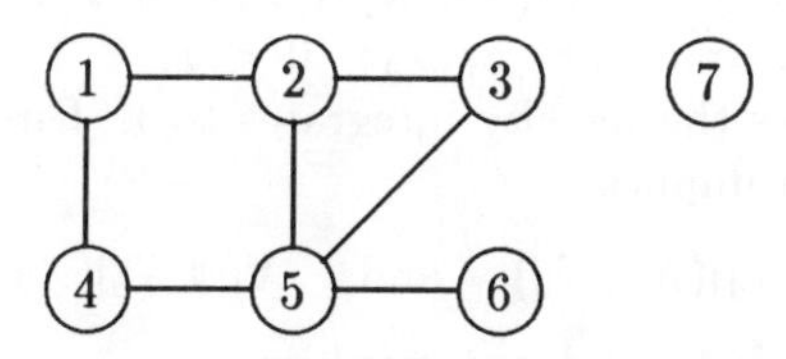

There are many paths from 1 to 6, and 1, 2, 5, 6 is one, but the graph is not connected as there is no path between vertex 7 and the other vertices. The boundary of vertex 1, $\text{bd}(1)$, is the set of neighbours $\{2,4\}$, and of $\{1,2\}$ is the set $\{3,4,5\}$. The cycle 1, 2, 5, 4, 1 is chordless. The cycle 1, 2, 3, 5, 4, 1 is not. The cliques of this graph are the subsets $\{1,2\}, \{1,4\}, \{4,5\}, \{2,3,5\}, \{5,6\}$

and {7}. The subgraphs induced by $\{1,2,3\}$ and by $\{1,2,6\}$ have diagrams

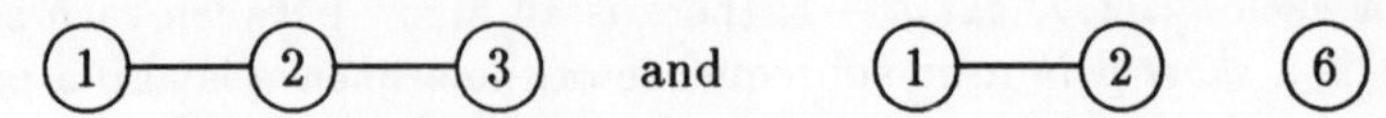

The first is connected, the second is not. □

3.2 Independence Graphs

Let $X = (X_1, X_2, \ldots, X_k)$ denote a vector of random variables, and $K = \{1, 2, \ldots, k\}$ the corresponding set of vertices. The graph is an independence graph, or more precisely a conditional independence graph, if there is no edge between two vertices whenever the pair of variables is independent given all the remaining variables. The vector of the remaining variables are referred to as the *rest*. We use the shorthand $1 \perp\!\!\!\perp 2 | \{3,4\}$ for $X_1 \perp\!\!\!\perp X_2 | (X_3, X_4)$, so that the independence of X_i and X_j given the rest can be written as $i \perp\!\!\!\perp j | K \backslash \{i,j\}$; occasionally and rather loosely we write $X_i \perp\!\!\!\perp X_j |\text{rest}$. The resulting undirected graph gives a picture of the pattern of dependence or association between the variables.

Definition. The *conditional independence graph* of X is the undirected graph $G = (K, E)$ where $K = \{1, 2, .., k\}$ and (i,j) is *not* in the edge set E if and only if $X_i \perp\!\!\!\perp X_j | X_{K \backslash \{i,j\}}$. □

Because of its Markov properties, a better name might be a *Markov* graph, but unfortunately this term is extensively used in the theory of random graphs. We remark on the importance of conditioning; there is no suitable theory for graphs constructed from pairwise marginal independences. It is often easy to construct the independence graph if we are given the joint density function of X by repeated application of the factorisation criterion for conditional independence.

EXAMPLE 3.2.1 Take $k = 4$ and consider the density function of $X = (X_1, X_2, X_3, X_4)$, $f_X(x) = \exp(u + x_1 + x_1 x_2 + x_2 x_3 x_4)$, on the four dimensional cube, $\{x; x = (x_1, x_2, x_3, x_4),\ 0 < x_i < 1,\ i = 1,2,3,4\}$, where the constant u ensures the density integrates to 1. Direct application of the factorisation criterion implies

$$X_1 \perp\!\!\!\perp X_4 | (X_2, X_3) \quad \text{and} \quad X_1 \perp\!\!\!\perp X_3 | (X_2, X_4),$$

and consequently the independence graph is

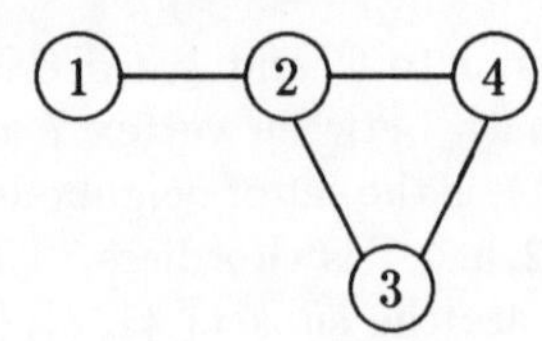

The graph uses the fact that vertex 1 is not adjacent to either 3 or 4 for construction but highlights the fact that the cliques of the graph are $\{1,2\}$ and $\{2,3,4\}$. □

EXAMPLE 3.2.2 Some independence graphs in three and four dimensions. The graphs are undirected but, for simplicity, only one ordering for an edge pair is included in the edge set.

independences	edge set E	diagram	comments
Three dimensions			
$1 \perp\!\!\!\perp 3 \mid 2$	$\{(1,2),(2,3)\}$		one independence
none	$\{(1,2),(1,3),$ $(2,3)\}$		complete inter-dependence
$1 \perp\!\!\!\perp 2 \mid 3$ $1 \perp\!\!\!\perp 3 \mid 2$	$\{(2,3)\}$		independent subsets
$1 \perp\!\!\!\perp 2 \mid 3$ $1 \perp\!\!\!\perp 3 \mid 2$ $2 \perp\!\!\!\perp 3 \mid 1$	$\{\}$		mutual independence
Four dimensions			
$1 \perp\!\!\!\perp 3 \mid \{2,4\}$ $1 \perp\!\!\!\perp 4 \mid \{2,3\}$ $2 \perp\!\!\!\perp 4 \mid \{1,3\}$	$\{(1,2),(2,3),$ $(3,4)\}$		a Markov chain
$1 \perp\!\!\!\perp 2 \mid \{3,4\}$ $1 \perp\!\!\!\perp 3 \mid \{2,4\}$ $2 \perp\!\!\!\perp 3 \mid \{1,4\}$	$\{(1,4),(2,4),$ $(3,4)\}$		possible latent structure
$1 \perp\!\!\!\perp 3 \mid \{2,4\}$ $2 \perp\!\!\!\perp 4 \mid \{1,3\}$	$\{(1,2),(1,4),$ $(2,3),(3,4)\}$		a chordless 4-cycle

The assumption generally made in latent structure models is that the observed variables are independent conditionally upon the unobserved latent variable: so the example requires that variable 4 is not observed. □

All possible graphs

In the exploratory analysis of data it may well be necessary to examine all possible graphs in order to choose one that may fit the data. In k dimensions there are $2^{\binom{k}{2}}$ different graphs, so that with $k = 4$ there are 64 such graphs. They can be classified in groups according to the number of edges together with the number of permutations:

Edges	0	1	2	3	4	5	6
	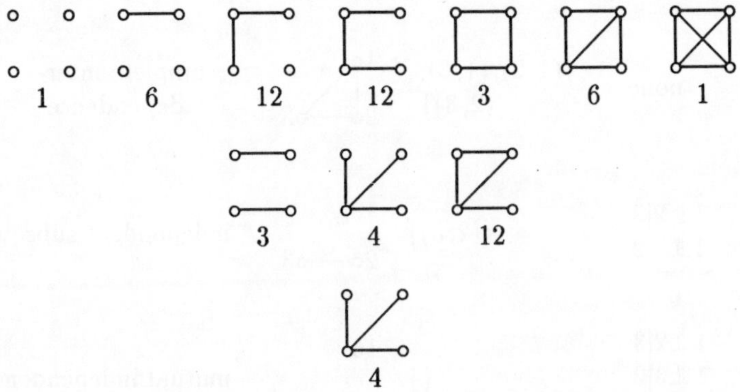 1	6	12	12	3	6	1
			3	4	12		
				4			

All the graphs above have exactly 4 vertices. If the enumeration is extended to include subgraphs, for example, by including graphs based on

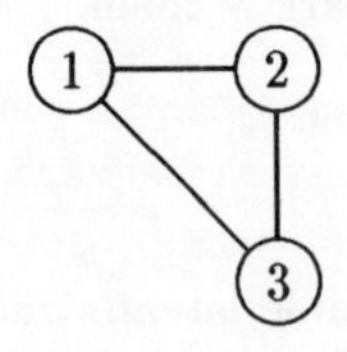

as well as those based on

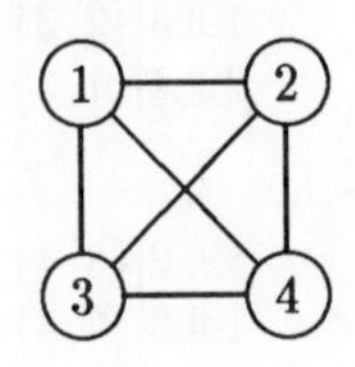

then this figure climbs to

$$\sum_{i=0}^{k} \binom{k}{i} 2^{\binom{i}{2}}$$

When $k = 4$ this is 113.

3.3 Separation

An independence graph highlights certain aspects of information about the interdependence of the variables, in particular, whether or not two variables are adjacent, and if not, how they are separated. It is tempting to conclude that non-adjacent variables are independent given the separating set alone. The separation theorem provides the theoretical justification for this interpretation.

EXAMPLE 3.3.1 Consider the independence statements that can be made from the graph

The manner in which an independence graph is defined means that

$$X_2 \perp\!\!\!\perp X_3 | (X_1, X_4)$$

but we would like to conclude that

$$X_2 \perp\!\!\!\perp X_3 | X_1,$$

where X_4 has been dropped from the conditioning set because the graph suggests that it is redundant in explaining the dependence between X_2 and X_3. This is a statement about the marginal distribution of (X_1, X_2, X_3) and its truth is a consequence of the separation theorem. □

Before proceeding to a proof of the theorem it is worth pondering over what the term separation implies. Recall that two vertices, i and j, are separated by a subset a if and only if all paths connecting the two pass through at least one member of the subset. Consider the following example.

EXAMPLE 3.3.2 Separating subsets. The vertices i and j are distinct and are not members of the separating subset. In each of the following graphs the ringed subsets are separating subsets for i and j. Not all possible separating subsets are explicitly indicated.

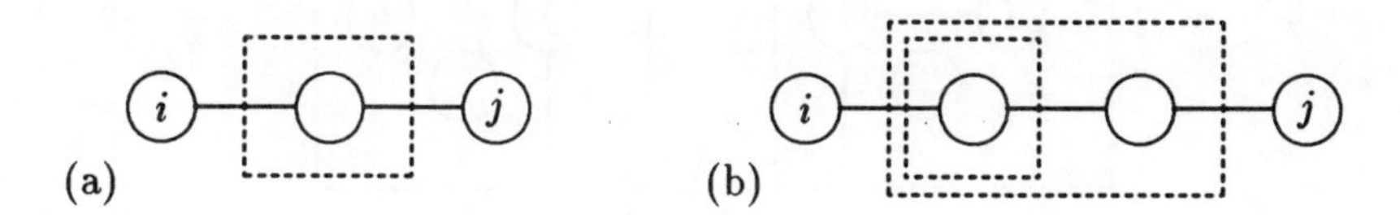

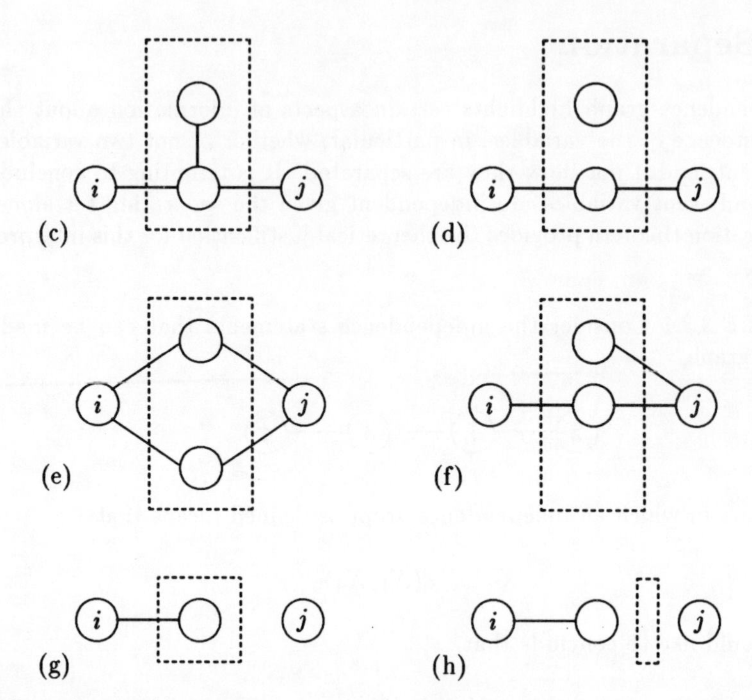

When the vertices are not connected then any subset not containing i and j is separating, including the empty set. □

The argument used here to establish the separation theorem depends upon successive application of block independence and reduction. To reveal this structure explicitly we first establish a version of the separation theorem for a graph that is composed of two disconnected subgraphs.

Lemma 3.3.1 *Suppose that the vertex set $K = \{1, 2, \ldots, k\}$ can be partitioned into two sets b and c where in the independence graph of K there is no path between any vertex in c with any vertex in b. Then*

$$i \perp\!\!\!\perp j \quad \text{for all } i \in b \text{ and } j \in c.$$

Proof: Bear this picture in mind:

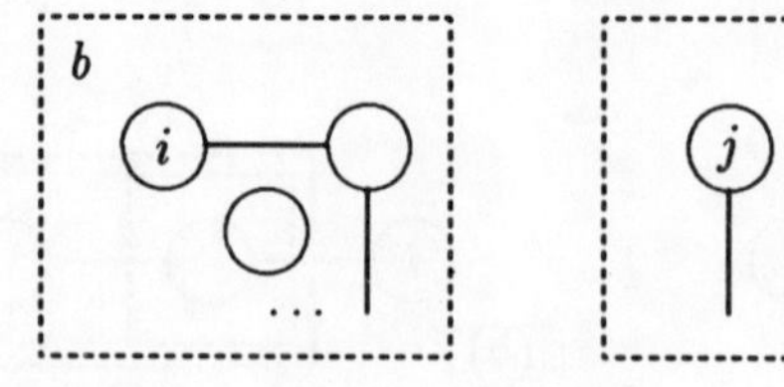

Fix vertices i and j, select an arbitrary vertex, say p, in c and consider integrating out X_p. By construction we know that

$$i \perp\!\!\!\perp j | K \backslash \{i, j\} \quad \text{and} \quad i \perp\!\!\!\perp p | K \backslash \{i, p\}.$$

By the block independence lemma it follows that

$$i \perp\!\!\!\perp \{j, p\} | K \backslash \{i, j, p\}$$

and by the reduction lemma

$$i \perp\!\!\!\perp j | K \backslash \{i, j, p\}.$$

The vertex p has been eliminated from the conditioning set. As this independence statement $i \perp\!\!\!\perp j | K \backslash \{i, j, p\}$ holds for all i in b and for all $j \neq p$ in c, we can conclude that in the independence graph for $K \backslash \{p\}$ with k-1 vertices, there is no edge between any vertex in b with any vertex in c. This graph again has two disconnected subgraphs.

Repetition of this argument eliminates further vertices from c until c is empty. The same process is applied to b, and the process terminates as K is finite, leaving $i \perp\!\!\!\perp j$, as required. □

We remarked above that any subset is a separating set for unconnected vertices so we need the somewhat more general:

Lemma 3.3.2 *Under the same conditions as in the preceding lemma*

$$i \perp\!\!\!\perp j | a \quad \textit{for all } i \in b \textit{ and } j \in c$$

and any subset a of K not containing i or j.

Proof: Modify the proof of the lemma above by choosing to integrate out those variables, X_p, only for p not in a. Then generalise the argument to an arbitrary finite graph. □

Lemma 3.3.3 *If a is any subset of vertices of K that separates two vertices i and j then $i \perp\!\!\!\perp j | a$, or more precisely, $X_i \perp\!\!\!\perp X_j | X_a$.*

Proof: We suppose that i and j are connected, for if not the preceding lemma applies. The first part of the proof blocks certain vertices together; to help understand the construction hold the following diagram in mind:

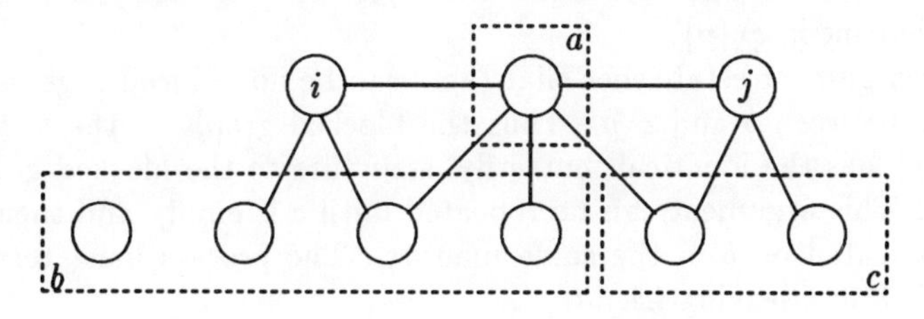

By hypothesis vertices i and j are separated by a. The remaining vertices can be partitioned in the following manner: any other vertex is either connected or is not connected to $a \cup \{i, j\}$. If connected, it is either separated from i by a or separated from j by a or both. So partition these remaining vertices into subsets b and c where

$$\begin{aligned} b &= \{l;\ l \text{ is not connected to } a \cup \{i,j\} \textit{ or } l \text{ is separated from } j \text{ by } a\}, \\ c &= \{l;\ l \text{ is not in } b \textit{ and } l \text{ is separated from } i \text{ by } a\ \}. \end{aligned}$$

Now map the independence graph of X into the blocked graph with 5 vertices $\{i, j, a, b, c\}$, and draw an edge between any pair if an edge can exist between the elements of that pair in the original independence graph. The blocked graph must have the following 'butterfly' structure:

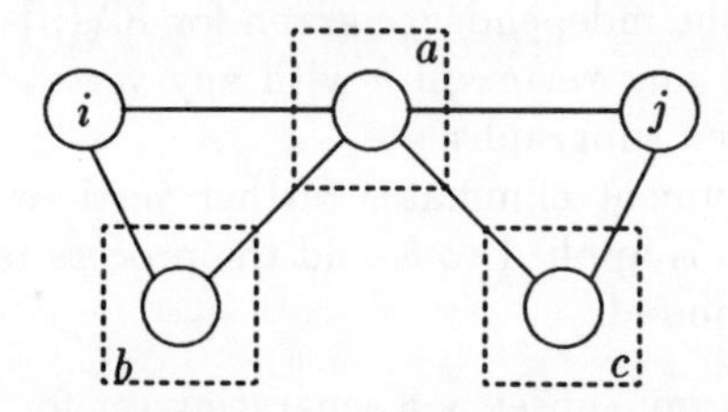

because, by hypothesis, there can be no edge between i and j, and by construction, no edges between i and c, j and b, and finally, between b and c.

The second part of the proof is to repeat the block independence and reduction argument. First consider two vertices p and q in c, and integrate out X_p. The construction of an independence graph implies that

$$i \perp\!\!\!\perp j | K \backslash \{i, j\} \quad \text{and} \quad i \perp\!\!\!\perp p | K \backslash \{i, p\}$$

so that application of block independence and reduction implies

$$i \perp\!\!\!\perp j | K \backslash \{i, j, p\}.$$

Thus, in the independence graph on $K \backslash \{p\}$, there is no edge between i and j. Furthermore because q is in c and all vertices in c are separated from i by a, a similar argument reveals

$$i \perp\!\!\!\perp q | K \backslash \{i, q, p\} \quad \text{for all } q \text{ in } c.$$

Thus integrating out X_p has not induced an edge between i and j, nor between i and any element in $c \backslash \{p\}$.

A similar argument establishes that there can be no induced edge between j and b nor between b and $c \backslash p$. Thus the blocked graph on the 5 vertices $\{i, j, a, b, c \backslash p\}$ has the identical 'butterfly' structure to the blocked graph on $\{i, j, a, b, c\}$. This argument can be repeated until c is empty and then reapplied and repeated on b in the same manner. The process must terminate because K is finite, leaving $i \perp\!\!\!\perp j | a$. □

The construction used for this lemma immediately generalises to random vectors and the same argument goes through with X_b and X_c replacing X_i and X_j. Under the condition that the density function be positive and with the pairwise definition of an independence graph, we have proved

Theorem 3.3.4 The separation theorem. *If X_a, X_b and X_c are vectors containing disjoint subsets of variables from X, and if, in the independence graph of X, each vertex in b is separated from each vertex in c by the subset a, then*

$$X_b \perp\!\!\!\perp X_c | X_a.$$

The converse of the separation theorem is immediate: if it is true that whenever a separates i and j that $i \perp\!\!\!\perp j | a$ then it is true that, whenever i and j are not adjacent, $i \perp\!\!\!\perp j |$rest. However it is trivial because the 'rest' is always a separating set for non-adjacent vertices and hence the conclusion forms part of the premiss to the assertion.

Minimal separating subsets

One consequence of the separation theorem is that some of the conditioning variables in an independence relation may be redundant. A conditional independence between a pair of variables is said to be *minimal* if it is not possible to apply the separation theorem to eliminate any variable from the conditioning set.

EXAMPLE 3.3.3 The conditional independences embodied in the graph

are $1 \perp\!\!\!\perp 3 | \{2, 4\}$, $1 \perp\!\!\!\perp 4 | \{2, 3\}$ and $2 \perp\!\!\!\perp 4 | \{1, 3\}$. The minimal independences are $1 \perp\!\!\!\perp 3 | 2$, $1 \perp\!\!\!\perp 4 | 2$, $1 \perp\!\!\!\perp 4 | 3$ and $2 \perp\!\!\!\perp 4 | 3$. In this example the set of minimal independences is larger than the original set of pairwise conditional independences. □

One might conjecture that a strong converse to the separation theorem holds: if, for all non-adjacent vertices i and j and for all minimal separators a, it is true that $X_i \perp\!\!\!\perp X_j | X_a$, then it is true that $X_i \perp\!\!\!\perp X_j |$rest. However the following is a counter-example.

EXAMPLE 3.3.4 Consider the graph on three vertices

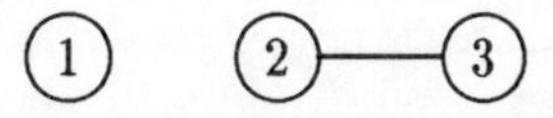

The set of all pairwise conditional independences given the rest is

$$I_{pair} = \{ 1 \perp\!\!\!\perp 2 | 3,\ 1 \perp\!\!\!\perp 3 | 2 \},$$

while the set of minimal conditional independences is

$$I_{min} = \{ 1 \perp\!\!\!\perp 2,\ 1 \perp\!\!\!\perp 3 \}.$$

Now block independence establishes that I_{pair} implies $1 \perp\!\!\!\perp \{2, 3\}$ and hence that I_{pair} implies I_{min}, but the converse is false, as marginal independence does not imply joint independence. For instance, see the counterexample in Example 2.1.3. □

3.4 Markov Properties

In this section we take up the issue of equivalent Markov properties.

Separation and prediction

The following may be obvious but its importance makes it worth stating explicitly. Consider predicting the value of one random vector, say Y, from information on another, say Z. We take it to be axiomatic that the structural information relating Z to Y is entirely contained in the conditional density function of Y given Z, $f_{Y|Z}$, which thereby determines the form of the optimal predictor. Now suppose we are given the independence graph of a random vector X, that includes the subset, X_a $(= Y)$, which we wish to predict. What information does this graph have about the form of an optimal predictor? Let X_b denote the vector containing exactly those variables that are adjacent to at least one element of X_a, that is the boundary of a, and let X_c denote the remaining variables in X. By construction the vectors X_a and X_c are separated by X_b and so by the corollary to the separation theorem $X_a \perp\!\!\!\perp X_c | X_b$. But this independence statement is equivalent to $f_{a|b \cup c} = f_{a|b}$. Hence, given X_b, no more information for predicting X_a can be extracted from X_c.

EXAMPLE 3.4.1 Suppose X has the graph

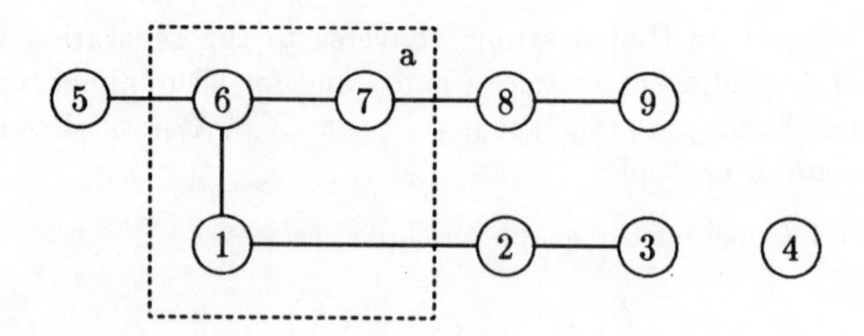

and X_a corresponds to the set $\{1, 6, 7\}$, then the boundary of a is $b = \{2, 5, 8\}$, and the remaining vertices are $c = \{3, 4, 9\}$. The information on (X_3, X_4, X_9) is irrelevant for predicting (X_1, X_6, X_7). □

Local Markov property

The local Markov property is closely related to prediction, and expresses the independence statements in the graph in terms of a specified vertex and its nearest neighbours in the boundary set. It turns out, in treatments of graphs with infinite numbers of vertices, that this is the easiest property to generalise. A random vector with graph G has the *local Markov property*, if, for every vertex i, with boundary $a = \mathrm{bd}\,(i)$, and b the set of remaining vertices, then

$$X_i \perp\!\!\!\perp X_b | X_a, \quad \text{where } b = K \backslash (\{i\} \cup a).$$

A mnemonic formulation is that $i \perp\!\!\!\perp$rest|boundary. Each of the k independence statements are statements about the joint distribution of all k variables, because $\{i\} \cup a \cup b = K$. Note that the list of independences may contain some redundancies.

EXAMPLE 3.4.2 For the graph with five vertices

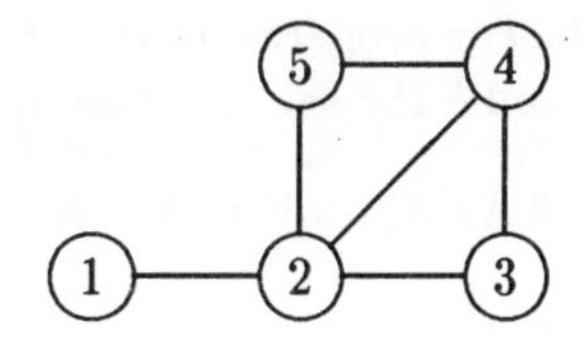

the full list of independences specified by the local Markov property is

$$1 \perp\!\!\!\perp \{3,4,5\}|2, \quad 3 \perp\!\!\!\perp \{1,5\}|\{2,4\}, \quad 4 \perp\!\!\!\perp 1|\{2,3,5\}, \quad 5 \perp\!\!\!\perp \{1,3\}|\{2,4\}.$$

There is redundancy here as the independence between X_3 and X_5 given (X_2, X_4) occurs twice in the list; and the first independence implies $X_1 \perp\!\!\!\perp X_5 | X_2$ so that part of the final independence, namely $X_1 \perp\!\!\!\perp X_5 | (X_2, X_4)$, is made redundant. □

To check that a set of variables is locally Markov with respect to a given graph means checking that the distribution satisfies, for each vertex i, the condition $i \perp\!\!\!\perp$rest|boundary. The converse problem, of constructing the graph given the information that the distribution of the set is locally Markov, is rather more difficult than the equivalent exercise based on the pairwise conditional independence. This is because, for any given vertex, there are 2^{k-1} ways of partitioning the remaining variables into the two sets, the boundary and the rest. Solving this problem of determining the neighbours of a given vertex, by checking for conditional independence with each of its k-1 potential neighbours, presupposes the equivalence of the pairwise independence and local Markov properties.

Equivalent Markov properties

We have used the independence of a pair of variables given the rest to define the conditional independence graph, G, for the random vector X. However there are alternative formulations:

- Firstly, the *pairwise Markov* property, that for all non-adjacent vertices i and j,
$$X_i \perp\!\!\!\perp X_j | X_a, \text{ where } a = K \backslash \{i, j\}.$$

- Secondly, the *global Markov* property, that, for all disjoint subsets a, b and c, of K, whenever b and c are separated by a in the graph, then X_b and X_c are independent given X_a alone:
$$X_b \perp\!\!\!\perp X_c | X_a.$$
It is global in the sense that the subsets, and in particular the separating set, are potentially arbitrary subsets of vertices.

- Thirdly, the *local Markov* property, that, for every vertex i, if $a = bd(i)$ is its boundary set, and b the set of remaining vertices, then
$$X_i \perp\!\!\!\perp X_b | X_a, \text{ where } b = K \backslash (\{i\} \cup a).$$

It is a remarkable fact that these are equivalent.

Theorem 3.4.1 *The three Markov properties: pairwise Markov, local Markov and global Markov, are equivalent.*

Proof: To summarise the proof of these equivalences we argue

1. The global Markov property implies the local Markov property, because the boundary set is always a separating subset.

2. The local Markov property implies the pairwise independence property, because, if the graph with vertices $K = \{1, 2, \ldots, k\}$ satisfies the local Markov property then for every vertex i, with boundary set $a = \text{bd}(i)$,
$$X_i \perp\!\!\!\perp X_b | X_a,$$
where b denotes the remaining vertices, i.e. $b = K \backslash (\{i\} \cup a)$. Select any vertex j not adjacent to i, then j is in b; put $c = b \backslash \{j\} = K \backslash (\{i, j\} \cup a)$ and rewrite the independence as
$$X_i \perp\!\!\!\perp (X_j, X_c) | X_a.$$
By the block independence lemma this is equivalent to
$$X_i \perp\!\!\!\perp X_j | (X_a, X_c) \quad \text{and} \quad X_i \perp\!\!\!\perp X_c | (X_a, X_j).$$

But $a \cup c = K \backslash \{i, j\}$ is the 'rest', and the first independence is exactly the pairwise independence property.

3. The separation theorem asserts that the pairwise conditional independence property implies the global Markov property. □

3.5 Directed Acyclic Independence Graphs

In many, if not most, studies of several interacting variables there is a striking lack of symmetry in the roles played by the variables that corresponds to a notion of causality and the premiss that if X causes Y then Y cannot cause X. We will not digress into the philosophy of causation, with its notions of time and time irreversibility, but we certainly wish to incorporate methods for dealing with such assymmetries. Now a neat way to portray the relationship that 'X effects Y' is by means of a directed graph and its diagram

together with the conditional probability density function, $f_{Y|X}$, a natural object of study for the probability modeller.

For example, suppose that in an educational study, measures are made on on the social class, X_1, and income X_2, of the head of a family, and also on the educational achievement of the eldest child, Y. The variables are not symmetric but satisfy a partial ordering. Our approach is firstly, to inquire if Y depends on both of X_1 and X_2, by assessing the independence statements $Y \perp\!\!\!\perp X_1 | X_2$ and $Y \perp\!\!\!\perp X_2 | X_1$ in the conditional distribution of Y given X; and secondly to assess the interaction between X_1 and X_2 in the distribution of (X_1, X_2) alone, without reference to Y. The conditional independence $X_1 \perp\!\!\!\perp X_2 | Y$ is of no interest.

We investigate how to represent such structures using directed independence graphs. In a directed graph a subgraph with edge (1,2) and a subgraph with the edge (2,1) have the respective diagrams

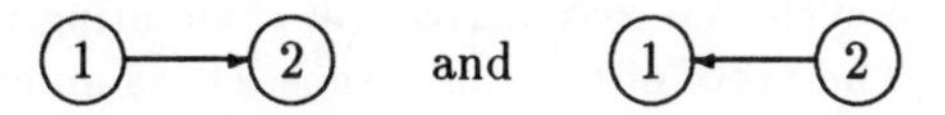

We start by requiring all edges to be directed. When later we do allow a mix of directed and undirected edges then we suppose

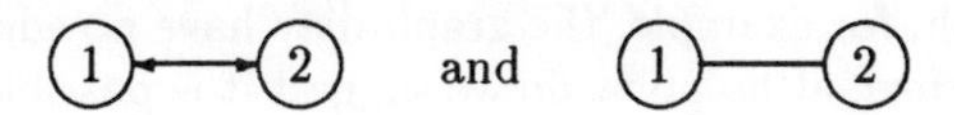

to be equivalent.

Directed cycles and orderings

Extending independence graphs to include directed edges immediately faces the problem of directed cycles such as

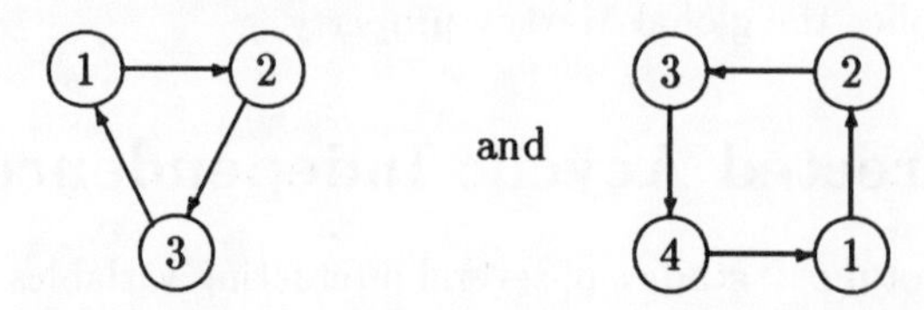

At first sight allowing directed cycles appears to have good possibilities of modelling 'feed-back', so that X_1 effects X_2 which effects X_3 which in turn effects X_1. But unfortunately, there is no suitable joint probability to model this situation; for instance, in the directed 3-cycle pictured here, we would like to express the joint density function as $f_{3|2}f_{2|1}f_{1|3}$, but apart from very special instances, this is not a well defined probability density function. Hence, by decree, we shall not allow a graph that contains a directed cycle to represent a directed version of an independence graph.

It turns out that an equivalent assumption to excluding directed cycles is to suppose that the vertices are completely ordered, that is, there exists a relation $\prec$ on the elements of $K = \{1, 2, \ldots, k\}$ such that: for all i and j in the set, (i) either $i \prec j$ or $j \prec i$, (ii) $\prec$ is irreflexive, and (iii) $\prec$ is transitive, so that if $i \prec j$ and $j \prec l$ then $i \prec l$. In this case we can write $1 \prec 2 \prec \ldots \prec k$, and think that each variable has a well defined past and future. Applied to the directed graph, an ordering means that any edge in the graph can have only one possible direction. By restricting attention to acyclic graphs, we may maintain the notion of parenthood, in which the direct antecedents of vertex i are known as the parents of i, and represented as $\mathrm{pa}(i)$.

Lemma 3.5.1 Ordering an acyclic directed graph. *In a directed graph, the conditions that: (i) there is no directed cycle, and (ii) there exists a complete ordering of the vertices that is respected in the graph, are equivalent.*

Proof: To go one way: if there is a directed cycle in the graph then for any element, i, in the cycle, we may derive $i \prec i$, violating the irreflexive nature of the ordering. The proof of the converse is left as an exercise. □

We have just seen that such an assumption is equivalent to supposing the existence of a complete ordering for the vertices. Note that it is not usually possible to deduce the underlying complete ordering of the vertices from a particular graph, for example, the graph may have no edges at all. The order is assumed *a priori*; at no point do we suggest it is possible for the probability modeller or data analyst to determine the direction of an edge, or the implicit order underlying the graph, other than by presupposition.

Directed independence graphs

The *a priori* ordering of the vertices endows each variable with a past, present and future. We can now define a directed independence graph, termed a *recursive* graph, by Wermuth and Lauritzen (1983): the natural conditioning set for each pairwise independence statement is the past, so define $K(i) = \{1, 2, \ldots, i\}$, to be the set which comprises the past and present for the i-th variable.

Definition. The *directed independence graph* of X is the directed graph $G^{\prec} = (K, E^{\prec})$, where $K = \{1, 2, \ldots, k\}$, $K(j) = \{1, 2, \ldots, j\}$ and the edge (i, j), with $i \prec j$, is *not* in the edge set $E^{\prec}$ if and only if $j \perp\!\!\!\perp i | K(j) \setminus \{i, j\}$. □

This is the same definition used for the undirected independence graph with the conditioning set modified, from the 'rest' comprising all past and future variables, to just the past. This one crucial difference between directed and undirected independence graphs means that for an undirected graph the independence statements are statements about a single joint distribution while for a directed graph they are statements about a sequence of marginal distributions. Though, we must remark that this sequence has the property that enough information is given to define the joint distribution by virtue of the *recursive factorisation identity*:

$$f_{12\ldots k} = f_{k|K(k)\setminus\{k\}} f_{k-1|K(k-1)\setminus\{k-1\}} \cdots f_{2|1} f_1.$$

Because we have an implicit *a priori* ordering of the vertices, application of the independences to evaluate the joint density function is entirely straightforward.

EXAMPLE 3.5.1 Given a seven dimensional vector X and the following pairwise conditional independences

$2 \perp\!\!\!\perp 1$	$5 \perp\!\!\!\perp 3 \| \{1,2,4\}$	$6 \perp\!\!\!\perp 4 \| \{1,2,3,5\}$	$7 \perp\!\!\!\perp 4 \| \{1,2,3,5,6\}$
$3 \perp\!\!\!\perp 1 \| \{2\}$	$5 \perp\!\!\!\perp 1 \| \{2,3,4\}$	$6 \perp\!\!\!\perp 2 \| \{1,3,4,5\}$	$7 \perp\!\!\!\perp 3 \| \{1,2,4,5,6\}$
	$5 \perp\!\!\!\perp 2 \| \{1,3,4\}$	$6 \perp\!\!\!\perp 1 \| \{2,3,4,5\}$	$7 \perp\!\!\!\perp 2 \| \{1,3,4,5,6\}$
			$7 \perp\!\!\!\perp 1 \| \{2,3,4,5,6\}$,

the independence graph has the diagram

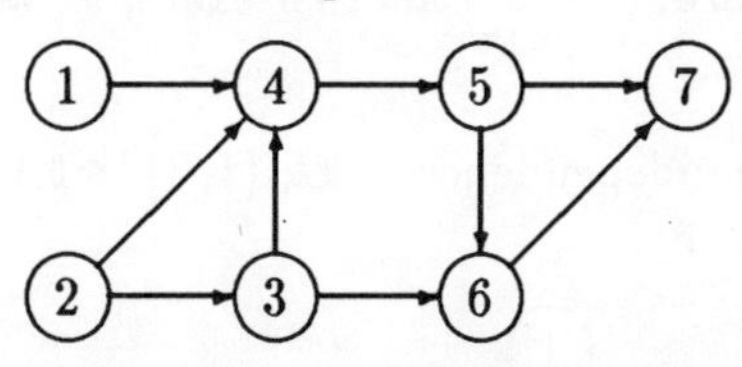

Each pairwise independence can be immediately applied to the recursive factorisation identity to derive the form of the joint density as

$$f_{12\ldots 7} = f_{7|56} f_{6|53} f_{5|4} f_{4|123} f_{3|2} f_2 f_1.$$

It is the ordering of the vertices, here specified numerically, that determines the conditioning set. □

Wermuth condition

We wish to elucidate the Markov properties of a directed independence graph, but hope to avoid much further work in proving such properties by exploiting the relationship of the directed graph to independence statements elicited from its associated undirected graph. If $G^{\prec} = (K, E^{\prec})$ then the *associated undirected graph* is defined as $G^u = (K, E^u)$ with the same vertex set and an undirected edge replacing each directed edge.

EXAMPLE 3.5.2 Consider the simple Markov chain defined from the independences $4 \perp\!\!\!\perp 2 | \{1,3\}$, $4 \perp\!\!\!\perp 1 | \{2,3\}$, and $3 \perp\!\!\!\perp 2 | \{1\}$. Its diagram is

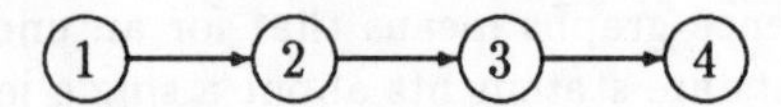

Our intuition suggests that as the path from 1 to 4 is cut by vertex 2, then $4 \perp\!\!\!\perp 1 | 2$, and variable 2 alone is sufficient in the conditioning set for this independence to hold. But how can we prove this? It is not directly obvious: for instance, applying block independence to the two statements $4 \perp\!\!\!\perp 2 | \{1,3\}$ and $4 \perp\!\!\!\perp 1 | \{2,3\}$ and deriving $4 \perp\!\!\!\perp \{1,2\} | 3$ does not help, because we cannot integrate out 3 from the conditioning set.

Now the associated undirected graph, G^u, has the diagram

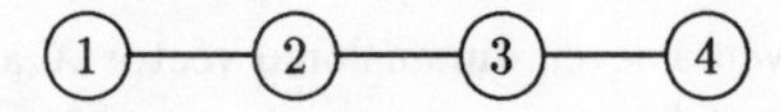

If we interpret this as an independence graph then firstly, three independence statements hold: $4 \perp\!\!\!\perp 2 | \{1,3\}$, $4 \perp\!\!\!\perp 1 | \{1,3\}$ and $3 \perp\!\!\!\perp 1 | \{2,4\}$; and secondly, these three statements imply the three statements in the list defining the directed independence graph. Will this insight do the trick? □

Before launching forward, consider another example, which indicates we need to be careful.

EXAMPLE 3.5.3 The independences $4 \perp\!\!\!\perp 2 | \{1,3\}$, $4 \perp\!\!\!\perp 1 | \{2,3\}$ and $1 \perp\!\!\!\perp 2$ define the recursive graph

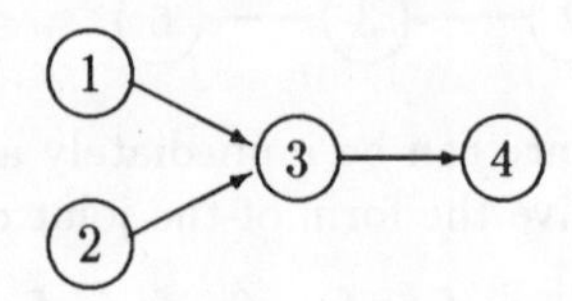

If we interpret the associated undirected graph as an independence graph we could conclude that $1 \perp\!\!\!\perp 2|3$; but this independence statement cannot hold in the directed graph, because as we saw in Section 2.1, marginal independence does not imply conditional independence. □

Thus we have two examples, one where the directed and the undirected graph might well have the same independence interpretations, and another where the interpretations are certainly different. It transpires that there is an easy way to classify this behaviour:

Definition. A directed graph satisfies the ***Wermuth condition*** if no subgraph has the configuration

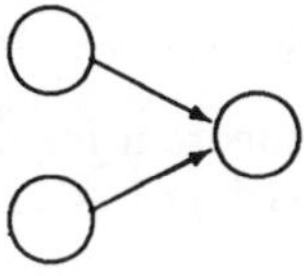

□

Thus the graph of Example 3.5.2 is Wermuth, while the graph of Example 3.5.3 is not.

EXAMPLE 3.5.4 Any ordering of the (acyclic) chordless four cycle fails the Wermuth condition. There are only three distinct labellings:

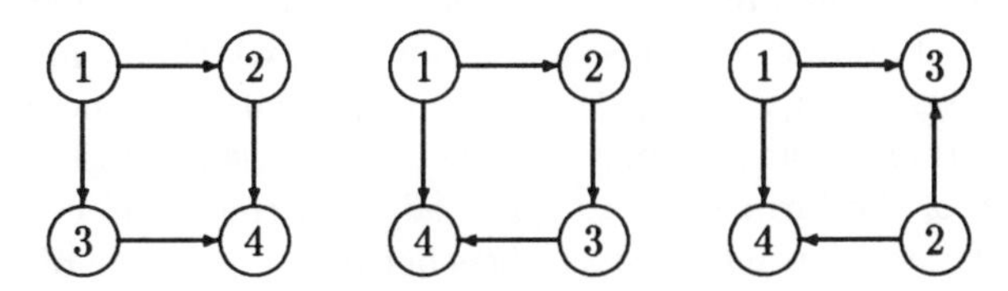

and each contains a forbidden configuration. □

Moral graphs and Markov properties

We return to the question of when the directed and the associated undirected graph have the same independence intepretations.

Definition. The ***moral graph*** associated with the directed graph $G^{\prec} = (K, E^{\prec})$ is the undirected graph $G^m = (K, E^m)$ on the same vertex set and with an edge set obtained by including all edges in $E^{\prec}$ together with all edges necessary to eliminate forbidden Wermuth configurations from $G^{\prec}$. □

It is termed a moral graph because it 'marries parents'; the adjective is due to Lauritzen and Spiegelhalter (1988).

EXAMPLE 3.5.5 A directed acyclic graph and its associated moral graph:

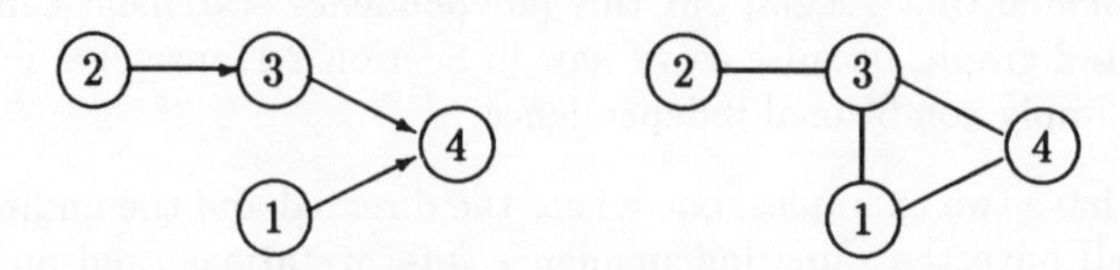

The Wermuth condition fails because variables 1 and 3, the parents of 4, are unmarried. They are married with an undirected edge and all other directions are dropped, giving the associated moral graph. This marriage does not make variable 1 a parent of 3, and so does not introduce another forbidden configuration on the subgraph $\{1,2,3\}$. □

Theorem 3.5.2 A Markov theorem for directed independence graphs. *The directed independence graph $G^{\prec}$ possesses the Markov properties of its associated moral graph, G^m.*

Proof: The recursive factorisation identity is simplified by application of each pairwise independence statement in the directed graph:

$$\begin{aligned} f_K &= \prod_{j=2}^{k} f_{j|K(j)\setminus\{j\}} f_1 && \text{is an identity,} \\ &= \prod_{j=2}^{k} f_{j|\mathrm{pa}(j)} f_1 && \text{application of independences,} \\ &= \prod_{j=2}^{k} g_{j\cup \mathrm{pa}(j)} g_1 \end{aligned}$$

by appropriately defining the functions g. We thus have an expansion for the joint density function in terms of functions, g_a, which are functions of x_a for $a = \{1\}, \{2\} \cup \mathrm{pa}(2), \ldots, \{k\} \cup \mathrm{pa}(k)$. By application of the factorisation criterion to this expansion we may deduce all pairwise conditional independence statements of the form $i \perp\!\!\!\perp j | \text{rest}$. The edges of the undirected independence graph for f_K are characterised as edges between j and each of its parents, and edges between the members of each pair of parents of j. That is, the edge set of the moral graph, G^m. □

EXAMPLE 3.5.6 Example 3.5.1 continued: the seven dimensional vector X with directed independence graph given in the example above, factorised down to

$$f_{12\ldots7} = f_{7|56} f_{6|53} f_{5|4} f_{4|123} f_{3|2} f_2 f_1.$$

Its independences can be read off the moral graph, G^m, with diagram

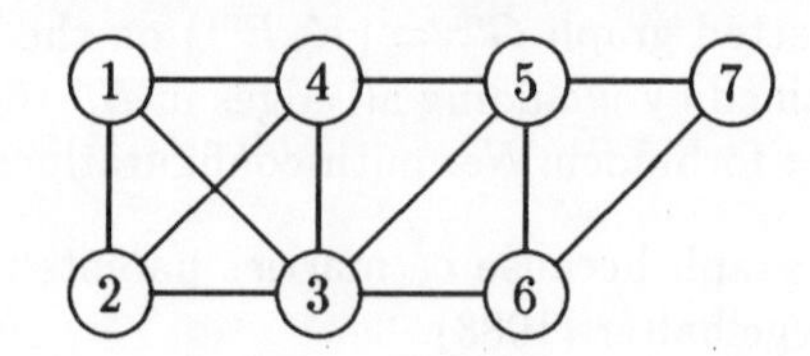

For example $7 \perp\!\!\!\perp \{1,2\} | \{3,4\}$, a fact not easy to deduce from the original defining independences. □

The full moral graph can obscure certain independences. In this example, $G^{\prec}$ implies $\{3,2\} | \perp\!\!\!\perp 1$, but the subgraph $\{1,2,3\}$ is complete in G^m. The fact may be deduced, though, by strengthening the assertion of the equivalence of directed and moral graphs to refer to the graph on any initial segment $K(i)$ of the vertex set, rather than just the one on the full vertex set K itself.

The following is a reformulation of a result from Wermuth (1980).

Corollary 3.5.3 *If $G^m = G^u$ then the Markov properties of the directed graph, $G^{\prec}$, are exactly identical to those of G^m.*

When the moral graph is identical to the undirected graph, so that no 'marrying' is required, then the Markov properties of the directed graph are exactly identical to those of the moral graph. We shall leave the proof as an exercise to the reader. Furthermore they will be the same as those of any directed graph whose undirected graph is identical to this moral graph; an application of this is the reversal of direction in a simple Markov chain.

EXAMPLE 3.5.7 The Markov properties of

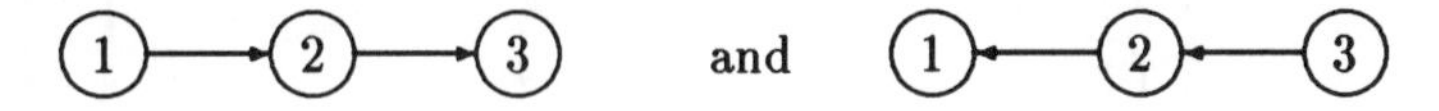

are identical. □

3.6 Chain Independence Graphs

We want to extend the theory to cover independence graphs with a mixture of directed and undirected edges. Though this is a generalisation, for it must include the theory of both these special cases, its motivation is entirely practical. For example, consider an experiment on crop yield, in which the response vector records measures of plant size during the growing period, final yield and fruiting quality, all made under a variety of treatment combinations. The graph must allow the possibility of some directed edges, because it is unrealistic to imagine that final fruiting quality, say, effects the mid-term growth rate, while the converse effect is eminently reasonable. It must also allow the possibility of undirected edges, because any interaction between final fruiting quality and yield, say, may be a two-way affair.

Mathematically, the way ahead is to assume that the vertex set satisfies a particular type of partial ordering, $\preceq$, instead of the complete ordering imposed on directed graphs. The partial ordering is derived by supposing the rather strong condition that the vertex set K can be partitioned into subsets

$b_1, b_2, \ldots, b_m$, called ***blocks***, which are completely ordered, that is, the blocks form a *chain*. The induced partial order, $\preceq$, on the individual vertices of K is that $i \prec j$, whenever $i \in b_r$, and $j \in b_s$ and $r < s$; and $i \preceq j$, whenever $i, j \in b_r$. The parents of i in b_r are drawn from the 'past', $b_1 \cup b_2 \cup \ldots \cup b_{r-1}$, and are joined to i by directed edges or arrows. The elements in b_1 are potential causes of the elements in b_2, the elements in $b_1 \cup b_2$ are potential causes of the elements in b_3, and so on.

EXAMPLE 3.6.1 Consider an eight variable system $K = \{1, 2, \ldots, 8\}$, partitioned into subsets $b_1 = \{1, 2, 3\}$, $b_2 = \{4\}$, $b_3 = \{5, 6\}$, and $b_4 = \{7, 8\}$, with an edge set defined by the edges in the diagram:

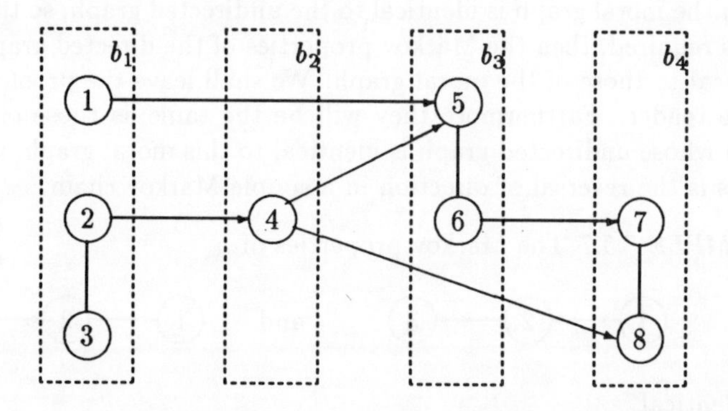

Any two elements from different blocks are only joined by an arrow; and two from the same block are only joined by a line. Consider vertex $5 \in b_3$. The parents of 5 are $\{1, 4\}$ and lie in blocks preceding b_3, while 6 is a neighbour of 5. The past and present for each vertex are the sets $K(1) = b_1$, $K(2) = b_1$, $K(3) = b_1$, $K(4) = b_1 \cup b_2$, and so on, until $K(8) = K$. Note $K(5) = K(6) = \{1, 2, 3, 4, 5, 6\}$. □

The essential property satisfied by this construction is that any edge is undirected for intra-block vertices, and directed for inter-block vertices with direction determined by the ordering on the blocks. This block formulation excludes graphs with directed cycles, as did our formulation of completely directed independence graphs, and, in addition, disallows graphs with cycles containing at least one directed edge, such as

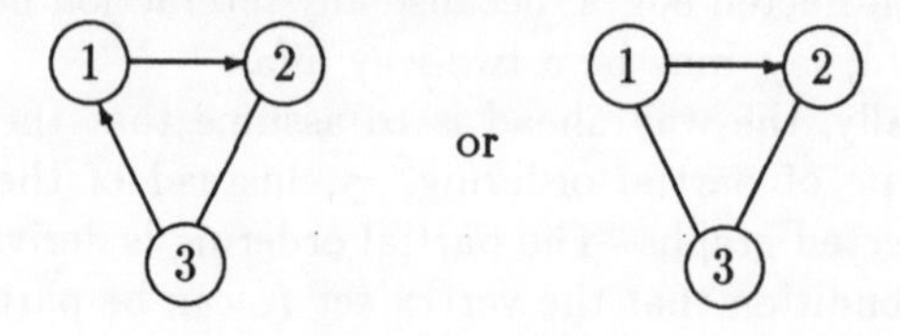

Graphs such as these violate the partial ordering assumption, by requiring that at least one vertex must exist in more than one block. Such cycles as these, containing at least one directed edge, like directed cycles consisting only of directed edges, do not admit a reasonable factorisation of the joint probability density function.

We may now define the block independence graph from pairwise conditional independences taking the conditioning set for each statement as $K(i)$,the set of all 'past' and 'present' variables defined with respect to vertex i, that is, $K(i) = b_1 \cup b_2 \cup \ldots \cup b_{r(i)}$ where $r(i)$ is the index of the block containing i.

Definition. The *chain independence graph* of X is the graph $G^{\preceq} = (K, E^{\preceq})$, where $K = \{1, 2, \ldots, k\}$, $K(i) = \cup_{l \leq r(i)} b_l$ and the edge (i, j) with $i \preceq j$, is *not* in the edge set $E^{\preceq}$ if and only if $j \perp\!\!\!\perp i | K(j) \backslash \{i, j\}$. If this condition fails, and $i \prec j$, then the edge is directed and only $(i, j) \in E^{\preceq}$; otherwise it is undirected and both $(i, j) \in E^{\preceq}$ and $(j, i) \in E^{\preceq}$. □

Such graphs are also called *block recursive* by Lauritzen and Wermuth (1984, 1989), and $K(i)$ is the *concurrent* past.

Instead of specifying a block structure, an apparently more general definition is to allow a mixture of directed and undirected edges subject to the constraint that no cycle containing one or more directed edges is permissible. Under this hypothesis it is fairly straightforward to demonstrate the existence of a block representation which separates vertices into blocks such that inter-block edges are arrows, intra-block edges are lines, and such that the blocks are completely ordered. However, the matter is a little more complicated, because the block representation need not be unique. For example, with vertex set $K = \{1, 2, 3\}$ and an empty edge set, $E = \phi$, three alternative block structures are: (i) $b_1 = \{1, 2\}$, $b_2 = \{3\}$ or (ii) $b_1 = \{1\}$, $b_2 = \{2\}$, $b_3 = \{3\}$ or (iii) $b_1 = \{1, 2, 3\}$. Clearly the initial list of pairwise independences is different, for instance: as neither $(1, 2)$ nor $(2, 1)$ is in the edge set, then under (i), $1 \perp\!\!\!\perp 2$; while under (iii), $1 \perp\!\!\!\perp 2 | 3$. It is shown by Frydenberg (1986) that all consistent block representations are probabilistically equivalent, and Frydenberg (1989) gives a fairly definitive answer to the question of when two chain graphs are equivalent.

Markov properties of chain graphs

The argument of the previous section, deriving the Markov properties of directed independence graphs extends with little difficulty to chain independence graphs. Again, the crucial ingredient that allows attribution of the Markov properties of the associated undirected graph, G^u, is the verification of the Wermuth condition; and, if the condition fails, substitution of the moral graph, G^m, constructed by marrying parents. The definition of the Wermuth condition requires a slight extension. It must still prohibit graphs with unmarried parents, but the set of possible parents is enlarged to include all parents of a connected subset of 'children', as illustrated in the next example.

EXAMPLE 3.6.2 The vertex set $K = \{1,2,3,4,5\}$, is partitioned into two blocks, $b_1 = \{1,2\}$, and $b_2 = \{3,4,5\}$ and the defining independences are $1 \perp\!\!\!\perp 2$, $3 \perp\!\!\!\perp 2 | \{1,4,5\}$, $4 \perp\!\!\!\perp 1 | \{2,3,5\}$, $5 \perp\!\!\!\perp 1 | \{2,3,4\}$, and $5 \perp\!\!\!\perp 2 | \{1,3,4\}$. The diagram of the graph is given on the left. The graph fails the Wermuth condition because there are two vertices, 3 and 4, in the same block which are connected, but with parents that are not joined.

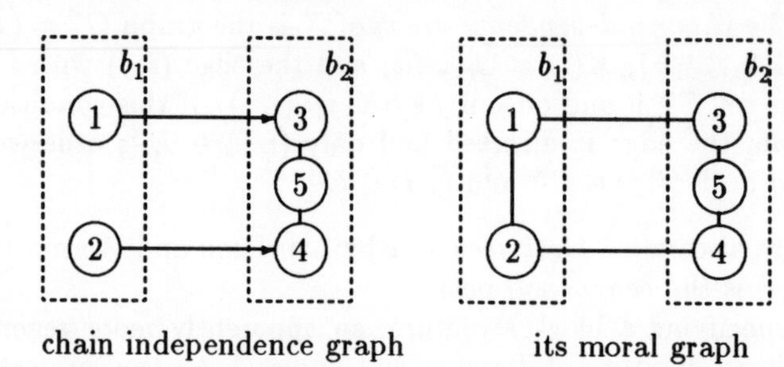

chain independence graph its moral graph

The moral graph on the right joins these parents. Note that the diagonal vertices are not joined: even though vertex 5 is required in the conditioning set for the independence of 1 and 4, this does not make 5 a parent of 4. The blocks, of course, are redundant for reading the moral graph. □

EXAMPLE 3.6.3 A second example gives a different moral graph. The vertex set $K = \{1,2,3,4,5\}$, is now partitioned into three blocks, $b_1 = \{1,2\}$, $b_2 = \{3,4\}$ and $b_3 = \{5\}$ with diagram

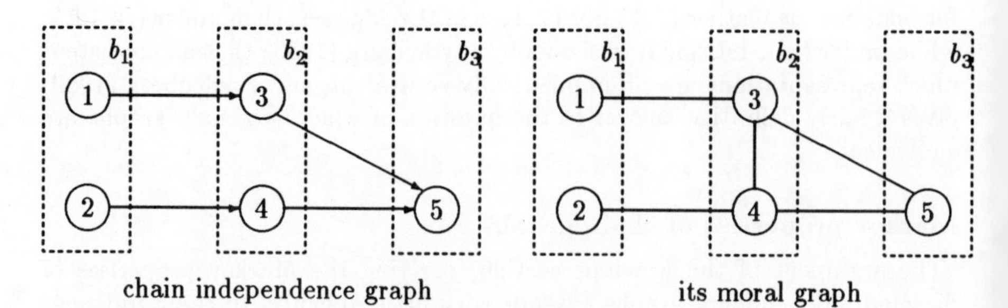

chain independence graph its moral graph

As vertices 3 and 4 are not connected in this block independence graph, vertices 1 and 2 are not joined in the moral graph. □

We now derive the Markov properties of the block independence graph. Fundamental is the recursive factorisation identity expressed in the terms of the blocks:

$$f_K = f_{b_1} \prod_{r=2}^{m} f_{b_r|b_1 \cup b_2 \cup \ldots \cup b_{r-1}}.$$

The independence properties of the undirected graph are characterised by its pairwise conditional independence statements. Our method of attack is to apply the factorisation criterion to this identity for the joint density function, and elicit all such statements. The proof provides a precise definition of the Wermuth condition.

Theorem 3.6.1 A Markov theorem for chained block independence graphs. *The directed independence graph $G^{\preceq}$ posseses the same independence interpretations as its associated moral graph, G^m.*

Proof: Consider two arbitrary vertices i and j of K with $i < j$. There are two cases to consider: either $i \in b_r$ and $j \in b_s$ with $r < s$, or both $i, j \in b_r$; when an edge exists, these cases correspond to an arrow and a line, respectively.

We argue that

1. All edges present in $G^{\preceq}$ will be preserved in G^m. This is because if the structure of $f_{b_r|\text{past}}$ does not permit a factorisation of j from i then, by virtue of the recursive factorisation identity, such a factorisation is also prohibited for f_K.

2. Suppose now that the edge between i and j is missing in $G^{\preceq}$. The edge can only be reinstated if, in the factorisation of f_K, there is some block r such that the factor $f_{b_r|\text{past}}$ prevents a factorisation of i and j. That is, if and only if, either, i and j are the parents of some future vertex $p \in b_r$, or, i is the parent of $p \in b_r$ and j is the parent of vertex $q \in b_r$ that are connected.

To establish this last point, suppose that p and q are not connected. Then b_r may be partitioned into two sets, c_1 containing p and c_2 containing q, so that

$$f_{b_r|\text{past}} = f_{c_1|\text{past}} f_{c_2|\text{past}}.$$

As i is only a parent to p and j only a parent to q, the term $f_{b_r|\text{past}}$ factorises into the product of a function of i and a function of j and hence does not induce an edge in G^m.

But the moral graph is constructed from $G^{\preceq}$ exactly according to the recipe outlined in 1 and 2. □

EXAMPLE 3.6.4 Consider the chained block graph of Example 3.6.1 in this section. Its defining independence statements are

$1 \perp\!\!\!\perp 3 \mid 2$	$5 \perp\!\!\!\perp 3 \mid \{1,2,4,6\}$	$7 \perp\!\!\!\perp 1 \mid \{2,3,4,5,6,8\}$	$8 \perp\!\!\!\perp 1 \mid \{2,3,4,5,6,7\}$
$1 \perp\!\!\!\perp 2 \mid 3$	$6 \perp\!\!\!\perp 1 \mid \{2,3,4,5\}$	$7 \perp\!\!\!\perp 2 \mid \{1,3,4,5,6,8\}$	$8 \perp\!\!\!\perp 2 \mid \{1,3,4,5,6,7\}$
$4 \perp\!\!\!\perp 1 \mid \{2,3\}$	$6 \perp\!\!\!\perp 2 \mid \{1,3,4,5\}$	$7 \perp\!\!\!\perp 3 \mid \{1,2,4,5,6,8\}$	$8 \perp\!\!\!\perp 3 \mid \{1,2,4,5,6,7\}$
$4 \perp\!\!\!\perp 3 \mid \{1,2\}$	$6 \perp\!\!\!\perp 3 \mid \{1,2,4,5\}$	$7 \perp\!\!\!\perp 4 \mid \{1,2,3,5,6,8\}$	$8 \perp\!\!\!\perp 5 \mid \{1,2,3,4,6,7\}$
$5 \perp\!\!\!\perp 2 \mid \{1,3,4,6\}$	$6 \perp\!\!\!\perp 4 \mid \{1,2,3,5\}$	$7 \perp\!\!\!\perp 5 \mid \{1,2,3,4,6,8\}$	$8 \perp\!\!\!\perp 6 \mid \{1,2,3,4,5,7\}$

The recursive factorisation identity expressed in the terms of the blocks is

$$f_K = f_{b_4|b_1 \cup b_2 \cup b_3} f_{b_3|b_1 \cup b_2} f_{b_2|b_1} f_{b_1},$$

and simplifies to

$$\begin{aligned} f_{12345678} &= f_{87|46} f_{56|14} f_{4|2} f_1 f_{23} \\ &= f_{87|46} f_{6|5} f_{5|14} f_{4|2} f_1 f_{23}. \end{aligned}$$

Regarding the right hand side as a factorisation into coordinate projection functions, we have

$$f_K = g_{4678} g_{456} g_{145} g_{24} g_{23}.$$

There are two independences, that between 8 and 6 and that between 7 and 4, that have not been applied to this factorisation. Taking these into account generates the independence graph:

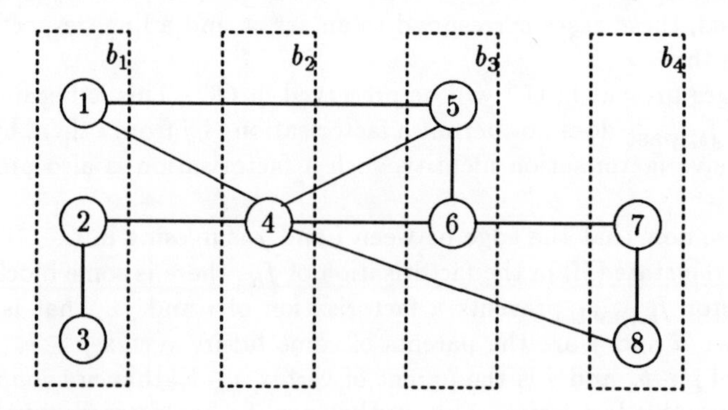

which is, of course, just the moral graph G^m. □

3.7 Exercises

1: Draw the undirected graph on 9 vertices that has edges between the following pairs of vertices $(1,2)$, $(2,3)$, $(4,5)$, $(5,6)$, $(2,5)$, $(8,9)$. Find the neighbours (the adjacency set) of vertex 2.

2: Find all paths from A to F in the graph

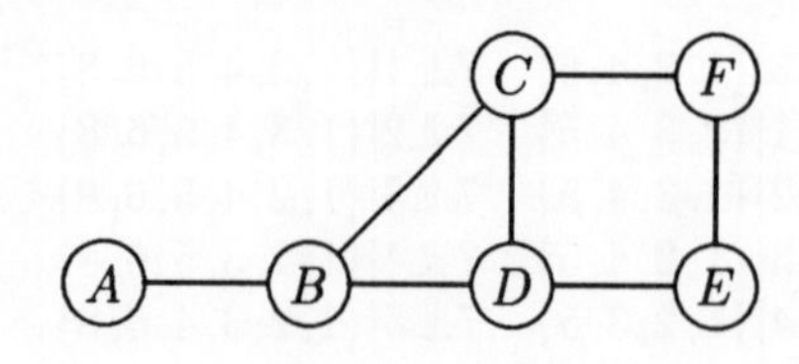

Which of these are short paths? Find all possible separating sets for these two vertices. Find the boundary set of $\{A, E\}$. Find the subgraph induced by $\{A, B, C, E\}$.

3: The *complementary graph* of the graph $G = (V, E)$ is the graph with the same vertex set, V, and with an edge set E' containing only those edges that do not occur in E. Find the complementary graphs of

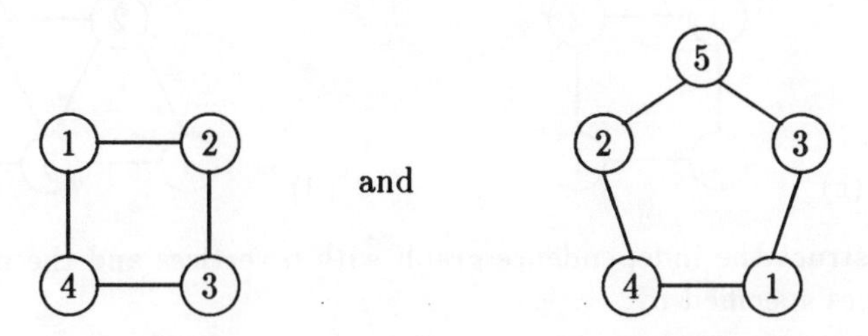

4: Find all chordless cycles and cliques in the graphs

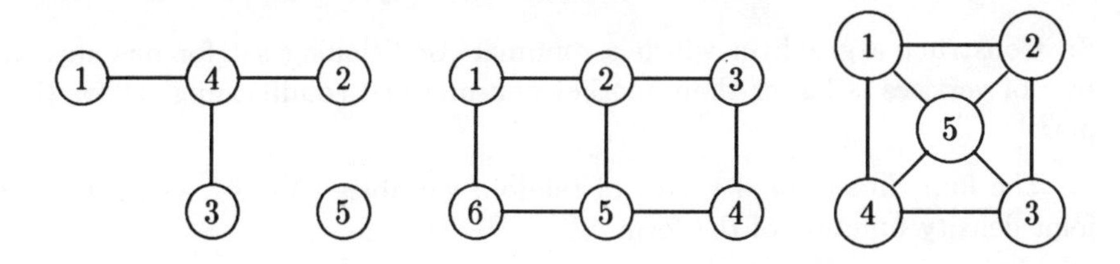

Are there any separating cliques?

5: Write down the conditional independences that define the independence graph

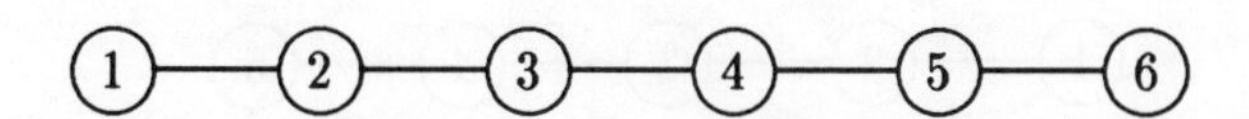

6: A conditional independence between a pair of variables is said to be minimal if it is not possible to use the separation theorem to eliminate any variable from the conditioning set. Write down the minimal conditional independences

embodied in the graphs

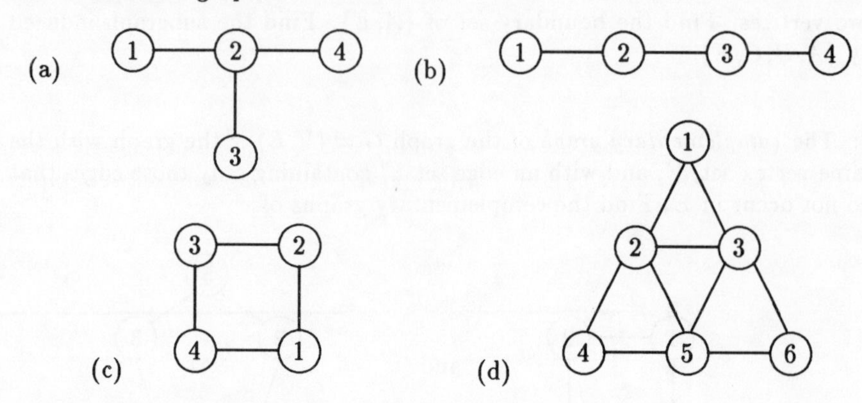

7: Construct the independence graph with 6 vertices and the minimal independences specified by:

given 1 the pairs (2,6), (3,6), (4,6), (5,6);
given 3 the pairs (1,5), (2,5), (4,5), (5,6);
given {1,3} the pair (2,4);
and given {2,4} the pairs (3,6), (5,6), (3,1), (5,1).

8: Construct a graph in which a minimal conditioning set for one specific pair of vertices is larger than another non-minimal conditioning set for that pair.

9: The four dimensional vector of random variables (X_1, X_2, X_3, X_4) has a joint density function of the form

$$f_{1234}(x) = \exp(c + x_1x_2 + x_2x_3 + x_3x_4 + x_1x_4),$$

on the four dimensional cube, $0 < x_i < 1, i = 1, 2, 3, 4$. Construct its independence graph.

10: The five random variables $(X_1, X_2, \ldots, X_5)$ satisfy $\{1,2\} \perp\!\!\!\perp \{3,4\} | 5$; prove that their independence graph is a butterfly.

11: If the graph

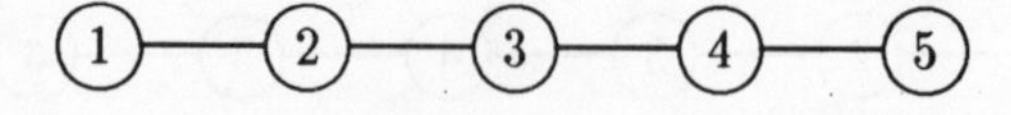

is constructed from the local Markov property, list the defining independence statements.

12: Derive the formula $\sum_{i=0}^{k} \binom{k}{i} 2^{\binom{i}{2}}$ for the number of different graphs in k dimensions when vertices are allowed to die. Verify that if $k = 4$ this is 113.

13: Draw all the distinctly different unlabelled (undirected) graphs with 5 vertices.

14: Without recourse to the global Markov property prove that the pairwise conditional independence property implies the local Markov property.

15: Prove the separation theorem by induction on the number of vertices in a graph, k. Again, start by considering a graph with disconnected subgraphs.

16: Prove that if a directed graph has no directed cycle then there exists at least one complete ordering of the vertices. Give an example for a graph with 7 vertices.

17: The following is the diagram of a directed independence graph.

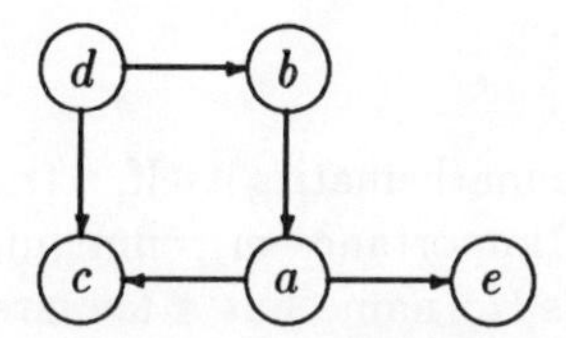

Show there are two possible implicit orderings for the variables, and thus two possible sets of defining pairwise independences. Are they equivalent?

18: The parents of 4 are 2 and 3, and 1 is the only parent of 2 and of 3. Find the associated moral graph of $\{1, 2, 3, 4\}$ and of $\{1, 2, 3\}$.

19: Explain why the defining independences in the following directed graphs are not the same.

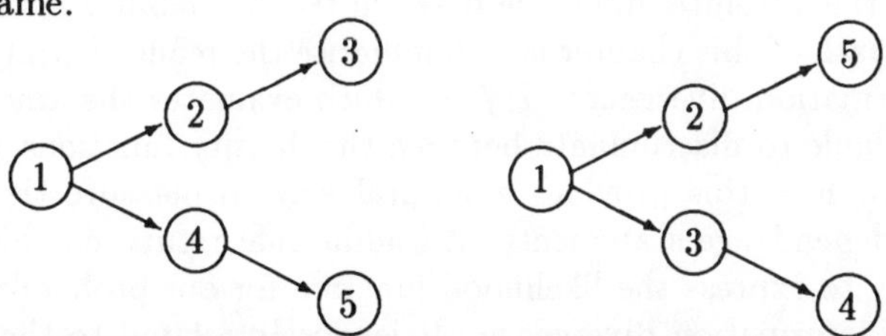

20: Consider a graph $G = (K, E^{\preceq})$ with a mixture of directed and undirected edges, but restrained to have no cycle that contains one or more arrows. Prove that a chain block representation exists, i.e. one for which the blocks are completely ordered and inter-block edges are arrows and intra-block edges are lines. Prove that the Markov properties of any chain block representation are the same.

Chapter 4

Information Divergence

Information theory is, like mathematics itself, a truly inter-disciplinary study, and occupies positions of importance in communication theory, engineering, psychology and linguistics, to name but a few areas of scientific endeavour. Not surprisingly it has a place in applied statistics as well, and arises in graphical modelling because of two important applications. The first is to measure the information contained in one random variable about the value of another, a primary objective of classical information theory, and clearly of interest in a study of independence structures; the second is to measure how well one density function can be approximated by another with a simpler structure, for example, some form of independence. We show that the divergence gives a bound on the absolute difference between two probability distributions.

The first task of this chapter is to familiarise the reader with the Kullback-Leibler information divergence $I(f;g)$ which evaluates the amount of information available to discriminate between the density functions f and g; and then to show how this provides a natural way to measure the strength of specified independence statements. An additional return on this investment is the ability to express the likelihood function for our probability models in terms of an information divergence. It is closely related to the information matrix of statistics introduced by Fisher (1925), though we do not make any use of this connection here.

The divergence $I(f;g)$ itself can be regarded as a function of either f or of g. Minimising with respect to f leads to the theory of minimum discrimination information and is discussed at length in Kullback (1968). On the other hand, it transpires that likelihood ratio tests for independence based on samples from the cross-classified Multinomial and from the multivariate Normal distributions can be expressed in terms of information divergence, and maximising the likelihood is equivalent to minimising the divergence $I(f;g)$ with respect to g. In general, if g is an arbitrary density function this can be a formidable task, and any solution rests heavily on the nature of constraints placed on g. The matter is further complicated because the divergence is not a proper measure of distance. For instance it is not symmetric between f and g, nor does it satisfy the triangle inequality apart from some special cases, see Csiszar (1975). Surprisingly then, it turns out that we can get some sensible answers. The key properties required to generate the results are, on the one hand, the positive definite nature of the divergence, and on the other, the convexity of certain subsets of density functions.

We begin with a definition and some examples of its evaluation, and then we show how it can be derived from consideration of the classical measures of information theory.

4.1 Kullback-Leibler Information Divergence

Definition. The *Kullback-Leibler information divergence* between two probability density functions f and g for the random vector X is

$$I(f;g) = E\log\frac{f(X)}{g(X)}$$

where the expectation is taken with respect to the density f. Closely related is the *Shannon entropy* defined as $-E\log f(X)$ and sometimes shortened to $-E\log f$. □

To make the notation clearer the random variable or vector X can be explicitly mentioned, thus we write $I(f_X;g_X)$ or, more succinctly, $I(f;g\|X)$. Note that it is not symmetric between f and g as is clear when writing out the expectation as the integral or the sum,

$$I(f;g) = \int \log\frac{f(x)}{g(x)}f(x)dx \quad \text{or} \quad I(f;g) = \sum_x \log\frac{f(x)}{g(x)}f(x),$$

as the random variable X is continuous or discrete. The generalisation to higher dimensions is immediate. The information to discriminate between f and g based on a partitioned random vector, (X,Y), of dimension 2 or more, is just

$$I(f;g\|X,Y) = E\log f(X,Y)/g(X,Y).$$

Here are some examples of its evaluation.

EXAMPLE 4.1.1 The divergence between two Bernoulli densities. Suppose that the random variable X is Bernoulli under both f and g and that f is given by $f(0) = 1-p$, $f(1) = p$, while g is given by $g(0) = 1-q$, $g(1) = q$, then, by taking expectations

$$I(f;g\|X) = E\log\frac{f(X)}{g(X)} = (1-p)\log\frac{(1-p)}{(1-q)} + p\log\frac{p}{q}.$$

To emphasise that this is a comparison of the values p and q the right hand side may be referred to as $I(p;q)$. If $p = 0.3$ and $q = 0.2$ the $I(p;q) = 0.7\log(0.7/0.8) + 0.3\log(0.3/0.2) = 0.028$; other values are given in the table. The plot of the divergence $I(p,q)$ against p for a fixed value of q is given

Table 4.1.1: The divergence I(p;q) between two Bernoulli densities.

p	q				
	0.1	0.2	0.3	0.4	0.5
0.1	.000	.037	.116	.226	.368
0.2	.044	.000	.026	.092	.193
0.3	.154	.028	.000	.022	.082
0.4	.311	.105	.023	.000	.020
0.5	.511	.223	.087	.020	.000
0.6	.751	.382	.192	.081	.020
0.7	1.033	.583	.339	.184	.082
0.8	1.363	.832	.534	.335	.193
0.9	1.758	1.146	.794	.551	.368

in Figure 4.1.1. Note that $I(p;q)$ is not symmetric between p and q. It is symmetric in p about $p = .5$ when $q = .5$ but not otherwise. Figure 4.1.1 suggests several conjectures about the Bernoulli divergence:

- $I(p;q)$ is always non-negative,
- the further p and q are apart the larger the divergence,
- the divergence is convex for variation in p,
- $I(p;q)$ has a unique zero at $p = q$ which is a global minimum, and there is no information to discriminate between the two densities,
- if $I(p;q) = \infty$, then discrimination between p and q is certain, and finally
- for values of p close to q, the divergence $I(p;q)$ is approximately quadratic.

Each of these assertions can be proved directly though some care has to be taken about defining $I(p;q)$ on the boundary of the square $0 \leq p,q \leq 1$. The

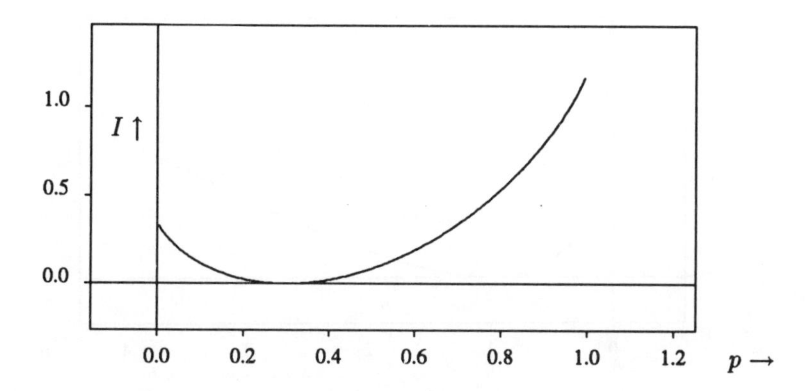

Figure 4.1.1: The diagram of $I(p;q)$ for fixed q.

function $t\log t$ is continuous for t in $(0,1)$ and as $t \to 0$ its limiting value is 0, so it is natural to define $0\log 0 = 0$. Then $I(p;q)$ is well defined for $0 \leq p \leq 1$ and $0 < q < 1$. Consider the case when $q = 0$. If $p = 0$ as well then $f(x) = g(x)$ for all x and there is no information to discriminate between f and g; so that $I(p;q) = 0$ when $p = 0 = q$. But if $p > 0$ then $I(p;q) = +\infty$; consequently $I(p;q)$ is not continuous at the point $(p,q) = (0.0)$. □

EXAMPLE 4.1.2 The bivariate random vector (X,Y) has the Bernoulli distribution with probabilities $p(x,y)$ under f and probabilities $q(x,y)$ under g where p and q are given by

$p(x,y)$	$y=0$	$y=1$
$x=0$	0.1	0.2
$x=1$	0.3	0.4

$q(x,y)$	$y=0$	$y=1$
$x=0$	0.05	0.15
$x=1$	0.20	0.60

We again write $I(f;g)$ as $I(p;q)$ though p and q are now tables of probabilities. So

$$\begin{aligned} I(p;q) &= 0.1\log(.1/.05) + 0.2\log(.2/.15) + 0.3\log(.3/.2) + 0.4\log(.4/.6) \\ &= 0.069 + 0.058 + 0.122 - 0.162 = 0.086 \end{aligned}$$

Though some of the components of this sum are negative the sum itself is positive. When p and q are arbitrary the expression does not readily simplify and in particular it is not a simple function of the cross-product ratio or of the coefficients in the log-linear expansion of the bivariate Bernoulli density function described in Section 2.4. □

EXAMPLE 4.1.3 The divergence between Normal distributions with different variances. The random variable X is Normally distributed under both f and

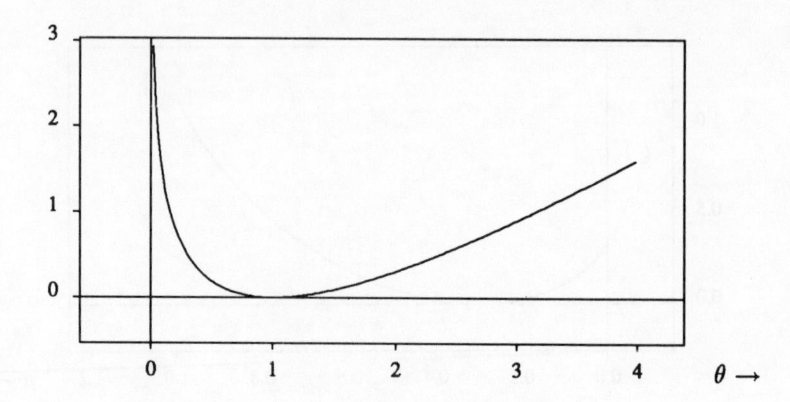

Figure 4.1.2: The diagram of $\theta - \log\theta - 1$.

g where under $f : X \sim N(0, \sigma^2)$ and under $g : X \sim N(0, \tau^2)$. Then

$$\begin{aligned} I(f;g) &= E\log f(X)/g(X) \\ &= E\log f(X) - E\log g(X) \\ &= \tfrac{1}{2}(-\log\sigma^2 - EX^2/\sigma^2) - \tfrac{1}{2}(-\log\tau^2 - EX^2/\tau^2) \end{aligned}$$

As the expectation is taken with respect to f, $EX^2 = \sigma^2$ and the divergence simplifies to

$$I(f;g) = I(\sigma^2;\tau^2) = \frac{1}{2}(\frac{\sigma^2}{\tau^2} - \log\frac{\sigma^2}{\tau^2} - 1),$$

a function of the variance ratio, $\theta = \sigma^2/\tau^2$, whose diagram against θ is shown in Figure 4.1.2. It has similar properties to the Bernoulli divergence, $I(p;q)$, in Figure 4.1.1 above, though here there is no upper bound on I as θ goes to 0. □

EXAMPLE 4.1.4 The divergence between standard bivariate Normal distributions with differing correlation coefficients. Suppose that under f_{XY} the correlation between X and Y is ρ while under g_{XY}, the correlation is zero. The bivariate Normal density function is given in Section 2.6, but in this example a certain amount of hard work can be spared by using Normal theory to simplify the expression for the divergence. By Proposition 2.6.3, and under both f_{XY} and g_{XY}, the marginal distributions of X and of Y are standard univariate Normal; under g_{XY}, $X \perp\!\!\!\perp Y$ and consequently g_{XY} is the product of its marginals: $g_{XY} = f_X f_Y$; so that

$$\begin{aligned} I(f_{XY};g_{XY}) &= E\log f_{XY}/g_{XY} \\ &= E\log f_{XY}/f_X f_Y \\ &= E\log f_{Y|X}/f_Y. \end{aligned}$$

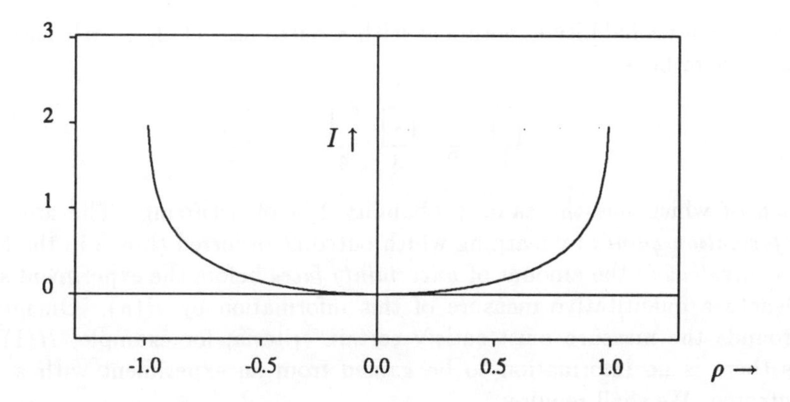

Figure 4.1.3: The diagram of $I(\rho;0) = -\frac{1}{2}\log(1-\rho^2)$.

The conditional distribution of $Y|X = x$ is $N(\rho x, 1-\rho^2)$, so

$$E\log f_{Y|X} - E\log f_Y = -\frac{1}{2(1-\rho^2)}\{\log(1-\rho^2) - E(Y-\rho X)^2\} - \frac{1}{2}EY^2.$$

Expectations are taken with respect to f_{XY} and this simplifies to $-\frac{1}{2}\log(1-\rho^2)$. Hence

$$I(f_{XY}; g_{XY}) = I(\rho;0) = -\frac{1}{2}\log(1-\rho^2).$$

In contrast to the bivariate Bernoulli, this information divergence between two bivariate Normal densities is directly expressible in terms of its usual measure of dependence, the correlation coefficient. Expressions similar to this frequently arise in our analysis of the multivariate Normal distribution. This divergence is plotted in Figure 4.1.3. Quite clearly if $\rho = 0$ then $I(\rho;0) = 0$ but, and more importantly, the converse: $I(\rho;0) = 0$ implies $\rho = 0$, is also true. □

4.2 Divergence: a Heuristic Derivation

A full axiomatic treatment of information measures is given in several standard texts, for example, Renyi (1970) has a good discussion, and all we require here is a flavour of the argument leading up to the definition of the divergence. The following informal derivation gives substantial insight into the underlying structure of the Kullback-Leibler information measure.

Step 1: Hartley's (1928) information number, H.

Consider a probability experiment with n outcomes, $\{1, 2, \ldots, n\}$, listed here as a row of boxes:

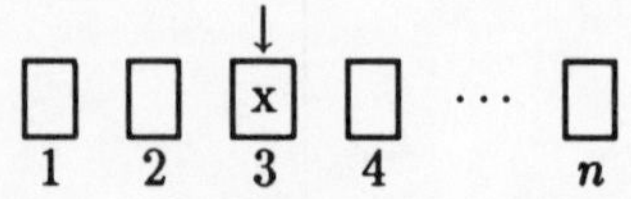

each of which has the same probability $1/n$ of occurring: The amount of *information gained* by learning which outcome occurred (box 3 in the figure) is equivalent to the amount of *uncertainty faced* before the experiment starts. Denote a quantitative measure of this information by $H(n)$. On intuitive grounds the measure must satisfy certain criteria; for example, $H(1) = 0$, as there is no information to be gained from an experiment with a single outcome. We shall require:

(a) $H(2) = 1$, to calibrate the measure on the simplest possible experiment;

(b) *monotonicity:* $H(n) \leq H(n+1)$ for $n = 1, 2, \ldots$, as the uncertainty faced increases with the difficulty of guessing the outcome; and most importantly

(c) *additivity:* $H(nm) = H(n) + H(m)$ for $n, m = 1, 2, \ldots$, because if the boxes are arranged in rows and columns, to know the row and the column in which the outcome occurs is exactly equivalent to knowing which outcome occurs, as the figure makes clear.

	1	2	3 ↓	4	…	n
1	□	□	□	□	…	□
→ 2	□	□	x	□	…	□
⋮	⋮	⋮	⋮	⋮	⋱	⋮
m	□	□	□	□	…	□

A function that satisfies these three properties is $\log_2 n$. If the monotonicity and additivity conditions are strengthened by requiring they hold for all positive rationals n and m then $\log_2 n$ is also the unique solution. We shall modify the first condition slightly by taking natural logarithms and define

$$H(n) = -\log \frac{1}{n}.$$

Step 2: Shannon's (1948) measure of entropy, h.
In an experiment with equally probable outcomes the 'uncertainty faced' and the 'information gained' can be identified. However these concepts must be

distinguished in an experiment in which the outcomes are no longer equi-probable because there is more information gained from the occurrence of an outcome with a relatively low probability of occurrence. Shannon defined the *entropy* or *average uncertainty faced* as the average gain of information.

To evaluate the entropy for an experiment in which the outcomes are no longer equi-probable first consider an experiment with two outcomes, $\{A, A^c\}$, where A has probability p of occurring and the other outcome A^c has probability $1-p$. Denote the entropy in the experiment by $h(p, 1-p)$. Then

$$h(p, 1-p) = -p\log p - (1-p)\log(1-p).$$

Its plausibility follows from the argument that if there are integers m and n such that $p = m/(m+n)$, then this experiment is equivalent to an experiment $E = \{1, 2, \ldots, m, m+1, \ldots, m+n\}$ with $m+n$ equally probable outcomes where it is recorded which of the two events $A = \{1, 2, \ldots, m\}$ or $A^c = \{m+1, \ldots, m+n\}$ is recorded, but not which particular outcome within A or A^c. The fine detail is suppressed. The tree diagram in Figure 4.2.1 illustrates.

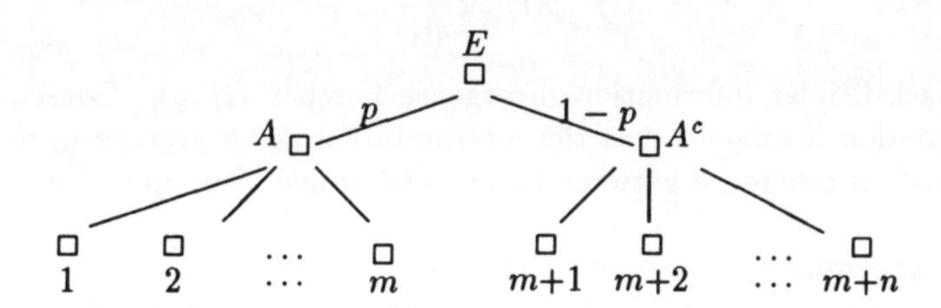

Figure 4.2.1: A tree diagram for the suppression of information.

The Hartley information in a completely recorded experiment with $m+n$ equi-probable outcomes is $\log(m+n)$. The Hartley information in the sub-experiment $A = \{1, 2, \ldots, m\}$ is $\log(m)$ and on average this information is suppressed a fraction p of the time. Similarly the information in learning the outcome A^c is $\log(n)$ and is suppressed a fraction $1-p$ of the time. As information is additive, the uncertainty or the average information gained by knowing which of A or A^c occurred is

$$\begin{aligned} h(p, 1-p) &= \log(m+n) - p\log(m) - (1-p)\log(n) \\ &= -p\log p - (1-p)\log(1-p) \end{aligned}$$

as we wanted to show.

In an experiment with n rather than two outcomes and with probabilities given by $p(1), p(2), \ldots, p(n)$ the same argument leads to the expression

$$h(p(1), p(2), \ldots, p(n)) = -\sum_i p(i)\log p(i),$$

for the entropy.

It is only slightly less rigorous to argue that, by continuity, the Hartley information gained from the experiment if the i-th outcome occurred is $-\log p(i)$ rather than $-\log(1/n)$. (So an experiment in which a rare event occurs has a higher information content that one in which a rather probable event occurs.) This leads to the conclusion that the average value of this information is just $-\sum p(i)\log p(i)$. Note that the Shannon entropy equals Hartley's number when each outcome has the same probability.

Step 3: Kullback-Leibler's (1951) information divergence, I.
Now suppose that the true probability of the n outcomes $\{1,2,\ldots,n\}$ is $p = (p(1),p(2),\ldots,p(n))$ but that $q = (q(1),q(2),\ldots,q(n))$ is an initial assessment or guess of the value. The imagined Hartley information gained from the i-th outcome is, using continuity, $-\log q(i)$, and on average the imagined uncertainty in the experiment is $-\sum p(i)\log q(i)$. The difference between this and the true uncertainty $-\sum p(i)\log p(i)$ is just

$$\sum p(i)\log\frac{p(i)}{q(i)},$$

the Kullback-Leibler information divergence number $I(p;q)$. Consequently this information divergence has the interpretation as an average measure of the information difference between two sets of probabilities, p and q.

Step 4: A generalisation to random variables.
The divergence $I(p;q)$ for a finite probability experiment treated here easily generalises to more abstract sample spaces, for example, to that for a continuous random variable. The outcome of an experiment $E = \{1,2,\ldots,n\}$ can be recorded by a random variable X that takes these integer values with probabilities $p = (p(1),p(2),\ldots,p(n))$. The density function $f_X(x)$ of X is given by $p(X = x) = p(x)$, for $x = 1,2,\ldots,n$, and the Shannon entropy $\sum_x p(x)\log p(x)$ is $E\log f_X(X)$. With q replaced by $g(x)$, the divergence $\sum p(x)\log p(x)/q(x)$ becomes $E\log f_X(X)/g_X(X)$. We now use this expression as the definition of divergence for a continuous random variable, and of course, the two notions coincide when X is a random variable on a finite sample space.

4.3 Properties of Information Divergence

The divergence is well defined, the divergence is additive over independent repeated samples and the divergence is positive definite. These are three of the most important properties of the Kullback-Leibler information divergence and the proof of the last is quite demanding; on the way we prove an inequality which explicitly shows how the divergence can be used to give an indication of how well one distribution approximates another. A comprehensive account of the properties of the divergence is given in the text by Kullback (1959, 1968), which in particular, emphasises an area that is not discussed here, that of minimum discrimination information which solves the problem of minimising the divergence $I(f;g)$ for variation in f.

On a finite sample space the divergence, $I(f;g)$, is a finite sum of terms of the form $p\log(p/q)$, each of which may be positive or negative. If p is 0 then the contribution to the sum is 0 but if q is zero then the contribution is $+\infty$. As long as the negative terms contribute only a finite quantity the divergence, though infinite, will be well defined; that is, it can never lead to expressions of the form $+\infty - \infty$. To see this define

$$r(x) = \frac{f(x)}{g(x)} \log \frac{f(x)}{g(x)},$$

so that the divergence can be rewritten as

$$\begin{aligned} I(f;g) &= \int r(x)g(x)dx \\ &= \int_{r<0} r(x)g(x)dx + \int_{r\geq 0} r(x)g(x)dx \\ &\geq -e^{-1} + \int_{r\geq 0} r(x)g(x)dx \end{aligned}$$

because the function r is of the form $t\log t$, and this function has a minimum value of $-e^{-1}$ in the interval $0 < t < 1$ where r is negative. The integrand in the second term on the right hand side is always positive.

Proposition 4.3.1 Additivity under repeated sampling. *If X and Y are independent random variables under both f and g then*

$$I(f;g\|X,Y) = I(f;g\|X) + I(f;g\|Y).$$

The proof is a straightforward exercise in algebraic manipulation.

By definition, a real valued function, h, is *convex* on a region of the real line R^1 if whenever $a < b$ in the region and $0 \leq t \leq 1$ then

$$h(ta + (1-t)b) \leq th(a) + (1-t)h(b).$$

One simple check of convexity is to test if the second derivative is always positive; thus $-\log t$ is convex for positive t because its second derivative is $1/t^2$. If X is a discrete random variable taking the value a with probability t and the value b with probability $1-t$ this definition of convexity can be recast as

$$h(EX) \leq Eh(X).$$

If X is an arbitrary random variable with a finite expectation and h is an integrable convex function this is known as *Jensen's inequality*, and a proof may be found in any text on real analysis.

Proposition 4.3.2 *The divergence is non-negative.*

Proof: Note that the function $h = -\log$ is convex and $I(f;g) = E\{-\log(g/f)\}$. Consequently Jensen's inequality gives $I(f;g) \geq -\log E(g/f) = -\log 1 = 0$. □

Positive definiteness

Let $[A]$ denote a finite partition, $\{A_1, A_2, \ldots, A_k\}$, of the sample space into a collection of disjoint sets, and put

$$F(A_i) = \int_{A_i} f(x)dx \quad \text{and} \quad G(A_i) = \int_{A_i} g(x)dx.$$

Define the divergence between f and g based on the partition $[A]$ by

$$I(f;g\|[A]) = \sum_{i=1}^{k} F(A_i)\log\frac{F(A_i)}{G(A_i)}.$$

Note that $I(f;g\|[A])$ is a finite sum. There are two results of importance.

Lemma 4.3.3 *The divergence, $I(f;g\|[A])$, on the partition $[A]$, increases under the refinement of $[A]$.*

Proof: Consider the simple refinement $[B]$ of the partition $[A]$ given by $\{B_1, B_2, A_2, \ldots, A_k\}$ obtained by splitting A_1 into $A_1 = B_1 \cup B_2$. Then

$$\begin{aligned} I(f;g\|[B]) &= I(f;g\|[A]) \\ &\quad -F(A_1)\log\frac{F(A_1)}{G(A_1)} + F(B_1)\log\frac{F(B_1)}{G(B_1)} + F(B_2)\log\frac{F(B_2)}{G(B_2)}. \end{aligned}$$

By setting $p = F(B_1)/F(A_1)$ and $q = G(B_1)/G(A_1)$ we have

$$I(f;g\|[B]) = I(f;g\|[A]) + I(p;q),$$

where $I(p;q)$ is the Bernoulli divergence in Example 4.1.1. But $I(p;q)$ is always non-negative and so $I(f;g\|[B]) \geq I(f;g\|[A])$. As any refinement of a finite partition can be obtained by repeatedly dividing a member of the collection in this fashion the assertion follows. □

Lemma 4.3.4 *The divergence satisfies*

$$I(f;g) \geq I(f;g\|[A])$$

whatever the partition $[A]$.

Proof: Now

$$\begin{aligned} I(f;g) &= E\log f(X)/g(X) \\ &= \sum_i \int_{A_i} \log \frac{f(x)}{g(x)} f(x)dx \\ &= \sum_i F(A_i) \int_{A_i} -\log \frac{g(x)}{f(x)} \frac{f(x)}{F(A_i)} dx. \end{aligned}$$

But $f(x)/F(A_i)$ is a probability density function, that of X conditional on the event that X lies in A_i, so Jensen's inequality applies and gives

$$I(f;g) \geq \sum_i F(A_i)\{-\log \int_{A_i} g(x)/F(A_i)dx\}.$$

The right hand side immediately simplifies to $I(f;g\|[A])$. □

Proposition 4.3.5 Dobrushin's theorem. *The divergence* $I(f;g)$ *is given by*

$$I(f;g) = \sup I(f;g\|[A])$$

where the supremum is taken over all partitions of the sample space.

The proof, which is too demanding for this text, proceeds by constructing a partition $[A]$ that has the property that $I(f;g) \leq I(f;g\|[A]) + \epsilon$ for ϵ arbitrarily small. In fact, in an advanced treatment, it is the right hand side, $\sup I(f;g\|[A])$, rather than the left, which is taken as the definition of the divergence.

An information inequality for the divergence

The inequality gives an explicit upper bound on the uniform metric for comparing probability distributions in terms of the information divergence. This inequality appears to have been independently discovered by several writers in the in the late 1960's; and we mention Csiszar (1967), Kemperman (1967) and Kullback (1967). It is used here to give a proof that the divergence is positive definite. The proof used in this text is essentially an exercise in Taylor series expansions; the standard proof relies on Jensen's inequality applied to strictly convex functions.

Lemma 4.3.6 *The Bernoulli divergence satisfies*

$$I(p;q) = p\log \frac{p}{q} + (1-p)\log \frac{(1-p)}{(1-q)} \geq 2(p-q)^2$$

for all $0 \leq p, q \leq 1$.

Proof: If $q = 0$ or 1 the inequality is trivial. Otherwise look at the divergence as a function of p and set

$$r(p) = p\log p/q + (1-p)\log(1-p)/(1-q).$$

Then $r(q) = 0$ and the first derivatives of g are

$$\begin{aligned} r'(p) &= \log p - \log q - \log(1-p) - \log(1-q), \qquad \text{and} \\ r''(p) &= 1/p + 1/(1-p). \end{aligned}$$

Consequently $r'(q) = 0$ and $r''(p) \geq 4$ for all p. From a Taylor series expansion with remainder there exists a number θ between p and q and such that

$$r(p) = r(q) + (p-q)r'(q) + \frac{1}{2}(p-q)^2 r''(\theta).$$

Substituting for the derivatives gives the inequality. □

The proof is much harder if $p\log p/q + (1-p)\log(1-p)/(1-q)$ is regarded as a function of q.

Proposition 4.3.7 An information inequality. *For any event A in the sample space*

$$\sup_A |F(A) - G(A)| \leq \{\frac{1}{2}I(f;g)\}^{\frac{1}{2}}.$$

Proof: Consider the partition, $\{A, A^c\}$, of the sample space; then, as the divergence on the partition is bounded by the divergence

$$F(A)\log\frac{F(A)}{G(A)} + (1-F(A))\log\frac{1-F(A)}{1-G(A)} \leq I(f;g).$$

Now use the preceding lemma with $p = F(A)$ and $q = G(A)$ to get a lower bound of $2(F(A) - G(A))^2$ for the left hand side. Consequently $I(f;g) \geq 2(F(A) - G(A))^2$ for any event A and hence for the supremum over all A. This gives the stated inequality. □

Table 4.3.1 gives an idea of how close the bound given by this inequality comes to equality. The value of $|p-q|$ is constant on the diagonals of this table. For example, when $p = 0.7$ and $q = 0.2$ the absolute difference is 0.5 and the bound is 0.540, not too bad. The bound is better for values of p and q near 0.5; and it improves as p and q move down a diagonal. The worst value in the table occurs when $p = 0.9$ and $q = 0.1$ with a difference of 0.8 bounded by 0.937.

We can at last prove that the divergence is positive definite.

Table 4.3.1: An information inequality, $\{\frac{1}{2}I(p;q)\}^{\frac{1}{2}}$ with $|p-q|$ in brackets.

p	q				
	0.1	0.2	0.3	0.4	0.5
0.1	.000 (.0)	.135 (.1)	.241 (.2)	.336 (.3)	.429 (.4)
0.2	.149 (.1)	.000 (.0)	.113 (.1)	.214 (.2)	.310 (.3)
0.3	.277 (.2)	.119 (.1)	.000 (.0)	.104 (.1)	.203 (.2)
0.4	.394 (.3)	.229 (.2)	.106 (.1)	.000 (.0)	.100 (.1)
0.5	.505 (.4)	.334 (.3)	.209 (.2)	.101 (.1)	.000 (.0)
0.6	.613 (.5)	.437 (.4)	.310 (.3)	.201 (.2)	.100 (.1)
0.7	.719 (.6)	.540 (.5)	.412 (.4)	.303 (.3)	.203 (.2)
0.8	.825 (.7)	.645 (.6)	.517 (.5)	.409 (.4)	.310 (.3)
0.9	.937 (.8)	.757 (.7)	.630 (.6)	.525 (.5)	.429 (.4)

Proposition 4.3.8 *The divergence is* positive definite; *that is*

$$\begin{cases} I(f;g) \geq 0, & \textit{and} \\ I(f;g) = 0 & \textit{if and only if} \quad f(x) = g(x) \end{cases}$$

for all values of x (save possibly on a set of probability measure zero).

Proof: We have seen that $I \geq 0$ and that $I = 0$ when $f = g$. To go the other way, if $I = 0$, then the information inequality gives $F(A) = G(A)$ for all events A in the sample space so $f(x) = g(x)$, save for a set with probability measure zero. □

At the conclusion of this argument we argue that $F(A) = G(A)$ for all events A implies $f(x) = g(x)$, for all x apart from a set of values of x with measure zero. What is meant is the following. Choose an arbitrary small number $\epsilon > 0$ and let $A = \{x; g(x)/f(x) - 1 \geq \epsilon\}$. Then $F(A) = G(A)$ implies

$$0 = G(A) - F(A) = \int_A \{g(x) - f(x)\}dx \geq \int_A \epsilon f(x)dx = F(A)\epsilon$$

because $g - f \geq f\epsilon$ on A. This is only possible if $F(A) = 0$ so that the set A has probability 0. Further discussion of this point requires a more meticulous measure theoretic approach.

4.4 Independence and Information Proper

In addition to the Shannon entropy and the Kullback-Leibler divergence there is a third fundamental measure of information. Suppose that X and Y are two random variables (or vectors) with a joint density function f_{XY} and marginal densities f_X and f_Y, respectively.

Definition. The information in one random vector about another, the ***information proper***, is

$$\text{Inf}(X \perp\!\!\!\perp Y) = I(f_{XY}; f_X f_Y),$$

where I is the Kullback-Leibler information divergence. □

The information proper, $\text{Inf}(X \perp\!\!\!\perp Y)$, is symmetric in X and Y and is 0 when $X \perp\!\!\!\perp Y$. It measures the divergence between the joint density of (X, Y) and the 'independent' density given by the product of the marginal densities of X and Y. In this sense it quantifies the information contained in one random variable about the value of another. The information proper is closely related to prediction and the problem of predicting the value of one random variable from the value of another; it measures average predictability. That it is a natural measure of the information held by one random variable about another is argued here by a simple derivation from the entropy.

Information proper: a derivation

Let the bivariate random vector (X, Y) denote the outcome of an experiment with a finite number of outcomes where $P(X = i, Y = j)$ is the probability that the (i, j)-th outcome occurs. The Shannon entropy, or uncertainty, in the partial experiment recorded by X alone is, from Section 4.1,

$$h(X) = -\sum_i P(X = i) \log P(X = i).$$

Similarly, the conditional uncertainty in X given $Y = j$ is

$$h(X|Y = j) = -\sum_i P(X = i|Y = j) \log P(X = i|Y = j),$$

and on ***average*** this conditional uncertainty is

$$h(X|Y) = \sum_j h(X|Y = j) P(Y = j).$$

Satisfyingly, this satisfies the identity

$$h(X, Y) = h(Y) + h(X|Y),$$

or mnemonically: the uncertainty in (X, Y) is the uncertainty in Y together with the average conditional uncertainty in X given knowledge of Y.

Now the *information in Y about X* is, by presumption, the average uncertainty in X removed by a knowledge of Y, that is,

$$h(X) - h(X|Y).$$

A straightforward evaluation shows that

$$h(X) - h(X|Y) = \sum_{i,j} P(X = i, Y = j) \log \frac{P(X = i, Y = j)}{P(X = i)P(Y = j)}$$

which is just the Kullback-Leibler information divergence between the joint probability and the product of the two marginal probabilities. The derivation is complete.

Rather than derive the information proper from the Shannon entropy as is done here, the argument can be reversed to obtain the uncertainty $h(X)$ from the information $\text{Inf}(X \perp\!\!\!\perp Y)$. It needs the identity $h(X) = \text{Inf}(X \perp\!\!\!\perp X)$, valid if it is accepted that $f_{XX} = f_X$.

Diagrams, convexity and extrema

Our interest in independence naturally leads to attempts to draw Venn diagrams displaying subsets of density functions possessing various independence properties. In order to draw better pictures we need to know a little more about the structure of these subsets; for instance, is the *independence set* $\{f_{XY}; X \perp\!\!\!\perp Y\}$ convex or is it just a blob? Unfortunately, the answer is the latter, because, given a real number θ in the interval $(0, 1)$ and any two densities f_{XY} and g_{XY} for which $X \perp\!\!\!\perp Y$, the density on the 'line' joining f to g is the mixture

$$h_{XY} = \theta f_{XY} + (1 - \theta) g_{XY};$$

and, there is no reason to suppose that $X \perp\!\!\!\perp Y$ under h.

In fact, this question of convexity can be cast in the framework of the relationship between conditional and marginal independence. Let Z denote a Bernoulli random variable, taking the values 0 and 1 with probabilities $1 - \theta$ and θ, and suppose the conditional density of (X, Y) given Z is f when $z = 1$ and g when $z = 0$; then the mixture h above is just the marginal density of (X, Y). The condition that f is in the independence set is $X \perp\!\!\!\perp Y | Z = 1$ and that g is in the independence set is $X \perp\!\!\!\perp Y | Z = 0$. Hence the independence set is convex if the conditional independence $X \perp\!\!\!\perp Y | Z$ implies the marginal independence $X \perp\!\!\!\perp Y$. This we know to be false, as witnessed by Simpson's paradox, and consequently it is misleading to portray the independence set as approximately oval.

But all is not lost, there are some important subsets that are convex.

Lemma 4.4.1 *The set of density functions for the partitioned random vector* (X, Y) *with a fixed* X *margin,* $\{f_{XY}; f_X = f_X^0\}$ *where* f^0 *is given, is convex.*

Proof: Suppose that f and g have the same fixed X margin. Then integrating the mixture density, h_{XY} above, over y, the marginal of X is

$$h_X = \theta f_X^0 + (1-\theta) f_X^0 = f_X^0,$$

whatever the value of θ. □

The next result is an important technical lemma that proves useful later on.

Lemma 4.4.2 Minimisation with arbitrary marginals. *If the partitioned random vector (X,Y) has a fixed density f_{XY}^0, and if g is any other density for (X,Y) in which the marginal density of X is arbitrary then the divergence, $I(f^0;g\|X,Y)$, is minimised with respect to variation in g, by setting $g_X = f_X^0$.*

Proof: Simple manipulation gives

$$\begin{aligned} I(f^0;g\|X,Y) &= E\log f_{XY}^0/g_{XY} \\ &= E\log f_{Y|X}^0/g_{Y|X} + E\log f_X^0/g_X \\ &= EI_x(f_{Y|X}^0;g_{Y|X}\|Y) + I(f_X^0;g_X\|X), \end{aligned}$$

where the subscript on the divergence I_x signifies that X is fixed at x. Since neither I_x, nor $EI_x = \int I_x f_X^0(x)dx$, depend on g_X, the divergence $I(f^0;g\|X,Y)$ is minimised by minimising the second term $I(f^0;g\|X)$. As the divergence is positive definite this implies that $g_X(x) = f_X^0(x)$ for all x. □

A picture, Figure 4.4.1, helps. We shall now show that the information proper

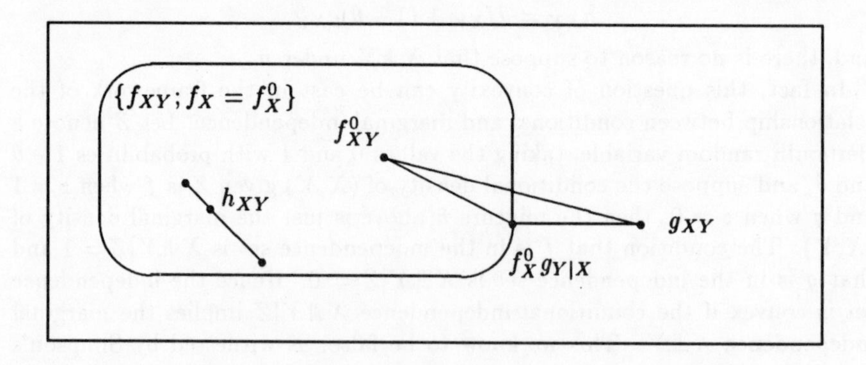

Figure 4.4.1: Sets with given margins are convex.

is a solution to an extremal problem. To simplify notation, we drop the superscript on the density f^0 though it is still held fixed.

Proposition 4.4.3 Minimising the divergence against independence. *If g_{XY} is any density function for which $X \perp\!\!\!\perp Y$, the divergence $I(f;g||X,Y)$ is minimised with respect to variation in g, by choosing $g_X = f_X$ and $g_Y = f_Y$, and the attained minimum is the information proper between X and Y,* $\text{Inf}(X \perp\!\!\!\perp Y)$.

Proof: As $X \perp\!\!\!\perp Y$, $g_{XY} = g_X g_Y$, so consider the identity

$$\frac{f_{XY}}{g_X g_Y} = \frac{f_{XY}}{f_X f_Y}\frac{f_X f_Y}{g_X g_Y}.$$

Taking expected values of the logarithm of this identity gives

$$I(f;g||X,Y) = \text{Inf}(X \perp\!\!\!\perp Y) + I(f;g||X) + I(f;g||Y).$$

Now $\text{Inf}(X \perp\!\!\!\perp Y)$ does not depend on g at all and minimising the left hand side with respect to g devolves to separately minimising $I(f;g||X)$ and $I(f;g||Y)$ of the right hand side. Application of the previous lemma concludes the argument. □

The minimal value of the divergence $I(f;g||X,Y)$ is just $\text{Inf}(X \perp\!\!\!\perp Y)$, the information contained in X about Y. We can thus sketch the space containing all bivariate density functions and the subspace for which $X \perp\!\!\!\perp Y$, in a Venn diagram such as Figure 4.4.2. Given a fixed point $f = f_{XY}$ the point in the subspace nearest to f is $g_{XY} = f_X f_Y$. The two subsets with fixed

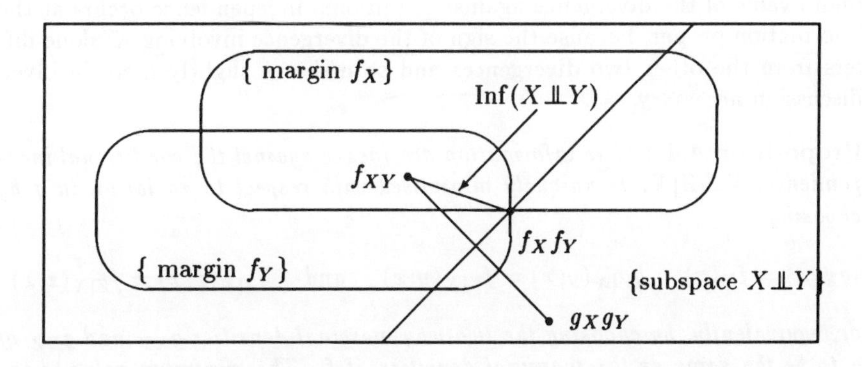

Figure 4.4.2: The information against independence.

margins are convex. So their intersection consisting of all densities with given X and given Y margins is also convex. The intersection of this subset with all densities exhibiting independence consists of the single point $f_X f_Y$.

The information against conditional independence

In order to use the divergence against independence as a numerical measure for the strength of a connection in an independence graph, we have to extend the above discussion to conditional independence.

Definition. The *information in Y about Z conditional on X* is

$$\text{Inf}(Y \perp\!\!\!\perp Z|X) = I(f_{XYZ}; f_{Y|X}f_{Z|X}f_X).$$

where I is the information divergence. □

Note that the expectation in the divergence is an average taken over X as well as Y and Z, so that this is truly a real valued scalar. It is a measure of the average amount of information in f against the independence of Y and Z conditional on X because

$$\text{Inf}(Y \perp\!\!\!\perp Z|X) = 0 \quad \text{if and only if} \quad Y \perp\!\!\!\perp Z|X,$$

which follows from the positive definite nature of the divergence. Consider the following identity. If the random vectors X, Y and Z have a joint probability density function f_{XYZ} and g_{XYZ} is some other density for which $Y \perp\!\!\!\perp Z|X$, then

$$\text{Inf}(Y \perp\!\!\!\perp Z|X) = I(f;g||X,Y,Z) - I(f;g||X,Y) - I(f;g||X,Z) + I(f;g||X).$$

Unfortunately one cannot directly use this expression to argue that the minimum value of the divergence against conditional independence occurs at the information proper, because the sign of the divergence involving X alone differs from the other two divergences and so makes a slightly more involved discussion necessary.

Proposition 4.4.4 *The information divergence against the conditional independence, $Y \perp\!\!\!\perp Z|X$, is uniquely minimised with respect to variation in g by choosing*

$$g_X(x) = f_X(x), \quad g_{Y|X}(y;x) = f_{Y|X}(y;x) \quad \text{and} \quad g_{Z|X}(z;x) = f_{Z|X}(z;x)$$

or, equivalently, by choosing the two-way marginal densities g_{XY} and g_{XZ} of g to be the same as the marginal densities of f. The minimum value is the information in Y about Z conditioned on X, $\text{Inf}(Y \perp\!\!\!\perp Z|X)$.

Proof: First establish the identity

$$\begin{aligned} I(f;g||X,Y,Z) &= \text{Inf}(Y \perp\!\!\!\perp Z|X) \\ &+ EI_x(f_{Y|X}; g_{Y|X}) + EI_x(f_{Z|X}; g_{Z|X}) + I(f_X; g_X) \end{aligned}$$

where X is held fixed in the divergences I_x, and the expectation is taken over X with respect to f_X. Each of the terms on the right is non-negative, and so the minimisation with respect to $g_{XYZ} = g_{Y|X}g_{Z|X}g_X$ decomposes into three separate minimisations. The last term is minimised by choosing $g_X = f_X$. The term $I_x(f_{Y|X}; g_{Y|X})$ is minimised by choosing $g_{Y|X} = f_{Y|X}$. Uniqueness holds with the caveat regarding sets of probability measure zero as in the proposition about positive definiteness. □

Hence the minimum divergence occurs when certain marginal distributions of f and g are equal, for independence: $g_X = f_X$ and $g_Y = f_Y$; for conditional independence: $g_{XY} = f_{XY}$ and $g_{XZ} = f_{XZ}$.

4.5 Information in an Independence Graph

The independence graph of a set of variables conveys a vivid but terse description of their pattern of interaction. One way to augment its rather modular nature is to attach a number to each edge in the graph to indicate the strength of the bond. As independence graphs are constructed from pairwise independences conditional on the remaining variables, the natural candidate is the divergence against conditional independence. The information for contrasting the graph

is $\mathrm{Inf}(X_2 \perp\!\!\!\perp X_3 | X_1)$ and this number can be associated with the edge between vertices 2 and 3. With just two variables the information for the graph

(2)—(3) against (2) (3)

is the information in X_3 about X_2, $\mathrm{Inf}(X_2 \perp\!\!\!\perp X_3)$. With k variables, $X_K = (X_1, X_2, \ldots, X_k)$, the *strength* of the edge between vertices i and j is measured by

$$\mathrm{Inf}(X_i \perp\!\!\!\perp X_j | X_{K \setminus \{i,j\}})$$

for i and j in $K = \{1, 2, \ldots, k\}$, or more succintly $\mathrm{Inf}(i \perp\!\!\!\perp j | K \setminus \{i, j\})$. A familiar scale of measurement is inherent in the *equivalent Normal correlation coefficient* defined by

$$\rho = +\{1 - \exp(-2\mathrm{Inf})\}^{\frac{1}{2}}$$

since this has a maximum of 1, and a minimum of $\rho = 0$, when Inf $= 0$. It is derived by inverting the expression for the information divergence between two correlated Normal random variables in Example 4.1.4, see Kent (1983) for further discussion. Of course ρ should be interpreted as a partial correlation coefficient when it is used to rescale the divergence against conditional independence.

The information measures are best numerically evaluated from

$$\begin{aligned}\mathrm{Inf}(X_2 \perp\!\!\!\perp X_3 | X_1) &= E\log f_{123}/f_{2|1}f_{3|1}f_1 \\ &= E\log f_{123} - E\log f_{12} - E\log f_{13} + E\log f_1\end{aligned}$$

and similarly for the other two divergences. In a full analysis of the joint distribution we may well need to numerically evaluate the Shannon entropy, $-E\log f_a$, for each of the $2^k - 1$ non-empty subsets of $(X_1, X_2, \ldots, X_k)$.

EXAMPLE 4.5.1 The three dimensional random vector (X_1, X_2, X_3) has a trivariate Bernoulli distribution, with a table of probabilities

$p(x_1, x_2, x_3)$		$x_3 = 0$	$x_3 = 1$
$x_1 = 0$	$x_2 = 0$	0.03 (1/35)	0.06 (2/35)
	$x_2 = 1$	0.11 (4/35)	0.23 (8/35)
$x_1 = 1$	$x_2 = 0$	0.34 (12/35)	0.09 (3/35)
	$x_2 = 1$	0.11 (4/35)	0.03 (1/35)

The variables X_3 and X_2 are independent at each level of X_1. The entropies

$$\begin{aligned}-E\log f_{123} &= 1.78, & -E\log f_{23} &= 1.33,\\ -E\log f_{12} &= 1.22, & -E\log f_{2} &= 0.69,\\ -E\log f_{13} &= 1.24, & -E\log f_{3} &= 0.67,\\ -E\log f_{1} &= 0.68, & -E\log f_{\phi} &= 0.00,\end{aligned}$$

are computed directly from the full table. For example

$$-E\log f_{123} = -0.03\log 0.03 - 0.06\log 0.06 - \cdots - 0.03\log 0.03 = 1.78,$$

and all the marginal tables, for example,

$$-E\log f_{12} = -0.09\log 0.09 - 0.34\log 0.34 - 0.43\log 0.43 - 0.14\log 0.14 = 1.22.$$

The conditional information numbers are

$$\begin{aligned}\mathrm{Inf}(X_2 \perp\!\!\!\perp X_3 | X) &= E\log f_{123} - E\log f_{13} - E\log f_{12} + E\log f_1\\ &= -(1.78 - 1.22 - 1.24 + 0.68) = 0.0,\\ \mathrm{Inf}(X_1 \perp\!\!\!\perp X_3 | X_2) &= -(1.78 - 1.22 - 1.33 + 0.69) = 0.08, \quad \text{and}\\ \mathrm{Inf}(X_2 \perp\!\!\!\perp X_1 | X_3) &= -(1.78 - 1.24 - 1.33 + 0.67) = 0.12.\end{aligned}$$

We can use these numbers to measure the strength of association on the independence graph of (X_1, X_2, X_3)

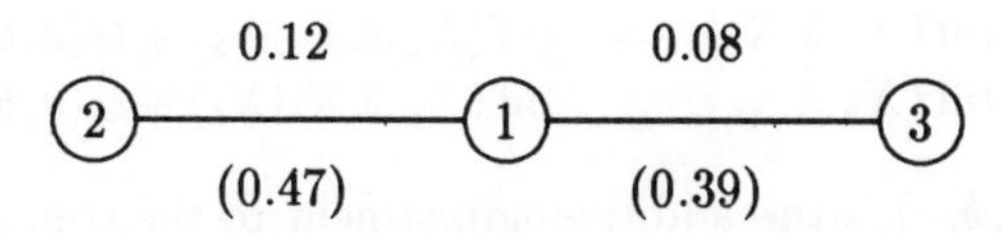

with the equivalent Normal partial correlation coefficients of 0.47 and 0.39 given in brackets. There is no edge between X_3 and X_2 because $X_2 \perp\!\!\!\perp X_3 | X_1$ and $\text{Inf}(X_2 \perp\!\!\!\perp X_3 | X_1) = 0$. The marginal information numbers are

$$\begin{aligned}
\text{Inf}(X_2 \perp\!\!\!\perp X_3) &= E \log f_{23} - E \log f_3 - E \log f_2 \\
&= -(1.33 - 0.69 - 0.67) = 0.033, \\
\text{Inf}(X_1 \perp\!\!\!\perp X_3) &= -(1.24 - 0.67 - 0.68) = 0.114, \text{ and} \\
\text{Inf}(X_2 \perp\!\!\!\perp X_1) &= -(1.22 - 0.68 - 0.69) = 0.157,
\end{aligned}$$

so that no pair of variables is marginally independent. It is intriguing that the difference between the marginal and the conditional information numbers is the same whatever the pair; that is

$$\begin{aligned}
\text{Inf}(X_2 \perp\!\!\!\perp X_3) - \text{Inf}(X_2 \perp\!\!\!\perp X_3 | X_1) &= \text{Inf}(X_1 \perp\!\!\!\perp X_3) - \text{Inf}(X_1 \perp\!\!\!\perp X_3 | X_2) \\
&= \text{Inf}(X_2 \perp\!\!\!\perp X_1) - \text{Inf}(X_2 \perp\!\!\!\perp X_1 | X_3)
\end{aligned}$$

and are all equal to 0.033. This is not a coincidence. □

We shall defer examples based on Normally distributed random variables until after our discussion of the multivariate Normal distribution.

The information in marginal distributions

The numerical example above suggested that the information divergences against marginal and conditional independence satisfy certain identities.

Proposition 4.5.1 *Suppose that the partitioned random vector (X_a, X_b, X_c) has a joint density function f_{abc}. The difference between the divergence against the conditional independence of X_b and X_c given X_a and the marginal independence of X_b and X_c is invariant to permuting the symbols X_a, X_b and X_c.*

Proof: Subtracting $\text{Inf}(X_b \perp\!\!\!\perp X_c | X_a) = I(f_{abc}; f_{ab} f_{ac} / f_a)$ from $\text{Inf}(X_b \perp\!\!\!\perp X_c) = I(f_{bc}; f_b f_c)$ gives

$$\text{Inf}(X_b \perp\!\!\!\perp X_c) - \text{Inf}(X_b \perp\!\!\!\perp X_c | X_a) = -E \log \frac{f_{abc} f_a f_b f_c}{f_{ab} f_{ac} f_{bc}};$$

and the right hand side is invariant to permuting the labels a, b and c. □

If this common difference is denoted by $G(a,b,c)$, we have the identities:

$$\begin{aligned}\mathrm{Inf}(X_b \perp\!\!\!\perp X_c) &= \mathrm{Inf}(X_b \perp\!\!\!\perp X_c | X_a) + G(a,b,c),\\ \mathrm{Inf}(X_a \perp\!\!\!\perp X_c) &= \mathrm{Inf}(X_a \perp\!\!\!\perp X_c | X_b) + G(a,b,c),\\ \mathrm{Inf}(X_a \perp\!\!\!\perp X_b) &= \mathrm{Inf}(X_a \perp\!\!\!\perp X_b | X_c) + G(a,b,c).\end{aligned}$$

The term $G(a,b,c)$ is the additive adjustment to the conditional information in X_b about X_c, given X_a, necessary to calculate the marginal information in X_b about X_c. Note that G can be positive or negative. When it is positive there is more information for deciding against marginal independence than against conditional independence, and remarkably this is true simultaneously for all pairs of variables.

The information measures for assessing the strengths of the independence relationships between three random vectors are

$$\begin{array}{ccc}\mathrm{Inf}(X_a \perp\!\!\!\perp X_b | X_c), & \mathrm{Inf}(X_a \perp\!\!\!\perp X_c | X_b), & \mathrm{Inf}(X_b \perp\!\!\!\perp X_c | X_a),\\ \mathrm{Inf}(X_a \perp\!\!\!\perp X_b), & \mathrm{Inf}(X_a \perp\!\!\!\perp X_c), & \mathrm{Inf}(X_b \perp\!\!\!\perp X_c),\end{array}$$

and a consequence of this proposition is that two of the terms in the list are redundant. Because independence graphs are characterised by conditional independence it is natural to retain $\mathrm{Inf}(X_b \perp\!\!\!\perp X_c | X_a)$, $\mathrm{Inf}(X_a \perp\!\!\!\perp X_c | X_b)$, $\mathrm{Inf}(X_a \perp\!\!\!\perp X_b | X_c)$ and for completeness $G(a,b,c)$.

Proposition 4.5.2 *The information proper satisfies the identities*

$$\begin{aligned}&(i)\mathrm{Inf}(X_a \perp\!\!\!\perp (X_b, X_c)) = \mathrm{Inf}(X_a \perp\!\!\!\perp X_c | X_b) + \mathrm{Inf}(X_a \perp\!\!\!\perp X_b); \quad \textit{and}\\ &(ii)\mathrm{Inf}(X_a \perp\!\!\!\perp (X_b, X_c)) = \mathrm{Inf}(X_a \perp\!\!\!\perp X_c | X_b) + \mathrm{Inf}(X_a \perp\!\!\!\perp X_b | X_c) + G(a,b,c).\end{aligned}$$

The proof is just simple algebraic manipulation. The term $G(a,b,c)$ measures the lack of 'orthogonality' in the three variables (or vectors) and provides another interpretation for this quantity. Note that if $\mathrm{Inf}(X_a \perp\!\!\!\perp (X_b, X_c))$ is zero then $\mathrm{Inf}(X_a \perp\!\!\!\perp X_c | X_b)$ and $\mathrm{Inf}(X_a \perp\!\!\!\perp X_b | X_c)$ and consequently $G(a,b,c) = 0$. By the positive definite nature of the divergence, the first decomposition provides an alternative proof of the equivalence of $X_a \perp\!\!\!\perp (X_b, X_c)$ and the pair, $X_a \perp\!\!\!\perp X_c | X_b$, $X_a \perp\!\!\!\perp X_b$, a version of which was proved in Proposition 2.2.6.

4.6 Divergence and Collapsibility

Some further discussion about the use of the information proper as an indicator of the strength of association is needed.

EXAMPLE 4.6.1 Reconsider the three-way table in Example 4.5.1 above. The information strengths of the three edges are (1,2): 0.12, (1,3): 0.08 and finally, (2,3): 0.0 as $2 \perp\!\!\!\perp 3 | 1$ for this table. Consider evaluating the information strength for each edge in the corresponding marginal distribution. A straightforward numerical calculation gives the information numbers $\text{Inf}(X_2 \perp\!\!\!\perp X_1) = 0.157$ and $\text{Inf}(X_1 \perp\!\!\!\perp X_3) = 0.114$. The strength on the $(1,3)$ edge has changed from 0.081 to 0.114. Is this reasonable?

Firstly, this information measure of strength makes an implicit comparison of two independence graphs. Thus $\text{Inf}(X_1 \perp\!\!\!\perp X_3 | X_2)$ compares

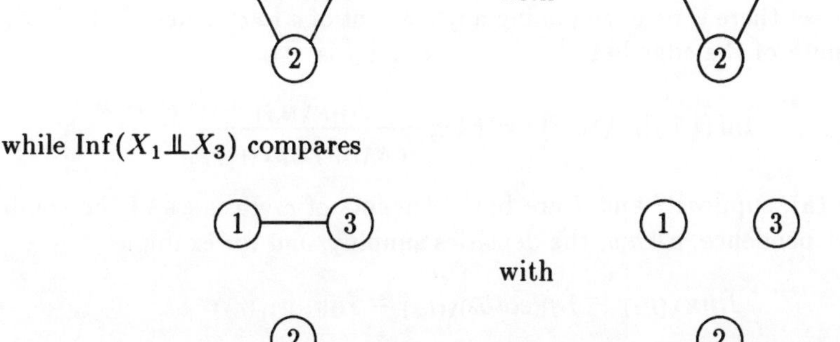

while $\text{Inf}(X_1 \perp\!\!\!\perp X_3)$ compares

so it is only in the exceptional case when the term $G(1,2,3) = 0$ that these comparisons are the same. Because independence graphs are derived from conditional independence constraints it would be wrong to use the latter measures.

Secondly, in this example, there is a feeling that vertex 1 'blocks' the (1,3) interaction off from vertex 2; and hence, when the graph is collapsed over 2, the (1,3) strength should be invariant under this marginalisation. Now manipulating identities in information establishes the equivalence of the assertions

(a) $X_3 \perp\!\!\!\perp X_2 | X_1$,

(b) $\text{Inf}(X_3 \perp\!\!\!\perp X_1) = \text{Inf}(X_3 \perp\!\!\!\perp X_1 | X_2) + \text{Inf}(X_3 \perp\!\!\!\perp X_2)$, and

(c) $\text{Inf}(X_3 \perp\!\!\!\perp X_1) = \text{Inf}(X_3 \perp\!\!\!\perp (X_1, X_2))$.

In the context of the independence (a), the marginal information $\text{Inf}(X_3 \perp\!\!\!\perp X_1)$ in (c) is a measure not only of the dependence of X_3 on X_1 but also of X_3 on (X_1, X_2). Accepting the independence (a) increases the ability to discriminate between graphs, so by (b), $\text{Inf}(X_3 \perp\!\!\!\perp X_1)$ is always larger than $\text{Inf}(X_3 \perp\!\!\!\perp X_1 | X_2)$ by an amount equal to $\text{Inf}(X_3 \perp\!\!\!\perp X_2)$. □

The next proposition formalises these ideas.

Proposition 4.6.1 *If the random vector X is collapsed over the vertices in X_a, and if $b = \mathrm{bd}\,(a)$ is the boundary set and c is the set of remaining vertices, then the information proper between any two variables in the (X_b, X_c) margin*

(a) is invariant to collapsing if neither element is in the boundary of a;

(b) increases if exactly one element is in the boundary of a; but

(c) may increase or decrease if both elements are in the boundary of a.

Proof: The subsets a, b and c partition $K = \{1, 2, \ldots, k\}$, and, as b is the boundary set there is no edge joining any element of a to c; hence $X_a \perp\!\!\!\perp X_c | X_b$. The strength of the edge between vertex i and j is

$$\mathrm{Inf}\,(i \perp\!\!\!\perp j | K\backslash\{i,j\}) = E\log \frac{f_{ij|K\backslash\{i,j\}}}{f_{i|K\backslash\{i,j\}} f_{j|K\backslash\{i,j\}}}.$$

To prove (a), suppose i and j are both elements of c. Because of the conditional independence, $c \perp\!\!\!\perp a | b$, the densities simplify, and for example

$$f_{ij|K\backslash\{i,j\}} = f_{ij|(a\cup b\cup c)\backslash\{i,j\}} = f_{ij|(b\cup c)\backslash\{i,j\}}.$$

The subset a vanishes from each conditioning set, leaving

$$\mathrm{Inf}\,(i \perp\!\!\!\perp j | (b \cup c)\backslash\{i,j\}).$$

But this is just the edge strength in the marginal graph, as is to be shown.

For part (b), suppose that i is in b and j is in c.

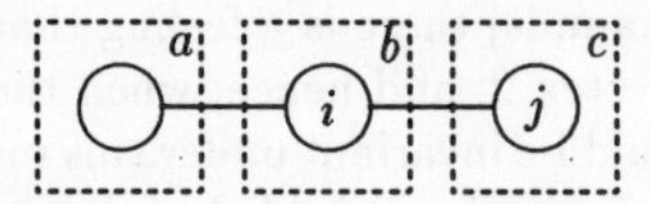

The original edge strength can be expressed in the form

$$E\log \frac{f_{j|K\backslash\{j\}}}{f_{j|K\backslash\{i,j\}}} = E\log \frac{f_{j|K\backslash\{j\}}}{f_{j|a\cup(b\backslash\{i\})\cup(c\backslash\{j\})}}$$

which, using conditional independence, becomes

$$E\log \frac{f_{j|b\cup(c\backslash\{j\})}}{f_{j|(b\backslash\{i\})\cup(c\backslash\{j\})}} + E\log \frac{f_{j|(b\backslash\{i\})\cup(c\backslash\{j\})}}{f_{j|a\cup(b\backslash\{i\})\cup(c\backslash\{j\})}}$$

on manipulating the last expression. Hence we have established the identity

$$\mathrm{Inf}\,(i \perp\!\!\!\perp j | (a \cup b \cup c)\backslash\{i,j\}) = \mathrm{Inf}\,(i \perp\!\!\!\perp j | (b \cup c)\backslash\{i,j\}) - \mathrm{Inf}\,(j \perp\!\!\!\perp a | (b \cup c)\backslash\{i,j\})$$

and as the second term is negative the assertion follows. Finally the third part, part (c), follows by constructing examples. □

In Example 4.5.1, one vertex is in the boundary of the set over which the table is collapsed, and hence the information proper must increase. Interestingly, in proving part (b) of the proposition, we calculated the difference between the marginal measure and the full measure as $\mathrm{Inf}(j \perp\!\!\!\perp a|(b \cup c)\backslash\{i,j\})$; this is just the information against the independence of j from a when the element i is removed from the separating set. In some cases this may be small, and in particular, it will be zero if there exists another element in c separating a from j.

Invariance of the mixed derivative

In contrast to the information measure of edge strength the mixed derivative measure of interaction discussed in Section 2.3 is invariant to collapsing as long as at least one of the interacting variables lies outside the boundary set. This is easy to show, because, with the same notation as Proposition 4.6.1 above, if e is a subset of K containing the p vertices of interacting variables, the p-th order cross partial derivative

$$D_e^p \log f_{a \cup b \cup c} = D_e^p \log f_{b \cup c},$$

due to the independence $X_a \perp\!\!\!\perp X_c | X_b$. In Examples 4.5.1 and 4.6.1 above,

$$D_{13}^2 \log f_{123} = D_{13}^2 \log f_{13},$$

due to the conditional independence between X_2 and X_1; and is the log cross-product ratio between X_1 and X_3. It is easily verified from the table that

$$\log \mathrm{cpr}\,(X_1, X_3) = -2.079 = \log \mathrm{cpr}\,(X_1, X_3|X_2).$$

EXAMPLE 4.6.2 Consider a five-way table of probabilities generated from the formula

$$p(x) = \exp(u + 2x_1x_2 - 2x_2x_4 + x_1x_3 + x_3x_4 - 4x_4x_5),$$

with $u = -(x_1 + x_2 + x_3 + x_4 + x_5)$, containing only linear terms, and where each x_i, $i = 1, 2, \ldots, 5$, takes the values 0 or 1. In standard order, with x_1 varying quickest, these 2^5 probabilities are

.129 .047 .047 .129 .047 .047 .017 .129 .047 .017 .002 .006 .047 .047
.002 .017 .047 .017 .017 .047 .017 .017 .006 .047 .000 .000 .000 .000
.000 .000 .000 .000

The independence graph of X is

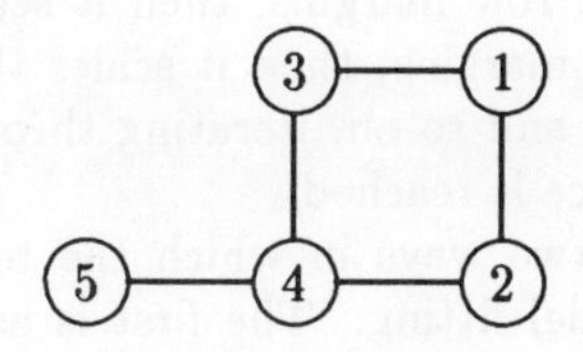

and the edge strengths are

	full table		2345 marginal table	
edge	strength	log-cpr	strength	log-cpr
(1,2)	.091	2.00		
(1,3)	.023	1.00		
(1,4)	.000	0.00		
(1,5)	.000	0.00		
(2,3)	.000	0.00	.005	0.43
(2,4)	.044	-2.00	.052	-2.00
(2,5)	.000	0.00	.000	0.00
(3,4)	.015	1.00	.015	1.00
(3,5)	.000	0.00	.000	0.00
(4,5)	.042	-4.00	.042	-4.00

Now it is seen that if the table is collapsed over vertex 1, the edge strengths in the four-way table alter. The $(2,4)$ edge strength increases substantially, though the $(3,4)$ edge strength remains practically the same. All strengths for edges including vertex 5 remain the same because 5 is separated from the boundary set of 1, $\mathrm{bd}(1) = \{2,3\}$, by vertex 4; as does the strength for edge $(4,5)$ because neither vertex is in the boundary set of the collapsed vertices. The logs of the cross-product ratios calculated from the log-linear expansion above generating the table are given for comparison. □

Measures of interaction based on the correlation coefficient behave in a similar manner to the information measures of edge strength because they are not mixed derivative measures.

4.7 Iterative Proportional Fitting

The iterative proportional fitting (ipf) algorithm has rather a long history. Its inception into the statistical literature is due to an early paper by Deming and Stephan (1940), but the same Deming-Stephan technique can be traced even earlier in other disciplines, for instance, back to the 1930's in fitting gravity models for transport studies. In its simplest form, the algorithm provides a method of adjusting one two-way contingency table to conform to the margins specified by another two-way table. It begins by scaling the rows of the first table to have the correct row margins, then it scales the resulting table to have the correct column margins, then it scales the resulting table to have the correct row margins, and so on, iterating through the cycle of rows and columns, until convergence is reached.

There are essentially two ways in which the technique is applied in the context of log-linear model fitting. The first is as described, to adjust the margins of a given table so as to better explore the higher order interactions, and an example of its use is given in the discussion of the cross-product ratio in Section 2.4. The second application is as a way to solve the likelihood equations. For example, the fitted values under the model of row and column independence can be calculated by scaling any independent table by the ipf algorithm to have row and column margins equal to the observed margins. An interesting and easily accessible discussion is given in Bishop *et al.* (1975). Our interest is stimulated by the use of the ipf algorithm in fitting graphical models. The application to covariance selection models is considered by Speed and Kiiveri (1986) and the procedure is implemented by Edwards (1987) for fitting hierarchical mixed interaction models.

The problem we consider is, given a density function f, of finding another density function, g, that is close to f, but which has a specified interaction structure.

EXAMPLE 4.7.1 Consider a fixed 2×2 table of probabilities, determining the density function f_{12} with margins given by

$$f_{12}: \quad \begin{array}{cc|c} & & .5 \\ & & .5 \\ \hline .6 & .4 & 1.0 \end{array} \qquad \text{and consider adjusting} \quad g_{12}^0 = \begin{array}{cc|c} .1 & .2 & \\ .3 & .4 & \\ \hline & & \end{array}$$

which has a cross-product ratio of 4/6=.667 to have the same margins as f_{12}. The ipf algorithm first adjusts the elements of g_{12}^0 to have the same row margins as f_{12} and holds the result in g_{12}^1 :

$$g_{12}^1 = \begin{array}{cc|c} (.1/.3).5 & (.2/.3).5 & \\ (.3/.7).5 & (.4/.7).5 & \\ \hline & & \end{array} = \begin{array}{cc|c} .167 & .333 & \\ .214 & .286 & \\ \hline & & \end{array}.$$

Now g_{12}^1 has the same row margins as f_{12}, that is $g_1^1 = f_1$. Then g_{12}^1 is adjusted to have the same column margins as f_{12}:

$$g_{12}^2 = \begin{array}{cc|c} (.167/.381).6 & (.333/.619).4 & \\ (.214/.381).6 & (.286/.619).4 & \\ \hline & & \end{array} = \begin{array}{cc|c} .263 & .215 & \\ .338 & .185 & \\ \hline & & \end{array},$$

so that, $g_2^2 = f_2$. However this disturbs the row margins and so g_{12}^2 is now adjusted to have the same row margins as f_{12}. The procedure continues cycling between row and column adjustments until convergence, for instance, until there is no change in the third decimal place. Finally

$$g_{12}^\infty = \begin{array}{cc|c} .276 & .224 & .5 \\ .324 & .176 & .5 \\ \hline .6 & .4 & 1.0 \end{array}.$$

The cross-product ratio in this table is still .667 and the interaction structure of g_{12}^0 is preserved in g_{12}^∞ while the margins of g_{12}^∞ are those of f_{12}. □

EXAMPLE 4.7.2 Take the same values for f_{12}, but start the iteration off with a table in which the row and column variables are independent

$$f_{12}: \quad \begin{array}{cc|c} & & 0.5 \\ & & 0.5 \\ \hline .6 & .4 & 1.0 \end{array} \quad \text{and adjust} \quad g_{12}^0 = \begin{array}{cc|c} 2/9 & 4/9 & \\ 1/9 & 2/9 & \\ \hline & & \end{array}$$

which has a cross-product ratio of 1. Then

$$g_{12}^1 = \begin{array}{cc|c} (2/6).5 & (4/6).5 & \\ (1/3).5 & (2/3).5 & \\ \hline & & \end{array} = \begin{array}{cc|c} 1/6 & 2/6 & \\ 1/6 & 2/6 & \\ \hline & & \end{array}.$$

and g_{12}^1 has the same row margins as f_{12}. Then g_{12}^1 is adjusted to have the same column margins as f_{12}:

$$g_{12}^2 = \begin{array}{cc|c} (1/2).6 & (1/2).4 & \\ (1/2).6 & (1/2).4 & \\ \hline & & \end{array} = \begin{array}{cc|c} 3/10 & 2/10 & \\ 3/10 & 2/10 & \\ \hline & & \end{array}.$$

But this table, g_{12}^2, has exactly the right margins, and any further iteration will just return g_{12}^2. Convergence has occurred in a finite number of steps. □

Another source of examples can be constructed by adjusting a correlation matrix so that sub-matrices corresponding to pre-specified marginal distributions are equal to those of another correlation matrix. For instance, we may wish to find the closest independent approximation, to a standardised bivariate Normal density. However, a satisfactory treatment of even this simple case involves linear least squares prediction theory, and so we defer the topic until we need it to construct algorithms for the evaluation of maximum likelihood estimates in graphical Gaussian models.

Convergence

The convergence of the ipf procedure has been considered by several authors, and the connection with minimum information divergence was made by Kullback (1968), however it is not until the important theoretical contribution of Csiszar (1975), that the problem was completely solved. Though the proof is too difficult for this text, we hope to indicate the connection with the divergence, and give some indication of why it works.

In the first instance suppose that the full vector is partitioned into just two sub-vectors so that $X_K = (X_a, X_b)$. The interaction structure is predetermined by an appropriate choice for g^0_{ab} and the margins by a choice of f_{ab}. The ipf procedure scales the rows according to

$$g^1_{ab} = g^0_{ab} f_a / g^0_a = g^0_{b|a}\, f_a,$$

and the columns according to

$$g^2_{ab} = g^1_{ab} f_b / g^1_b = g^1_{a|b}\, f_b.$$

The first transformation is then re-applied to g^2_{ab}, giving g^3_{ab}, and so on, the iteration continuing until some convergence criterion is attained. The n-th term, n even, of this sequence of iterations is the updating equation

$$g^{n+1}_{ab} = g^n_{b|a}\, f_a.$$

We can now show that the ipf algorithm preserves the higher order interaction structure determined by the density g^0, when interaction is measured by the mixed derivative as discussed in Section 2.3.

Let e be any subset of K that contains at least one vertex element that is not in a and one that is not in b. Suppose the cardinality of e is p. The mixed derivative is derived by taking logarithms:

$$\log g^{n+1}_{ab} = \log g^n_{ab} - \log g^n_a + \log f_a,$$

and differentiating, to get

$$D^p_e \log g^{n+1}_{ab} = D^p_e \log g^n_{ab},$$

because, by presumption, e has an element that is not in a so that the term containing a is annihilated by differentiating with respect to the corresponding variable. Any terms involving b vanish for the same reason. This expression iterates to give

$$D^p_e \log g^{n+1}_{ab} = D^p_e \log g^0_{ab},$$

so preserving the interaction structure of the initial density function. Thus if a limiting density function exists it has the 'right' interaction structure.

Now consider the convergence of the procedure. The divergence between f and g^{n+1} is

$$\begin{aligned} I(f; g^{n+1} \| X_a, X_b) &= \mathrm{E} \log f_{ab}/g^{n+1}_{ab} \\ &= \mathrm{E} \log f_{ab}/g^n_{ab} + \mathrm{E} \log g^n_{ab}/g^{n+1}_{ab} \\ &= I(f; g^n \| X_a, X_b) - \mathrm{E} \log f_a/g^n_a \\ &= I(f; g^n \| X_a, X_b) - I(f; g^n \| X_a). \end{aligned}$$

This is highly suggestive. The sequence of divergences between f and g^{n+1} is monotonically decreasing, and as the divergence is bounded below by zero, it must have a limit. If the sequence of divergences has a limit, intuition suggests that this implies the existence of a limiting density function, g^{∞}_{ab}; and furthermore, because the increments in this sequence must go to zero:

$$\lim_n I(f; g^n \| X_a) = 0 \quad \text{and} \quad \lim_n I(f; g^n \| X_b) = 0,$$

the limiting density must have the correct margins, as well as the right interaction structure. A rigorous proof whihc crucially depends on the notion of convex sets, is rather more difficult, and we refer the interested reader to Csiszar (1975).

A more general set-up

Slightly more generally the ipf algorithm operates in the following set-up: the partition $K = \{a, b\}$ determining the margins is generalised to a set of subsets, $\{a_1, a_2, \ldots, a_m\}$ which need not be disjoint but must cover K, so that $\cup_i a_i = K$, and are pairwise incomparable: that is no one set can be contained in another.

The aim is to find a joint density g for the k random variables that agrees with the density f_K on the m margins corresponding to the subsets a_i but with a higher order interaction structure determined by g^0. Let $\{a, b\}$ denote a partition of K, so that $b = a^c$. The algorithm proceeds by cycling through the list of subsets,

$$a = a_i, \quad i = 1, 2, \ldots, m,$$

using the *ipf updating equation*

$$g^{n+1}_{ab} = g^n_{b|a} f_a,$$

to make an adjustment to each margin in turn, until convergence is reached. In our discussion above, the list contained only two elements, but our reflections on the existence of a limit, g^{∞}, its interaction structure and the convergence of the algorithm, are easily generalised.

EXAMPLE 4.7.3 A chordless 4-cycle. Consider the four-way table of probabilities, p_{1234} :

		$x_2 = 0$		$x_2 = 1$	
		$x_1 = 0$	$x_1 = 1$	$x_1 = 0$	$x_1 = 1$
$x_4 = 0$	$x_3 = 0$	1	7	1	2
	$x_3 = 1$	2	3	4	5
$x_4 = 1$	$x_3 = 0$	1	2	1	2
	$x_3 = 1$	2	3	4	3
		6	15	10	12

multiplied by 1/43 so as to add to 1. The two-way marginal table, p_{12}, is given in the last line. Consider finding the table of probabilities, closest to a table for which all variables are independent, but with two-way margins on the subsets $\{12, 23, 34, 41\}$ identical to those for p_{1234} given here. We can use the ipf procedure to find the solution. A choice of the the initial table is $g^0(x) = 1/16$ for all x. Successively adjust this table to have the same $\{1,2\}$, $\{2,3\}$, $\{3,4\}$, $\{1,4\}$, margins, in the manner described above. It converges to the table g^∞

		$x_2 = 0$		$x_2 = 1$	
		$x_1 = 0$	$x_1 = 1$	$x_1 = 0$	$x_1 = 1$
$x_4 = 0$	$x_3 = 0$	1.75	5.44	1.52	2.29
	$x_3 = 1$	1.34	4.17	3.38	5.10
$x_4 = 1$	$x_3 = 0$	1.33	2.47	1.15	1.04
	$x_3 = 1$	1.57	2.92	3.95	3.57
		6	15	10	12

which has exactly the right margins but is by no means identical to p_{1234}. In fact this table, g^∞_{1234}, has the log-linear expansion

$$\log g^\infty(x) = -3.20 + 1.13x_1 - 0.145x_2 - 0.266x_3 - 0.276x_4 \\ - 0.721x_1x_2 + 1.07x_2x_3 - 0.512x_1x_4 + 0.43x_3x_4$$

The non-zero u-terms correspond to the adjusted margins. All higher order u-terms are zero, as in the log-linear expansion g^0. □

In implementing the algorithm for practical use, certain technical details have to be considered, in particular, choosing the order for the marginal sets a_i, $i = 1, 2, \ldots, m$, poses an intriguing problem. One result in this direction is that if the density is decomposable, see Chapter 12, then there exists an order for the marginal sets such that the algorithm converges in a finite number of steps.

4.8 Exercises

1: The random variable X has a Poisson distribution if its density function is $f(x) = \exp(-\lambda + x \log \lambda)/x!$ for $x = 0, 1, 2, \ldots$, and we say $X \sim P(\lambda)$. Find the information divergence between $f : X \sim P(\lambda)$ and $g : X \sim P(\nu)$.

2: Find the divergence between $f : X \sim N(\mu, \sigma^2)$ and $g : X \sim N(0, 1)$. Show that the divergence to discriminate (μ, σ^2) against $(\mu = 0, \sigma^2 = 1)$ is additive in the sense that is it is the sum of the divergences to discriminate μ against $\mu = 0$ and σ^2 against $\sigma^2 = 1$.

3: The random variable X has a Gamma distribution if its density function $f(x) = \exp(-x/\alpha + (p-1)\log x - p\log\alpha - \log\Gamma(p))$ for $x \geq 0$ where $\Gamma(p)$ is the gamma function. We say $X \sim$ Gamma(p, α). Find the divergence between $f : X \sim$ Gamma(p, α) and $g : X \sim$ Gamma(p, β).

4: Plot the function $t\log t$ for $0 < t < 1$. Show that $t\log t \to 0$ as $t \to 0$. Find the first two derivatives of $t\log t$ and prove that $t\log t$ is convex for $t \geq 0$.

5: Show that $\log t \leq t - 1$ for $t > 0$.

6: By writing $-\log(1-t)$ as an integral, or otherwise, show that

$$t \leq -\log(1-t) \leq t + t^2/2(1-t) \text{ for } 0 \leq t < 1.$$

7: Denote $-t\log t - (1-t)\log(1-t)$ by $h(t)$. Plot $h(t)$ for $0 \leq t \leq 1$. Find its first two derivatives and show that it is convex for $t \geq 0$. Compare the behaviour of $h(t)$ with the function $t\log 2t + (1-t)\log 2(1-t)$.

8: Show that the Bernoulli divergence

$$I(p; q) = p\log p/q + (1-p)\log(1-p)/(1-q)$$

satisfies the inequalities
(a) $I(p; 1/2) \leq 2\log 2|p - 1/2|$ for $0 \leq p \leq 1$, and
(b) $2(p-q)^2 \leq I(p; q) \leq 2(p-q)^2/q(1-q)$ for $0 \leq p \leq 1$, and $0 < q < 1$.

9: Show that the Shannon entropy $h(p, 1-p)$ is maximised at $p = 1/2$ and, more generally, $h(p(1), p(2), \ldots, p(k)) = -\sum p(i)\log p(i)$ is maximised at $p(i) = 1/k$.

10: If the mixed derivative interaction measure for the random variables X and Y is $i(x, y)$ show that $i(x, y) = 0$ if and only if $\mathrm{Inf}(X \perp\!\!\!\perp Y) = 0$. If (X, Y) is bivariate Bernoulli use this to show that the condition $\mathrm{cpr}(X, Y) = 0$ implies the divergence is zero.

11: Write out the proof of the proposition asserting the additivity of the information divergence under repeated sampling in Proposition 4.3.1.

12: Verify the identity $\mathrm{Inf}(X \perp\!\!\!\perp Y) = h(X) + h(Y) - h(X, Y)$ relating Shannon's entropies to the divergence against independence (information proper).

13: For jointly distributed Bernoulli random variables X, Y and Z show that

$$0 \leq -E\log f_{Z|X,Y} \leq \log 2.$$

and find the largest attainable value of $\mathrm{Inf}(Y \perp\!\!\!\perp Z|X)$.

14: Plot the function relating the information divergence against independence to the equivalent Normal correlation coefficient. What is the largest attainable equivalent correlation between Bernoulli random variables?

15: Generalise the identity

$$\text{Inf}(X \perp\!\!\!\perp (Y,Z)) = \text{Inf}(X \perp\!\!\!\perp Z|Y) + \text{Inf}(X \perp\!\!\!\perp Y)$$

by replacing (Y, Z) by $(Y_1, Y_2, \ldots, Y_q)$.

16: Define

$$V(f;g) = \int |f(x) - g(x)|dx \quad \text{and} \quad P(f;g) = \int \frac{(f(x) - g(x))^2}{g(x)} dx$$

in honour of Pearson's chi-squared statistic.
(a) Show that $V(f;g) \leq \{P(f;g)\}^{\frac{1}{2}}$.
(b) Show that $V(f;g) \leq \{2I(f;g)\}^{\frac{1}{2}}$. (Difficult)

Chapter 5

The Inverse Variance

The multivariate Normal distribution is a uniquely tractable probability model for continuous random variables, and among many other remarkable properties, its conditional independences are simply characterised in terms of zeros in the inverse variance. In consequence, an adequate treatment of graphical models based on the Normal distribution needs a full discussion of this entity.

The chapter begins by defining what is meant by a random vector, by its expectation and by the covariance between two random vectors. The text highlights the bilinearity of the covariance, before moving onto the fundamentals of linear least squares prediction and a discussion of the partial covariance. These concepts provide the basic technical machinery for the interpretation of the elements of the inverse variance. The prediction problem is of interest in its own right, and it is presented here to emphasise the parallels with the regression techniques of applied statistics. The partial covariance provides the theoretical generalisation of the covariance needed to consider conditional independence in the multivariate Normal distribution.

The important results are summarised in the inverse variance lemma and its corollaries. The lemma, the centrepiece of this chapter, evaluates the inverse in the same way as the inverse is determined in a Cholesky decomposition of a matrix into the product of triangular and diagonal matrices. It shows that the diagonal elements of the inverse are related to multiple correlation coefficients and the off diagonal elements to partial correlation coefficients. Other corollaries to the lemma provide additional tools required for the analysis of the multivariate Normal distribution: completing the square in a quadratic form and the evaluation of a determinant. The latter topic makes the connection between the Schur complement of a partitioned matrix and the partial variance.

We use the language of and several concepts from matrix and linear algebra, see Lang (1970)) for example, but we do not depend on any detailed knowledge. The general treatment has been influenced by Dempster (1969).

5.1 Random Vectors, Expectation and Covariance

The notions of a scalar random variable, its expectation and variance, and of the covariance between two random variables are taken as understood. Linearity is the essential property. If X and Y are scalar random variables and α is a fixed scalar then the *expectation* of X is denoted by $E(X)$ and satisfies

1. Linearity: $\begin{cases} E(X+Y) &= E(X)+E(Y), \\ E(\alpha X) &= \alpha E(X), \end{cases}$

2. $E(1) = 1$, and if X is non-negative then $E(X)$ is non-negative.

We usually drop the brackets and write EX for $E(X)$. The *covariance* of X and Y is denoted by $\text{cov}(X, Y)$ and defined as $\text{cov}(X,Y) = E(X - EX)(Y - EY)$. The *variance* of X is defined as $\text{var}(X) = \text{cov}(X,X)$. Note that $\text{var}(X) = E(X - EX)^2$. This definition, together with the linearity of the expectation operator implies that covariance is a symmetric bilinear form, so that

$$\begin{aligned} \text{cov}(X,Y) &= \text{cov}(Y,X), \\ \text{cov}(X,Y+Z) &= \text{cov}(X,Y) + \text{cov}(X,Z), \text{ and} \\ \text{cov}(X,\alpha Y) &= \alpha\text{cov}(X,Y). \end{aligned}$$

In extending the theory to higher dimensions it is natural to focus on an operator with simple mathematical properties, and one such, due to its built-in bilinearity, is the covariance. For instance, it is easier to manipulate $\text{cov}(X,X)$ rather than $\text{var}(X)$ as illustrated in deriving the following well known formulae for the variance of a sum of random scalars:

$$\begin{aligned} \text{var}(X+Y) &= \text{cov}(X+Y, X+Y) \\ &= \text{cov}(X, X+Y) + \text{cov}(Y, X+Y) \\ &= \text{var}(X) + 2\text{cov}(X,Y) + \text{var}(Y). \end{aligned}$$

Although the *correlation:* $\text{corr}(X,Y) = \text{cov}(X,Y)/\{\text{var}(X)\text{var}(Y)\}^{1/2}$, is perhaps the natural scale in which to measure association it is almost never true that $\text{corr}(X, Y + Z)$ simplifies to $\text{corr}(X,Y) + \text{corr}(X,Z)$.

Random vectors

We consider column vectors in p-dimensional Euclidean space. The *random vector* X is the vector

$$X = \begin{pmatrix} X_1 \\ X_2 \\ \vdots \\ X_p \end{pmatrix}$$

constructed from the p random variables $X_1, X_2, \ldots, X_p$. Similarly, random matrices of order $p \times q$ are just matrices with random variables as elements. The *expectation* of the random vector X is the vector $E(X)$ consisting of the expected values of the elements of X :

$$E(X) = \begin{pmatrix} EX_1 \\ EX_2 \\ \vdots \\ EX_p \end{pmatrix}.$$

Proposition 5.1.1 *The expectation E is linear over R^p, in the sense that*

$$E(X+Y) = E(X) + E(Y) \quad \textit{and} \quad E(\alpha X) = \alpha E(X)$$

for any scalar α, and all p-dimensional random vectors X and Y. In consequence

$$E(AX) = AE(X)$$

where A is any fixed linear transformation of X from R^p to R^q.

Proof: If X and Y are both p-dimensional, then the vector $E(X+Y)$ has elements $E(X_i + Y_i) = E(X_i) + E(Y_i)$ and the result follows from vector addition. The second result follows in the same manner. The last is derived by writing out the i-th element of AX as

$$\sum_{j=1}^{p} a_{ij} X_j$$

and then taking expectations. □

Because of the linearity of E it is rarely confusing to drop the brackets and write EX for $E(X)$.

The *covariance* between the p-dimensional random vector X and the q-dimensional random vector Y is the matrix

$$\operatorname{cov}(X,Y) = E(X - EX)(Y - EY)^T,$$

which, when written out in full, is

$$\operatorname{cov}(X,Y) = \begin{pmatrix} \operatorname{cov}(X_1,Y_1) & \operatorname{cov}(X_1,Y_2) & \ldots & \operatorname{cov}(X_1,Y_q) \\ \operatorname{cov}(X_2,Y_1) & \operatorname{cov}(X_2,Y_2) & \ldots & \operatorname{cov}(X_2,Y_q) \\ \ldots & \ldots & \ldots & \ldots \\ \operatorname{cov}(X_p,Y_1) & \operatorname{cov}(X_p,Y_2) & \ldots & \operatorname{cov}(X_p,Y_q) \end{pmatrix}.$$

Properties of the covariance

The bilinearity of the covariance on random variables underlies the bilinearity of the covariance on random vectors.

Proposition 5.1.2 *The covariance operator,* cov (.,.), *is bilinear, so that it is linear in the first argument:* $\text{cov}(\alpha X, Y) = \alpha \text{cov}(X, Y)$ *for* α *scalar and* $\text{cov}(X + Z, Y) = \text{cov}(X, Y) + \text{cov}(Z, Y)$ *when* X *and* Z *are both p-dimensional; similarly linear in the second argument; and, whenever* A *and* B *are fixed conformable linear transformations*

$$\text{cov}(AX, BY) = A\text{cov}(X, Y)B^T.$$

Furthermore, the covariance satisfies

$$\text{cov}(Y, X) = \text{cov}(X, Y)^T,$$

the transposed covariance matrix.

Proof: Write $\text{cov}(X, Y)$ as $E(X - EX)(Y - EY)^T$, and use the linearity of the expectation. The symmetry of the covariance between random variables, $\text{cov}(X_i, Y_j) = \text{cov}(Y_j, X_i)$, gives the last result. □

The *variance* var (X) is defined in terms of the covariance by

$$\text{var}(X) = \text{cov}(X, X),$$

which is, of course, symmetric. Now

$$\text{var}(AX) = \text{cov}(AX, AX) = A\text{cov}(X, X)A^T = A\text{var}(X)A^T$$

as the covariance is bilinear. By taking the transformation A to be of order $1 \times p$, AX is one dimensional. As the variance of a random variable cannot be negative, we have shown that the matrix var (X) is non-negative definite. In fact, unless otherwise specified, we will always assume that var (X) is positive definite, which implies that the inverse $\text{var}(X)^{-1}$ exists.

EXAMPLE 5.1.1 If A has the matrix $\begin{pmatrix} 1 & 2 \\ 3 & 4 \end{pmatrix}$ and var (X) the matrix $\begin{pmatrix} 2 & 1 \\ 1 & 2 \end{pmatrix}$ then the variance matrix of AX is

$$\begin{pmatrix} 1 & 2 \\ 3 & 4 \end{pmatrix}\begin{pmatrix} 2 & 1 \\ 1 & 2 \end{pmatrix}\begin{pmatrix} 1 & 2 \\ 3 & 4 \end{pmatrix}^T = \begin{pmatrix} 14 & 32 \\ 32 & 74 \end{pmatrix}.$$

□

EXAMPLE 5.1.2 Find the covariance matrix of AX and BX if

$$AX = \begin{pmatrix} 2X_1 + X_3 \\ X_2 - X_3 \end{pmatrix}, \quad BX = (X_1 + X_2 + X_3)$$

and the variance matrix of X is

$$\mathrm{var}(X) = \mathrm{var}\left(\begin{pmatrix} X_1 \\ X_2 \\ X_3 \end{pmatrix}\right) = \begin{pmatrix} 5 & 2 & 1 \\ 2 & 3 & 3 \\ 1 & 3 & 6 \end{pmatrix}.$$

Now $A = \begin{pmatrix} 2 & 0 & 1 \\ 0 & 1 & -1 \end{pmatrix}$ and $B = \begin{pmatrix} 1 & 1 & 1 \end{pmatrix}$ so the covariance is

$$\mathrm{cov}(AX, BY) = \begin{pmatrix} 2 & 0 & 1 \\ 0 & 1 & -1 \end{pmatrix} \begin{pmatrix} 5 & 2 & 1 \\ 2 & 3 & 3 \\ 1 & 3 & 6 \end{pmatrix} \begin{pmatrix} 1 \\ 1 \\ 1 \end{pmatrix} = \begin{pmatrix} 26 \\ -2 \end{pmatrix}.$$

□

We draw the reader's attention to the remarks on notation made towards the end of the first chapter, in particular, the convention that all vectors are column vectors, but are often displayed as row vectors for typographical reasons. Hence if the random vector T, say, is partitioned into two components X and Y then we write $T = (X, Y)$ when we should write $\begin{pmatrix} X \\ Y \end{pmatrix}$. Similarly $\mathrm{var}\left(\begin{pmatrix} X \\ Y \end{pmatrix}\right)$ is written as $\mathrm{var}(X, Y)$. We crave the reader's indulgence.

Conditional expectation and variance

The *conditional expectation and variance* and of Y given X is the ordinary expectation or variance operator applied to Y but evaluated with reference to the conditional distribution of Y given X rather than the marginal distribution of Y. The conditional expectation of Y given X is written as $E_{Y|X}(Y)$, and the conditional variance as $\mathrm{var}_{Y|X}(Y)$, so that the subscript indicates the distribution over which the expectation operations are evaluated. Thus the conditional variance of a function of Y, $g(Y)$ say, is $\mathrm{var}_{Y|X}(g(Y))$. Certain consequences of this convention are easily checked, for example, $E_{Y|X}(X) = X$ and $\mathrm{var}_{Y|X}(X) = 0$. The identity $\mathrm{var}_Y(Y) = \mathrm{var}_{XY}(Y)$ just states that the variance of Y evaluated in the joint distribution (X, Y) is the same as the variance evaluated in the marginal distribution. When we need to distinguish between a conditional expectation evaluated at a point, we write $E_{Y|X=x}(Y)$; note that the result is no longer random.

An advantage of the notation is to distinguish the conditional variance of Y given X, $\mathrm{var}_{Y|X}(Y)$, from the partial variance of Y given X, $\mathrm{var}(Y|X)$, defined later in this chapter as the ordinary variance of the least squares residual. In general, the partial and the conditional variances are not the same.

The conditional covariance of Y and Z given X satisfies an interesting identity

$$\operatorname{cov}_{YZ}(Y, Z) = E_X \operatorname{cov}_{YZ|X}(Y, Z) + \operatorname{cov}_X(E_{Y|X}(Y), E_{Z|X}(Z)).$$

It can be specialised by setting $Y = Z$ to give an expression for the conditional variance,

$$\operatorname{var}_Y(Y) = E_X \operatorname{var}_{Y|X}(Y) + \operatorname{var}_X(E_{Y|X}(Y));$$

and generalised by considering the covariance of $g(Y, Z)$ and $h(Y, Z)$. It also extends to higher moments and cumulants.

5.2 Linear Least Squares Prediction

Predicting the unknown from uncertain information is a central concern of statistical theory and practice. The problem takes many forms and has many variations: forecasting in time series analysis, estimating parameters of probability models, discrimination and classification, growth curve analysis; these are just a few examples from applied statistics. One can find numerous applications of these techniques in all areas of science.

The statistical theory of prediction developed here is concerned to find linear predictors that minimise a quadratic objective function. The mathematical tractability of linear least squares as a criterion for prediction makes it an extremely attractive technique, which works well in practice, and leads to more general techniques based on weighted least squares. The solution to the linear least squares prediction problem is an important stepping stone on the way to the analysis of the multivariate Normal distribution.

The problem is to predict a scalar random variable Y (in R^1) from an observation on a p-dimensional random vector X (in R^p). This task is simplified by requiring the predictor to be a linear function of X, and then to choose that one with the smallest mean square error of prediction. The result is known as the *linear least squares predictor* (llsp) of Y from X.

We may assume that the expected values of X and Y are zero because, if not, we only have to replace X by $X - EX$ and Y by $Y - EY$. Hence the linear least squares prediction problem for (X, Y) in R^{p+1} with $E(X, Y) = 0$ is to choose the vector of prediction coefficients, b, so that the predictor of Y, $\hat{Y} = b^T X$, minimises $E(Y - \hat{Y})^2$. The predictor is written as $\hat{Y}(X)$ to make its dependence on X explicit.

We shall soon see that Pythagoras knew the answer; the residual from the predictor must lie at right angles to the plane spanned by the variables in the predictor. This is the content of the next result.

Proposition 5.2.1 The normal equations. *The vector b minimises $E(Y - \hat{Y})^2$ if and only if b satisfies*

$$\text{cov}\,(Y - b^T X, X) = 0.$$

Proof: As $E(X, Y) = 0$, note that $E(Y - \hat{Y}) = 0$ and $E(Y - \hat{Y})^2 = \text{var}\,(Y - \hat{Y})$. Any vectors b and c in R^p satisfy the identity

$$\text{var}\,(Y - c^T X) = \text{var}\,(Y - b^T X + (b - c)^T X)$$

Expanding the right hand side gives

$$\text{var}\,(Y - b^T X) + 2\text{cov}\,(Y - b^T X, (b - c)^T X) + \text{var}\,((b - c)^T X)$$

which simplifies to give

$$\text{var}\,(Y - c^T X) = \text{var}\,(Y - b^T X) + 2\text{cov}\,(Y - b^T X, X)(b - c) + \text{var}\,((b - c)^T X).(*)$$

Now if b is chosen such that $\text{cov}\,(Y - b^T X, X) = 0$ then $\text{var}\,(Y - c^T X) \geq \text{var}\,(Y - b^T X)$ for all c, because $\text{var}\,((b - c)^T X) \geq 0$. Hence b is the minimising value.

Conversely let b denote the minimising value of $\text{var}\,(Y - b^T X)$ and c be any other vector so that $\text{var}\,(Y - c^T X) \geq \text{var}\,(Y - b^T X)$. The identity $(*)$ gives

$$\begin{aligned} 0 &\leq 2\text{cov}\,(Y - b^T X, X)(b - c) + \text{var}\,((b - c)^T X) \\ &= 2\text{cov}\,(Y - b^T X, X)(b - c) + (b - c)^T \text{var}\,(X)(b - c). \end{aligned}$$

Putting $c = b + \alpha v$ for some real scalar α and arbitrary v in R^p yields the inequality

$$0 \leq 2\text{cov}\,(Y - b^T X, X)v\alpha + v^T \text{var}\,(X)v\alpha^2.$$

But the right hand side is a simple quadratic function of α that passes through the origin, so the non-negativity constraint can only be satisfied if the coefficient of α is zero. Hence

$$\text{cov}\,(Y - b^T X, X)v = 0 \text{ for all vectors } v \text{ in } R^p,$$

or equivalently, $\text{cov}\,(Y - b^T X, X) = 0$, as had to be shown. □

An explicit expression for the coefficients is possible.

Corollary 5.2.2 *The linear least squares* prediction coefficients *are*

$$b^T = \text{cov}\,(Y, X)\text{var}\,(X)^{-1}$$

and the linear least squares predictor of Y from X is

$$\hat{Y}(X) = \text{cov}\,(Y, X)\text{var}\,(X)^{-1} X.$$

Proof: The normal equations $\text{cov}(Y - b^T X, X) = 0$ can be solved for b. Two applications of the bilinearity of $\text{cov}(.,.)$ give, firstly

$$\text{cov}(Y - b^T X, X) = \text{cov}(Y, X) - \text{cov}(b^T X, X),$$

and secondly,

$$0 = \text{cov}(Y, X) - b^T \text{cov}(X, X).$$

This is a system of simultaneous linear equations which can be solved to give $b^T = \text{cov}(Y, X)\text{var}(X)^{-1}$ as $\text{cov}(X, X) = \text{var}(X)$ is invertible. □

EXAMPLE 5.2.1 If X is a zero mean two dimensional random vector and Y is a random variable with $EY = 0$ and

$$\text{var}(X, Y) = \begin{pmatrix} 1 & r & a \\ r & 1 & b \\ a & b & 1 \end{pmatrix}$$

the linear least squares predictor of Y is given by selecting the appropriate part of the partition of $\text{var}(X, Y)$:

$$\begin{aligned} \hat{Y}(X) &= \text{cov}(Y, X)\text{var}(X)^{-1}X \\ &= \begin{pmatrix} a & b \end{pmatrix} \begin{pmatrix} 1 & r \\ r & 1 \end{pmatrix}^{-1} \begin{pmatrix} X_1 \\ X_2 \end{pmatrix} \\ &= \frac{(a - br)}{(1 - r^2)} X_1 + \frac{(b - ar)}{(1 - r^2)} X_2 \end{aligned}$$

This is an instructive example. The predictor is a linear combination of the coordinates of X whose coefficients depend on both $\text{cov}(X, Y)$ and $\text{var}(X)$. The coefficients are not determined by $\text{cov}(Y, X_1)$ and $\text{cov}(Y, X_2)$ alone, and in fact, a zero coefficient of X_1 neither implies nor is implied by, $\text{cov}(Y, X_1) = 0$; the necessary and sufficient condition for this is $a - br = 0$. The predictor simplifies when $\text{cov}(X_1, X_2) = 0$, so that $r = 0$ and $\text{var}(X)$ is diagonal, to $\hat{Y} = aX_1 + bX_2$; or equivalently $\hat{Y}(X_1, X_2) = \hat{Y}(X_1) + \hat{Y}(X_2)$. □

On to q-dimensions

Minimising the mean square error of prediction only makes sense when Y is scalar; however the normal equations for the predictor can be just as easily solved when Y is q-dimensional. Thus we use the definition

Definition. The *linear least squares predictor of Y from X* when (X, Y) is $p+q$-dimensional with variance $\text{var}(X, Y)$ and $E(X, Y) = 0$ is

$$\hat{Y}(X) = \text{cov}(Y, X)\text{var}(X)^{-1}X.$$

We occasionally write $\hat{Y}(X)$ as $\hat{Y}$ and often as BX with

$$B = \text{cov}\,(Y, X)\text{var}\,(X)^{-1},$$

where B is the $q \times p$ matrix of ***prediction coefficients.*** When the expected values are non-zero then replacing Y by $Y - EY$ and X by $X - EX$ gives

$$\hat{Y}(X) = EY + \text{cov}\,(Y, X)\text{var}\,(X)^{-1}(X - EX).$$

□

EXAMPLE 5.2.2 Suppose $p = 2 = q$.
(a) If $\text{var}\,(X) = I$ and $\text{cov}\,(X, Y) = I$ are both given by the identity matrix then

$$\hat{Y} = I(I)^{-1}X = X,$$

(b) If $\text{var}\,(X) = I$ and $\text{cov}\,(X, Y) = 11^T$ is a matrix of ones, then

$$\hat{Y} = (11^T)I^{-1}X = \begin{pmatrix} X_1 + X_2 \\ X_1 + X_2 \end{pmatrix},$$

(c) If $\text{var}\,(X) = \begin{pmatrix} 1 & r \\ r & 1 \end{pmatrix}$ and $\text{cov}\,(X, Y) = I$ then

$$\hat{Y} = \begin{pmatrix} \hat{Y}_1 \\ \hat{Y}_2 \end{pmatrix} = I\delta \begin{pmatrix} 1 & -r \\ -r & 1 \end{pmatrix} X = \begin{pmatrix} \delta(X_1 - rX_2) \\ \delta(X_2 - rX_1) \end{pmatrix}$$

where $\delta = 1/(1 - r^2)$. Hence the llsp of Y_1 depends upon X_2 as well as X_1 even though Y_1 is uncorrelated with X_2. □

5.3 Properties of the Predictor

The linear least squares predictor of Y from X, $\hat{Y}(X)$, has some remarkable properties. Because Y is q-dimensional the derivations begin with definition of the predictor, and do not use the least squares derivation explicitly. It is assumed that $\text{var}\,(X, Y)$ of the partitioned vector (X, Y), consisting of the variances of X, of Y, and the covariance of X and Y, is given.

Orthogonality

The first property shows that the residual $Y - \hat{Y}$ is orthogonal, in the sense of zero covariance, to X and to any linear transformation of X. Pictures are always helpful, so we include Figure 5.3.1 However the purist may remark, how can we represent Y which is one dimensional and X which is p-dimensional in the same space? The answer is that the 'vector' representing X in the diagram repesents the subspace generated by the p variables $X_1, X_2, \ldots, X_p$.

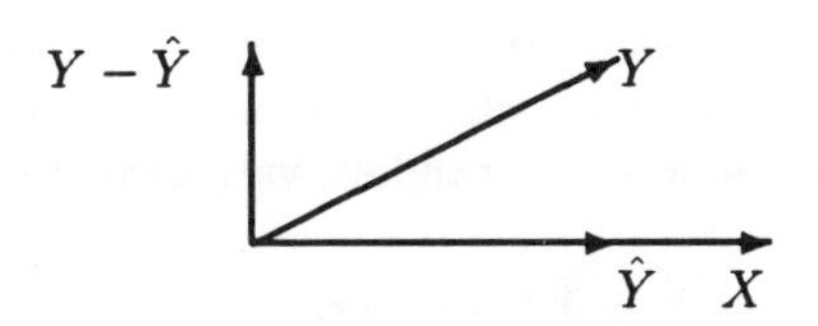

Figure 5.3.1: The orthogonality of X and $Y - \hat{Y}$

Proposition 5.3.1 Orthogonality of the residual. *The predictor satisfies*

$$\text{cov}\,(Y - \hat{Y}, X) = 0.$$

Proof: Direct evaluation gives

$$\begin{aligned}
\text{cov}\,(Y - \hat{Y}, X) &= \text{cov}\,(Y, X) - \text{cov}\,(\hat{Y}, X) && \text{bilinearity of cov,} \\
&= \text{cov}\,(Y, X) - \text{cov}\,(BX, X) && \text{definition of } B, \\
&= \text{cov}\,(Y, X) - B\text{var}\,(X) && \text{bilinearity of cov,} \\
&= 0 && \text{on substituting for } B.
\end{aligned}$$

□

The next corollary is immediate.

Corollary 5.3.2 *The predictor satisfies*

$$\begin{aligned}
\text{cov}\,(\hat{Y}, X) &= \text{cov}\,(Y, X), \\
\text{cov}\,(Y - \hat{Y}, \hat{Y}) &= 0, \quad \textit{and} \\
\text{cov}\,(Y - \hat{Y}, AX) &= 0 \textit{ for all } A.
\end{aligned}$$

The next two propositions concern the decomposition of the variance of Y. One might imagine that the second was well known to Pythagoras.

Proposition 5.3.3 Predictor variance. *The variance of the linear least squares predictor is*

$$\text{var}\,(\hat{Y}) = \text{cov}\,(Y, X)\text{var}\,(X)^{-1}\text{cov}\,(X, Y).$$

Proof: Writing $\hat{Y}$ as BX and using the corollary above gives

$$\text{var}\,(\hat{Y}) = \text{cov}\,(\hat{Y}, \hat{Y}) = B\text{cov}\,(X, \hat{Y}) = B\text{cov}\,(X, Y),$$

now substitute for B. □

Proposition 5.3.4 Variance decomposition. *The variance of Y decomposes into the sum*

$$\text{var}\,(Y) = \text{var}\,(\hat{Y}) + \text{var}\,(Y - \hat{Y}).$$

Proof: Write $\text{var}\,(Y) = \text{var}\,(\hat{Y} + Y - \hat{Y})$, expand, and use the orthogonality relations in Corollary 5.3.2 above. □

Invariance and additivity

Here we look at the predictor as a function of the explanatory variables, X. A distinction has to be made as to whether Y is to be predicted at a new realisation x of X, or on a new set of random variables. In the former case, we write

$$\hat{Y} = \hat{Y}(x) = Bx,$$

and as B does not change with x it is clear that $\hat{Y}(x)$ is a linear transformation of x. This formulation would be appropriate if we were to predict Y from new values of the variables in X. Alternatively, if we are to predict from a different set of random variables, Z, say, then the appropriate prediction coefficients are $\text{cov}(Y,Z)\text{var}(Z)^{-1}$. In this case the appropriate formulation of the linear least squares predictor is

$$\hat{Y}(X) = \text{cov}(Y,X)\text{var}(X)^{-1}X,$$

which is not at all linear in X.

Proposition 5.3.5 Invariance and additivity. *The predictor satisfies*

$$\begin{array}{rrcll} \textit{invariance:} & \hat{Y}(Y) & = & Y, & \\ & \hat{Y}(AX) & = & \hat{Y}(X) & \textit{if} \quad A \textit{ is invertible; and} \\ \textit{additivity:} & \hat{Y}(X,Z) & = & \hat{Y}(X)+\hat{Y}(Z) & \textit{if} \quad \text{cov}(X,Z)=0. \end{array}$$

Proof: Direct substitution gives the first: $\hat{Y}(Y) = \text{cov}(Y,Y)\text{var}(Y)^{-1}Y = Y$. For the second, use the bilinearity of $\text{cov}(.,.)$ to simplify

$$\begin{aligned} \hat{Y}(AX) &= \text{cov}(Y,AX)\text{var}(AX)^{-1}AX \\ &= \text{cov}(Y,X)A^T(A\text{var}(X)A^T)^{-1}AX, \end{aligned}$$

which simplifies to $\hat{Y}$ as A is invertible. For additivity note that if $\text{cov}(X,Z) = 0$ then $\text{var}(X,Z)$ is block diagonal and so

$$\hat{Y}(\begin{pmatrix} X \\ Z \end{pmatrix}) = \text{cov}(Y, \begin{pmatrix} X \\ Z \end{pmatrix})\text{var}(\begin{pmatrix} X \\ Z \end{pmatrix})^{-1}\begin{pmatrix} X \\ Z \end{pmatrix}$$

and as $\text{cov}(X,Z) = 0$, the predictor is

$$\begin{pmatrix} \text{cov}(Y,X) & \text{cov}(Y,Z) \end{pmatrix}\begin{pmatrix} \text{var}(X)^{-1} & 0 \\ 0 & \text{var}(Z)^{-1} \end{pmatrix}\begin{pmatrix} X \\ Z \end{pmatrix}$$

which simplifies to $\hat{Y}(X) + \hat{Y}(Z)$. □

The sweep operator

Consider the orthogonalising mapping L, known as the *sweep operator*,

$$L = \begin{pmatrix} I & 0 \\ -\mathrm{cov}\,(Y,X)\mathrm{var}\,(X)^{-1} & I \end{pmatrix} = \begin{pmatrix} I & 0 \\ -B & I \end{pmatrix}$$

that transforms the partitioned vector (X, Y) to the vector

$$L\begin{pmatrix} X \\ Y \end{pmatrix} = \begin{pmatrix} X \\ Y - \hat{Y} \end{pmatrix}$$

which has orthogonal components. Clearly L is invertible. A corollary to the last proposition, and one that is heavily used in the sequel, is the following.

Corollary 5.3.6 Generalised additivity. *The llsp of Y from the partitioned vector (X, Z) is the sum of the llsp of Y from X and of Y from the residual, $Z - \hat{Z}(X)$, after predicting Z from X:*

$$\hat{Y}(X, Z) = \hat{Y}(X) + \hat{Y}(Z - \hat{Z}).$$

Proof: By the invariance of the llsp to invertible transformations of the explanatory vector, the predictor $\hat{Y}(L(X, Z)) = \hat{Y}(X, Z)$ as L is invertible, so that $\hat{Y}(X, Z) = \hat{Y}(X, Z - \hat{Z})$. The covariance between X and $Z - \hat{Z}$ is zero so that the previous proposition applies. □

Prediction as linear projection

The behaviour of $\hat{Y}(X)$ is examined here as a function of the argument Y.

Proposition 5.3.7 Prediction as projection. *The predictor $\hat{Y}(X)$ is a linear projection. That is $\hat{Y}(X)$ is linear in Y:*

$$\begin{aligned} (\widehat{Y + Z})(X) &= \hat{Y}(X) + \hat{Z}(X), \\ (\widehat{\alpha Y})(X) &= \alpha\hat{Y}(X) \text{ for } \alpha \text{ scalar, and} \\ \widehat{AY} &= A\hat{Y}; \end{aligned}$$

and $\hat{Y}(X)$ is idempotent, so that $\hat{\hat{Y}} = \hat{Y}$.

Proof: As $\hat{Y}(X) = \mathrm{cov}\,(Y, X)\mathrm{var}\,(X)^{-1}X$, the linearity of $\hat{Y}$ is equivalent to the linearity of the covariance operator in the first argument. To show that $\hat{Y}$ is idempotent, put $Z = \hat{Y}(X)$ so that we have to show that $\hat{Z}(X) = Z$. Now

$$\begin{aligned} \hat{Z}(X) &= \mathrm{cov}\,(Z, X)\mathrm{var}\,(X)^{-1}X \\ &= \mathrm{cov}\,(BX, X)\mathrm{var}\,(X)^{-1}X \\ &= B\mathrm{cov}\,(X, X)\mathrm{var}\,(X)^{-1}X = BX \text{ by orthogonality,} \end{aligned}$$

and this of course is just Z. □

The reader may well feel at this point that the term linearity is somewhat overused if not positively abused. If we begin with the simple linear regression model $EY = a + bx$, then linearity refers to the linearity in x and so, $EY = a + bx + cx^2$, for example, is not linear but quadratic. But in the usual treatment of the multiple linear regression model where $EY = a + b_1x_1 + b_2x_2 + \ldots + b_px_p$, which (paradoxically) includes the quadratic, linearity refers to the coefficients $(a, b_1, \ldots, b_p)$. Then, at the beginning of our treatment of prediction, the predictor $\hat{Y} = b^T x$ appears to be linear in both the prediction coefficients b and the explanatory variables X; but finally when it is viewed as $\hat{Y}(X) = \text{cov}\,(Y, X)\text{var}\,(X)^{-1}X$ it turns out to be linear in Y and non-linear in X!

The least squares property of the q-dimensional predictor, which motivated its construction, is left as an exercise.

The multiple correlation coefficient

If Y is scalar ($q = 1$) the correlation between Y and its predictor $\hat{Y}$ is defined as the *multiple correlation coefficient,* R . With this definition and the help of the properties above it is easy to show that

$$R^2 = \text{var}\,(\hat{Y})/\text{var}\,(Y),$$

see the exercises to this chapter. This coefficient is a popular one to report for it can be interpreted as the proportion of the variation in Y that can be predicted or 'explained' by X. We sometimes write it as $R^2(Y; X)$.

5.4 Predicting the Mathematics Marks

The variance matrix of the five mathematics examination marks, given in Table 1.1.2, provides a fine illustration of how the linear least squares predictor is calculated and interpreted. It is possible to find the predictor of any one mark given information on the others, or indeed to find the predictor of a group of marks from information on some or all of the others. For didactic purposes this variance is supposed to be known without error and no allowance is made here for sampling variation. The sequence of operations starts with $\text{var}\,(X, Y)$ given numerically and then calculates $\text{var}\,(X)^{-1}, B = \text{cov}\,(Y, X)\text{var}\,(X)^{-1}, \hat{Y} = BX, \text{var}\,(\hat{Y}) = B\text{cov}\,(X, Y)$ and $\text{var}\,(Y - \hat{Y}) = \text{var}\,(Y) - \text{var}\,(\hat{Y})$, in order.

We start with the one dimensional problem of predicting the statistics mark, so $Y = (\text{stat})$, when information is given on none, some and all of the other marks. For comparison, these predictors are evaluated numerically at the point

$$x = (\text{mech,vect,alg,anal}) = (20, 20, 60, 40)$$

in the X space.

No other information: If there is no information about the other marks then the best predictor of $Y - EY$ is zero; which leads to $\hat{Y} = 42.31$ as a base from which to compare other predictions. The variance 'to explain' is $\text{var}(Y) = 294.37$ and a 95% prediction interval (assuming Normality) is

$$42.31 \pm 1.96(294.37)^{1/2} = 42.31 \pm 33.63 = (8.68, 75.94).$$

Prediction from one variable: With information available on just analysis, $X = (\text{anal})$ and the linear least square predictor of Y, at $X = x$,

$$\hat{Y} = EY + \text{cov}(Y, X)\text{var}(X)^{-1}(x - EX)$$

becomes

$$\begin{aligned}\hat{Y} &= 42.31 + 153.77/217.88(x - EX) \\ &= 42.31 + 0.71(x - 46.68) = 9.36 + 0.71x.\end{aligned}$$

The predictor of the statistics mark is 9.36 + 0.71anal. A rise of one unit in the analysis mark leads to a rise of 0.71 in the predicted statistics mark. When the analysis mark is $x = 40$ the predicted statistics mark is 37.59, some five marks below the statistics average.

The variance accounted for by this predictor is

$$\text{var}(\hat{Y}) = B\text{cov}(X, Y) = (0.71)153.77 = 108.52,$$

which is some 36.9% of the variance of Y. The residual variance is

$$\text{var}(Y - \hat{Y}) = \text{var}(Y) - \text{var}(\hat{Y}) = 294.37 - 108.52 = 185.85,$$

so that

$$37.59 \pm 1.96(185.5)^{1/2} = 37.59 \pm 26.69 = (10.90, 64.29)$$

is a 95% prediction interval for the statistics mark.

Prediction from two variables: When information is given on both algebra and analysis the relevant chunk of the variance matrix is

alg	111.60		
anal	110.84	217.88	
stat	120.49	153.77	294.37

and the prediction equation is given by

$$\hat{Y} - 42.31 = \begin{pmatrix} 120.49 & 153.77 \end{pmatrix} \begin{pmatrix} 111.60 & 110.84 \\ 110.84 & 217.88 \end{pmatrix}^{-1} \begin{pmatrix} \text{alg} - 50.60 \\ \text{anal} - 46.68 \end{pmatrix}.$$

Inversion and simplification leads to

$$\hat{Y} = -11.19 + 0.77\text{alg} + 0.32\text{anal}.$$

Note how the coefficient of the analysis mark has changed: from 0.71, when it is the only variable in the predictor, down to 0.32. Clearly care is necessary to interpret the value of the coefficients in the predictor as they can be very sensitive to which of the other variables are included in the predictor. The relatively larger size of the algebra coefficient means that the statistics predictor responds more to a change in the algebra mark than in the analysis mark. At (alg,anal) = $(60, 40)$ the predicted statistics mark is 47.81, some 5 marks above the statistics average.

The variation in the statistics mark accounted for by this predictor is

$$\text{var}(\hat{Y}) = \text{cov}(Y, X)B^T = \begin{pmatrix} 120.49 & 153.77 \end{pmatrix}\begin{pmatrix} 0.77 & 0.32 \end{pmatrix}^T = 140.86,$$

which is 47.9% of the variance of Y; thus the inclusion of algebra on top of analysis accounts for an extra 10.0%. The residual variance given algebra and analysis is $\text{var}(Y - \hat{Y}) = 153.51$. The 95% prediction interval for statistics is now

$$47.81 \pm 1.96(153.51)^{1/2} = 47.81 \pm 24.2842 = (23.53, 72.09).$$

Prediction from all variables: The predictor based on marks from all four subjects requires the inversion of a 4×4 matrix; it simplifies to

$$\hat{Y} = -11.38 + 0.02\text{mech} + 0.03\text{vect} + 0.73\text{alg} + 0.31\text{anal}.$$

The small values of the mechanics and vectors coefficients means that they hardly contribute in predicting the statistics mark, and furthermore, their inclusion has hardly altered the algebra and analysis coefficients. This concurs with the conjecture made from the independence graph in the introduction. The predicted value at $(20, 20, 60, 40)$, which signals poor performances in both mechanics and vectors, is 45.86, only just down on the predicted value when these poor performances are ignored.

The variance of Y accounted for by the predictor is 141.10 (some 47.9% of the total) and the residual variance is 153.27. The 95% prediction interval is

$$45.86 \pm 1.96(153.27)^{1/2} = 45.86 \pm 24.27 = (21.59, 70.13)$$

which is almost identical to the one based on predicting statistics from algebra and analysis alone.

5.5 The Partial Covariance

The residual variance of the linear least squares predictor of the statistics mark is a useful way of summarising the effect of incorporating different information in the prediction equation. From now on we shall refer to it as a partial variance. Associated with a partial variance is a partial covariance, and, interestingly, this is directly related to the measure of conditional independence for the trivariate Normal distribution discussed in Section 2.6. Furthermore it transpires that the conditional expectation of a multivariate Normal random vector is just a linear least squares predictor and its conditional covariance is just a partial covariance. So, not surprisingly, the partial covariance and the elements of the inverse variance are intimately related. The mathematical approach chosen here corresponds directly to the method of finding the inverse of a matrix by developing a formula for the inverse of a partitioned matrix, and a byproduct of this approach is that the partial variance is just the Schur complement of a positive definite matrix.

Consider the partitioned random vector (X, Y, Z) where X is p-dimensional, Y is q-dimensional and Z is r-dimensional. Suppose that $E(X, Y, Z) = 0$ and that the variance, $\text{var}(X, Y, Z)$, is partitioned as

$$\begin{pmatrix} \text{var}(X) & \text{cov}(X,Y) & \text{cov}(X,Z) \\ \text{cov}(Y,X) & \text{var}(Y) & \text{cov}(Y,Z) \\ \text{cov}(Z,X) & \text{cov}(Z,Y) & \text{var}(Z) \end{pmatrix}.$$

The predictors of Y from X, of Z from X, and of Z from X and Y are

$$\begin{aligned} \hat{Y}(X) &= \text{cov}(Y,X)\text{var}(X)^{-1}X, \\ \hat{Z}(X) &= \text{cov}(Z,X)\text{var}(X)^{-1}X \text{ and} \\ \hat{Z}(X,Y) &= \text{cov}(Z,(X,Y))\text{var}(X,Y)^{-1}(X,Y). \end{aligned}$$

Definition. The *partial covariance* of Y and Z given X is

$$\text{cov}(Y, Z|X) = \text{cov}(Y - \hat{Y}(X), Z - \hat{Z}(X)),$$

the ordinary covariance of the two residual vectors, $Y - \hat{Y}$ and $Z - \hat{Z}$, obtained by predicting Y and Z from X. □

Its properties are easy to deduce and not more than a hint of a proof is given. The partial variance and correlation are then defined in terms of the partial covariance.

Proposition 5.5.1 Bilinearity. *The partial covariance,* $\text{cov}(Y, Z|X)$, *is bilinear in its arguments* (Y, Z) *and satisfies the formula*

$$\text{cov}(Y, Z|X) = \text{cov}(Y, Z) - \text{cov}(Y, X)\text{var}(X)^{-1}\text{cov}(X, Z).$$

Proof: As $\text{cov}(Y, Z|X) = \text{cov}(Y - \hat{Y}(X),\ Z - \hat{Z}(X))$, bilinearity follows from the linearity of the linear least squares predictors, $\hat{Y}$ in Y and $\hat{Z}$ in Z, together with the bilinearity of the covariance operator. By Proposition 5.3.1 the residual $Y - \hat{Y}$ and any transform of X have zero covariance so that

$$\text{cov}(Y, Z|X) = \text{cov}(Y - \hat{Y}, Z) = \text{cov}(Y, Z) - \text{cov}(\hat{Y}, Z).$$

The formula follows by substituting BX for the predictor $\hat{Y}$. □

The proof of the next corollary is immediate.

Corollary 5.5.2 *The partial variance,* $\text{var}(Y|X) = \text{cov}(Y, Y|X)$, *satisfies*

$$\begin{aligned}\text{var}(Y|X) &= \text{var}(Y) - \text{var}(\hat{Y}) \\ &= \text{var}(Y) - \text{cov}(Y,X)\text{var}(X)^{-1}\text{cov}(X,Y).\end{aligned}$$

When Y and Z are both one dimensional their *partial correlation* adjusted for X is

$$\text{corr}(Y, Z|X) = \text{cov}(Y, Z|X)/\{\text{var}(Y|X)\text{var}(Z|X)\}^{1/2}.$$

EXAMPLE 5.5.1 Find the partial covariance and partial correlation of Y and Z given X when (X, Y, Z) is three dimensional and $\text{var}(X, Y, Z)$ is given by

$$\text{(a)} \quad \begin{pmatrix} 1 & a & b \\ a & 1 & ab \\ b & ab & 1 \end{pmatrix} \qquad \text{(b)} \quad \begin{pmatrix} 1 & a & a \\ a & 1 & b \\ a & b & 1 \end{pmatrix}$$

For (a), $\text{cov}(Y, Z|X) = ab - a1^{-1}b = 0$, so the partial covariance and consequently the partial correlation are both zero. For (b), $\text{cov}(Y, Z|X) = b - a1^{-1}a = b - a^2$ while $\text{var}(Y|X)$, and by symmetry $\text{var}(Z|X)$, are both equal to $1 - a1^{-1}a = 1 - a^2$. Thus the partial correlation is $(b - a^2)/(1 - a^2)$. It is zero if and only if $b = a^2$; it equals the ordinary correlation between Y and Z only in the trivial case that $a = 0$ or $b = 1$. □

EXAMPLE 5.5.2 The mathematics marks again. Consider evaluating the partial covariance of the statistics and the analysis mark given the remaining variables X. Rather than put Y = (stat), Z = (anal) and compute $\text{cov}(Y, Z|X)$ directly, we can put Y = (stat,anal) and X = (mech,vect,alg) and then compute the partial variance $\text{var}(Y|X)$. From Section 5.4, the variance of the marks partitions into

$$\begin{aligned}\text{var}(X) &= \begin{pmatrix} 302.29 & & \\ 125.78 & 170.88 & \\ 100.43 & 84.19 & 111.60 \end{pmatrix} \\ \text{cov}(Y, X) &= \begin{pmatrix} 105.07 & 93.60 & 110.84 \\ 116.07 & 97.89 & 120.49 \end{pmatrix} \quad \text{var}(Y) = \begin{pmatrix} 217.88 & \\ 153.77 & 294.37 \end{pmatrix}.\end{aligned}$$

Now $\text{var}(\hat{Y}(X)) = \text{cov}(Y, X)\text{var}(X)^{-1}\text{cov}(X, Y)$ and this expression simplifies numerically to give

$$\text{var}(\hat{Y}(X)) = \begin{pmatrix} 111.01 & \\ 120.33 & 130.63 \end{pmatrix}.$$

The partial variance, $\text{var}(Y|X) = \text{var}(Y) - \text{var}(\hat{Y})$, comes by subtraction

$$\text{var}(Y|X) = \begin{pmatrix} 106.86 & \\ 33.44 & 163.74 \end{pmatrix}.$$

The partial covariance between statistics and analysis after adjustment for the other three marks is 33.44. The associated partial correlation is 0.25, down from the value of 0.61 before adjustment but nowhere near zero. Note that it is identical to the negative of the corresponding entry in the scaled inverse correlation matrix of the mathematics marks in Section 1.1. □

5.6 Invariance, Additivity and Recurrence

The partial covariance, $\text{cov}(Y, Z|X)$, can be analysed from several points of view: for variation in Y, in Z, or in X. Here, the properties of invariance and additivity, and a recurrence relationship satisfied by the partial covariance, concern variations in the conditioning set, X. The proofs hinge on expressing the partial covariance in terms of the original covariance.

Invariance

The partial covariance between Y and Z is invariant to replacing X by any invertible linear transformation of X. So if A is an invertible linear transformation then

$$\text{cov}(Y, Z|AX) = \text{cov}(Y, Z|X),$$

which is easily proved by writing $\text{cov}(Y, Z|AX) = \text{cov}(Y - \hat{Y}(AX), Z - \hat{Z}(AX))$ and using the corresponding property for the linear least squares predictor. This conforms to our intuitive notions, and reflects an invariance, for example, to the rescaling of the variables in X.

Additivity

We saw in Proposition 5.3.5 that the predictor of Y is additive when $\text{cov}(X, Z) = 0$, so that

$$\hat{Y}(X, Z) = \hat{Y}(X) + \hat{Y}(Z),$$

and that more generally,

$$\hat{Y}(X, Z) = \hat{Y}(X) + \hat{Y}(Z - \hat{Z}(X)).$$

The last term on the right hand side is easily expressed in terms of the partial covariance:

$$\begin{aligned} \hat{Y}(Z - \hat{Z}(X)) &= \text{cov}(Y, Z - \hat{Z}(X))\text{var}(Z - \hat{Z}(X))^{-1}(Z - \hat{Z}(X)) \\ &= \text{cov}(Y, Z|X)\text{var}(Z|X)^{-1}(Z - \hat{Z}(X)); \end{aligned}$$

which explicitly demonstrates that it has the same structure as an ordinary predictor. This last formula deserves highlighting because, by choosing the dimension of Z to be $r = 1$, we may extract the prediction coefficient of a single explanatory variable from the prediction equation. It is $\text{cov}\,(Y, Z|X)\text{var}\,(Z|X)^{-1}$, and explicitly shows that all the coefficients in a predictor are *partial* prediction coefficients, and are adjusted for the other variables in the predictor.

This leads to an elegant statement of variance additivity.

Proposition 5.6.1 Additivity of the explained variation. *If* $\text{cov}\,(X, Z) = 0$ *then*

$$\text{var}\,(\hat{Y}(X, Z)) = \text{var}\,(\hat{Y}(X)) + \text{var}\,(\hat{Y}(Z)),$$

while in general

$$\text{var}\,(\hat{Y}(X, Z)) = \text{var}\,(\hat{Y}(X)) + \text{var}\,(\hat{Y}(Z - \hat{Z}(X))),$$

in which $\text{var}\,(\hat{Y}(Z - \hat{Z}(X)))$ *simplifies to*

$$\text{cov}\,(Y, Z|X)\text{var}\,(Z|X)^{-1}\text{cov}\,(Z, Y|X).$$

Proof: If $\text{cov}\,(X, Z) = 0$ then $\hat{Y}(X, Z) = \hat{Y}(X) + \hat{Y}(Z)$, and $\hat{Y}(X)$ and $\hat{Y}(Z)$ have zero covariance. Taking the variance of $\hat{Y}(X) + \hat{Y}(Z)$ gives the result. If not, write $\hat{Y}(X, Z) = \hat{Y}(X) + \hat{Y}(Z - \hat{Z}(X))$ and note $\hat{Y}(X)$ and $\hat{Y}(Z - \hat{Z}(X))$ have zero covariance because X and $Z - \hat{Z}(X)$ have zero covariance. The additivity result then follows, and the formula for $\text{var}\,(\hat{Y}(Z - \hat{Z}(X)))$ follows by taking the variance of

$$\hat{Y}(Z - \hat{Z}(X)) = \text{cov}\,(Y, Z|X)\text{var}\,(Z|X)^{-1}(Z - \hat{Z}(X)).$$

□

The result is important for applied work because it shows that the difference, $\text{var}\,(\hat{Y}(X, Z)) - \text{var}\,(\hat{Y}(X))$, which is the difference between the amount of variation in Y 'explained' by including X and Z in the predictor and by including X alone, is itself interpretable as an amount of 'explained' variation. It is that amount 'explained' by including the residual $Z - \hat{Z}(X)$, in the predictor for Y, or equivalently the amount 'explained' by including Z adjusted for X.

EXAMPLE 5.6.1 A variance decomposition for the mathematics marks. In Section 5.4 on predicting the mathematics marks we evaluated the explained and residual variance of the statistics mark for various choices of predictor. To summarise:

choice of X	explained variance $\mathrm{var}(\hat{Y})$	residual variance $\mathrm{var}(Y\|X)$
empty	0.00	294.37
(anal)	108.52	185.85
(anal,alg)	140.86	153.51
(mech, vect, alg, anal)	141.10	153.27

Clearly some of the information in this table is redundant. It is most useful to report the changes in the explained (or residual) variance when the X-set varies. The table that gives these corresponds to the classical analysis of variance table: Implicit in the construction of this decomposition is the variance additivity proposition above: to be able to refer to the difference

$$\mathrm{var}\,(\widehat{\mathrm{stat}}(\mathrm{anal,alg})) - \mathrm{var}\,(\widehat{\mathrm{stat}}(\mathrm{anal})) = 140.86 - 108.52 = 32.34,$$

as the variance of algebra adjusted for analysis is the direct consequence of this proposition which asserts that

$$\mathrm{var}\,(\widehat{\mathrm{stat}}(\mathrm{anal} - \widehat{\mathrm{alg}}(\mathrm{anal}))) = \mathrm{var}\,(\widehat{\mathrm{stat}}(\mathrm{anal,alg})) - \mathrm{var}\,(\widehat{\mathrm{stat}}(\mathrm{anal})).$$

Most of the explained variance of $Y = (\mathrm{stat})$ is accounted for at the first step when analysis enters the predictor. Algebra still makes a substantial contribution given that analysis is already in the predictor but the contribution of the last two marks, mechanics and vectors, is negligible.

If the variables enter into the predictor in a different order then this table would alter correspondingly. □

A recurrence relation for the partial covariance

The next assertion actually requires some work to prove. It underlies many interesting formulae concerning partial correlation coefficients.

Table 5.6.1: Decomposing the variance of the statistics mark.

variance explained by	variance	percent
(anal)	108.52	36.87
(alg) adjusted for (anal)	32.34	10.99
(mech, vect) adjusted for (alg, anal)	0.24	0.08
residual given (mech, vect, alg,anal)	153.27	52.07
total	294.37	100.00

Proposition 5.6.2 *Consider the partitioned random vector (T,X,Y,Z). The partial covariance satisfies the recurrence relationship*

$$\mathrm{cov}\,(Y,Z|X,T) = \mathrm{cov}\,(Y,Z|X) - \mathrm{cov}\,(Y,T|X)\mathrm{var}\,(T|X)^{-1}\mathrm{cov}\,(T,Z|X).$$

Proof: First assume that $\mathrm{cov}\,(X,T) = 0$. From Proposition 5.3.5 the predictors $\hat{Y}$ and $\hat{Z}$ are additive, that is

$$\hat{Y}(X,T) = \hat{Y}(X) + \hat{Y}(T) \quad \text{and} \quad \hat{Z}(X,T) = \hat{Z}(X) + \hat{Z}(T).$$

Now

$$\begin{aligned}
\mathrm{cov}\,(Y,Z|X,T) &= \mathrm{cov}\,(Y-\hat{Y}(X,T),Z) && \text{orthogonality}\\
&= \mathrm{cov}\,(Y-\hat{Y}(X)-\hat{Y}(T),Z) && \text{additivity}\\
&= \mathrm{cov}\,(Y-\hat{Y}(X),Z) - \mathrm{cov}\,(\hat{Y}(T),Z) && \text{bilinearity of cov}\\
&= \mathrm{cov}\,(Y,Z|X) - \mathrm{cov}\,(\hat{Y}(T),Z) && \text{def partial cov}\\
&= \mathrm{cov}\,(Y,Z|X) - \mathrm{cov}\,(Y,T)\mathrm{var}\,(T)^{-1}\mathrm{cov}\,(T,Z)
\end{aligned}$$

as asserted. More generally, when $\mathrm{cov}\,(X,T)$ is not zero, replace (X,T) by $(X,T-\hat{T}(X))$. These vectors do have zero covariance and, by invariance,

$$\mathrm{cov}\,(Y,Z|X,T) = \mathrm{cov}\,(Y,Z|X,T-\hat{T}(X)).$$

Substituting $T-\hat{T}(X)$ for T in the expression above for $\mathrm{cov}\,(Y,Z|X,T)$ gives

$$\begin{aligned}
\mathrm{cov}\,(Y,Z|X,T) = {} & \mathrm{cov}\,(Y,Z|X) - \\
& \mathrm{cov}\,(Y,T-\hat{T}(X))\mathrm{var}\,(T-\hat{T}(X))^{-1}\mathrm{cov}\,(T-\hat{T}(X),Z)
\end{aligned}$$

and the right hand side reduces to

$$\mathrm{cov}\,(Y,Z|X) - \mathrm{cov}\,(Y,T|X)\mathrm{var}\,(T|X)^{-1}\mathrm{cov}\,(T,Z|X),$$

as had to be shown. □

EXAMPLE 5.6.2 This recurrence relationship underlies many well known identities for partial variances and correlations. For instance, if X is p-dimensional and Y, Z and T are all one dimensional then

$$\mathrm{var}\,(Y|X,T) = \mathrm{var}\,(Y|X)\{1-\mathrm{corr}^2(Y,T|X)\}$$

is a famous formula that relates the change in the partial variance to the squared partial correlation coefficient. The formula

$$\mathrm{corr}\,(Y,Z|X,T) = \frac{\mathrm{corr}\,(Y,Z|X) - \mathrm{corr}\,(Y,T|X)\mathrm{corr}\,(T,Z|X)}{\{(1-\mathrm{corr}^2(Y,T|X))(1-\mathrm{corr}^2(T,Z|X))\}^{1/2}}$$

is even more famous. Further simplification follows if it is assumed that the variance of X is diagonal. □

5.7 The Inverse Variance

The inverse variance $\mathrm{var}(X)^{-1}$ occurred in the expression for the linear least squares predictor of Y from X and in the resulting partial variance. It appears in the quadratic form for the multivariate Normal density and as part of the criterion for the conditional independence of Normal random variables. We now come to grips with it directly and summarise the results in the inverse variance lemma and its corollaries:

- The diagonal elements of the inverse are directly related to the corresponding multiple correlation coefficients.
- The scaled off diagonal elements are the negatives of partial correlation coefficients.
- The corollaries provide the other tools required to analyse the multivariate Normal distribution: in particular, the techniques of completing the square in a quadratic form and the evaluation of a determinant.

A Cholesky decomposition of a matrix factorises the matrix into the product of a triangular matrix, a diagonal matrix and the triangular one transposed. The inverse variance lemma evaluates the inverse in the same way as finding the inverse from the Cholesky decomposition. A block version of this decomposition is basic to the proof.

Cholesky factorisation

The random vector $T = (X, Y)$ is $p+q$-dimensional with variance $\mathrm{var}(X, Y)$. The linear least squares predictor of Y is $\hat{Y} = \hat{Y}(X)$ which can be written as $\hat{Y} = BX$ where $B = \mathrm{cov}(Y, X)\mathrm{var}(X)^{-1}$. The linear mapping L, the sweep operator, introduced in Section 5.3, which orthogonalises (X, Y), is

$$L = \begin{pmatrix} I & 0 \\ -\mathrm{cov}(Y, X)\mathrm{var}(X)^{-1} & I \end{pmatrix}.$$

It takes (X, Y) to $L(X, Y) = (X, Y - \hat{Y})$ which has the property that its components are orthogonal. Clearly L is invertible.

Proposition 5.7.1 A block Cholesky factorisation of the variance. *The variance can be expressed as*

$$\mathrm{var}(X, Y) = L^{-1}\mathrm{var}(X, Y - \hat{Y})L^{-T}$$

or equivalently as

$$\mathrm{var}(X, Y) = \begin{pmatrix} I & 0 \\ \mathrm{cov}(Y, X)\mathrm{var}(X)^{-1} & I \end{pmatrix}$$
$$\begin{pmatrix} \mathrm{var}(X) & 0 \\ 0 & \mathrm{var}(Y|X) \end{pmatrix} \begin{pmatrix} I & \mathrm{var}(X)^{-1}\mathrm{cov}(X, Y) \\ 0 & I \end{pmatrix}.$$

Proof: Now $(X, Y - \hat{Y}) = L(X, Y)$, so $(X, Y) = L^{-1}(X, Y - \hat{Y})$. Taking variances and simplifying gives

$$\text{var}(X, Y) = L^{-1}\text{var}(X, Y - \hat{Y})L^{-T}.$$

But $\text{cov}(X, Y - \hat{Y}) = 0$ and $\text{var}(Y - \hat{Y}) = \text{var}(Y|X)$ so that $\text{var}(X, Y - \hat{Y})$ is block diagonal. □

This is only a block version of the Cholesky decomposition because while L and so L^{-1} are lower triangular, $\text{var}(X, Y - \hat{Y})$ is only block diagonal. A corollary of this Cholesky factorisation is a proof of the following.

Corollary 5.7.2 *Any variance matrix can be factorised into the product AA^T where A is lower triangular.*

Proof: Consider $\text{var}(T)$ where $T = (T_1, T_2, \ldots, T_k)$. Write $T = (X, Y)$ where $Y = T_k$ is one dimensional and apply the block Cholesky factorisation to $\text{var}(X, Y)$ giving $\text{var}(X)$ and $\text{var}(Y|X)$ on the diagonal. Repeat the procedure on $\text{var}(X)$ until T is exhausted. Finally the diagonal matrix remaining is

$$\begin{pmatrix} \text{var}(T_1) & 0 & \cdots & 0 \\ 0 & \text{var}(T_2|T_1) & \cdots & 0 \\ \vdots & \vdots & \ddots & 0 \\ 0 & \cdots & \cdots & \text{var}(T_k|T_1, T_2, .., T_{k-1}) \end{pmatrix}.$$

All elements on the diagonal are strictly positive and taking the square root of each completes the proof. □

Note that the product of lower triangular matrices is also lower triangular. This proof could have supposed that X rather than Y is one dimensional.

The inverse variance lemma

Denote the inverse variance matrix $\text{var}(X, Y)^{-1}$ by D, which is sometimes known as the *concentration* or *precision* matrix, and denote its corresponding partition by

$$D = \begin{pmatrix} D_{XX} & D_{XY} \\ D_{YX} & D_{YY} \end{pmatrix}.$$

The main import of the inverse variance lemma is that $D_{YY} = \text{var}(Y|X)^{-1}$ where $\text{var}(Y|X)$ is the partial variance of Y given X.

Proposition 5.7.3 Inverse Variance Lemma. *The inverse of the partitioned variance* $\text{var}(X, Y)$ *is given by*

$$\text{var}(X, Y)^{-1} = \begin{pmatrix} \text{var}(X)^{-1} + B^T\text{var}(Y|X)^{-1}B & -B^T\text{var}(Y|X)^{-1} \\ -\text{var}(Y|X)^{-1}B & \text{var}(Y|X)^{-1} \end{pmatrix}.$$

Proof: Inverting the block Cholesky factorisation in Proposition 5.7.1 gives

$$\mathrm{var}\,(X,Y)^{-1} = L^T \mathrm{var}\,(X,Y-\hat{Y})^{-1} L,$$

and as $\mathrm{var}\,(X,Y-\hat{Y})$ is block diagonal it is easily invertible, so that

$$\mathrm{var}\,(X,Y)^{-1} = \begin{pmatrix} I & -B^T \\ 0 & I \end{pmatrix} \begin{pmatrix} \mathrm{var}\,(X)^{-1} & 0 \\ 0 & \mathrm{var}\,(Y|X)^{-1} \end{pmatrix} \begin{pmatrix} I & 0 \\ -B & I \end{pmatrix}.$$

Multiplying out completes the proof. □

Note that there is no need to invert L explicitly.

5.8 Inverse Variance Lemma: Corollaries

The corollaries to the lemma clarify the relationship between the elements of the inverse variance and partial correlation coefficients. The first states that each element on the diagonal of the inverse variance matrix is the reciprocal of the partial variance of the corresponding variable predicted from the rest. The second states that if the inverse variance is scaled to have a unit diagonal then the off diagonal elements are the negatives of the corresponding partial correlations. The proofs depend on varying the original partition so that while $T = (X,Y)$ is fixed, Y, and consequently X, varies.

Corollary 5.8.1 *Each diagonal element of the inverse variance is the reciprocal of a partial variance.*

Proof: Set $q = 1$. Then D_{YY} is scalar and from the lemma

$$D_{YY} = \mathrm{var}\,(Y|X)^{-1} = 1/\mathrm{var}\,(Y|X_1, X_2, \ldots, X_p).$$

As any diagonal element can be selected by first permuting the original random vector, the proposition is true for all diagonal elements. □

Consequently the inverse of the correlation matrix has elements of the form $\mathrm{var}\,(Y)/\mathrm{var}\,(Y|X_1, X_2, \ldots, X_p)$ on the diagonal, which is just $1/(1-R^2)$ where R is the multiple correlation coefficient between Y and the rest of the other variables, X.

Corollary 5.8.2 *Each off diagonal element of the inverse variance, scaled to have a unit diagonal, is the negative of the partial correlation between the two corresponding variables, partialled on all the remaining variables, the rest.*

Proof: Set $q = 2$ and denote the elements of the 2×2 matrix D_{YY} by $\{d_{ij}\}$, then

$$D_{YY} = \begin{pmatrix} d_{11} & d_{12} \\ d_{21} & d_{22} \end{pmatrix} = \mathrm{var}\,(Y|X)^{-1},$$

or turning this around, we have the identity

$$\begin{pmatrix} \mathrm{var}\,(Y_1|X) & \mathrm{cov}\,(Y_1, Y_2|X) \\ \mathrm{cov}\,(Y_2, Y_1|X) & \mathrm{var}\,(Y_2|X) \end{pmatrix} = \begin{pmatrix} d_{11} & d_{12} \\ d_{21} & d_{22} \end{pmatrix}^{-1}.$$

Now invert the 2×2 matrix on the right hand side and scale both sides of the identity to have unit diagonal. The identity becomes

$$\begin{pmatrix} 1 & * \\ \mathrm{cov}\,(Y_1, Y_2|X)/\{\mathrm{var}\,(Y_1|X)\mathrm{var}\,(Y_2|X)\}^{1/2} & 1 \end{pmatrix} = \begin{pmatrix} 1 & * \\ -d_{12}/\{d_{11}d_{22}\}^{1/2} & 1 \end{pmatrix}.$$

Consequently, $\mathrm{corr}\,(Y_1, Y_2|X) = -d_{12}/\{d_{11}d_{22}\}^{1/2}$. Permutation allows any two variables to be selected in the partition and so the result is true for all off diagonal elements. □

Thus the elements of the scaled inverse correlation matrix are the negative partial correlations of the corresponding elements given the rest. We have finally begun to prove some of the assertions made in our discussion of the mathematics marks in the first section of the first chapter.

There was a fashion set by Cramer (1946) to take this latter property of partial correlations as their definition, see Kendall and Stuart (1961). That is to define them in terms of the elements of the inverse matrix; it leads to some interesting identities in determinants and cofactors but to a certain opacity as to the meaning of a partial correlation.

Zero partial covariance

We now give conditions for zero covariance and partial covariance in terms of the inverse variance.

Corollary 5.8.3 Zero covariance. *The off diagonal block D_{XY} of the inverse variance $D = \mathrm{var}\,(X,Y)^{-1}$ is zero if and only if $\mathrm{cov}\,(X,Y) = 0$.*

Proof: The inverse variance lemma gives

$$\begin{aligned} D_{XY} &= -B^T \mathrm{var}\,(Y|X)^{-1} \\ &= -\mathrm{var}\,(X)^{-1}\mathrm{cov}\,(X,Y)\mathrm{var}\,(Y|X)^{-1}. \end{aligned}$$

As $\mathrm{var}\,(X,Y)$ is positive definite, both $\mathrm{var}\,(X)$ and $\mathrm{var}\,(Y|X)$ are positive definite, and the result follows. □

The proof to the next corollary has a recursive flair.

Corollary 5.8.4 Zero partial covariance. *If the random vector is partitioned into $T = (X, Y, Z)$ with inverse variance $D = \mathrm{var}\,(X, Y, Z)^{-1}$ then the term $D_{YZ} = 0$ if and only if $\mathrm{cov}\,(Y, Z|X) = 0$.*

Proof: Replace the vector Y by the partitioned vector (Y, Z) so that D_{YY} becomes

$$\begin{pmatrix} D_{YY} & D_{YZ} \\ D_{ZY} & D_{ZZ} \end{pmatrix}.$$

By the inverse variance lemma this equals $\operatorname{var}(Y, Z|X)^{-1}$. The condition that $D_{YZ} = 0$ is now that the corresponding term in $\operatorname{var}(Y, Z|X)^{-1}$ be zero. But now apply the inverse variance lemma again, this time to the partitioned partial variance $\operatorname{var}(Y, Z|X)^{-1}$: then by the previous corollary the off diagonal block is zero if and only if $\operatorname{cov}(Y, Z|X) = 0$. □

Cunning!

Block zero covariance

The next result is the analogue of the block independence theorem applied to the covariance operator. Recall the following picture to mind:

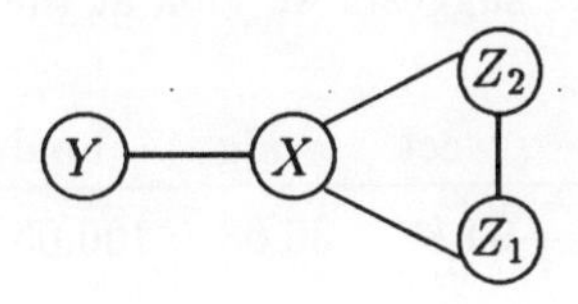

Corollary 5.8.5 Block zero covariance. *If the random vector Z in (X, Y, Z) is further partitioned into Z_1 and Z_2 then the following assertions are equivalent:*

(a) $\operatorname{cov}(Y, Z|X) = 0$, *and*

(b) $\operatorname{cov}(Y, Z_1|X, Z_2) = 0$ *and* $\operatorname{cov}(Y, Z_2|X, Z_1) = 0$.

Proof: If (b) holds then by Corollary 5.8.2 the inverse variance D has the following structure:

$$\begin{array}{r|cccc} X & * & & & \\ Y & * & * & & \\ Z_1 & * & 0 & * & \\ Z_2 & * & 0 & * & * \\ \hline & & & & \end{array}$$

and consequently $D_{YZ} = 0$. Applying Corollary 5.8.2 again gives $\operatorname{cov}(Y, Z|X) = 0$. Conversely, if $\operatorname{cov}(Y, Z|X) = 0$ then $D_{YZ} = 0$ so that $D_{YZ_1} = 0$ and $D_{YZ_2} = 0$. □

An important application of these results is to prove a parallel to the separation theorem, in which pairwise zero correlation partialled on the 'rest' implies zero correlation partialled on only the separating set. That is, to prove a version of the global Markov property without assuming Normality.

EXAMPLE 5.8.1 Application to the Mathematics marks. This last corollary establishes the third assertion made concerning the mathematics marks: that the covariance of (mech,vect) and (anal,stat) given (alg) is zero. Inverting the variance matrix of the mathematics marks reported in Table 1.1.1, the inverse variance is

mech	0.00530				
vect	-0.00246	0.01055			
alg	-0.00277	-0.00476	0.02726		
anal	0.00001	-0.00080	-0.00713	0.01000	
stat	-0.00014	-0.00017	-0.00476	-0.00204	0.00652
	mech	vect	alg	anal	stat

As it stands the elements of this matrix have no obvious significance. However Corollary 5.8.1 above suggests we look at the reciprocals of the diagonal elements. These are

mech	vect	alg	anal	stat
188.51	94.82	36.68	100.03	153.27

and are the residual variances having predicted from all other variables. Note that the figure 153.27 already occurred in Table 5.6.1 giving the residual variances for different statistics predictors. Algebra has by far the smallest residual variance, mechanics has the greatest. The raw variances are

mech	vect	alg	anal	stat
302.29	170.88	111.60	217.88	294.37

so that the proportion of the variance of the mechanics mark that remains unexplained is

$$\text{var}(\text{mech}|\text{rest})/\text{var}(\text{mech}) = 188.51/302.29 = 0.62.$$

To illustrate that the same calculation may be reached by inverting the correlation matrix, the reciprocals of the diagonal elements (1.60,1.80,3.04,2.18,1.92) of the inverse correlation evaluated in Section 1.1 are

mech	vect	alg	anal	stat
0.62	0.55	0.33	0.46	0.52

These are each of the form $1 - R^2$, and are interpretable as the proportion of the variance that remains unexplained. For example, 52% of the statistics mark is 'unexplained' but only 33% of the algebra mark.

Scaling either the inverse variance or the inverse correlation gives

mech	1.00				
vect	-0.33	1.00			
alg	-0.23	-0.28	1.00		
anal	0.00	-0.08	-0.43	1.00	
stat	-0.02	-0.02	-0.36	-0.25	1.00
	mech	vect	alg	anal	stat

The partial correlation of the statistics and the analysis mark having adjusted for the rest is 0.25 while that between statistics and mechanics is 0.02. If we approximate the near zero partial correlations by zero then, for example, mechanics and analysis are uncorrelated given the remaining variables (vect,alg,stat); similarly mechanics and statistics are independent given the remaining variables, (vect,alg,anal). The block zero covariance proposition implies that (mech) and (anal,stat) are uncorrelated given (vect,alg). Further application of the block zero theorem implies that (mech,vect) and (anal,stat) are uncorrelated given algebra alone. □

Completing the square

The final corollary to the inverse variance lemma allows us to obey the familiar high school admonition, 'when in doubt complete the square'. It is needed to derive conditional distributions from the multivariate Normal. Rather than write $u^T v$ for the inner-product, $\sum u_i v_i$, of two p-dimensional vectors, u and v, we use the notation $[.,.]$, so that $[u,v] = u^T v$. This notation stresses the bilinearity of the inner-product, in line with our stress on the bilinearity of the covariance operator.

Corollary 5.8.6 Completing the square. *The identity*

$$[(X,Y), \text{var}(X,Y)^{-1}(X,Y)] = [X, \text{var}(X)^{-1}X] + [Y-\hat{Y}, \text{var}(Y|X)^{-1}(Y-\hat{Y})]$$

in quadratic forms based on the variance $\text{var}(X,Y)$ *is satisfied.*

Proof: In the proof of the inverse variance lemma we established the expressions

$$(X,Y) = L^{-1}(X, Y-\hat{Y}) \text{ and } \text{var}(X,Y)^{-1} = L^T \text{var}(X, Y-\hat{Y})^{-1} L.$$

Hence

$$\begin{aligned} \text{var}(X,Y)^{-1}(X,Y) &= L^T \text{var}(X,Y-\hat{Y})^{-1} L L^{-1}(X,Y-\hat{Y}) \\ &= L^T \text{var}(X,Y-\hat{Y})^{-1}(X,Y-\hat{Y}) \end{aligned}$$

and taking inner products gives

$$\begin{aligned} [(X,Y), \text{var}(X,Y)^{-1}(X,Y)] &= [(X,Y), L^T \text{var}(X,Y-\hat{Y})^{-1}(X,Y-\hat{Y})] \\ &= [(X,Y-\hat{Y}), \text{var}(X,Y-\hat{Y})^{-1}(X,Y-\hat{Y})]. \end{aligned}$$

As $\text{var}(X, Y-\hat{Y})$ is block diagonal the proof is finished. □

5.9 Variance: Trace and Determinant

The determinant of a matrix crops up in many branches of applied mathematics and so we should not be surprised to see it occurring in the study of conditional independence. We shall see in the next chapter that the determinant of a variance arises in the derivation of the multivariate Normal density function as the Jacobian of a linear transformation. It then enters as an important ingredient of the information divergence between two Normal distributions, and similarly enters into the likelihood function. We briefly summarise some of its more important properties together with those of a similarly useful quantity, the trace of a matrix.

Matrix properties

The definition of a trace and a determinant and proofs of the following assertions can be found in most texts on matrix or linear algebra. See, for example, Lang (1970).

The *trace* of a $p \times q$ matrix A with elements $\{a_{ij}\}$ is the sum

$$\mathrm{tr}\,(A) = \sum_i a_{ii},$$

of the diagonal elements of A. The trace is linear and satisfies the following

$$\begin{array}{rrcll} \text{linearity}: & \mathrm{tr}\,(\alpha A) & = & \alpha \mathrm{tr}\,(A) & \text{for } \alpha \text{ scalar, and} \\ & \mathrm{tr}\,(A+B) & = & \mathrm{tr}\,(A) + \mathrm{tr}\,(B) & \text{for } A \text{ and } B \text{ conformable;} \\ \text{transpose}: & \mathrm{tr}\,(A^T) & = & \mathrm{tr}\,(A) & \\ \text{product}: & \mathrm{tr}\,(AB) & = & \mathrm{tr}\,(BA) & \text{for } A \text{ and } B \text{ are conformable.} \end{array}$$

If A, B, C and D are real matrices of orders $p \times p$, $q \times p$, $p \times q$ and $q \times q$ respectively, and α is scalar then the *determinant*, which is scalar, satisfies the following properties

$$\begin{array}{rrcl} \text{scalar multiplication}: & \det(\alpha A) & = & \alpha^p \det(A), \\ \text{transpose}: & \det(A^T) & = & \det(A), \\ \text{product}: & \det(AD) & = & \det(A)\det(D), \text{ if } p = q, \\ \text{inverse}: & \det(A^{-1}) & = & 1/\det(A) \text{ and } \det(I) = 1, \\ \text{block triangular}: & \det\begin{pmatrix} A & C \\ 0 & D \end{pmatrix} & = & \det(A)\det(D), \\ \text{partition}: & \det\begin{pmatrix} A & C \\ B & D \end{pmatrix} & = & \det(D - BA^{-1}C)\det(A) \end{array}$$

if A is invertible.

The *Schur complement* of A in the partitioned matrix $\begin{pmatrix} A & C \\ B & D \end{pmatrix}$ is defined as $D - BA^{-1}C$, and first appears in Schur's (1917) proof of the determinantal partition rule. Though the name, the Schur complement, is a relatively recent acquisition, this matrix has arisen in many fields of applications and its role in statistics is discussed in review articles by Ouellette (1981) and Styan (1985). The partial variance $\text{var}(Y|X)$ is the operator that corresponds to the Schur complement of the operator $\text{var}(X)$ in $\text{var}(X,Y)$.

Proposition 5.9.1 Schur's identity. *The determinant of* $\text{var}(X,Y)$ *factorises according to*

$$\det \text{var}(X,Y) = \det \text{var}(X) \det \text{var}(Y|X).$$

Proof: The proof of the inverse variance lemma established the decomposition

$$\text{var}(X,Y) = L^{-1}\text{var}(X, Y - \hat{Y})L^{-T}.$$

By the rule for the determinant of a product

$$\det \text{var}(X,Y) = \det(L^{-1}) \det \begin{pmatrix} \text{var}(X) & 0 \\ 0 & \text{var}(Y|X) \end{pmatrix} \det(L^{-T}).$$

As L is lower triangular with ones on the diagonal $\det(L) = 1$, and the result follows. □

We can now draw a picture. A geometrical representation of the covariance is to represent a p-dimensional random vector X by p fixed vectors in a vector space V say, with inner product given by the covariance, $\text{cov}(.,.)$. It turns out that the square root of the determinant of $\text{var}(X)$ is just the *volume* of the p-dimensional parallelopiped generated by the p vectors representing X. If X is one dimensional then

$$\det \text{var}(\alpha X) = \alpha^2 \det \text{var}(X) = \alpha^2 \text{var}(X)$$

and is a squared length. If Y is also one dimensional then

$$\{\det \text{var}(X,Y)\}^{1/2} = \{\det \text{var}(X)\}^{1/2}\{\det \text{var}(Y|X)\}^{1/2}$$

and the equation expresses the equality of the areas

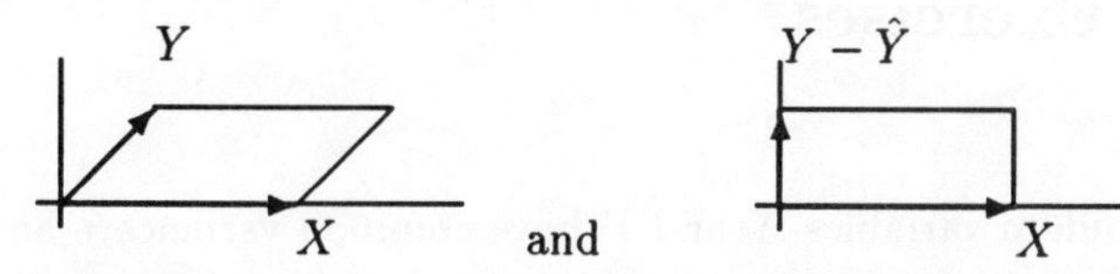

When X is two dimensional and Y is one dimensional, the identity $\det \text{var}(X,Y)^{1/2} = \det \text{var}(X)^{1/2} \det \text{var}(Y|X)^{1/2}$ shows how to compute the volume of a 'squashed' cube; and this generalisation to three dimensions is sufficient to show that the determinant represents squared volume in any number of dimensions.

The expected value of a quadratic form

Finding the expected value of a quadratic form $[X, AX] = X^T AX$ in Normal random variables often reduces to evaluating the trace of a matrix.

Proposition 5.9.2 *If the random vector* X *has expectation* $EX = 0$ *and variance* $\text{var}(X)$ *then*

$$E[X, AX] = \text{tr}(A\text{var}(X))$$

for all fixed linear transformations A.

Proof: First assume that $\text{var}(X) = I$, so that as $EX = 0$, $E(X_i X_j) = 1$ if $i = j$ and 0 otherwise. Then

$$E[X, AX] = E\sum_i \sum_j X_i X_j a_{ij} = \sum a_{ii} = \text{tr}(A)$$

by the definition of the trace. More generally, suppose that we can find a factorisation of the variance, so that $\text{var}(X) = CC^T$ where C is positive definite, for example, one way is the Cholesky decomposition in Section 5.7. A corollary of such a factorisation is that if $Y = C^{-1}X$ then $\text{var}(Y) = I$. Hence

$$\begin{aligned} E[X, AX] &= E[CY, ACY] = E[Y, C^T ACY] \\ &= \text{tr}(C^T AC) \text{ since } \text{var}(Y) = I \\ &= \text{tr}(ACC^T) = \text{tr}(A\text{var}(X)). \end{aligned}$$

in consequence. □

A rather more subtle proof uses the fact that because $[X, AX] = X^T AX$ is a scalar it is identical to $\text{tr}(X^T AX)$, and so we have $E[X, AX] = E\text{tr}(X^T AX)$ which in turn equals $E\text{tr}(AXX^T)$. But the trace is linear and so this in turn is just $\text{tr}(EAXX^T)$ and equals $\text{tr}(A\text{var}(X))$ since $EX = 0$.

5.10 Exercises

1: The random variables X and Y have common variance v and correlation r,
(a) find $\text{cov}(X, X + Y)$, $\text{cov}(X - Y, X + Y)$ and $\text{cov}(X, Y - rX)$;
(b) by evaluating $\text{var}(X + Y)$ and $\text{var}(X - Y)$ show that r lies in $[-1, 1]$;
(c) minimise $\text{var}(Y - bX)$ with respect to the real scalar b;
(d) maximise $\text{var}(aX + bY)$ with respect to a and b subject to the constraint that $a^2 + b^2 = 1$.

2: The vectors u and v, and var(X) are given by

$$u = \begin{pmatrix} 1 \\ 2 \\ 3 \end{pmatrix}, v = \begin{pmatrix} 1 \\ -2 \\ 1 \end{pmatrix}, \text{var}(X) = \begin{pmatrix} 3 & 2 & 1 \\ 2 & 4 & 1 \\ 1 & 1 & 2 \end{pmatrix}$$

then evaluate the correlation between $u^T X$ and $v^T X$.

3: The random variables $X_0, X_1, \ldots, X_p$ and Z are uncorrelated with common variance 1, find the variance matrix of the random vector $Y = (Y_1, \ldots, Y_p)$, when Y_i is given by
(a) $Y_i = Z + X_i$,
(b) $Y_i = X_i + \alpha X_{i-1}$,
(c) $Y_i = X_i + Y_{i-1}, \quad Y_0 = 0$,
(d) $Y_i = X_i + \alpha Y_{i-1}, \quad Y_0 = (1-\alpha^2)^{-1/2} X_0$
for $i = 1, \ldots, p$ and α scalar. Try $p = 4$ to start with.

4: If var$(X) = I$ and $Y_1 = X_1$, $Y_2 = X_1 + X_2$, $Y_3 = X_1 + X_3$, $Y_4 = X_2 + X_4$ find var(Y) and by writing X in terms of Y find var$(Y)^{-1}$.

5: If X is a zero mean p-dimensional random vector and Y is a random scalar then evaluate the predictor of Y when var(X, Y) is given by

$$\text{(a)} \begin{pmatrix} 1 & 0 & a \\ 0 & 1 & b \\ a & b & 1 \end{pmatrix} \quad \text{(b)} \begin{pmatrix} 1 & a & a^2 \\ a & 1 & a \\ a^2 & a & 1 \end{pmatrix} \quad \text{(c)} \begin{pmatrix} 1 & r & 0 & a \\ r & 1 & 0 & b \\ 0 & 0 & 1 & c \\ a & b & c & 1 \end{pmatrix}$$

6: Find the predictor of $Y_1 + Y_2$ when

$$\text{var}(X_1, X_2, Y_1, Y_2) = \begin{pmatrix} 1 & r & r & -r \\ r & 1 & r & r \\ r & r & 1 & r \\ -r & r & r & 1 \end{pmatrix}.$$

7: Write out the steps in the derivation of var$(\hat{Y}) = B$cov(X, Y). Since B has already been computed in the computation of $\hat{Y}$ this formula is a quick way to compute var$(\hat{Y})$.

8: Show that the linear least squares predictor of AY is $A\hat{Y}$ where A is of order $1 \times q$. Hint: re-derive the normal equations.

9: If $\hat{X} = \hat{X}(Y)$ is the predictor of X from Y and $\hat{Y} = \hat{Y}(X)$ then show
(a) $\text{cov}(\hat{Y}, \hat{X}) = \text{cov}(Y, X)\text{var}(X)^{-1}\text{var}(\hat{X}) = \text{var}(\hat{Y})\text{var}(Y)^{-1}\text{cov}(Y, X)$
(b) $\text{cov}(Y - \hat{Y}, X - \hat{X}) = -\text{cov}(Y, X)\text{var}(X)^{-1}\text{var}(X - \hat{X})$,
(c) $\text{cov}(Y, X) = \text{cov}(\hat{Y}, \hat{X}) - \text{cov}(Y - \hat{Y}, X - \hat{X})$.

10: Prove that $\hat{Z}(Y,X) = \hat{Z}(X,Y)$.

11: Write out the definition of the multiple correlation coefficient explicitly and deduce that $R^2 = \text{var}(\hat{Y})/\text{var}(Y)$.

12: Show that $\text{corr}(Y, b^T X)^2 \leq \text{var}(\hat{Y})/\text{var}(Y)$ with equality if and only if $b^T X$ is a scalar multiple of $\hat{Y}$. Thus $\hat{Y}$ is a maximal correlation predictor as well as a least squares predictor. (Use the Cauchy-Schwarz inequality.)

13: If (X,Y) is a bivariate Bernoulli random variable show that the linear least squares predictor of Y from X is

$$\hat{Y}(x) = (1-x)P(Y=1|X=0) + xP(Y=1|X=1) \quad \text{for} \quad x = 0,1.$$

14: Find the predictor when algebra alone is used to predict the statistics mark. Show that the residual variance is 164.3 and that the variance accounted for is 130.1. Compare the value of R^2 with the diagonal of the inverse correlation matrix given in Section 1.1 when $Y = (\text{stat})$.

15: Confirm that
(a) the predictor of algebra from the other marks is

$$\widehat{\text{alg}} = 18.22 + 0.10 \text{ mech} + 0.17 \text{ vect} + 0.26 \text{ anal} + 0.17 \text{ stat},$$

and that the percentage of variance accounted for is 67.1%;
(b) the predictor of mechanics from (vect,alg,anal) is

$$\widehat{\text{mech}} = -12.33 + 0.47\text{vect} + 0.54\text{alg} + 0.01\text{anal},$$

and that the variance decreases from 302.3 to 188.6 so the percentage variance accounted for is 37.4%.
(c) Comment on the values of the coefficients in relation to the independence graph.
(d) Find the predictor of (mech,vect) given information on the other three variables.

16: The vector (X,Y,Z) is three dimensional. Find the partial covariance and partial correlation of X and Y given Z and the multiple correlation coefficient between X and (Y,Z) when $\text{var}(X,Y,Z)$ is given by

$$\begin{pmatrix} 3 & 2 & 1 \\ 2 & 4 & 1 \\ 1 & 1 & 2 \end{pmatrix}.$$

17: It is said that Y is *perfectly predictable* from X if $\text{var}(Y|X) = 0$. If Y is one dimensional and X is three dimensional with

$$\text{var}(Y,X) = \begin{pmatrix} 1 & r & 0 & -r \\ r & 1 & 0 & r \\ 0 & 0 & 1 & 0 \\ -r & r & 0 & 1 \end{pmatrix}$$

what values of r make Y perfectly predictable?

18: If $p = 2$, $q = 1$ and the predictor $\hat{Y}$ of Y reduces to $\hat{Y}(X) = bX_2$ where b is scalar (i.e. the coefficient of X_1 is zero) then show that $\text{cov}(Y, X_1|X_2) = 0$. Does this mean that Y and X_1 are uncorrelated?

19: Find the partial correlation between (a) X_1 and X_3 given X_2, and (b) X_1 and X_3 given (X_2, X_4), when $\text{var}(X)$ is given by

$$\begin{pmatrix} 1 & s & r & s \\ s & 1 & s & r \\ r & s & 1 & s \\ s & r & s & 1 \end{pmatrix}$$

with $r = 0.6$, $s = \sqrt{0.48}$.

20: Generalise the formula for partial correlations

$$\text{corr}(Y, Z|X) = \frac{\text{corr}(Y,Z) - \text{corr}(Y,X)\text{corr}(X,Z)}{\{(1 - \text{corr}^2(Y,X))(1 - \text{corr}^2(Z,X))\}^{\frac{1}{2}}},$$

for X, Y and Z scalar, to the case when X is p-dimensional and $\text{var}(X) = I$.

21: Show that the partial variance var((mech,stat)|(vect,alg,anal)) based on the mathematics mark is $\begin{pmatrix} 188.62 & 4.18 \\ 4.18 & 153.37 \end{pmatrix}$ and hence that the partial correlation between mechanics and statistics having adjusted for the rest is 0.02.

22: If $\text{var}(X) = \begin{pmatrix} 2 & 1 \\ 1 & 3 \end{pmatrix}$ find A such that $\text{var}(X) = AA^T$.

23: The random vector X is p-dimensional and has variance $\text{var}(X) = I + \theta 11^T$, for θ scalar. Verify that its inverse, $\text{var}(X)^{-1}$, is of the form $aI + b11^T$ for some scalars a, b. Find a and b and interpret the elements of this matrix.

24: Show that scaling $\text{corr}(X)^{-1}$ to have unit diagonals leads to the same matrix as scaling $\text{var}(X)^{-1}$. Show that the diagonal elements of the inverse of a positive definite matrix with unit diagonals are always greater than unity.

25: Show that the variance satisfies the identity

$$\text{var}(X, Y) = \begin{pmatrix} V_X & V_X B^T \\ BV_X & V + BV_X B^T \end{pmatrix}$$

where $V_X = \text{var}(X)$ and $V = \text{var}(Y|X)$, and hence develop a symmetric expression for the inverse variance lemma.

26: By reversing the roles of X and Y show explicitly that the following identities follow from the inverse variance lemma:

$$\text{var}\,(X)^{-1}\text{cov}\,(X,Y)\text{var}\,(Y|X)^{-1} = \text{var}\,(X|Y)^{-1}\text{cov}\,(X,Y)\text{var}\,(Y)^{-1}$$

27: Show that $\text{var}\,(X,Y)$ is positive definite if and only if $\text{var}\,(X)$ and $\text{var}\,(Y|X)$ are both positive definite. If $Z = (X_1, X_2, \ldots, X_p)$ and $\text{var}\,(X_i) = 1$ for all i, show that $\det\text{var}\,(Z)$ is maximised for variation in $\text{cov}\,(X_i, X_j)$ with $i \neq j$ by $\text{var}\,(Z) = I$.

Chapter 6

Graphical Gaussian Models

The independence graph of a set of random variables is defined by a set of pairwise conditional independence relationships that determine the edge set of the graph. While no more is required to establish the Markov properties of the graph, this is usually a too general framework for data analysis and we need to make further assumptions about the joint distribution of the variables. The payoff is the development of specific measures and tests for independence and conditional independence. Here we consider graphical models based on the Normal distribution, and remark that, for historical reasons, some authors refer to a Gaussian graphical model as a covariance selection model.

The simplest assumption for continuous random variables is that they have a jointly Normal distribution and, after a short discussion of approaches to graphical modelling, the next two sections of this chapter review the basic theory associated with the multivariate Normal distribution. Of especial importance are the results that the multivariate Normal distribution is closed under marginalisation and conditioning; the generalistion of the elementary result that uncorrelated Normal random variables are independent; and, that the independence graph of a jointly Normal set of random variables is determined by the location of zeros in the inverse variance.

Explanation of the maximum likelihood procedure applied to fitting a graphical Gaussian model takes up the rest of the chapter. Given a random sample of multivariate Normal random vectors, and an independence graph, the likelihood equations for determining the best fitting parameters, are derived and their solution discussed. The representation of the likelihood function as an information divergence, expressed in terms of the variance, provides a useful tool for deriving the general equations; though specific examples can be tackled using straightforward calculus. A version of the iterative fitting procedure is used to solve the likelihood equations and some discussion is given of when direct estimates exist. The generalised log-likelihood ratio test statistic, herein referred to as the deviance, is defined, and used to test goodness of fit and to compare different graphical models. The equations determining the second derivatives of the log-likelihood function are rather unpleasant, and so explicit expressions for the asymptotic approximations to the standard errors of the parameter estimates are reluctantly ommitted. Numerical illustrations of fitting graphical Gaussian models were produced using the program MIM, discussed in Appendix A.

The final section provides a scant introduction to more specialised topics: the sampling distribution of the sufficient statistic is the Wishart distribution; Bartlett correction factors are introduced as one way to improve the distribution of the test statistic based on the deviance; and the computation of Mahalanobis deviates is used to illustrate the important role of diagnostic procedures.

6.1 Graphical Models and Modelling

Consider a k-dimensional random vector $X = (X_1, X_2, \ldots, X_k)$ and an independence graph $G = (K, E)$. A *graphical model* for X is a single family of probability distributions for X, that satisfy the pairwise conditional independence restrictions inherent in G, but are otherwise arbitrary. When the distributions are multivariate Normal then we speak of the *graphical Gaussian* model. In this case we shall see that such conditional independence constraints are equivalent to specifying zeros in the inverse variance.

We turn to the issue of statistical inference for graphical models: the activity of relating a sample of observations to a putative probability model. We suppose we have a random sample of N observations on X and we compute the sample mean and sample variance.

A naive graphical modelling procedure

A naive graphical modelling procedure, used to analyse the mathematics marks in the introduction, goes as follows:

1. Estimate the variance matrix $V = \text{var}(X)$ by the sample variance matrix, S. (The variance operator $\text{var}_N(.)$ corresponds to S in the same way as $\text{var}(.)$ corresponds to V.)

2. Compute its inverse, S^{-1}, or the inverse of the sample correlation matrix. The diagonal elements are interpretable in terms of partial variances: $1/\text{var}_N(X_i|\text{rest})$ or $\text{var}_N(X_i)/\text{var}_N(X_i|\text{rest})$ if the sample corre-

lation matrix is used, and indicate how well variable X_i can be predicted from the other variables.

3. Scale either inverse to have a unit diagonal and so compute the sample partial correlations, $\text{corr}_N(X_i, X_j|\text{rest})$.

4. Set any sufficiently small element of the scaled inverse to zero. Draw the resulting independence graph according to the rule that no edge is included in the graph if the partial correlation coefficient is zero.

Firstly, ignoring the problems due to sampling, we must establish the implicit claim that the graph produced is an independence graph. Secondly, we have to face the many questions of detail raised by the procedure, for instance, how small is small? and of methodology, such as, is the implied estimation procedure best possible? Thirdly, and more importantly, the general question of how a new procedure could be constructed to cope with different, or more complicated, sampling situations has to be addressed. We can do better than this.

Likelihood inference

The observations in the sample are directly related to the probability model under consideration by the likelihood function. The standard results and techniques of maximum likelihood estimation and likelihood ratio tests (Cox and Hinkley, 1974) are then available to the graphical modeller. The initial steps of such an approach are described by:

1. First specify the family of distributions for X: the family of k-dimensional Normal distributions.

2. Select the parameter set. For the multivariate Normal distribution, independence is characterised by the variance, $V = \text{var}(X)$, or its inverse D, and as the correspondence between V and D is one to one, we are at liberty to use either parameterisation. There are $k(k+1)/2$ parameters of which k are concerned with scale and $k(k-1)/2$ with interaction. At this stage, we are not interested in the mean value, μ, and so allow it to be entirely arbitrary.

3. Determine the graphical model of interest. Each pairwise independence constraint specified by the independence graph generates a constraint on the parameters. For the multivariate Normal distribution the constraint is a zero in the inverse variance; note, the same constraint expressed in terms of the variance parameter is substantially more complicated.

4. Construct the likelihood function. We assume simple random sampling, so that the sample observations are mutually independent and each has the same density function.

5. Estimate the unknown parameters. The likelihood is maximised over the unknown parameters, subject to the constraints that determine the particular graphical model.

6. Check goodness of fit. The deviance of a fitted graphical model is twice the generalised log-likelihood ratio test of that model against the saturated model with the complete graph in which all parameters are entirely unconstrained. The test compares the calculated deviance to percentage point with the relevant chi-squared distribution, and so gives an assessment of whether a particular graphical model provides an adequate representation of the data.

6.2 The Multivariate Normal Distribution

We begin with a definition.

Definition. The k-dimensional random vector X has the multivariate *Normal distribution* if and only if its density function is of the form

$$f_X(x) = (2\pi)^{-k/2} \det(D)^{1/2} \exp\{-\frac{1}{2}[x-\mu, D(x-\mu)]\}, \quad \text{for } x \text{ in } R^k$$

where μ is a fixed k-dimensional vector and the $k \times k$ matrix D is symmetric and positive definite. □

The vector μ and the matrix D are parameters of this distribution. We shall see that μ is the expected value of X, and that if V is the variance matrix of X, then $D = V^{-1}$ the inverse variance. We write $X \sim N(\mu, V)$ to denote that X has this multivariate Normal distribution and $n(x; \mu, V)$ for the actual function displayed in this definition. We continue to write the quadratic form in the exponent as $[x-\mu, D(x-\mu)]$ rather than the equivalent expression $(x-\mu)^T D(x-\mu)$, to stress the bilinearity of the term.

The following direct construction of a multivariate Normal random vector from a set of independent Normal random variables shows that such a distribution exists and establishes the meaning of the parameters μ and D. Suppose that μ and D are given.

Step 1: Let $Z_1, Z_2, \ldots, Z_k$ be k independent scalar random variables each with the standard $N(0,1)$ density function, and put

$$Z = \begin{pmatrix} Z_1 \\ Z_2 \\ \vdots \\ Z_k \end{pmatrix}.$$

Note that $E(Z) = 0$ and $\text{var}(Z) = I$.

Step 2: Identify the density function of Z with the joint density function of $Z_1, Z_2, \ldots, Z_k$ which, by mutual independence, is

$$f_Z(z) = f_{Z_1,Z_2,..,Z_k}(z_1, z_2, .., z_k) = (2\pi)^{-k/2} \exp(-\frac{1}{2}[z, z]),$$

where we use the fact that $[z, z] = \sum_i z_i^2$.

Step 3: Given a symmetric positive definite matrix we can always find a positive definite matrix A to express it as AA^T, for example, by a Cholesky decomposition as in Proposition 5.7.1. As D^{-1} is symmetric and positive definite we can therefore find A such that $D^{-1} = AA^T$.

Step 4: Put $Y = AZ$. The Jacobian of this tranformation is $\det(A)^{-1} = \det(D)^{1/2}$ and so the density function of Y is

$$f_Y(y) = (2\pi)^{-k/2} \det(D)^{1/2} \exp(-\frac{1}{2}[y, Dy]).$$

Step 5: Finally put $X = Y + \mu$ to give f_X as in the definition.

In fact we have proved the following:

Proposition 6.2.1 *Any multivariate Normal random vector X can be represented as a linear transformation of a* standard *multivariate Normal random vector Z, that is, $Z \sim N(0, I)$, together with a shift: $X = AZ + \mu$.*

The next result relates the work of the previous chapters on covariance and diagonalisation to the multivariate Normal.

Corollary 6.2.2 *The expected value and the variance of X are $E(X) = \mu$, and* $\text{var}(X) = D^{-1}$.

Proof: Now $X = AZ + \mu$ and so by the linearity of expectation, $E(X) = AE(Z) + \mu$. But $E(Z) = 0$. As $\text{var}(Z) = I$, it follows that $\text{var}(X) = \text{var}(AZ) = A\text{var}(Z)A^T = AA^T$ and, by construction, the last term is just D^{-1}. □

Corollary 6.2.3 *If $X \sim N$ then the linear combination $[v, X] \sim N$ for all fixed k-dimensional vectors v.*

Proof: As $X = AZ + \mu$ then

$$[v, X] = [v, AZ + \mu] = [A^T v, Z] + [a, \mu] = \sum \alpha_j Z_j + [v, \mu].$$

But this is just a linear combination of Normal random scalars and hence is Normal by the assertion of Proposition 2.6.1 in Chapter 2. □

There is a more powerful theorem along these lines which states that $X \sim N$ if and only if $[v, X] \sim N$ for all v in R^k. Following Rao (1973) several authors have used this as a definition of multivariate Normality; and in consequence there is a very direct proof of the result that the marginal distributions of the multivariate Normal are also multivariate Normal. However, from our viewpoint this approach suffers two disadvantages: it makes characteristic function theory a prerequisite to the theory; and, as it does not present the density function directly, the likelihood function is not immediately accessible.

Corollary 6.2.4 *If $EX = 0$ then $[X, \text{var}(X)^{-1}X]$ has the chi-squared distribution with k degrees of freedom.*

Proof: We have $X = AZ$ and $\text{var}(X) = AA^T$ so

$$[X, \text{var}(X)^{-1}X] = [AZ, (AA^T)^{-1}AZ] = [Z, Z] = \sum_{j=1}^{k} Z_j^2,$$

is a sum of k independent chi-squared random variables each with 1 degree of freedom; hence it has a chi-squared distribution with k degrees of freedom. □

Variance: operator and parameter

We wish to establish if the family of multivariate Normal density functions is closed under the operations of marginalisation and conditioning. Our program is clear: we need to partition the full vector into two sub-vectors, to integrate out one sub-vector from the joint density to get its marginal density, and then to use this to derive the conditional density function. But first we must clarify the notation.

There are three inter-related notational distinctions, to make: firstly, there is a choice of partitioning notation for the random vector; secondly, a distinction is to be drawn between the variance operator and the variance parameter; and finally, we must distinguish the partial variance from the conditional variance.

Firstly, the choice of partitioning notation: we will partition the full vector into (X_a, X_b) in accordance with the theoretical exposition of independence graphs, even though this places a heavy reliance on subscripts unlike the alternative (X, Y) notation employed in the previous chapter. We write the full vector as X_K where K is the vertex set $\{1, 2, \ldots, k\}$. If a is a subset of K, X_a is well defined, as is, $X_{a \cup b} = (X_a, X_b)$ for disjoint subsets a and b, of K.

Secondly: by definition, the covariance is $\text{cov}(X_a, X_b) = E(X_a - EX_a)(X_b - EX_b)^T$, where E is the expectation operator. This defines the variance of X_a from $\text{var}(X_a) = \text{cov}(X_a, X_a)$ and the variance of $X_{a \cup b} = (X_a, X_b)$ is the partitioned matrix

$$\text{var}(X_{a\cup b}) = \begin{pmatrix} \text{var}(X_a) & \text{cov}(X_a, X_b) \\ \text{cov}(X_b, X_a) & \text{var}(X_b) \end{pmatrix}.$$

Note that $\text{var}(X_a, X_b)$ and $\text{cov}(X_a, X_b)$ are different objects. The linear least squares predictor of X_b from X_a is

$$\hat{X}_b(X_a) = EX_b + \text{cov}(X_b, X_a)\text{var}(X_a)^{-1}(X_a - EX_a),$$

and the partial variance of X_b given X_a is $\text{var}(X_b|X_a)$.

The value taken by the variance operator is used to parameterise the distribution. The *variance parameter* V is the $k \times k$ matrix of variances and covariances given from $V = \text{var}(X)$. The partitioning notation extends to the variance parameters: $V_{aa} = \text{var}(X_a)$, $V_{ab} = \text{cov}(X_a, X_b)$, and more generally,

$$V_{a\cup b, a\cup b} = \begin{pmatrix} V_{aa} & V_{ab} \\ V_{ba} & V_{bb} \end{pmatrix}.$$

Some authors write V_{aa} as V_a. If we want to make the dimension of V explicit, we write V_{KK}. To reiterate, V, the variance matrix is a parameter of the density function, while $\text{var}(.)$ is an operator on a random vector.

Finally there is a difference between the conditional variance and the partial variance, which is manifest for both the variance operator and variance parameter. The conditional variance of X_b given X_a is the ordinary variance operator applied to X_b evaluated with reference to the conditional distribution of X_b given X_a, and written as $\text{var}_{b|a}(X_b)$ as in Section 5.1 of the previous chapter. The subscript indicates the distribution over which the implicit expectations are evaluated. We can now easily distinguish the conditional variance of X_b given X_a, $\text{var}_{b|a}(X_b)$, from the partial variance of X_b given X_a,

$$\text{var}(X_b|X_a) = \text{var}(X_b) - \text{cov}(X_b, X_a)\text{var}(X_a)^{-1}\text{cov}(X_a, X_b),$$

which is defined in the joint distribution as the ordinary variance of the least squares residual, and so is more explicitly written as $\text{var}_{a\cup b}(X_b|X_a)$. The corresponding notation for the partial variance parameter is

$$V_{bb|a} = V_{bb} - V_{ba}V_{aa}^{-1}V_{ab}.$$

Though, in general, the partial and the conditional variances are not the same, it turns out that they are identical in the multivariate Normal distribution. Hence there is no need to introduce special notation for the conditional variance parameters.

The *inverse variance parameter* is $D = V^{-1}$, or making the underlying random vector more explicit, $D_{KK} = V_{KK}^{-1}$. It is partitioned conformably with V, so that D_{aa} is the corresponding diagonal block of $D = V^{-1}$,

$$D = \begin{pmatrix} D_{aa} & D_{ab} \\ D_{ba} & D_{bb} \end{pmatrix} = \begin{pmatrix} V_{aa} & V_{ab} \\ V_{ba} & V_{bb} \end{pmatrix}^{-1}.$$

The main thrust of the inverse variance lemma is that

$$D_{bb} = V_{bb|a}^{-1}.$$

At times we need to refer to the individual elements of the variance or inverse variance parameter; we simplify $V_{\{i\},\{j\}}$ to v_{ij} and note that $v_{ij} = \operatorname{cov}(X_i, X_j)$. The corresponding element of D is d_{ij}.

We denote the sample variance-covariance matrix by S and use the same partitioning conventions for S as for V. Thus we talk of S_{aa} and S_{ab} and in detailed examples of its elements s_{ij}. It is sometimes convenient to view the sample variance as an operator, so we define a variance operator, $\operatorname{var}_N(.)$ with the property $\operatorname{var}_N(X_a) = S_{aa}$ for all $a \subseteq K$. For maximum likelihood estimates we use $\hat{V}$ and $\hat{v}_{ij}$.

We need to distinguish between the multivariate Normal density with variance matrix V and another with variance W, and so write f^V and f^W; the empirical multivariate Normal density function $f^S(x)$ is the function $n(x; 0, S)$.

6.3 Marginal and Conditional Distributions

It transpires that both the marginal and the conditional density functions of the multivariate Normal are multivariate Normal. That is, the class of multivariate Normal density functions is closed under the operations of marginalisation and conditioning. The proof proceeds along the same lines as the corresponding proof for the bivariate Normal given in Section 2.6, using the additional machinery developed in the previous chapters on covariance and inverse variance.

Proposition 6.3.1 Marginal and conditional density functions. *If the partitioned vector (X_a, X_b) has a Normal distribution parameterised by the mean vector (μ_a, μ_b) and variance*

$$V_{a\cup b, a\cup b} = \begin{pmatrix} V_{aa} & V_{ab} \\ V_{ba} & V_{bb} \end{pmatrix}$$

then

(i) the marginal distribution of X_a is Normal with mean μ_a and variance V_{aa}; and

(ii) the conditional distribution of X_b given $X_a = x_a$ is Normal with mean

$$E_{b|a}(X_b) = \mu_b + B_{b|a}(x_a - \mu_a), \quad \textit{where } B_{b|a} = V_{ba}V_{aa}^{-1},$$

and variance

$$\text{var}_{b|a}(X_b) = V_{bb|a}, \quad \textit{where } V_{bb|a} = V_{bb} - V_{ba}V_{aa}^{-1}V_{ab}.$$

Proof: Without loss of generality we can assume that $E(X_a, X_b) = (\mu_a, \mu_b) = 0$. Suppose X_a is p-dimensional and X_b is q-dimensional. The joint density function of (X_a, X_b) is

$$f_{ab}(x_a, x_b) = (2\pi)^{-(p+q)/2}\det(V)^{-1/2}\exp\{-\frac{1}{2}[(x_a, x_b), V^{-1}(x_a, x_b)]\}.$$

writing V for $V_{a\cup b, a\cup b}$. The two corollaries to the inverse variance lemma required here are Schur's identity, Proposition 5.9.1, that

$$\det V = \det V_{aa} \det V_{bb|a},$$

and Corollary 5.8.6 that the quadratic form in the exponent of the density function can be completed to give

$$[(x_a, x_b), V^{-1}(x_a, x_b)] = [x_a, V_{aa}^{-1}x_a] + [(x_b - B_{b|a}x_a), V_{bb|a}^{-1}(x_b - B_{b|a}x_a)].$$

Substituting for the determinant and the inner product in the joint density gives

$$n((x_a, x_b); 0, V) = n(x_a; 0, V_{aa})\, n(x_b; B_{b|a}x_a, V_{bb|a})$$

as an identity in three functions. Both factors in this product are structured as multivariate Normal densities and the vector x_b only occurs in the second factor. Integrating out x_b gives the marginal density function of X_a. It equals $n(x_a; 0, V_{aa})$ because the integral of $n(x_b; B_{b|a}x_a, V_{bb|a})$ over x_b must equal unity. The conditional density function then falls out of

$$\begin{aligned} f_{b|a}(x_b; x_a) &= f_{ab}(x_a, x_b)/f_a(x_a) \\ &= n(x_b; B_{b|a}x_a, V_{bb|a}) \end{aligned}$$

on substitution. □

By substituting for each expectation and variance in terms of the partitioned parameter V we may derive

Corollary 6.3.2 Conditional expectation and variance. *The conditional mean of X_b given $X_a = x_a$, and the linear least squares predictor of X_b from $X_a = x_a$, are identical; so to are the conditional variance and the partial variance. That is*

$$E_{b|a}(X_b) = \hat{X}_b(x_a) \quad \textit{and} \quad \text{var}_{b|a}(X_b) = \text{var}(X_b|X_a).$$

Hence the mean of X_b given X_a is just a linear transformation of X_a and even more startling, the conditional variance of X_b given $X_a = x_a$ is constant for all values of x_a.

Independence and conditional independence

Simple conditions for independence are derived in terms of the covariance and the inverse variance.

Corollary 6.3.3 Independence and zero covariance. *The Normal random vectors X_a and X_b are independent, $X_a \perp\!\!\!\perp X_b$, if and only if either*

(i) $\mathrm{cov}(X_a, X_b) = 0$, *or expressed parametrically,* $V_{ab} = 0$; *or*

(ii) $D_{ab} = 0$, *where D is the inverse variance.*

Proof: There are several ways to prove this; one is to apply the factorisation criterion to the joint density function. Again set means to zero without loss of generality. The log density function of (X_a, X_b) is

$$\log f_{ab}(x_a, x_b) = \text{const} - \frac{1}{2}[(x_a, x_b), D(x_a, x_b)].$$

where $D = V^{-1}$. By partitioning D we see

$$[(x_a, x_b), D(x_a, x_b)] = [x_a, D_{aa}x_a] + 2[x_a, D_{ab}x_b] + [x_b, D_{bb}x_b],$$

and an application of the factorisation criterion proves that $X_a \perp\!\!\!\perp X_b$ if and only if $D_{ab} = 0$. The corollary to the inverse variance lemma, Corollary 5.8.6, completes the argument. □

A similar argument using the factorisation criterion for conditional independence produces the most important result that relates zeros in the inverse variance and zero partial covariance to conditional independence.

Corollary 6.3.4 Conditional independence and zero partial covariance. *The Normal vectors X_b and X_c are independent conditional on X_a, $X_b \perp\!\!\!\perp X_c | X_a$, if and only if either*

(i) $\mathrm{cov}(X_c, X_b | X_a) = 0$ *or* $V_{bc|a} = V_{bc} - V_{ba}V_{aa}^{-1}V_{ac} = 0$; *or*

(ii) the block of the inverse variance $D_{bc} = 0$.

In particular, if X_b and X_c are both one dimensional then

$$X_i \perp\!\!\!\perp X_j | X_{K\setminus\{i,j\}} \text{ if and only if } d_{ij} = 0.$$

We have a simple parametric criterion for pairwise independence conditional on the rest.

Graphical Gaussian models

There is enough machinery to specify a graphical Gaussian model. Given an independence graph G, and a k-dimensional random vector X, a ***graphical Gaussian model*** is a family of Normal distributions for X constrained to satisfy the pairwise conditional independence restrictions inherent in the independence graph. Corollary 6.3.4 shows that such conditional independence constraints are equivalent to specifying zeros in the inverse variance parameter corresponding to the absence of an edge in G. Otherwise the distribution is arbitrary. It is sometimes called a covariance selection model, following Dempster (1972) who first suggested fitting models with zeros in the inverse variance. However, note it is really the partial covariances that are selected to be zero. There are $2^{\binom{k}{2}}$ graphical models corresponding to the $\binom{k}{2}$ off diagonal elements in the inverse variance matrix.

EXAMPLE 6.3.1 The vector X is four dimensional, $k = 4$. A graphical Gaussian model for X is that the inverse variance $D = \text{var}\,(X_1, X_2, X_3, X_4)^{-1}$ is of the form

$$D = \begin{pmatrix} d_{11} & d_{12} & 0 & 0 \\ d_{12} & d_{22} & d_{23} & 0 \\ 0 & d_{23} & d_{33} & d_{34} \\ 0 & 0 & d_{34} & d_{44} \end{pmatrix}$$

where the remaining d_{ij} are arbitrary, restrained only to ensure the matrix is symmetric and positive definite. The independence graph of X is

(1)——(2)——(3)——(4)

The global Markov property asserts that $X_1 \perp\!\!\!\perp X_3 | X_2$ and consequently, and rather remarkably, we may conclude that $\text{var}\,(X_1, X_2, X_3)^{-1}$ is of the form

$$\begin{pmatrix} * & * & 0 \\ * & * & * \\ 0 & * & * \end{pmatrix}$$

irrespective of values of the non-zero entries of D. □

The mean vector $\mu = E(X)$ may either be known, and so set to zero, or unknown and more complicated models may impose some structure on the mean, for example, some form of regression model relating X to some measured covariates. Similarly, to generalise in another direction, other models may incorporate additional modelling of the variance, for example, by imposing the symmetry restraint that $d_{ij} = d$ for all the non-zero elements of D.

Log-linear parameterisation

Graphical Gaussian models are based on the implication that

$$d_{ij} = 0 \text{ if and only if } X_i \perp\!\!\!\perp X_j \,|\text{rest}.$$

These quantities naturally arise because the multivariate Normal density function is a member of the exponential family of density functions, which is a large class of density functions of importance to statistical theory and practice, see Cox and Hinkley (1974). A characteristic of density functions in this class are the linearities they exhibit when expressed on a logarithmic scale, and an interest in such an expansion is motivated by a comparison with the corresponding 'log-linear model' for the Multinomial distribution. The log of the multivariate Normal density function is, directly from the definition, with $EX = \mu = 0$,

$$2\log f_X(x) = \text{const} - [x, Dx] + \log \det D.$$

It is a quadratic form in x with weights determined by D, the inverse variance. In terms of the coordinates of x this expands to

$$2\log f_X(x) = \text{const} - \sum_{i,j} d_{ij} x_i x_j + \log \det D,$$

and the interaction between x_i and x_j are entirely dictated by d_{ij}. By virtue of the corollaries to the inverse variance lemma the coefficients of this quadratic form are interpretable as partial variances and partial correlations:

$$\begin{aligned} d_{ii} &= 1/\text{var}\,(X_i|\text{rest}), \text{ and} \\ d_{ij} &= -\text{corr}\,(X_i, X_j|\text{rest})(d_{ii}d_{jj})^{1/2}. \end{aligned}$$

6.4 Divergence between Normal Distributions

The Kullback-Leibler information divergence between two density functions, f and g is defined by $I(f;g||X) = E\log f(X)/g(X)$, where the expectation is taken with respect to the density function f. Its applications have been discussed in general terms in Chapter 4 and here we consider its evaluation on multivariate Normal densities. The high points turn out to be, firstly, that the information in one random variable about another is a function of the correlation or partial correlation structure; and secondly, that the likelihood function for an unknown variance can be simply expressed as the divergence between an empirical Normal density with variance determined from the sample and a Normal density with variance determined by the parameters.

Evaluation of the divergence between sub-families of the multivariate Normal distribution is made a little easier by using the results from Chapter 5: if the random vector X has expected value $EX = \mu$ and variance $\mathrm{var}(X) = V$ then firstly,

$$E[X, AX] = \mathrm{tr}(AV) + [\mu, A\mu];$$

and secondly, $V = \mathrm{var}(X_a, X_b)$ satisfies Schur's identity

$$\det V = \det V_{aa} \det V_{bb|a}.$$

We shall calculate $E\log(f/g)$ by calculating $E\log f$ and $E\log g$ separately.

Lemma 6.4.1 Entropy evaluation. *If X has the k-dimensional Normal density $f_X(x) = n(x; \mu, V)$ then the entropy of X is*

$$-E\log f_X(X) = \frac{1}{2}k(1 + \log 2\pi) + \frac{1}{2}\log \det V.$$

Proof: From the definition of the Normal density function

$$2E\log f_X(X) = -k\log 2\pi + \log \det D - E[X - \mu, D(X - \mu)],$$

and, by the results above, $E[X - \mu, D(X - \mu)] = \mathrm{tr}(VD) = \mathrm{tr}(I) = k$. □

The entropy of the non-singular linear transformation of a Normal random vector $X \to AX + b$ satisfies

$$\mathrm{entropy}(AX + b) = \log \det(A) + \mathrm{entropy}(X),$$

which shows that the entropy depends on the scale to measure the variables, though is invariant to location.

Proposition 6.4.2 The divergence between two Normal distributions. *If $f(x) = n(x; \mu_1, V_1)$ and $g(x) = n(x; \mu_2, V_2)$ then*

$$I(f; g) = \frac{1}{2}[\mu_1 - \mu_2, V_2^{-1}(\mu_1 - \mu_2)] + \frac{1}{2}\mathrm{tr}(V_1 V_2^{-1}) - \frac{1}{2}\log \det(V_1 V_2^{-1}) - \frac{k}{2}.$$

To prove this, use Lemma 6.4.1 to calculate the entropy $E\log f$ and then find $E\log g$ in a similar way. The lack of symmetry between the subscripts 1 and 2 arises because all expectations are taken with respect to f. Unlike the entropy the divergence is invariant to changes in both location and scale, that is $I(f; g||AX + b) = I(f; g||X)$. The notation $I(f^{\mu_1, V_1}; f^{\mu_2, V_2})$ for this divergence proves useful.

The next two corollaries specialise the proposition by setting $V_1 = V_2 = V$ in the first and $\mu_1 = \mu_2 = \mu$ in the second.

Corollary 6.4.3 Comparing means. *The divergence for comparing two means when the variances are equal, $f : X \sim N(\mu_1, V)$ and $g : X \sim N(\mu_2, V)$, is*

$$I(f; g) = \frac{1}{2}[(\mu_1 - \mu_2), V^{-1}(\mu_1 - \mu_2)].$$

Twice the right hand side is the *Mahalanobis distance* between the points μ_1 and μ_2, it is symmetric between μ_1 and μ_2 and defines a proper metric. It goes to show that, even though the divergence is not a distance in general, it is rather 'near' (sic!) to being one at times.

Corollary 6.4.4 Comparing variances. *The divergence for comparing two variances when the means are equal, $f : X \sim N(\mu, V_1)$ and $g : X \sim N(\mu, V_2)$, is*

$$I(f;g) = \frac{1}{2}\mathrm{tr}\,(V_1V_2^{-1} - I_k) - \frac{1}{2}\log\,\det(V_1V_2^{-1})$$

where I_k is the identity matrix of order $k \times k$.

This is the essential expression needed for dealing with the likelihood function. Because expectation is taken with respect to f, var $(X) = V_1$, in the particular case that $V_2 = I_k$, the information divergence between $f : X \sim N(\mu, V)$ and $g : X \sim N(\mu, I_k)$ is

$$I(V; I_k) = \frac{1}{2}\{\mathrm{tr}\,(V - I_k) - \log\,\det V\}.$$

A corollary to this expression is the inequality, true for any positive definite symmetric matrix V of order k, $\mathrm{tr}\,(V) - \log\,\det(V) - k \geq 0$.

Corollary 6.4.5 The divergence against independence: information proper. *When*

$$f : (X_a, X_b) \sim N(\mu, V_{a\cup b, a\cup b}) \quad \textit{and} \quad g : g_{ab} = f_a f_b,$$

the information proper, defined as the divergence against independence, is

$$\mathrm{Inf}\,(X_a \perp\!\!\!\perp X_b) = -\frac{1}{2}\log\frac{\det V_{a\cup b, a\cup b}}{\det V_{aa}\,\det V_{bb}}$$

and simplifies to $\frac{1}{2}\log\det V_{bb|a}/\det V_{bb}$.

Proof: Suppose that X_a and X_b are p- and q-dimensional, respectively. From Lemma 6.4.1 the entropy of X_a is

$$2E\log f_a = -p\log 2\pi - \log\,\det V_{aa} - p;$$

by symmetry there is a similar formula with b replacing a; which furthermore holds with $a \cup b$ replacing a, giving

$$2E\log f_{ab} = -(p+q)\log 2\pi - \log\,\det V_{a\cup b, a\cup b} - (p+q).$$

Using the definition that $\mathrm{Inf}(X \perp\!\!\!\perp X_b) = E\log f_{ab}/f_a f_b$, subtraction gives the first expression and Schur's identity the second. □

Proposition 6.4.6 The divergence against conditional independence. *Suppose that*

$$f : (X_a, X_b, X_c) \sim N(\mu, V_{a\cup b\cup c, a\cup b\cup c}) \text{ and } g : g_{bc|a} = f_{b|a} f_{c|a},\ g_a = f_a.$$

Then the conditional information measure $\mathrm{Inf}(X_b \perp\!\!\!\perp X_c | X_a)$ *can be expressed as any one of*

$$-\frac{1}{2}\log\frac{\det V_{a\cup b\cup c, a\cup b\cup c} \det V_{aa}}{\det V_{a\cup b, a\cup b} \det V_{a\cup c, a\cup c}} = -\frac{1}{2}\log\frac{\det V_{b\cup c, b\cup c|a}}{\det V_{bb|a} \det V_{cc|a}} = -\frac{1}{2}\log\frac{\det V_{cc|a\cup b}}{\det V_{cc|a}}.$$

When $b = \{i\}$ *and* $c = \{j\}$ *are one dimensional, the conditional information simplifies to*

$$\mathrm{Inf}(X_i \perp\!\!\!\perp X_j | X_a) = -\frac{1}{2}\log(1 - \mathrm{corr}^2(X_i, X_j | X_a)).$$

Proof: The definition of the conditional information measure expressed in terms of marginal density functions is

$$\mathrm{Inf}(X_b \perp\!\!\!\perp X_c | X) = E\log\frac{f_{abc} f_a}{f_{ab} f_{ac}},$$

which is a linear combination of entropies of Normal distributions. Evaluating each from the entropy lemma proves the first statement in the proposition. The two other expressions can be deduced directly from Schur's lemma or from the conditional densities, by using $f_{abc} = f_a f_{b|a} f_{c|ab}$, and the result that Normal distributions are closed under conditioning. When $q = 1 = r$, the partial variance $V_{b\cup c, b\cup c|a}$ is just a 2×2 matrix, and the result transpires by evaluating its determinant. □

Thus in Normal distributions the information divergence for measuring the conditional independence of X_i and X_j given X_a is a simple function of the partial correlation between X_i and X_j given X_a. This expression gives us a measure of the strength of a connection in the independence graph of Normal random variables. It also allows us to generalise the notion of the equivalent Normal correlation coefficient discussed in Section 4.5 of Chapter 4 to an equivalent Normal partial correlation coefficient.

EXAMPLE 6.4.1 Suppose that the variance of X is

$$V = \begin{pmatrix} 1.0 & 0.5 & 0.4 \\ 0.5 & 1.0 & 0.2 \\ 0.4 & 0.2 & 1.0 \end{pmatrix}$$

and we wish to find the independence graph and its edge strengths. We need to find the three divergences against conditional independence. One way is to calculate the partial correlation coefficients directly from the recursion formula for partial covariance (tedious); alternatively, as with the mathematics marks, we may invert the variance matrix and scale the result to have unit diagonal. This gives all partial correlation coefficients at one pass:

$$\begin{pmatrix} 1.00 & 0.47 & 0.35 \\ 0.47 & 1.00 & 0.00 \\ 0.35 & 0.00 & 1.00 \end{pmatrix},$$

from which the divergences against conditional independence are directly calculated, according to Corollary 4.2.6, as

$$\mathrm{Inf}(X_1 \perp\!\!\!\perp X_2 | X_3) = 0.123, \mathrm{Inf}(X_1 \perp\!\!\!\perp X_3 | X_2) = 0.067, \text{ and } \mathrm{Inf}(X_2 \perp\!\!\!\perp X_3 | X_1) = 0.0.$$

The independence graph is

(2)——(1)——(3)
0.12 0.07

The variable X_1 is almost twice as informative about X_2 as it is about X_3. □

6.5 The Likelihood Function

The likelihood function is a fundamental tool of statistical inference. It relates the information in the observed sample to the unknown parameters of a putative probability model and so enables us to assess which values of the parameters are well supported, and which are not. For a general definition and a summary of its mathematical properties the reader is referred to Cox and Hinkley (1974). The concern of this section is solely to evaluate the function for samples from the multivariate Normal distribution; applications to estimation and testing of graphical models come later in the chapter.

In the chapter on information divergence, it was noted that information measures based on the Kullback-Leibler divergence naturally quantify the information in one random variable about another. Now we turn to a second application of divergence. The generalised likelihood ratio test statistics based on the multivariate Normal are information divergences which measure the distance between the observed sample variance and another estimated variance simplified by some form of independence structure. It turns out that, in a large family of probability models, the so-called exponential family, which includes the multivariate Normal and the Multinomial distributions, the log-likelihood function is expressible as an information divergence.

Before embarking on this enterprise, we remark that this is not quite so surprising as it may first appear. The *log-likelihood ratio test statistic* based on an observation, x, for testing the hypothesis that X has density f, against the alternative that X has density g, is

$$\log f(x)/g(x),$$

and the likelihood ratio for a sample of independent observations is the sum of such terms. Its expected value under either hypothesis is an information divergence, and the central limit theorem suggests that, in large samples, the sample average should be close to this expectation.

The likelihood function

Given observations $x^1, x^2, \ldots, x^N$, on a sample of N independent and identically distributed random vectors, each with the k-dimensional multivariate Normal density function $f(x;\theta)$, with parameter θ, the *log-likelihood function* is

$$l(\theta) = l(\theta; x^1, x^2, \ldots, x^N) = \sum_l \log f(x^l; \theta),$$

see Cox and Hinkley (1974). There are several ways to parameterise the distribution; of interest to us are the elements of the variance matrix V, denoted by $V = \{v_{ij}\}$, and the elements of the inverse variance $D = V^{-1} = \{d_{ij}\}$. The correspondence between the two sets is one to one and we are at liberty to choose either, but, being greedy, we will use both. The mean vector, $\mu = EX$, is of less interest and to simplify the exposition it could be set to zero, but for completeness is retained.

Substituting the formula for the multivariate Normal density function in the definition of the log-likelihood function

$$2\,l(\mu, V) = \text{const} - \sum_{l=1}^{N} [x^l - \mu, V^{-1}(x^l - \mu)] - N \log \det V$$

Define the *sample mean* vector by $\bar{x} = N^{-1}(\sum x^l)$ and the *sample variance* matrix by $S = \{s_{ij}\}$, where the element

$$s_{ij} = \frac{1}{N} \sum_{l=1}^{N} (x_i^l - \bar{x}_i)(x_j^l - \bar{x}_j)^T$$

is the sample covariance of the i-th and j-th coordinates of X. The constant in the log-likelihood function neither depends on the observations in the sample nor on the parameters V.

First note that

$$\sum [x^l - \mu, V^{-1}(x^l - \mu)] = \sum [x^l - \bar{x}, V^{-1}(x^l - \bar{x})] + N[\bar{x} - \mu, V^{-1}(\bar{x} - \mu)],$$

because the cross-product term, $\sum_l [x^l - \bar{x}, V^{-1}(\bar{x} - \mu)]$, is zero.

Now the first part of this sum of squares function can be substantially simplified using the following trick:

$$[v, V^{-1}v] = v^T V^{-1} v = \text{tr}\,(v^T V^{-1} v) = \text{tr}\,(V^{-1} v v^T),$$

which works because the inner product is a scalar and $\text{tr}\,(AB) = \text{tr}\,(BA)$. Hence

$$\begin{aligned} \sum [x^l - \bar{x}, V^{-1}(x^l - \bar{x})] &= \sum \text{tr}\,\{V^{-1}(x^l - \bar{x})(x^l - \bar{x})^T\} \text{ by this trick} \\ &= \text{tr}\,\{V^{-1} \sum (x^l - \bar{x})(x^l - \bar{x})^T\} \\ &\qquad \text{as } tr(A+B) = tr(A) + tr(B) \\ &= N\text{tr}\,(V^{-1}S), \text{ definition of } S. \end{aligned}$$

Putting this together by manipulating the constant, which may now depend on $\bar{x}$ or S, we have

$$2\,l(\mu, V) = \text{const} - N[\bar{x} - \mu, V^{-1}(\bar{x} - \mu)] - N\text{tr}\,(V^{-1}S) - N\log \det V.$$

We have proved the following proposition.

Proposition 6.5.1 The likelihood function and the divergence. *The log-likelihood function for the mean μ and variance V, of the multivariate Normal distribution based on a random sample of N observations, can be written as either*

(i) $2l(\mu, V) = const - N[\bar{x} - \mu, V^{-1}(\bar{x} - \mu)] - N\text{tr}\,(V^{-1}S) - N\log \det V$; *or*

(ii) $l(\mu, V) = const - NI(f^{\bar{x},S}; f^{\mu,V})$;

where $I(f^{\bar{x},S}; f^{\mu,V})$ is the information divergence given in Proposition 6.4.2 between two multivariate Normal density functions.

Both expressions for the likelihood function are valuable. The knowledge that the sample comes from a multivariate Normal distribution implies that the raw observations stored in a data matrix of order $k \times N$ may be condensed to the $k(k+1)/2$ elements of the sample mean and variance, which are the sufficient statistics for the parameters, μ and V. By closure of the distribution under addition, the sampling distribution of $\bar{X}$ is also multivariate Normal. The sampling distribution of S is Wishart, and we defer further discussion to a later section of this chapter.

With distance measured by information divergence, the closer the parameter V is to the sample variance S the larger is the log-likelihood.

Corollary 6.5.2 Unconstrained maximisation. *When V is unconstrained the unique maximum likelihood estimator of the mean μ and variance V from a sample of independent Normal observations is the sample mean $\bar{x}$ and variance S.*

Proof: As the divergence is non-negative, the minimum value of $I(f^{\bar{x},S}; f^{\mu,V})$ is 0. This occurs when $\mu = \bar{x}$ and $V = S$. It is unique because the divergence is positive definite so that if $I(f;g) = 0$ then $f = g$, and consequently $f^{\bar{x},S}(x) = f^{\mu,V}(x)$ for all x. From this it follows that $\mu = \bar{x}$ and $V = S$. □

In an initial discussion of graphical models based on the Normal distribution the mean μ is rather a nuisance, and our main interest is in models for V. There are two easy ways to eliminate μ from the likelihood function: we may either set $\mu = 0$ and rework the derivation, or we may set μ equal to its maximum likelihood estimator, and take the ***profile log-likelihood function*** defined by $l(V) = l(\hat{\mu}, V)$ see Cox and Hinkley (1974). We then write f^S for the multivariate Normal density function evaluated at $\mu = \bar{x}$ and $V = S$. We adopt the latter approach.

6.6 Maximum Likelihood Estimates

We can now write down the likelihood function for a graphical model given a random sample from the multivariate Normal distribution; our next task is to ascertain its maximum. In fact there are several related problems: to give a set of equations that define the estimate, to check that a solution exists, to solve them or at least to provide an algorithm for their solution, and finally to determine if the solution is unique. The general expression for the likelihood in terms of information divergence gives a great deal of theoretical and practical assistance and leads to a fairly general solution of the problem. But, before making such a broadside attack, substantial insight is gained by tackling a simple example with standard differential calculus. It goes to show that the maximisation problem is far from trivial.

A detailed example

Suppose $k = 3$ and consider the graphical model defined by the single independence constraint $X_2 \perp\!\!\!\perp X_3 | X_1$. Its independence graph is

and the element d_{23} of the inverse variance is zero. Subject to this and the constraint that the inverse variance matrix is symmetric and positive definite, the parameters are unrestricted. The log-likelihood function, from Proposition 6.5.1 and expressed as a function of the inverse variance D, is

$$2\,l(D) = \text{const} - N\text{tr}\,(SD) + N\log\det(D).$$

Now substituting for

$$D = \begin{pmatrix} d_{11} & d_{12} & d_{13} \\ d_{21} & d_{22} & 0 \\ d_{31} & 0 & d_{33} \end{pmatrix}.$$

shows that, apart from the constant,

$$\begin{aligned} -\text{tr}\,(SD) + \log\det(D) = & -s_{11}d_{11} - s_{22}d_{22} - s_{33}d_{33} \\ & - 2s_{12}d_{12} - 2s_{13}d_{13} + \log(d_{11}d_{22}d_{33} - d_{12}^2 d_{33} - d_{13}^2 d_{22}). \end{aligned}$$

Note that the sample covariance s_{23} does not appear in this likelihood. The trace is linear and is easy to differentiate directly, and the determinant of a 3×3 matrix is not too bad. Taking the partial derivatives with respect to each d_{ij} gives the set of likelihood equations

$$\begin{array}{lrcl} (1,1) & -s_{11} + d_{22}d_{33}/\det(D) & = & 0 \\ (2,2) & -s_{22} + (d_{11}d_{33} - d_{13}^2)/\det(D) & = & 0 \\ (3,3) & -s_{33} + (d_{11}d_{22} - d_{12}^2)/\det(D) & = & 0 \\ (1,2) & -2s_{12} - 2d_{12}d_{33}/\det(D) & = & 0 \\ (1,3) & -2s_{13} - 2d_{13}d_{22}/\det(D) & = & 0 \end{array}$$

These equations can be put together in the following way:

$$\begin{pmatrix} s_{11} & & \\ s_{12} & s_{22} & \\ s_{13} & * & s_{33} \end{pmatrix} = \frac{1}{\det(D)} \begin{pmatrix} d_{22}d_{33} & & \\ -d_{12}d_{33} & d_{11}d_{33} - d_{13}^2 & \\ -d_{13}d_{22} & * & d_{11}d_{22} - d_{12}^2 \end{pmatrix}$$

and the matrix equation is completed by symmetry. It looks as though the right hand side may be invertible if we can insert something suitable for the missing equation $* = *$, corresponding to the missing edge in the graph. Now we know that for this postulated graphical model the partial covariance of X_2 and X_3 given X_1 is zero, so that

$$\text{cov}\,(X_2, X_3) = \text{cov}\,(X_2, X_1)\text{cov}\,(X_1, X_3)/\text{var}\,(X_1).$$

Hence we replace the left hand $*$ by $s_{12}s_{13}/s_{11}$ and $* = *$ by the identity

$$s_{12}s_{13}/s_{11} = d_{12}d_{13}/\det(D).$$

Happily, the right hand side then inverts to D so that the likelihood equations are expressible as

$$\begin{pmatrix} s_{11} & & \\ s_{12} & s_{22} & \\ s_{13} & s_{12}s_{13}/s_{11} & s_{33} \end{pmatrix} = \hat{D}^{-1}.$$

The maximum likelihood estimator of D is given by inverting this equation, and as $\hat{V} = \hat{D}^{-1}$ the estimator of V for this graphical model is given directly by the left hand side of the equation.

The elements of the estimated variance are identical to those of the sample variance for the subsets $\{1,2\}$ and $\{1,3\}$. This is a general result: the estimated and the sample variance are identical for those subsets of the variables corresponding to cliques in the graph: that is

$$\hat{V}_{aa} = S_{aa} \quad \text{for } a = \{1,2\},\ \{1,3\}.$$

These equations, together with the condition

$$\hat{d}_{23} = 0,$$

where $\hat{D} = \hat{V}^{-1}$, are sufficient to uniquely determine the maximum likelihood estimators of the parameters.

Twice the value of the log-likelihood at the maximum, the deviance, is

$$\text{dev} = 2l(\hat{D}) = N\{\text{tr}\,(S\hat{D}) - \log\det(S\hat{D}) - k\},$$

which for this particular model simplifies to a function of the sample partial correlation coefficient

$$\text{dev} = -N\log\left(1 - \text{corr}_N(X_2, X_3|X_1)^2\right).$$

Just one parameter is constrained and consequently the deviance has one degree of freedom.

EXAMPLE 6.6.1 Suppose the sample variance calculated from the data is

$$S = \begin{pmatrix} 10 & & \\ 6 & 20 & \\ 5 & 7 & 30 \end{pmatrix}$$

based on a sample of size $N = 100$. The maximum likelihood estimate of V under the graphical model defined by $X_2 \perp\!\!\!\perp X_3 | X_1$ is

$$\hat{V} = \begin{pmatrix} 10 & & \\ 6 & 20 & \\ 5 & \hat{v}_{23} & 30 \end{pmatrix}$$

and $\hat{v}_{23} = (5)(6)/10 = 3$. The observed and fitted correlation matrices are identical apart from the (2,3) entry. The observed and the fitted inverse variance matrices are

$$\begin{pmatrix} .127 & & \\ -.033 & .063 & \\ -.013 & -.009 & .038 \end{pmatrix} \quad \text{and} \quad \begin{pmatrix} .131 & & \\ -.037 & .061 & \\ -.018 & -.000 & .036 \end{pmatrix}$$

respectively. The observed and fitted partial correlation matrices are

$$\begin{pmatrix} 1.0 & & \\ .372 & 1.0 & \\ .193 & .188 & 1.0 \end{pmatrix} \text{ and } \begin{pmatrix} 1.0 & & \\ .409 & 1.0 & \\ .263 & .000 & 1.0 \end{pmatrix}$$

respectively. Note that all off diagonal values change, and the 0 in entry $(2,3)$ is dictated by the postulated model. Twice the maximised log-likelihood ratio is

$$\text{dev} = -N\log\{1-(7-(5)(6)/10)^2/(20-6^2/10)(30-5^2/10)\} = N(0.036) = 3.612,$$

and is associated with 1 degree of freedom. Its observed P-value, $P(X^2 > 3.61)$, computed from the chi-squared distribution with 1df is 0.054, a value of borderline significance. □

The general case

We now consider the general case and the next proposition asserts that the maximum likelihood estimates are always of this form. Our aim is to outline a general argument, which can be made rigorous, but avoids the technical details. Such a proof can be found in Speed and Kiiveri (1986) using work of Csiszar (1975). In fact our discussion is somewhat incomplete: we derive the likelihood equations that determine the fitted variance matrix, but do not show that either a solution exists, nor that if it exists that it is unique. Deriving the equations is comparatively easy because of the positive definite nature of the information divergence. Recall that a clique is a maximally complete subset.

Theorem 6.6.1 Likelihood equations for graphical Gaussian models. *The maximum likelihood estimator of a graphical model with graph G, based on a random sample from the multivariate Normal distribution, satisfies the likelihood equations*

$$\hat{d}_{ij} = 0,$$

whenever vertices i and j are not adjacent in G, and,

$$\hat{V}_{aa} = S_{aa}$$

whenever the subset a of vertices in G form a clique. The estimated parameters D and V are related by $\hat{D} = \hat{V}^{-1}$, and are unique with probability one.

Proof: We already know that for every missing edge in the graph, the corresponding parameter in the inverse variance is zero. For the other part of the set of likelihood equations it is useful to first establish a little lemma.

Lemma 6.6.2 *If, in a graphical model for the k-dimensional random vector X, the vertices $a \subseteq K = \{1, 2, \ldots, k\}$ induce a complete subgraph in the independence graph of X, the marginal density, f_a, of X_a is entirely arbitrary.*

Proof: By definition a graphical model is arbitrary apart from conditional independences determined by the independence graph. With X partitioned into (X_a, X_b), the density $f_{ab} = f_a f_{b|a}$. As the subset a is complete all conditional independence statements concern $f_{b|a}$. Hence f_a is arbitrary. □

Recall Lemma 4.4.2 which showed that the divergence $I(f; g||(X_a, X_b))$ is uniquely minimised with respect to variation in g_a, the marginal distribution of X_a under g, by choosing $g_a = f_a$. Now the likelihood of the variance V is given by

$$l(S) - l(V) = NI(f^S; f^V)$$

where f^S denotes the empirical multivariate Normal density function with variance equal to the sample variance S, and f^V the density function with variance V. This is maximised when the divergence is minimised. If the vertices in a form a clique then, by this little lemma f_a^V is arbitrary and by Lemma 4.4.2 choosing f^V to agree with f^S on this margin uniquely minimises the divergence. Hence

$$f_a^S = f_a^V \quad \text{whenever } a \text{ induces a clique.}$$

But the multivariate Normal is closed under marginalisation, so that both these densities are multivariate Normal with mean zero. For these to be identical, the quadratic forms in the exponent of the density have to be the same:

$$[x_a, S_{aa}^{-1} x_a] = [x_a, V_{aa}^{-1} x_a] \quad \text{for all } x_a.$$

This implies that the empirical and the estimated variances are the same. Hence the result. □

6.7 Direct and Indirect Estimates

We have deduced the likelihood equations in the last theorem but it is not immediately obvious how one might solve these. At the end of this section we apply the iterative proportional fitting procedure, as described by Speed and Kiiveri (1986) and implemented in MIM, Edwards (1987) and see Appendix A. To begin we look at examples of graphical models with and without direct estimates, the first of which is an instance of the simplest irreducible model.

The chordless 4-cycle

The complexity is illustrated by fitting the seemingly straightforward graphical model determined by the independence graph

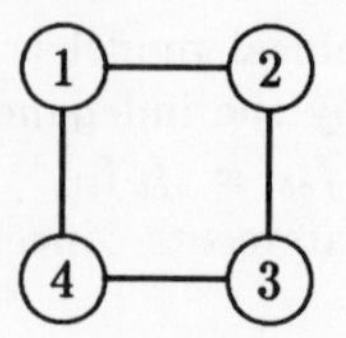

for the vector $X = (X_1, X_2, X_3, X_4)$. We wish to find the maximum likelihood estimate of V, the variance of X, under this model when the sample variance matrix is

$$S = \begin{pmatrix} 10 & & & \\ 1 & 10 & & \\ 5 & 2 & 10 & \\ 4 & 6 & 3 & 10 \end{pmatrix}.$$

The cliques in the graph are $\{1,2\}$, $\{2,3\}$, $\{3,4\}$ and $\{1,4\}$ and the likelihood equations imply that

$$\begin{pmatrix} \hat{v}_{11} & \\ \hat{v}_{12} & \hat{v}_{22} \end{pmatrix} = \begin{pmatrix} 10 & \\ 1 & 10 \end{pmatrix} \text{ and } \begin{pmatrix} \hat{v}_{22} & \\ \hat{v}_{23} & \hat{v}_{33} \end{pmatrix} = \begin{pmatrix} 10 & \\ 2 & 10 \end{pmatrix} \text{ and,} \ldots,$$

which can be collected together as

$$\begin{pmatrix} \hat{v}_{11} & & & \\ \hat{v}_{12} & \hat{v}_{22} & & \\ \hat{v}_{13} & \hat{v}_{23} & \hat{v}_{33} & \\ \hat{v}_{14} & \hat{v}_{24} & \hat{v}_{34} & \hat{v}_{44} \end{pmatrix} = \begin{pmatrix} 10 & & & \\ 1 & 10 & & \\ ? & 2 & 10 & \\ 4 & ? & 3 & 10 \end{pmatrix}.$$

Together with the equations corresponding to the independences in the graph namely $\hat{d}_{13} = 0$ and $\hat{d}_{24} = 0$, these determine the maximum likelihood estimators.

Consider their solution. Of the unknown variances, all apart from $\hat{v}_{13}$ and $\hat{v}_{24}$ can be extracted directly, and so there are in effect only two unknowns. By definition, if $\hat{d}_{13} = 0$, then element $(1,3)$ of the inverse of $\hat{V}$ is zero, equivalently its adjoint is zero, and equivalently the determinant

$$\det \begin{pmatrix} 1 & \hat{v}_{13} & 4 \\ 10 & 2 & \hat{v}_{24} \\ \hat{v}_{24} & 3 & 10 \end{pmatrix} = 0; \text{ also } \det \begin{pmatrix} 10 & \hat{v}_{13} & 4 \\ 1 & 2 & \hat{v}_{24} \\ \hat{v}_{13} & 10 & 3 \end{pmatrix} = 0$$

where the second determinant derives from $d_{24} = 0$. Expanding these determinants in terms of their elements gives the two non-linear simultaneous equations in two unknowns

$$\begin{aligned}\hat{v}_{13} &= (140 - 11\hat{v}_{24})/(100 - \hat{v}_{24}^2), \text{ and} \\ \hat{v}_{24} &= (100 - 11\hat{v}_{13})/(100 - \hat{v}_{13}^2).\end{aligned}$$

Substituting one equation into the other leads to a polynomial of degree 5, which guarantees the existence of at least one solution, albeit not analytic. Regarding these two equations as updating equations in an iterative algorithm leads to a procedure which converges rapidly from any feasible starting value. The numerical solution is

$$\hat{v}_{13} = 1.314 \text{ and } \hat{v}_{24} = 0.870.$$

The fitted variance matrix is then

$$\begin{pmatrix} 10.0 & & & \\ 1.0 & 10.0 & & \\ 1.314 & 2.0 & 10.0 & \\ 4.0 & 0.870 & 3.0 & 10.0 \end{pmatrix}.$$

However the model does not fit the data at all well. The observed and fitted partial correlations are

$$\begin{pmatrix} 1.0 & & & \\ -.225 & 1.0 & & \\ .448 & .124 & 1.0 & \\ .371 & .606 & .025 & 1.0 \end{pmatrix} \text{ and } \begin{pmatrix} 1.0 & & & \\ .070 & 1.0 & & \\ .000 & .183 & 1.0 & \\ .380 & .000 & .268 & 1.0 \end{pmatrix}$$

respectively; and the deviance of this model is 66.14 on 2df, which is computed from the expression for the deviance given in terms of the trace and log det.

Butterfly models

Suppose the vector X can be partitioned into three (X_a, X_b, X_c) and all conditional independence constraints can be summarised by the single statement $X_b \perp\!\!\!\perp X_c | X_a$.

Proposition 6.7.1 *The maximum likelihood estimate, $\hat{V}$, for the graphical model entirely specified by the conditional independence $X_b \perp\!\!\!\perp X_c | X_a$ is*

$$\hat{V} = \begin{pmatrix} S_{aa} & & \\ S_{ba} & S_{bb} & \\ S_{ca} & S_{ca}S_{aa}^{-1}S_{ab} & S_{cc} \end{pmatrix},$$

where S is the sample variance matrix of $X = (X_a, X_b, X_c)$ partitioned into conformable blocks. If X_a, X_b and X_c are p-, q- and r-dimensional respectively, then the deviance

$$\text{dev}\,(X_b \perp\!\!\!\perp X_c | X_a) = -N \log \frac{\det S \det S_{aa}}{\det S_{a \cup b, a \cup b} \det S_{a \cup c, a \cup c}}$$

has an asymptotic chi-squared distribution with qr degrees of freedom. When $q = 1 = r$ *this expression for the deviance simplifies to*

$$\text{dev}\,(X_b \perp\!\!\!\perp X_c|X_a) = -N\log(1 - \text{corr}\,_N^2(X_b, X_c|X_a))$$

where $\text{corr}\,_N(X_b, X_c|X_a)$ *is the sample partial correlation coefficient of* X_b *and* X_c *given* X_a.

Proof: The graph is of the form

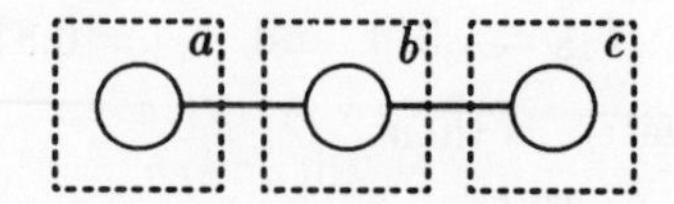

The zeros in the inverse variance form a rectangular array; that is, $D_{bc} = 0$. The subgraphs on $a \cup b$ and $a \cup c$ are both complete and as there can be no edge between an element of b and one of c, they are both maximally complete, that is cliques. Hence the likelihood equations are

$$\begin{aligned} \hat{V}_{a\cup b, a\cup b} &= S_{a\cup b, a\cup b}, \\ \hat{V}_{a\cup c, a\cup c} &= S_{a\cup c, a\cup c} \quad \text{and} \\ \hat{D}_{bc} &= 0. \end{aligned}$$

Corllary 6.3.4 established that $D_{bc} = 0$ if and only if $\text{cov}\,(X_b, X_c|X_a) = 0$ which, by the recurrence relationship for partial covariance, is equivalent to the condition

$$\text{cov}\,(X_b, X_c) = \text{cov}\,(X_b, X_a)\text{var}\,(X_a)^{-1}\text{cov}\,(X_a, X_c)$$

on the variance $V = \text{var}\,(X)$. Hence

$$\hat{V}_{bc} = \hat{V}_{ba}\hat{V}_{aa}^{-1}\hat{V}_{ac} = S_{ba}S_{aa}^{-1}S_{ac}$$

using the likelihood equations above. We have determined $\hat{V}$ completely.

The simple conditional independence structure of this model enables a simple derivation of the deviance. In Proposition 6.5.1 we may either simplify the general expression or use the representation of the deviance of a model M as a divergence:

$$\text{dev}\,(M) = 2N \min I(f^S; f^M)$$

where the minimum is taken over permissible f^M. Now when M corresponds to a conditional independence statement this minimum divergence exercise has been solved in Proposition 4.4.4; the result is

$$\text{dev}\,(M) = 2N\text{Inf}\,(X_b \perp\!\!\!\perp X_c|X_a),$$

where the information measure is computed with respect to f^S. Finally the evaluation of the divergence for multivariate Normal distributions at Corollary 6.4.4 allows us to express this as

$$\operatorname{dev}(M) = -N \log \frac{\det S \det S_{aa}}{\det S_{a\cup b, a\cup b} \det S_{a\cup c, a\cup c}}.$$

The statistic has an aysmptotic chi-squared distribution because of the results quoted in Section 6.1. When $q = 1 = r$ the simplification in terms of the sample partial corrrelation coefficient is given by the last part of Proposition 6.4.6. □

EXAMPLE 6.7.1 The mathematics marks again. We fit the model determined by the independence statement (mech,vect)⫫(anal,stat)|alg which has the graph

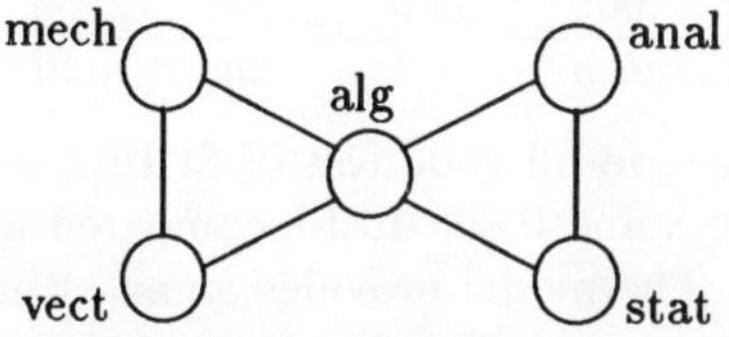

The sample size is $N = 88$ and the sample variance matrix is reproduced here from Table 1.1.2:

mech	302.29				
vect	125.78	170.88			
alg	100.43	84.19	111.60		
anal	105.07	93.60	110.84	217.88	
stat	116.07	97.89	120.49	153.77	294.37
	mech	vect	alg	anal	stat

Fitting this model produces the fitted variance matrix

mech	302.29				
vect	125.78	170.88			
alg	100.43	84.19	111.60		
anal	99.74	83.62	110.84	217.88	
stat	108.43	90.90	120.49	153.77	294.37
	mech	vect	alg	anal	stat

Only the covariances between (anal,stat) and (mech,vect) are different from the sample covariances. The sample and fitted correlations that differ are

anal	.409	.485
stat	.389	.436
	mech	vect

anal	.389	.433
stat	.363	.405
	mech	vect

The observed and fitted partial correlations are

mech	1.000				
vect	.329	1.000			
alg	.230	.281	1.000		
anal	- .002	.078	.432	1.000	
stat	.025	.020	.357	.253	1.000
	mech	vect	alg	anal	stat

and

mech	1.000				
vect	.332	1.000			
alg	.235	.327	1.000		
anal	- .000	.000	.451	1.000	
stat	- .000	.000	.364	.256	1.000
	mech	vect	alg	anal	stat

and compare well. The overall goodness of fit test for this model is the deviance which takes the value 0.895 to be compared against the chi-squared distribution with 4 df. The model provides an excellent fit to the data. □

Complete independence is a special case of the conditional independence model discussed above, where the graph splits into two disconnected complete subgraphs. We can derive the deviance for this model by supposing that a is empty in the expression for the conditional independence deviance: the deviance for testing the independence of X_c and X_b is

$$-N \log \det S / \det S_{bb} \det S_{cc},$$

where S_{bb} and S_{cc} are the diagonal blocks of the partition of S. This statistic has an asymptotic chi-squared distribution with qr df on the null hypothesis.

Iterative proportional fitting

When direct estimates are not available the iterative proportion fitting procedure discussed in Section 4.7 can be applied. Given two density functions, g^0 and f, for the k-dimensional random vector X, iterative proportional fitting is a procedure that finds a density function, g^∞, that has the same interaction structure as g^0, and the same marginal distributions as f on the subsets $\{a_1, a_2, \ldots, a_m\}$ of $K = \{1, 2, \ldots, k\}$. These subsets need not be disjoint but must not nest and must cover K. At the n-th step, the updating equation of the iteration is

$$g_{ab}^{n+1} = g_{b|a}^{n} f_a.$$

The subset b is the complement of a in K, $b = K \backslash a$, and at each step of the iteration, a cycles through the m marginal subsets so that $a = a_1$ when $n = 1$;

$a = a_m$ when $n = m$; $a = a_1$ when $n = m + 1$, and so on. So to be accurate, we should add the subscript n to the subset a in the updating equation.

Our discussion at the end of the chapter on information divergence proved that the ipf algorithm preserves all the higher order interactions of the density g^0, proved that if the algorithm converged then it would converge to a density function with the 'right' margins, and outlined a heuristic argument for its convergence. The basis of this argument is that at every step it decreases the divergence, and is thus an example of a cyclic descent algorithm. Here we discuss its application to solving the maximum likelihood equations for a Gaussian graphical model.

We have to evaluate the updating equation in the context of the multivariate Normal distribution. Most important are the closure properties of the distribution, needed at several points in the argument: (i) If f_{ab} is a Normal density function for (X_a, X_b) then the marginal density f_a is Normal and $f_{b|a}$ is Normal; and (ii) the converse, that if the marginal f_a is Normal and the conditional $g_{b|a}$ is Normal then the product, $f_a g_{b|a}$, is also Normal. Since application of the ipf algorithm only involves the operations of marginalising, conditioning and forming products, each member of the sequence $g^0, g^1, g^2, \ldots, g^n, \ldots$ is Normal.

To simplify the matter we suppose that both f and g^0 have mean 0; and we first wish to show that the expected value of X under g^{n+1} is also 0. The argument runs by induction. The updating equation implies

$$E^{n+1}_{ab}(X) = E^f_a[E^n_{b|a}(X)]$$

where the superscript on the expectation denotes which density is involved, and as before, the subscript indicates the marginal or the conditional density. Partition X into X_a and X_b. Firstly, $E^{n+1}_{ab}(X_a) = E^f_a[E^n_{b|a}(X_a)] = E^f_a(X_a) = 0$, by the assumption on f. Secondly, the conditional mean of the multivariate Normal distribution is the linear least squares predictor:

$$E^n_{b|a}(X_b) = \hat{X}_b(X_a) = B^n_{b|a}(X_a - E^n_a(X_a)),$$

where

$$B^n_{b|a} = \text{cov}^n(X_b, X_a)\text{var}^n(X_a)^{-1}$$

is the matrix of prediction coefficients. Hence $E^{n+1}_{ab}(X_b) = B^n_{b|a}(E^f_a(X_a) - E^n_a(X_a))$, and both these expectations are zero: the first by the assumption on f, and the second by the induction hypothesis.

Thus to specify g^{n+1} completely we only have to update the variance. Under f the variance matrix of X_a is V^f_{aa}, the leading block on the diagonal of V^f. At the n-th step the conditional distribution of X_b given X_a is $N(B^n_{b|a}X_a, V^n_{bb|a})$. We may combine this information to get the variance at the step $n + 1$. The formula discussed at the end of Section 5.1, expresses the variance in terms of conditional means and variances, heuristically: var = E(conditional var) + var(conditional E). Application here leads immediately to

$$V^{n+1} = \begin{pmatrix} V^f_{aa} & V^f_{aa}(B^n_{b|a})^T \\ B^n_{b|a}V^f_{aa} & V^n_{bb|a} + B^n_{b|a}V^f_{aa}(B^n_{b|a})^T \end{pmatrix},$$

and the problem is solved. To summarise, to calculate V^{n+1} from V^n:

(i) partition V^n according to a and $b = K\backslash a$;

(ii) compute the prediction coefficients $B^n_{b|a}$ and the partial variance $V^n_{bb|a}$ from V^n;

(iii) together with V^f_{aa}, substitute in the updating formula for the variance.

This formulation involves the inversion of V^n_{aa}, a matrix of order $|a|\times|a|$.

EXAMPLE 6.7.2 Suppose we wish to fit the chordless 4-cycle

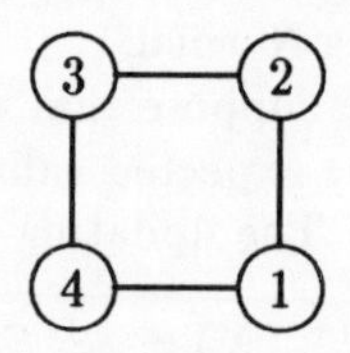

with cliques $a = \{1,2\}, \{2,3\}, \{3,4\}, \{1,4\}$, and that the sample variance matrix is

1	10.0			
2	1.0	10.0		
3	5.0	2.0	10.0	
4	4.0	6.0	3.0	10.0

We begin by choosing the variance matrix for g^0 to be the identity. Application of the updating equation cycling over the four cliques of the graph leads rapidly to V^∞.

1	10.000			
2	1.000	10.000		
3	1.314	2.000	10.000	
4	4.000	0.870	3.000	10.000

1	0.120			
2	−0.008	0.105		
3	−0.000	−0.020	0.114	
4	−0.047	−0.000	−0.032	0.129

The second matrix here, is the inverse D^∞. □

A second and more striking implementation of the procedure, is to update the inverse variance. The updating equation for D^{n+1} can be derived directly from the identity in quadratic forms implied by $g_{ab}^{n+1} = g_{b|a}^{n} f_a$, or by applying the inverse variance lemma to invert the updating equation for V^{n+1}. We choose this latter route. As

$$D^n = \begin{pmatrix} (V_{aa}^n)^{-1} + (B_{b|a}^n)^T (V_{bb|a}^n)^{-1} B_{b|a}^n & -(B_{b|a}^n)^T (V_{bb|a}^n)^{-1} \\ -(V_{bb|a}^n)^{-1} B_{b|a}^n & (V_{bb|a}^n)^{-1} \end{pmatrix}$$

inverting the updating equation gives

$$D^{n+1} = D^n + \begin{pmatrix} (V_{aa}^f)^{-1} - (V_{aa}^n)^{-1} & 0 \\ 0 & 0 \end{pmatrix}.$$

Surprisingly, at the n-th step only the block corresponding to a needs adjusting. However, this apparently more economical form of the updating equation is actually illusory, because at each stage we need to compute $(V_{aa}^n)^{-1}$ from the current estimate of the inverse variance D^n. Application of the inverse variance lemma to D^n shows that $(V_{aa}^n)^{-1} = D_{aa|b}^n$, and evaluation of this requires inverting D_{bb}, a matrix of order $|b| \times |b|$.

6.8 The Analysis of Deviance

The deviance is the essential way to to measure the overall goodness of fit of a graphical model, and the natural way to compare two models when one is nested in the other.

The deviance

The variance V is symmetric and positive definite, but otherwise arbitrary. The likelihood function assigns a relative ordering to each value of V; the higher the likelihood the more plausible the value. When V is unconstrained we have seen that the largest value occurs at $V = S$ and the difference between the log-likelihood at this point and at other values is

$$l(S) - l(V) = NI(f^S; f^V).$$

The constant term in the log-likelihood function has disappeared, suggesting this is a better scale of measurement because of the natural zero.

A graphical model, M, is the family of all multivariate Normal distributions constrained to have the same independence graph. The *saturated model* is the maximal graphical model with an independence graph that is complete, and contains all Normal distributions with k vertices.

Definition. The *deviance* of a model M, dev (M), is twice the difference between the the unconstrained maximum of the log-likelihood and the maximum taken over M. □

This generalised likelihood ratio test, the deviance, of a graphical model M can be expressed as a minimal divergence.

Proposition 6.8.1 Deviance and divergence for the multivariate Normal. *The deviance of a class of models M, determined by constraints on the variance V of the multivariate Normal distribution, is given by*

$$\text{dev}\,(M) = 2N \min I(f^S; f^V)$$

where $If^S; f^V)$ is the information divergence between two mean zero multivariate Normal distributions with variances S and V respectively, and the minimum is taken over all V subject to the constraint that $f^V \in M$.

Proof: From the definition dev (M) is

$$\begin{aligned} 2\max_V l(V) - 2\max_{V:f^V \in M} l(V) &= 2l(S) - 2\min_{V:f^V \in M} l(V) \\ &= 2N \min_{V:f^V \in M} I(f^S; f^V). \end{aligned}$$

□

The picture in Figure 6.8.1 gives an idea of the relationships.

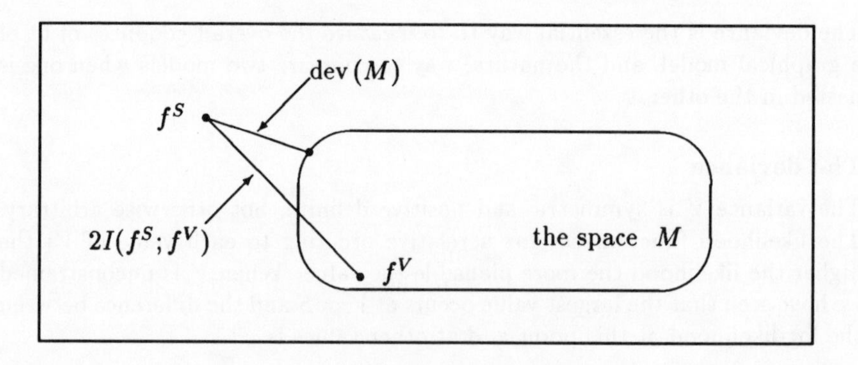

Figure 6.8.1: The space of values for V.

Maximum likelihood estimation is an important tool of statistical inference but a detailed description of its technical definition and of the sampling distribution properties would take us too far afield. Hence we only quote one sampling result of importance. For a proof, refer to Cox and Hinkley (1974) or Andersen (1980).

Proposition 6.8.2 The sampling distribution of the deviance. *Subject to certain regularity conditions on M, under the null hypothesis that f^V lies in the subclass M, the deviance has an asymptotic chi-squared distribution with degrees of freedom given by the number of constraints set on V to ensure f^V lies in M.*

If X is k-dimensional then the degrees of freedom for any graphical model M is just $\binom{k}{2}$ less the number of excluded edges, because excluding an edge corresponds to setting exactly one free parameter in the inverse variance matrix to zero.

An adjustment to the deviance made by multiplying by a scale factor, the *Bartlett correction factor*, improves the asymptotic chi-squared approximation to the exact sampling distribution of the deviance. The technique works for decomposable models, as these allow analytic expressions for the deviance and similar Bartlett corrections can be made for decomposable log-linear models. We shall postpone details until the last section of the chapter but just remark that the main conclusion is that observed P-values calculated without adjustment are slightly too high. In fact in certain graphical models, rather better results are possible, because a simple transformation of the deviance has an exact F-distribution, see Section 10.5, though this does not apply to the log-linear model. Overall it does not seem worthwhile to sacrifice the simplicity and unity of the procedures for borderline improvements.

Partitioning the deviance

When one model is nested within another the deviance difference between the two models is a generalised log-likelihood ratio and has an asymptotic chi-squared distribution on the null hypothesis.

EXAMPLE 6.8.1 A model for pitprops. Jeffers (1967) introduced a data set to the statistical literature concerning measurements taken on pit props and the relationship these variables had to the maximum compressive strength of the timber. The variables are

dia	top diameter of the prop;
lng	length of the prop;
wet	moisture content of the prop, as a percentage of the dry weight;
sg1	specific gravity of the timber at the time of the test;
sg2	oven-dry specific gravity of the timber;
rgt	number of annual rings at the top of the prop;
rgb	number of annual rings at the base of the prop;
bow	maximum bow;

lnt distance of the point in maximum bow from the top of the prop;
whl number of knot whorls;
lnc length of the clear prop from the top of the prop;
knt average number of knots per whorl;
dkn average diameter of the knots.

Table 6.8.1: The correlation matrix of 13 measurements on 180 Corsican pit props. From Jeffers (1967).

	dia	lng	wet	sg1	sg2	rgt	rgb	bow	lnt	whl
dia	1.0									
lng	.954	1.0								
wet	.364	.297	1.0							
sg1	.342	.284	.882	1.0						
sg2	-.129	-.118	-.148	.220	1.0					
rgt	.313	.291	.153	.381	.364	1.0				
rgb	.496	.503	-.029	.174	.296	.813	1.0			
bow	.424	.419	-.054	-.059	.004	.090	.372	1.0		
lnt	.592	.648	.125	.137	-.039	.211	.465	.482	1.0	
whl	.545	.569	-.081	-.014	.037	.274	.679	.557	.526	1.0
lnc	.084	.076	.162	.097	-.091	-.036	-.113	.061	.085	-.319
knt	-.019	-.036	.220	.169	-.145	.024	-.232	-.357	-.127	-.368
dkn	.134	.144	.126	.015	-.208	-.329	-.424	-.202	-.076	-.291

	lnc	knt	dkn
lnc	1.0		
knt	.029	1.0	
dkn	.007	.184	1.0

The deviance against complete independence is 1934.74 on 78 df and the independence graph has 13 vertices and 78 possible edges. An approach to models with large numbers of variables is to insist on a simple block structure, in which intra-block variables are adjacent and inter-block variables are adjacent if the blocks are adjacent. On the basis of our knowledge of the variables we divide them into the following five cognate groups: a_1={dia, lng, lnt}, a_2={bow, knt}, a_3={whl, lnc, dkn}, a_4={rgt, rgb}, and a_5={wet, sg1, sg2}. Consider the model $M_1 = a_1.a_3 + a_2.a_3 + a_3.a_4 + a_4.a_5$ corresponding to the graph

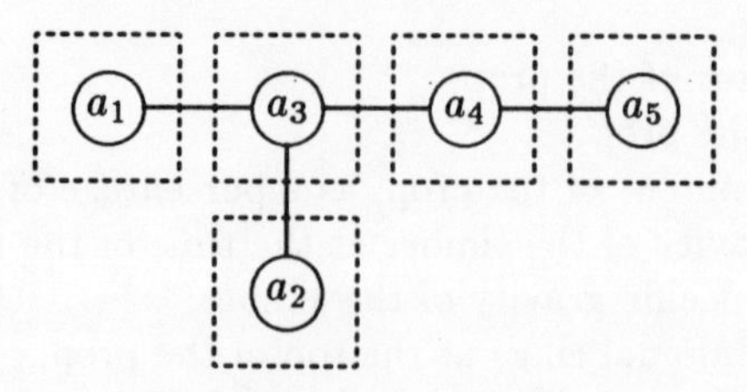

The deviance of this model is 157.21 with 40 df, so that we have fitted a model which uses 38 of the possible 78 parameters but accounts for 1-

157.21/1934.74= 91.9% of the total variation.

Consider the improvement made by including a further edge between groups a_4 and a_2, which has model formulae $M_2 = a_1.a_3+a_2.a_3+a_3.a_4+a_2.a_4+a_4.a_5$. It has a deviance of 144.34 on 36 df. The analysis of deviance table summarises these results:

attribution	dev	df
model M_1	1777.53	38
interaction $a_2.a_4$	12.87	4
residual from M_2	144.34	36
total	1934.74	78

Because M_1 is nested in M_2, the difference in deviance of 12.87 on 4 df can be compared to chi-squared tables; it has a P-value of 0.0119 . Though this and the residual from M_2 is significant compared to the critical value of the relevant chi-squared distribution, both terms are relatively small in relation to the rest of the variability. □

Edge exclusion deviances

When the random vectors X_b and X_c are both one dimensional the deviance given in Proposition 6.7.1 against their conditional independence is

$$-N\log(1 - \text{corr}_N(X_i, X_j \mid \text{rest})^2),$$

where $\text{corr}_N(X_i, X_j|\text{rest})$ is the observed partial correlation coefficient. It is an *edge exclusion deviance* as it is the appropriate deviance for testing if a single edge can be eliminated from the saturated model.

EXAMPLE 6.8.2 The mathematics marks. The edge exclusion deviances, computed directly from the partial correlation coefficients, are

	mech	vect	alg	anal	stat
mech	*				
vect	10.10	*			
alg	4.80	7.23	*		
anal	0.00	0.54	18.16	*	
stat	0.05	0.04	11.98	5.81	*

Each deviance sets one element of the inverse variance to zero, $d_{ij} = 0$, so the appropriate chi-squared distribution has one degree of freedom. If we exclude all edges with deviances that do not reach 3.84, the 5% point of the chi-squared distribution, then, fortunately, we arrive at the same conclusion as obtained by our first analysis of this data set in the introduction. □

There are several problems raised by this multiple testing procedure. Though each test is made at the nominal 5% significance level, the overall test has a much higher, and unknown, significance level. A second issue is that if the test for a particular $d_{ij} = 0$ is accepted, should this information be used in testing for other edges? Some of these issues are discussed in Chapter 9 on model selection.

We noted in the previous section that some models have direct estimates. These are entirely characterised as those graphical models with triangulated independence graphs, the decomposable models, and their theory is taken up in Chapter 12, where it is more generally shown that the estimation and testing problems are simplified whenever the model is reducible.

6.9 Wishart, Bartlett and Mahalanobis

The aim of this section is no more than to draw the reader's attention to some further specialised topics.

The Wishart Distribution

For completeness, we present without derivation, the density function of the Wishart distribution. Its importance rests in the fact that it is the exact sampling distribution of the sample variance matrix, the sufficient statistic for the parameter V in random samples from the multivariate Normal distribution. Pre-supposing a glance at the next chapter, we remark that while the multivariate Normal distribution is the continuous analogue of the cross-classified Multinomial distribution of size 1 which underlies graphical models for categorical observations, it is the Wishart distribution that parallels the cross-classified Multinomial distribution of size N; the latter is the sampling distribution of the table of counts, the sufficient statistic for the parameters of the graphical log-linear model.

In this section alone, the definition of the $k \times k$ sample variance matrix is modified to $S = n^{-1} \sum_{l=1}^{N} (X^l - \overline{X})(X^l - \overline{X})^T$, by changing the denominator to $n = N - 1$. The density function of the Wishart distribution is

$$f(S; D) = \frac{\det(D)^{n/2} \det(S)^{(n-k-1)/2}}{\Gamma_k\left(\frac{n}{2}\right)} \left(\frac{n}{2}\right)^{nk/2} \exp\{-\frac{n}{2}\mathrm{tr}\,(DS)\},$$

where $\Gamma_k(.)$ represents the multivariate gamma function, $D = V^{-1}$ and the support of f is determined by the constraint that the matrix S is positive definite, see Muirhead (1982) for more details.

Bartlett correction factors

Bartlett (1937, 1954) showed how the likelihood ratio statistic can be altered to improve the chi-squared approximation to its distribution. This was generalised by Lawley (1956) and applied by Williams (1976) to contingency tables. Our exposition is based on Porteous (1985ab) who developed the technique for graphical Gaussian models.

The method is based on a representation of the sampling distribution of the determinant of the sample variance matrix. Firstly, a series expansion for the expected value of the deviance in terms of the sample size is found; and then the leading two terms in this expansion are used to give the multiplying factor for the deviance, so chosen that the corrected statistic has an expected value that agrees with that of the corresponding chi-squared distribution to order N^{-1}. It only works for decomposable graphical models because only these admit an explicit expression for the deviance in terms of determinants, see Chapter 12.

We illustrate the derivation of the Bartlett correction for the model characterised by the conditional independence of X_b and X_c given X_a where, as usual, suppose that (X_a, X_b, X_c) is $p+q+r$-dimensional. The deviance of this model can be written as the logarithm of a ratio of sample variances.

We state two results without proof, the first is a property of the Wishart distribution:

Lemma 6.9.1 *(a) If S is the sample variance matrix of a k-dimensional Normal sample with variance V based on n df, then its determinant has the representation*

$$n \det(S)/\det(V) \stackrel{d}{=} \prod_{i=1}^{k} U_i,$$

where the U_i's are independently chi-squared distributed with $N-i+1$ degrees of freedom respectively, and $\stackrel{d}{=}$ indicates 'equal in distribution'. See Rao (1973, p540).

(b) If U is a chi-squared variate with n degrees of freedom then, for large n,

$$E\log(U) = \log(n) - 1/n + O(n^{-2}).$$

We can now prove

Proposition 6.9.2 *The Bartlett correction factor to apply to the deviance of the conditional independence model $X_b \perp\!\!\!\perp X_c | X_a$ is*

$$1 - \frac{1}{n}(p + 1 + \frac{1}{2}(q + r + 1)).$$

Proof: The deviance against conditional independence, $\operatorname{dev}(X_b \perp\!\!\!\perp X_c | X_a)$, is

$$-n\log \frac{\det S \det S_{aa}}{\det S_{a\cup b,a\cup b} \det S_{a\cup c,a\cup c}}.$$

Under the independence hypothesis

$$\frac{\det V \det V_{aa}}{\det V_{a\cup b,a\cup b} \det V_{a\cup c,a\cup c}} = 1,$$

so that we may ignore the true variance. Using the lemma the expected value is

$$(\sum_{i=1}^{p+q} + \sum_{i=1}^{p+r} - \sum_{i=1}^{p+q+r} - \sum_{i=1}^{p}) \log (N+i-1) - 1/(N-i+1) + O(N^{-2})$$

which simplfies to the expression given. □

The Bartlett factor ensures that the first moment of the corrected likelihood ratio test statistic agrees with that of the approximating chi-squared distribution to order N^{-1}. The fact that the correction factor is necessarily less than unity suggests that the uncorrected deviance is stochastically too large, and hence that the corresponding tests will be rejected more often than the stated significance level suggests. In the sampling experiment reported by Porteous, with $p = 2$, $q = 1 = r$, and a nominal 5% level, the actual level of the uncorrected statistic was 11.2% when the sample size was 10, and 7.1% when the sample size was 25. The corresponding figures for the corrected statistic were 4.9% and 5.0% respectively. Increasing the dimension of the conditioning set leads to larger correction factors.

The deviance for any decomposable model can be calculated in a similar manner. The above proof depends on the deviance factorising explicitly in terms of determinants of sample variances, and the postulated variance factorising in the same way.

Diagnostics: Mahalanobis deviates

At some point or other in the analysis of a data set the modeller is bound (even honour bound) to query the assumptions that underly the techniques of the statistical analysis. The graphical modeller is no different and, especially when substantive conclusions seem to hang on the interpretation of a fitted graph, the analyst requires to test the fundamental assumption of multivariate Normality that underpins graphical Gaussian models.

The study of residuals and more general diagnostic techniques have made great strides with the development of fast computing facilities. Reviews of currently available methods can be found in Barnett and Lewis (1978), Cook and Weisberg (1982), and Atkinson (1985). Though there is a great deal of material based on tests for univariate Normality related to diagnostics from the linear regression model there is less on multivariate generalisations relevant to graphical modelling. Clearly diagnostic tests for outliers, for contamination, and for skewness, as well as transformations to improve Normality, are needed. This section makes no pretence to survey the field; its aim is just to make the reader aware of the issue and to give an illustration of one type of check on multivariate Normality.

One technique is derived from the following standard argument from statistical distribution theory: if the random scalar $X \sim N(0,1)$ then X^2 has the chi-squared distribution with 1 df; and if X is a k-dimensional Normal random vector with mean μ and variance V then the weighted sum of squares

$$[(X-\mu), V^{-1}(X-\mu)]$$

has a chi-squared distribution with k df. With an independent random sample $X^1, X^2, \ldots, X^N$ the unknown mean and variance, (μ, V), is replaced by the sample mean and variance, $(\bar{X}, S)$, so that

$$Z^l = [(X^l - \bar{X}), S^{-1}(X^l - \bar{X})] \quad \text{for} \quad l = 1, 2, \ldots, N.$$

It is appropriate to refer to these quantities as *Mahalanobis deviates* because of the metric that goes under that name, and an index plot of these deviates may identify those observations that lie far from the central values.

This raises certain questions: though the marginal distribution of Z^l is approximately the same independent chi-squared distribution when N is large enough, on *average* the Z^l's will be smaller and have less variation than had the parameters μ and V be known exactly; thus tests based on the chi-squared distribution are thought to be somewhat conservative. Secondly, rather than the chi-squared distribution it is the distribution of the largest deviate that is needed for exact tests, see Barnett and Lewis (1978). Thirdly, if $\hat{V}$ is the fitted variance from a graphical model it is not clear if the Mahalanobis deviates should be computed with S or the estimate $\hat{V}$. There are arguments to support both contentions.

EXAMPLE 6.9.1 The mathematics marks. The index plot of the $N = 88$ Mahalanobis deviates for the mathematics marks using the observed sample mean and variance, is given in Figure 6.9.1.

There is noticeably less variation in the centre of the plot which is a consequence of the more or less ordered nature of the observations in Table 1.1.1. At first sight Figure 6.9.1 suggests that observation 81 appears be rather large.

However the histogram

interval	frequency	
0- 2	11	+++++++++++
- 4	28	++++++++++++++++++++++++++++
- 6	25	+++++++++++++++++++++++++
- 8	10	++++++++++
-10	7	+++++++
-12	5	+++++
-14	1	+
-16	0	
-18	1	+
total	88	

suggests the value 17.99 is entirely consistent with a chi-squared distribution on 5 degrees of freedom. Our analysis reaches the conclusion that this set of Mahalanobis deviates does not conflict with the assumption of multivariate Normality.

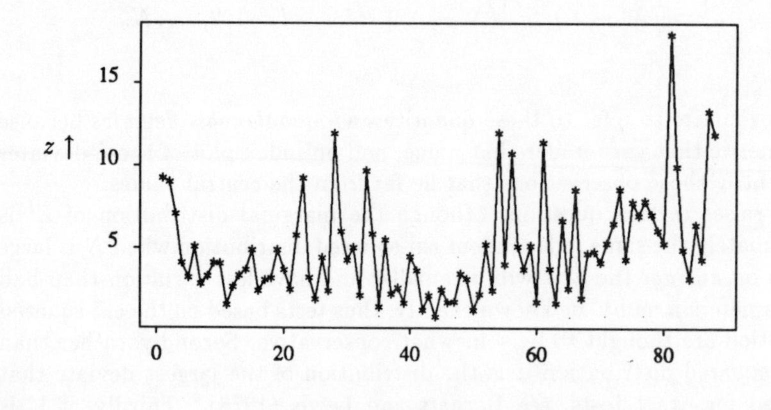

Figure 6.9.1: Index plot of Mahalanobis deviates for the mathematics marks.

6.10 Exercises

1: If $Y = \beta X + E$ where β is fixed and $E \sim N(0,1)$ independently of $X \sim N(0,1)$ then write down the conditional probability density function of Y given X, the joint probability density function of Y and X and finally the marginal probability density function of Y.

2: Write down the conditional probability density function of Y given X, when the partitioned random vector (X, Y) has a multivariate Normal distribution. The random Normal vector (X_1, X_2, Y_1, Y_2) is four dimensional with zero mean and variance

$$\begin{pmatrix} 1 & 0 & 0 & a \\ 0 & 1 & 0 & b \\ 0 & 0 & 1 & c \\ a & b & c & 1 \end{pmatrix}$$

Find the conditional distribution of $Y = (Y_1, Y_2)$ given $X = (X_1, X_2)$ and the partial correlation of Y_1 and Y_2 given X_1 and X_2.

3: The variance of the Normal random vector X is V; describe the structure of its inverse when the associated independence graph is

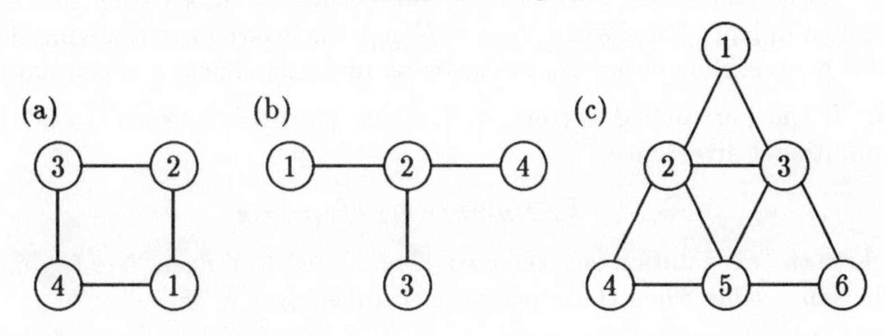

4: The scalar random variables X_1, X_2, X_3 and X_4, are independent, and Normal with zero mean and unit variance. The vector $Y = (Y_1, Y_2, Y_3, Y_4)$ is related to the X_i's by the following equations:

$$Y_1 = X_1,\ Y_2 = Y_1 + X_2,\ Y_3 = Y_2 + X_3 \text{ and } Y_4 = X_2 + X_4.$$

Construct and interpret the independence graph of Y. Find the least squares predictor of Y_3 from (Y_1, Y_2, Y_4).

5: The random vector Z is two dimensional Normal with mean zero and variance I. The vector Y is four dimensional and, conditional on Z, has a Normal distribution with mean BZ and variance I. Evaluate var(Y, Z) and show that

$$\text{var}\,(Y, Z)^{-1} = \begin{pmatrix} I & -B \\ -B^T & I + B^T B \end{pmatrix}.$$

If

$$B = \begin{pmatrix} 1 & 0 \\ 1 & 0 \\ 1 & 1 \\ 0 & 1 \end{pmatrix} \text{ show that var}\,(Y)^{-1} = 1/11 \begin{pmatrix} 8 & -3 & -2 & 1 \\ -3 & 8 & -2 & 1 \\ -2 & -2 & 6 & -3 \\ 1 & 1 & -3 & 7 \end{pmatrix}$$

given from the formula var$(Y)^{-1} = I - B(I + B^T B)^{-1} B^T$. Construct the independence graphs of Y and of (Y, Z) and comment.

6: The p-dimensional vector $X \sim N(0, V)$ and $Z(v) = [v, X]/[v, Vv]^{1/2}$ for a fixed vector v (note $Z \sim N(0,1)$). Use the Cauchy-Schwarz inequality to show that $P(|Z(v)| \leq c$ for all v in R^p) where c is any fixed scalar, can be computed from the chi-squared tables with p df.

7: The variance $V = \text{var}(X)$ has eigen values $\lambda_1, \lambda_2, \ldots, \lambda_k$; show that the divergence $2I(\text{var}(X); I_k)$ in Corollary 6.4.4 where I_k is the $k \times k$ identity matrix, can be expressed as $\sum_1^k \{\lambda_i - \log \lambda_i - 1\}$.

8: Prove that the divergence between two arbitrary multivariate Normal densities, f and g, is invariant to location or scale changes; that is

$$2I(f; g \| AX + b) = 2I(f; g \| X),$$

if A is non-singular. Furthermore, show that the information $\text{Inf}(X_a \perp\!\!\!\perp X_b)$ against independence when $X_{a \cup b} \sim N(\mu, V)$ is invariant to changing the value of μ or to scaling either X_a or X_b by an invertible linear transformation.

9: If the partitioned vector (X, Y, Z) is multivariate Normal evaluate the conditional divergence

$$E_{Y,Z|X} \log f_{Y,Z|X} / f_{Y|X} f_{Z|X},$$

where the expectation is taken over Y and Z with X fixed. Verify that it has the same value when the expectation is taken over X as well.

10: If (X, Y) is Normal and Y is one dimensional show that $2\text{Inf}(X \perp\!\!\!\perp Y) = -\log(1 - R^2)$ where R is the multiple correlation coefficient between X and Y.

11: Suppose that β and σ^2 are fixed scalars, that h is a density for X with mean 0 and finite variance, that Y is one dimensional and that the joint density of (X, Y) is specified by

$$f : Y|X = x \sim N(\beta x, \sigma^2) \text{ and } X \sim h_X, \text{ and } g : Y|X \sim N(0, \sigma^2) \text{ and } X \sim h_X.$$

Show that $2I(f; g) = R^2/(1 - R^2)$. Contrast this to the result of the previous question.

12: In the intraclass correlation model $X = Z1 + Y$ where Y is p-dimensional $N(0, I)$ and Z is 1-dimensional $N(0, \theta)$. Show that $\text{var}(X) = \theta 11^T + I$. Find the inverse variance D. Suppose that X alone is observed, write down the likelihood function of the parameter θ.

13: The sample variance matrix based on $N = 500$ observations is

$$S = \begin{pmatrix} 10 & 6 & 1 \\ 6 & 10 & 3 \\ 1 & 3 & 10 \end{pmatrix}.$$

Test if the third variable is independent of the previous two.

14: The mathematics marks. As a somewhat formal exercise verify that the deviance against the independence of X_a and X_b when a ={mech,vect,alg} and b ={anal,stat} is

15: Show that for the simple graphical model in the detailed example of Section 6.6 the maximum likelihood estimators of the elements of the inverse variance $D = \{d_{ij}\}$ are

$$\begin{aligned}
d_{23} &= 0, \\
d_{22} &= 1/(s_{22} - s_{12}^2/s_{11}), \\
d_{33} &= 1/(s_{33} - s_{13}^2/s_{11}), \\
d_{12} &= -s_{12}/s_{11}d_{22}, \\
d_{13} &= -s_{13}/s_{11}d_{33}, \\
d_{11} &= (1 - s_{12}d_{12} - s_{13}d_{13})/s_{11}.
\end{aligned}$$

16: Show that the Mahalanobis deviates can be derived as deletion residuals from the determinant of the sample variance. Hint: take $\bar{x} = 0$, for simplicity, and define S_j as the sample variance $1/N \sum_l x^l(x^l)^T$ where the sum excludes $l = j$. Then use Schur's identity to show that

$$\det(S_j)/\det(S) = 1 - 1/NS_j^T S^{-1} S_j.$$

17: Repeat the diagnostic test for the mathematics marks data based on the Mahalanobis deviates :

(a) for the first half of the sample;

(b) compare the results with that calculated from a random half of the sample;

(c) repeat the analysis on the sub-vector a ={mech,vect}.

Chapter 7

Graphical Log-linear Models

This chapter describes and fits graphical models to multi-way contingency tables based on sampling from the cross-classified Multinomial distribution. It turns out that these are a subclass of hierarchical log-linear models, specified by parameterising the density function in terms of the coefficients of its log-linear expansion, the u-terms. The development of this parameterisation for cross-classified tables of probabilities, colloquially known as log-linear models, was pioneered by Birch (1963), Goodman (1970), Haberman (1974), and Bishop, Fienberg and Holland (1975). Log-linear modelling has proved itself to be an extremely useful technique because of the way it arms the data analyst with a reliable method of extracting information from complex contingency tables.

Log-linear models: Firstly, the Bernoulli distribution introduced in Chapter 2, is generalised to the cross-classified Multinomial distribution of size 1. This generalisation goes in two directions: to an arbitrary number of variates rather than two or three, and to an arbitrary number of levels rather than two. From there, a similar route to our discussion of the Gaussian graphical model is taken: we examine the closure properties of the Multinomial distribution and then contrast two distinct parameterisations of the density function: the table of probabilities p and the u-terms. We relate conditions on these parameters to specifications on the independence graph, and consequently define a log-linear graphical model by setting certain u-terms to zero. The set of graphical models is located within the larger family of hierarchical log-linear models and the lattice structure of these models is briefly explored.

Naive modelling: A naive approach to parameter estimation and model selection might be to adopt the following program:

(i) reduce the full set of N observations to the k-way contingency table of interest by tabulation;

(ii) calculate the u-terms of the full log-linear expansion from an inversion of this observed table;

(iii) judge their size and decide which are negligible and which are not; and finally

(iv) draw conclusions as to the independence and strength of association between the variables cross-classifying the table.

In fact this is not so far off the mark, but it faces the difficulty that it is hard to decide how small a u-term has to be to be negligible, and even harder to decide if a whole collection of u-terms is negligible. Even should this be overcome, the procedure has to then decide how to estimate the retained u-terms; using the estimate from the empirical log-linear expansion is a rather poor method. The issues of model fitting and model selection have become entwined and need to be separated.

Likelihood inference: A systematic inferential procedure is based on the likelihood function composed from a random sample of independent and identically distributed observations on a cross-classified Multinomial distribution. Representation of the function as an information divergence, gives a neat way to find the maximum, an alternative is a straightforward application of Lagrange multipliers. Both methods lead to a set of likelihood equations summarised by the slogan 'observed = fitted', for all margins of the table corresponding to u-terms present in the model. The divergence derivation shows how the iterative proportional fitting algorithm can be used to compute parameter estimates.

The deviance, or more explicitly, twice the maximised log-likelihood ratio test statistic for the test of a specified model against the saturated model, is shown to simplify to

$$\text{dev}\,(M) = 2 \sum_{\text{cells}} \text{obs} \log \frac{\text{obs}}{\text{fv}\,(M)},$$

which directly measures the divergence between the observed values and the fitted values, $\text{fv}\,(M)$, throughout the whole table. The deviance attributable to excluding an edge from the complete graph, the edge exclusion deviance is an important tool for graphical model selection.

An essential difference between graphical Gaussian and graphical log-linear models is that pairwise conditional independence is dictated by a single parameter in the former while in the latter it is associated with several parameters. This allows possibly interesting ways of partitioning the deviance into meaningful components and two are discussed in this chapter: partitioning along chains in a lattice, and partitioning according to the values of the conditioning variable. Finally, the chapter ends with a brief discussion of diagnostic procedures.

Related material: We draw the reader's attention to related material discussed in other chapters: firstly, the sampling distribution of the sufficient statistic from a sample of size 1 is the cross-classified Multinomial distribution of size N. This is the correct parallel to the Wishart distribution under multivariate Normal sampling. It is discussed in Chapter 9 along with the Hypergeometric sampling distribution in our discussion of exact testing procedures.

Secondly, the issue of log-linear model selection is the same as the corresponding problem for graphical Gaussian models and the topics are treated together in Chapter 8.

Thirdly, the relation between logistic regression and log-linear models is an instance of the difference between modelling a joint distribution and modelling a conditional distribution, and is deferred to Chapter 10 on models with explanatory variables.

Finally, some rather more general issues are raised but are not resolved. For example: conditions for the existence of direct estimates, the decomposable models; conditions for collapsibility, so that inference can be conducted in a margin of the table, rather than the full table; questions concerning the general issue of maximum likelihood estimation: existence, uniqueness, consistency and efficiency. Explicit expressions for the asymptotic approximations to the standard errors of the parameter estimates are ommitted, see Lee (1977). Answers to all are known to some degree, but a concise treatment would place itself in the more general setting of exponential family models, and regard the log-linear and Gaussian models as special cases.

7.1 The Cross-classified Multinomial Distribution

A standard way to refer to an element of a table of probabilities classified by four factors is to denote it by

$$p_{ijkl}$$

but evidently this is not a sensible way to talk about k-dimensional tables where k is arbitrary. Our solution is to use the coordinate projection notation introduced in Chapter 1, and it is important that the reader feels comfortable with these conventions.

Let $K = \{1, 2, \ldots, k\}$ denote the set of indices that classify the table, and x_i denote the possible values taken by the i-th variable. Then $x =$

$(x_1, x_2, \ldots, x_k)$ denotes a particular cell in the table and $X = (X_1, X_2, \ldots, X_k)$ denotes a k-dimensional discrete observation or random vector. To emphasise the dimensionality both X and x are sometimes subscripted by the set K as X_K and x_K. The probability associated with each cell of the table is $p(x) = \text{Prob}(X = x)$. For instance, if an individual is randomly selected from a larger population then the probability that the individual is characteristised as a male, a smoker and university educated might be written as $p(1, 0, 4)$.

In parallel to our analysis of the multivariate Normal distribution results are conveniently expressed in terms of the partitioned observation $X = (X_a, X_b)$, of p and q dimensions respectively. The marginal subvector X_a has indices in the subset $a \subseteq K$ and is defined as the coordinate projection $X_a = (X_i; i \in a)$. The values, x_a, taken by this subvector are cells in a marginal table, and the associated marginal table of probabilities is $p_a(x_a)$ or p_a for short. If the full vector $X = X_K$ has the table of probabilities $p = p_K$ then the marginal table for X_a is

$$p_a(x_a) = \sum_{x_b} p_K(x_a, x_b).$$

We shall use the same convention for tables of counts, so that $n(x)$ denotes the count in cell x of the full table while $n_a(x_a)$ denotes the count in cell x_a of the marginal table.

When each variable is binary the cell structure of the table is simple: the Bernoulli random variable takes values in the set $\{0, 1\}$, and its k-dimensional generalisation, the multivariate Bernoulli random vector, $X = (X_1, X_2, \ldots, X_k)$ takes values in the Cartesian product $\{0, 1\}^k$ of the set $\{0, 1\}$ with itself. Geometrically, it is the set consisting of the 2^k vertices of a k-dimensional hypercube. Though a case could be made for referring to the multivariate Bernoulli distribution as the 'Multinoulli', it will not be espoused here.

Because many categorical variables take more than two responses we need to generalise to variables that take values in a finite set. Hence we suppose each coordinate, X_i, of the Multinomial random vector X takes a value in a set $\{0, 1, 2, \ldots, r_i - 1\}$ and the k-dimensional Multinomial random vector $X = (X_1, X_2, \ldots, X_k)$ takes values in the Cartesian product of these sets. When necessary, which is fortunately not often, we refer to this Cartesian product as H^k, even though the notation is slightly imprecise because each coordinate X_i takes a value in a different set.

Definition. The k-dimensional random vector has the *cross-classified Multinomial distribution* of size 1 if and only if its density function f_K is given by the non-zero table of probabilities p_K; that is,

$$f_K(x) = p_K(x)$$

where p_K is such that $p_K(x) > 0$ for all x, and $\sum_x p_K(x) = 1$. □

Note the requirement that p_K is strictly positive ensures the existence of a log-linear expansion and that all the conditional density functions exist.

EXAMPLE 7.1.1 A 2×3 table. In $k = 2$ dimensions, X takes the six possible values $(0,0), (0,1), (0,2), (1,0), (1,1)$ and $(1,2)$ of $H^2 = \{0,1\} \times \{0,1,2\}$ with the table of probabilities $p_K(x)$ given by

$p_{12}(x)$	x_2	0	1	2
x_1	0	0.1	0.2	0.3
	1	0.1	0.2	0.1

These are positive and sum to one. The values of the probability on the internal cells are, for example, $p_{12}(0,1) = 0.2$ and $p_{12}(1,2) = 0.1$.

The marginal table of probabilities for X_1 is a function of x_1 calculated by summing p_{12} over the values of x_2. We use p_1 to denote this table, so that $p_1(x_1)$ is specified by $p_1(0) = 0.6$ and $p_1(1) = 0.4$; similarly, p_2 denotes the marginal table for X_2 and $p_2(x_2)$ is specified by $p_2(0)$, $p_2(1)$ and $p_2(2)$. Later in this chapter we shall adopt the same notation for the observed counts, n.

An alternative notation is to let $p(0,+)$ and $p(1,+)$ denote the cells in the margin for X_1; however it suffers the same disadvantages as the notation p_{ij} : it crucially depends on the ordering of the variables, it does not easily generalise to k dimensions, and, it does not suggest notation for conditional tables. □

EXAMPLE 7.1.2 A second example is the family of probability distributions parameterised by α, β_1 and β_2 and given by

$p_{12}(x)$	x_2	0	1	2
x_1	0	$\alpha\beta_1$	$\alpha\beta_2$	$\alpha(1-\beta_1-\beta_2)$
	1	$(1-\alpha)\beta_1$	$(1-\alpha)\beta_2$	$(1-\alpha)(1-\beta_1-\beta_2)$

These probabilities are positive and sum to one whenever $0 < \alpha < 1$, $0 < \beta_1$, $0 < \beta_2$ and $\beta_1 + \beta_2 < 1$. For each permissible choice of the parameters this table of probabilities satisfies

$$p_{12}(x_1, x_2) = p_1(x_1)p_2(x_2) \text{ for all } x \text{ in } H^2;$$

so that the variables X_1 and X_2 are independent under every probability in the family of probabilities

$$\{p_{12}(x; \alpha, \beta_1, \beta_2);\ 0 < \alpha < 1,\ 0 < \beta_1, \beta_2,\ \beta_1 + \beta_2 < 1\}.$$

This example incorporates independence within a flexible family of distributions, in which the vagaries of real-life data can be accommodated by allowing the parameters to vary. In many applications, interest rests in neither the actual values of the probabilities nor of the parameters, but whether the assumption of independence is tenable.

Strict positivity is achieved by insisting that the parameters do not lie on the boundary of the parameter space, that is, α, β_1 and β_2 are bounded away from zero, and both α and the sum $\beta_1 + \beta_2$ are bounded below one. □

Marginal and conditional distributions

The family of cross-classified Multinomial density functions is closed under the operations of marginalisation and conditioning. Firstly, if the partitioned random vector (X_a, X_b) has the table of probabilities p_K then the marginal distribution of X_a is obtained by summing over the values of X_b:

$$f_a(x_a) = \sum_{x_b} f_K(x_a, x_b) = \sum_{x_b} p_K(x_a, x_b) = p_a(x_a),$$

for all x_a. As $p_K > 0$ and sums to unity then so does p_a and consequently is of the same form as in the definition of the Multinomial of size 1. Secondly, the conditional density function of X_b given X_a is

$$f_{b|a}(x_b; x_a) = \frac{p_{ab}(x_a, x_b)}{p_a(x_a)} \quad \text{for all } x_b.$$

The right hand side is strictly positive and sums to one for each fixed value of x_a, and so it is of the form of a table of probabilites, which we can denote by $p_{b|a}$. Hence the conditional distribution is also Multinomial of size 1.

Proposition 7.1.1 Conditional independence. *The partitioned Multinomial random vector (X_a, X_b, X_c) satisfies*

$$X_b \perp\!\!\!\perp X_c | X_a \quad \textit{if and only if} \quad p_{abc} = \frac{p_{ac}p_{ab}}{p_a}.$$

This is just a restatement of the definition of conditional independence.

Contrast to the multivariate Normal distribution

We make the following remarks:

- We derived the multivariate Normal distribution from a family of independent Normal random variables, via a linear transformation; it is not possible to construct a Multinomial random vector in this way.
- That the Multinomial random vector does not live in a vector space has the consequence that its density function cannot be expressed in terms of simple operations on vectors.

- An essential difference between the multivariate Multinomial and Normal density functions is that the latter has a log-linear expansion determined by a quadratic form and so only contains two-way interactions.
- When (X_a, X_b) has the joint Normal distribution $N(0, V_{a\cup b,a\cup b})$ the marginal distribution of X_a is $N(0, V_{a,a})$; and the parameters of the joint and marginal density functions, $V_{a\cup b,a\cup b}$ and $V_{a,a}$, are simply related by partitioning. The parameters $D_{a\cup b,a\cup b} = \text{var}(X_{a\cup b})^{-1}$ and $D'_{a,a} = \text{var}(X_a)^{-1}$, have a more complicated relationship, summarised by the inverse variance lemma. In the Multinomial distribution there is also a simple relation between the parameters $p_{ab} = p_{a\cup b}$ and p_a: summation; but relating the u-terms, which are the parameters of the expansion of $\log p_{ab}$ as a multilinear form, to the corresponding u-terms, u', of the expansion of $\log p_a$, is not so easy.

7.2 Log-linear Expansions and u-terms

An important advance in the analysis of categorical data was to write down the density function as a log-linear expansion rather than as a table of probabilities. We consider generalising this expansion from the three dimensional Bernoulli given in Section 2.5 to the cross-classified Multinomial distribution.

Recall that the expansion for a three dimensional density function of the random vector (X_1, X_2, X_3) is

$$\begin{aligned}\log f_{123}(x) &= u_\phi + u_1x_1 + u_2x_2 + u_3x_3 \\ &+ u_{12}x_1x_2 + u_{13}x_1x_3 + u_{23}x_2x_3 + u_{123}x_1x_2x_3;\end{aligned}$$

for $x = (x_1, x_2, x_3)$ in $\{0,1\}^3$. When $(X_1, X_2, \ldots, X_k)$ takes values in $\{0,1\}^k$, the multivariate Bernoulli, this generalises to

$$\begin{aligned}\log f_K(x) &= u_\phi + \sum_i u_i x_i \\ &+ \textstyle\sum_{i,j} u_{ij}x_ix_j + \sum_{i,j,l} u_{ijl}x_ix_jx_l + \cdots + u_{12\ldots k}x_1x_2\ldots x_k.\end{aligned}$$

This is a multi-linear form in the k variables, $x_1, x_2, \ldots, x_k$, with fixed coefficients, called the *u-terms*. The term u_{ij} is a *two-way interaction*, u_{ijl} is a *three-way interaction*, and $u_{12\ldots k}$ is a *k-way interaction*. There are 2^k u-terms which is exactly the same as the number of entries in the table of probabilities. The density function can be determined from either the table of probabilities, p_K, or the u-terms, $\{u\}$, in the log-linear expansion. The latter can be extracted from the table by substituting the values $x_i = 0, 1$ in the log-linear expansion, which gives rise to a triangular system of equations and so is easily solved.

When the coordinates of X take values in arbitrary finite sets this expansion is no longer appropriate. To see the way ahead we go back to the expansion for the bivariate Bernoulli:

$$\log f_{12}(x) = u_\phi + u_1 x_1 + u_2 x_2 + u_{12} x_1 x_2 \quad \text{for} \quad x \text{ in } \{0,1\}^2,$$

and consider how to modify it when X takes values in

$$\{0, 1, \ldots, r_1 - 1\} \times \{0, 1, \ldots, r_2 - 1\}.$$

The trick is to rewrite this expansion as

$$\log f_{12}(x) = u_\phi(x) + u_1(x) + u_2(x) + u_{12}(x)$$

where the u-terms u_ϕ, u_1, u_2 and u_{12} are now functions of $x = (x_1, x_2)$, rather than constants. To be useful, firstly, each u-term is constrained to be a function only of the coordinates of x indicated by the subscript. For example, u_1 is a function only of x_1, so that $u_1(x) = u_1(x_1)$; $u_{12}(x)$ is a function of both x_1 and x_2; and $u_\phi(x)$ is just a constant. As it stands this expression is over parameterised, so secondly, to ensure that the u-terms can be uniquely determined from the table of probabilities they are tied down at zero by insisting that

$$u_1(0, x_2) = 0, \; u_2(x_1, 0) = 0 \; \text{ and } \; u_{12}(0, x_2) = 0, \; u_{12}(x_1, 0) = 0,$$

for all x_1 and x_2. The correspondence between the table of probabilities, p, and the u-terms, $\{u\}$, is still one to one; the latter are determined as in the next example.

EXAMPLE 7.2.1 For the 2×3 table of Example 7.1.2 above substituting for $x_1 = 0, 1$ and $x_2 = 0, 1, 2$ gives

$$\begin{aligned}
\log p(0,0) &= u_\phi \\
\log p(0,1) &= u_\phi + u_2(1) \\
\log p(0,2) &= u_\phi + u_2(2) \\
\log p(1,0) &= u_\phi \qquad\qquad\;\; + u_1(1) \\
\log p(1,1) &= u_\phi + u_2(1) \;\; + u_1(1) + u_{12}(1,1) \\
\log p(1,2) &= u_\phi + u_2(1) \;\; + u_1(1) + u_{12}(1,2).
\end{aligned}$$

There are six equations in six unknowns. Back substitution gives $u_\phi = \log(0.1)$, $u_2(1) = \log(2)$, $u_2(2) = \log(3)$, $u_1(1) = 0$, $u_{12}(1,1) = 0$, and $u_{12}(1,2) = -\log(2)$.

In this 2×3 table there are now two parameters measuring two-way interaction, $u_{12}(1,1)$ and $u_{12}(1,2)$. To identify these note that if the equations for $\log p(0,2)$ and $\log p(1,2)$ are eliminated, the four remaining equations have an identical structure to the log-linear expansion for a 2×2 table. Solving these gives

$$u_{12}(1,1) = \log \frac{p(0,0)p(1,1)}{p(0,1)p(1,0)},$$

so that this parameter is just a log cross-product ratio in a 2×2 table. Similarly

$$u_{12}(1,2) = \log \frac{p(0,0)p(1,2)}{p(0,2)p(1,0)},$$

is also a log cross-product ratio in a 2×2 table. □

The generalisation to the k-dimensional cross-classified Multinomial distribution is the log-linear expansion

$$\log f_K(x) = u_\phi(x) + \sum u_i(x) + \sum u_{ij}(x) + \sum u_{ijl}(x) + \cdots + u_{12\ldots k}(x)$$

for all x in H^k. There are still 2^k u-terms, but these are now functions rather than scalars; each u-term is a coordinate projection function with the coordinates indicated by the suffix; and each u-term is constrained to be zero whenever one of its indicated coordinates is zero. There is a neater way to write this expansion in terms of subsets of $K = \{1, 2, \ldots, k\}$.

Definition. The *log-linear expansion* of the cross-classified Multinomial density function f_K is

$$\log f_K(x) = \sum_{a \subseteq K} u_a(x_a)$$

where the sum is taken over all possible subsets a of $K = \{1, 2, \ldots, k\}$ and where the *u-terms* $\{u_a\}$ are coordinate projection functions, so that $u_a(x) = u_a(x_a)$, and also satisfy the constraint that $u_a(x) = 0$ whenever $x_i = 0$ and $i \in a$. □

The expansion is linear in the u-terms rather than in their arguments, the coordinates of x. The importance of the multivariate Bernoulli as a special case is that the expansion is linear in the coordinates of x as well. We shall often drop the argument from the u-terms and write the expansion as $\log f_K = \sum u_a$; but we must remember that they are functions and not scalars. To compute $\{u_a\}$ from p proceeds exactly as in the example given, Example 7.2.1 The resulting system of equations is always linear and triangular. There are other ways to ensure the identifiability of the u-terms; for example, requiring that the sum of the u-terms be zero is often used, see Bishop *et al.* (1975).

Conditional independence

It is particularly easy to list the conditions for independence as conditions on the u-terms.

Proposition 7.2.1 Independence and the u-terms. *If* (X_a, X_b, X_c) *is a partitioned Multinomial random vector then* $X_b \perp\!\!\!\perp X_c | X_a$ *if and only if all u-terms in the log-linear expansion with one or more coordinate in b* and *with one or more in c, are zero.*

Proof: The proof is a direct application of the factorisation theorem for conditional independence. Let t be an arbitrary subset of $a \cup b \cup c = \{1, 2, \ldots, k\}$. First, verify that if all u-terms, u_t, are zero whenever $t \not\subseteq a \cup b$ or $t \not\subseteq a \cup c$, then

$$\log f_K = \sum_{t \subseteq a \cup b} u_t + \sum_{t \subseteq a \cup c} u_t - \sum_{t \subseteq a} u_t.$$

But this function is of the form $g(x_a, x_b) + h(x_a, x_c)$ and hence $X_b \perp\!\!\!\perp X_c | X_a$. The converse holds by reversing the argument. □

When a is empty, this is a proposition about independence. For pairwise independence of variables X_i and X_j conditioned on the remaining variables, the dimension of X_a is $k-2$ and $X_b = X_i$ and $X_c = X_j$ are both one dimensional. The proposition requires $u_{\{i,j\} \cup t} = 0$ whenever $t \subseteq a$. There are 2^{k-2} such terms. Contrast this with the corresponding condition on the Normal density function where only a single coefficient in the inverse variance is required to be 0.

7.3 Graphical Log-linear Models

Though a model, or more specificially a probability model, has a rather catch-all flavour, we use it to mean the specification of an arbitrary family of distributions simplified by a set of hypotheses or constraints.

The importance of the log-linear expansion rests in the fact that many interesting hypotheses can be generated by setting u-terms to zero; and Proposition 7.2.1 gives conditions on the u-terms for conditional independence. The independence graph $G = (K, E)$ of a k-dimensional random vector X, with k vertices in $K = \{1, 2, \ldots, k\}$ and edge set E, is defined in Chapter 3 to be the set of pairs (i, j) such that whenever (i, j) is not in E the variables X_i and X_j are independent conditionally on the rest. So, as with the graphical model for continuous observations with a Normal distribution, we use the independence graph and associated pairwise conditional independences to determine a graphical model for the k-dimensional categorical random vector X.

Definition. Given an independence graph G, the cross-classified Multinomial distribution for the random vector X is a *graphical model* for X if the distribution of X is arbitrary apart from constraints of the form that for all pairs of coordinates not in the edge set E of G, the u-terms containing the selected coordinates are identically zero. □

More explicitly, the density of a Multinomial graphical model is

$$\log f_K(x) = \sum_{a \subseteq K} u_a(x_a)$$

subject to the constraints that $u_a = 0$ if $\{i,j\} \subseteq a$ and (i,j) is not in the edge set E. The parameters of the graphical model are the remaining u-terms that are not set to zero.

EXAMPLE 7.3.1 The edge set $E = \{(1,2),(2,3),(2,5),(3,4),(3,5),(4,5)\}$ on five vertices generates the diagram

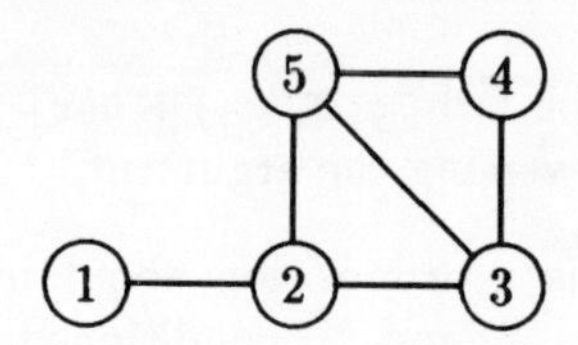

The corresponding Multinomial graphical model for $X = (X_1, X_2, \ldots, X_5)$ has the log-linear expansion

$$\begin{aligned}\log f_{12345} &= u_\phi + u_1 + u_2 + u_3 + u_4 + u_5 \\ &\quad + u_{12} + u_{23} + u_{25} + u_{35} + u_{34} + u_{45} + u_{235} + u_{345}.\end{aligned}$$

No u-term in this expansion is subscripted by pairs of variables not in the edge set; for example, the u-term u_{1234} is not in the expansion because $(2,4)$ is not in the edge set. □

EXAMPLE 7.3.2 Given $k = 5$ and the same edge set as the previous example, the model

$$\log f_X = u_\phi + u_1 + u_2 + u_3 + u_4 + u_5 + u_{12} + u_{23} + u_{25} + u_{35} + u_{34} + u_{45}$$

is not a purely graphical model. It has the same independence graph as in the previous example, but it is further restrained by requiring the three-way interactions, u_{235} and u_{345} are zero as well. □

Hierarchical log-linear models

A graphical model satisfies restraints of the form that all u-terms 'above' a fixed point have to be zero to get conditional independence. A larger class of models, the hierarchical models, is obtained by allowing more flexibility in setting u-terms to zero.

Definition. A log-linear model is *hierarchical* if, whenever one particular u-term is constrained to zero then all higher u-terms containing the same set of subscripts are also set to zero; that is, if $u_a = 0$ then $u_t = 0$ for all $a \subseteq t$. □

Before examples are given note that every distribution with a log-linear expansion has an *interaction graph* constructed by connecting any pair of vertices (i, j) by an edge if there is an interaction term u_a in the expansion in which the set a contains both i and j. Application of the factorisation criterion makes it evident that the interaction graph is an independence graph.

EXAMPLE 7.3.3 The Multinomial log-linear model for a 3-way table

$$\log f_{123} = u_\phi + u_1 + u_2 + u_3 + u_{12} + u_{13}$$

is hierarchical, its maximal u-terms are u_{12} and u_{13}, and it is purely graphical as well. Its independence graph is

The model

$$\log f_{123} = u_\phi + u_1 + u_2 + u_3 + u_{12} + u_{13} + u_{23}$$

is hierarchical; its maximal u-terms are u_{12}, u_{13} and u_{23}. Its independence graph is

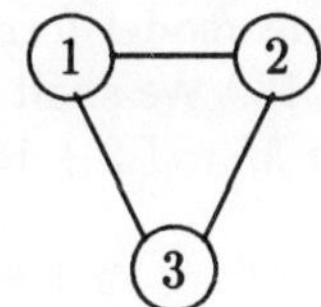

but it is not graphical, because the constraint $u_{123} = 0$ does not correspond to a pairwise conditional independence. The model

$$\log f_{123} = u_\phi + u_3 + u_{12}$$

is not hierarchical because, for instance, u_1 is set to zero but u_{12} is not. Its independence graph is

□

A non-hierarchical model is not necessarily uninteresting; it is just that the focus of interest is something other than independence.

All graphical models are hierarchical because they possess the property that whenever one u-term is set to zero then all higher u-terms are also set to zero. A purely graphical model puts no further constraints on the parameters other than the conditional independence constraints embodied in the graph. A consequence, which is easy to prove, is that a graphical model contains an interaction for each clique in the graph:

Proposition 7.3.1 *A hierarchical log-linear model is graphical if and only if its maximal u-terms correspond to cliques in the independence graph.*

Model formulae

The independence properties of a Multinomial random vector are directly determined by the presence or absence of particular u-terms in the log-linear expansion of the probability density function. If a model is hierarchical then this is just determined by the maximal u-terms in the expansion, and so the list of maximal u-terms generates the model. The list of subscripts of maximal u-terms present in the log-linear expansion is sometimes known as the *generating class* of the model and its *model formula* is one way of writing this list. The syntax for model formulae varies, but most are compact, algebraically tractable and suitable for machine implementation in model fitting software packages. A streamlined notation to discuss all models with a fixed generating class is illustrated by the next example.

EXAMPLE 7.3.4 Consider the hierarchical log-linear model

$$\log f_{123} = u_\phi + u_1 + u_2 + u_3 + u_{12} + u_{13}.$$

The maximal u-terms are u_{12} and u_{13} so that its generating class is the set of subsets $\{\{1,2\},\{1,3\}\}$. One model formula in popular use is $1.2 + 1.3$, and another alternative is $12, 13$. We shall use the former incorporating the symbols '.' and '+', and write $M = 1.2 + 1.3$. The saturated model

$$\log f_{123} = u_\phi + u_1 + u_2 + u_3 + u_{12} + u_{13} + u_{23} + u_{123},$$

has the formula $M = 1.2.3$; and the model of mutual independence,

$$\log f_{123} = u_\phi + u_1 + u_2 + u_3,$$

the formula $M = 1 + 2 + 3$. The model

$$\log f_{123} = u_\phi + u_1 + u_2 + u_{12},$$

for the three dimensional vector $X = (X_1, X_2, X_3)$ expresses the equi-probability of the values of X_3 for each of the possible values of (X_1, X_2). It has the model formula $M = 1.2$.

The variables need not be denoted by their subscripts and an entirely consistent way of writing $1.2 + 1.3$ is $X_1.X_2 + X_1.X_3$. □

Lattices of models

Use of model formulae allows a succinct description of the full set of hierarchical log-linear models. Setting subsets of the u-terms to zero creates a partial order $\preceq$ between model formulae M_1 and M_2, determined by $M_1 \preceq M_2$ if and only if M_1 contains all the maximal u-terms contained in M_2. It turns out that the set of model formulae together with this partial order is an algebraic lattice, Gratzer (1978). We shall not use any lattice theory results here apart from the insight given by the lattice diagram of the set of log-linear models. In two dimensions the lattice contains five model formulae

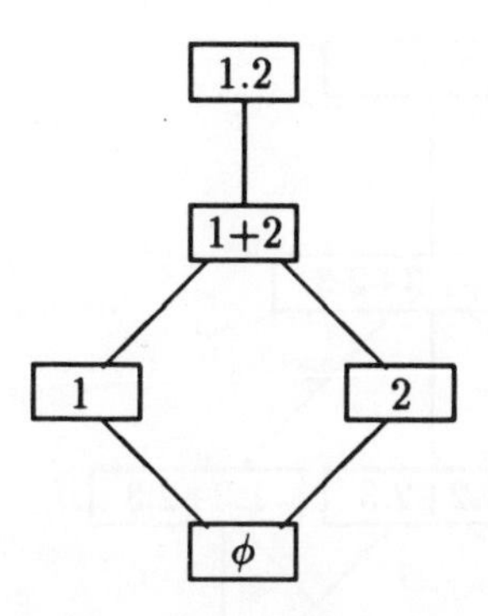

The model formula $M = 1+2$ denotes the family of independence models for a two way table, because the log-linear expansion is

$$\log f_{12} = u_\phi + u_1 + u_2,$$

and as $u_1(x) = u_1(x_1)$ is a function only of x_1, and u_2 of x_2, then, by the factorisation criterion, X_1 and X_2 are independent. The model formula $M = \phi$ represents the log-linear expansion

$$\log f_{12} = u_\phi$$

so that all values of $X = (X_1, X_2)$ have the same probability. The class $M = 1$ refers to the distributions with equi-probable values of X_2 for each value of X_1.

This finite set of five model formulae is a lattice because for every pair of model formulae there exists (i) a 'larger' model formula, the *join*, that contains all the u-terms contained in either; and (ii) a 'smaller' model formula, the *meet*, that contains all the u-terms contained in both. For instance, 1 and 2 join in $1+2$, and meet in ϕ.

In three dimensions the lattice is considerably more complicated, see Figure 7.3.1. The model $1.2+1.3$ is the graphical model corresponding to the independence of X_2 and X_3 conditional on X_1. The model $1.2+3$ is the graphical model corresponding to the independence of (X_1, X_2) and X_3. The model $1.2+1.3+2.3$ is the hierarchical model that contains all two-way interaction u-terms. The model 1.2 is the model of the conditional equi-probability of X_3 given (X_1, X_2). The minimal model ϕ contains just a single constant u-term and the maximal model 1.2.3 contains all u-terms and hence spans the set of all three dimensional Multinomial distributions.

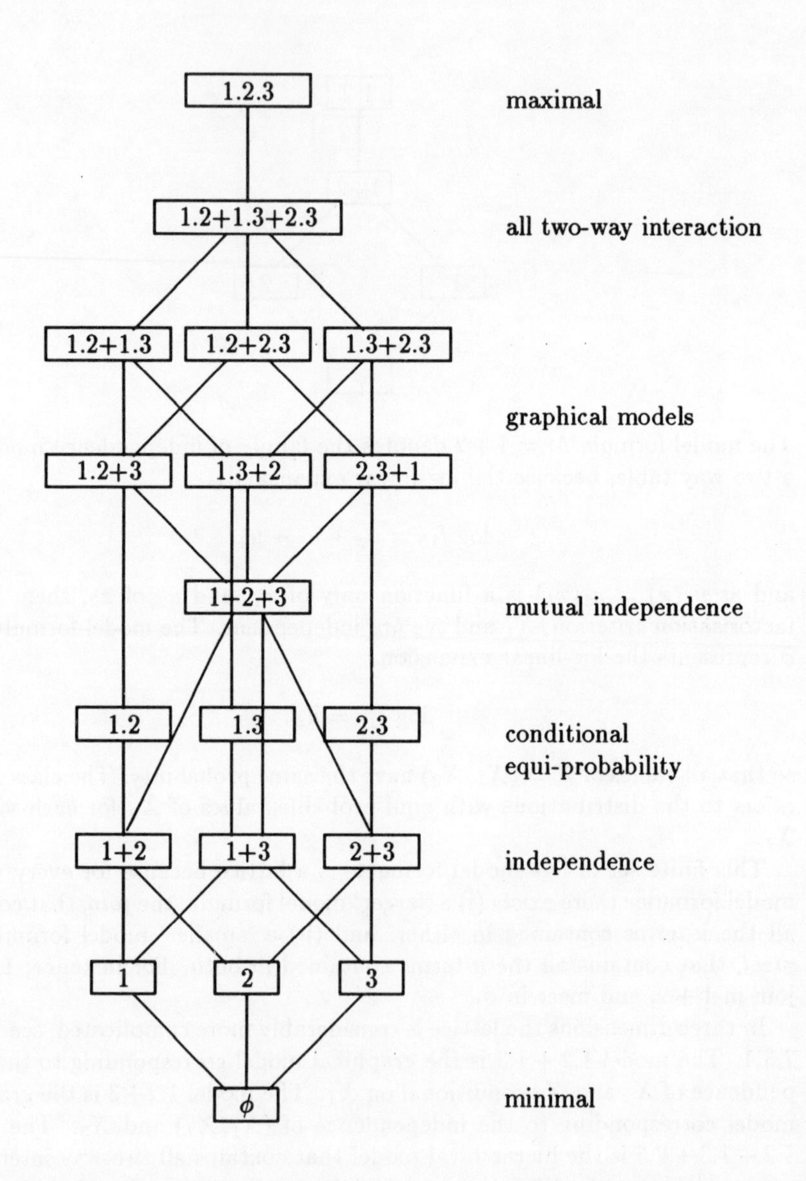

Figure 7.3.1: The lattice diagram of all three dimensional hierarchical log-linear models.

The lattice contains several sub-lattices of interest including the lattice of all graphical models.

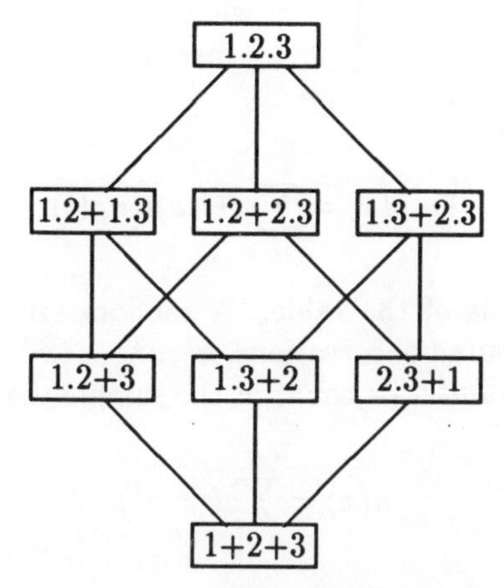

This is a Boolean sub-lattice in the sense that it can be generated by the presence or absence of a finite set of conditions: the presence or absence of the $\binom{k}{2}$ terms of the form $i.j$. The join for any two graphical models in this sub-lattice consists of those independences that both models have in common while the meet consists of those that occur in either one or both.

In four dimensions, the lattice of hierarchical log-linear models has 167 elements. More generally, it is possible to show that the lattice of all k-dimensional hierarchical log-linear models is a finite distributive lattice with minimal element ϕ and maximal element $1.2\ldots.k$, see Lauritzen, Speed and Vijayan (1984).

7.4 The Likelihood Function

The likelihood function is derived under the assumption of selecting a random sample of N observations on the k-dimensional random vector $X = (X_1, X_2, \ldots, X_k)$. The distribution of X is the cross-classified Multinomial distribution of size 1, parameterised by the associated table of probabilities, p, or by the u-terms in the associated log-linear expansion of the density function.

The likelihood for the parameters, p, based on a single observation, x, on the random vector X, is just $p(x)$. In order to get a mathematical handle consider the identity for p as the product over all the $r_1 \times r_2 \times \ldots \times r_k$ possible cells of the table:

$$p(x) = \prod_i p(i)^{\delta(i,x)},$$

where i ranges over the cells and δ is the indicator function defined as

$$\delta(i,x) = \begin{cases} 1 & \text{when } i = x \text{ and} \\ 0 & \text{otherwise.} \end{cases}$$

Taking logarithms gives

$$\log p(x) = \sum_i \delta(i,x) \log p(i),$$

as a sum over the cells of the table. A random sample of N independent and identically distributed observations, $x^1, x^2, \ldots, x^N$, is observed. Let $n(x)$ denote the ***observed number*** or count in the sample that fall in cell x, so that

$$n(x) = \sum_{l=1}^{N} \delta(x, x^l).$$

We adopt the same notation for the table of observed counts, as used for the table of probabilities: $n(x)$ is the value of the function n at each x (at each cell of the table); n_a is the marginal table of counts obtained by summing n over x_b where x is partitioned into (x_a, x_b); hence n_a is a function of x_a.

Proposition 7.4.1 The log-likelihood function.

1. *The log-likelihood function of the table of probabilities p based on a random sample of N Multinomial random observations is*
$$l(p; x^1, x^2, \ldots, x^N) = \sum_x n(x) \log p(x).$$

2. *The table of observed counts, n, is a sufficient statistic for the parameters p, and so $l(p; x^1, x^2, \ldots, x^N) = l(p; n)$.*

3. *The log-likelihood function expressed in terms of the information divergence, is*
$$l(p; n) = l(\frac{1}{N}n; n) - NI(\frac{1}{N}n; p).$$

4. *The log-likelihood expressed as a function of the u-terms, is*
$$l(u; n) = \sum_a \sum_{x_a} n_a(x_a) u_a(x_a).$$

Proof: Because of random sampling the log-likelihood function of p, based on the full sample, is just the sum of the log-likelihoods for each observation; simplifying this gives the first expression. The sufficiency of the cell count, n, follows by the standard factorisation of the likelihood. The Kullback-Leibler information measure for measuring the divergence between density functions f and g is the average value of the log-likehood ratio averaged with respect to f, so that $I(f;g) = E\log(f/g)$. If under f, the random vector X has the cross-classified Multinomial distribution with a table of probabilities q and under g the same, but with table p, then the divergence is the finite sum

$$I(q;p) = \sum_x q(x)\log\frac{q(x)}{p(x)},$$

taken over all cells, x, in the table. It is just a minor generalisation of Examples 4.1.1 and 4.1.2. Finally, the last expression comes from

$$l(u;n) = \sum_x n(x)\log p(x) = \sum_x n(x)\sum_a u_a(x)$$

and simplifies to $\sum_a \sum_x u_a(x_a)n(x) = \sum_a \sum_{x_a} u_a(x_a)n_a(x_a)$, where we use $u_a(x) = u_a(x_a)$. □

There is a remarkable unity between the likelihood functions for the Multinomial and multivariate Normal distributions, evidenced by comparing the functions expressed in terms of the divergence. This unity exists in spite of the disparate mathematical nature of the density functions and of the structure of the parameter sets, V and p.

The deviance

The difference in log-likelihoods at $p = n/N$ and an arbitrary point p is expressible as an information divergence and so maximising the likelihood is equivalent to minimising the divergence. This argument is later used to derive the general form of the likelihood equations. Here we first note that it provides a simple proof that the unrestricted maximum likelihood estimate of p is

$$\hat{p}(x) = \frac{1}{\mathrm{N}}n(x);$$

just the observed proportion in cell x. The assertion follows because the divergence is positive definite and so the minimum value of $I(\frac{1}{\mathrm{N}}n;p)$ is 0; this is attained only if $p = \frac{1}{\mathrm{N}}n$.

Now consider a hierarchical model parameterised by the table of probabilities p^M, that has an associated log-linear expansion with model formula M. We want to calculate the likelihood ratio of this model to the saturated model with the complete graph, and no restrictions on the parameters. The *deviance* of M is defined as twice the difference between the value of the unconstrained maximised log-likelihoods and the maximum under the constraints satisfied by M. We have shown

Proposition 7.4.2 Deviance and divergence. *For a random sample of N Multinomial random observations the deviance is*

$$\text{dev}\,(M) = 2\,N\,I(\frac{1}{\text{N}}n;\hat{p}^M) = 2\sum_x n(x)\log\frac{n(x)}{N\hat{p}^M(x)}.$$

where $\hat{p}^M$ is the maximum likelihood estimate of p^M.

Mnemonically this is twice the sum over cells of 'observed log (observed/fitted)' and of course to evaluate it one needs to have an expression for the fitted values.

Under the null hypothesis that M holds, the deviance has an asymptotic chi-squared sampling distribution with degrees of freedom given by the number of parameters set to zero. This is perhaps the single most important tool for the practical analysis of multi-dimensional contingency tables, and we make extensive use of it later in this text. It is however only an approximation to its exact sampling distribution, and strictly is valid only in large samples. It is derived from considerations concerning maximum likelihood procedures, because the deviance is twice a generalised likelihood ratio: see Cox and Hinkley (1974) or Rao (1973), and for results specialised to contingency tables, Andersen (1980). The validity of these general results requires certain regularity conditions to hold: in our case of simple random sampling from the Multinomial distribution these reduce to requiring that the parameters do not lie on on the boundary of the parameter space.

Though related, maximising the log-likelihood by minimising the information divergence is not equivalent to the minimum discrimination information procedure, which is concerned to minimise $I(p;\frac{1}{\text{N}}n)$ for variation in the first argument p, and as I is not symmetric leads to a different solution. See Kullback (1959, 1968) and Bishop *et al.* (1975) for an interesting discussion.

Pearson's goodness of fit statistic

A standard measure of goodness of fit for contingency tables is Pearson's statistic, sum (obs − fitted)^2/fitted, or more formally

$$X^2 = \sum_x (n(x) - N\hat{p}(x))^2/N\hat{p}(x),$$

where the sum is taken over all cells, x, in the table. It deserves mention because it is so well known, but if the observed and fitted tables, n and $N\hat{p}$, are not too far apart then the deviance and this goodness of fit statistic are almost identical. A heuristic argument is to write the deviance as

$$\begin{aligned}
\text{dev} &= 2NI(n/N;\hat{p}) \\
&= 2N\sum n\log(n/N\hat{p}) \\
&= -2N\sum N\hat{p}(n/N\hat{p})\log(n/N\hat{p}).
\end{aligned}$$

The Taylor series expansion for $t\log(t)$ for t in the neighbourhood of 1, containing just the first two terms, is $(t-1)+(t-1)^2/2$, so that if $n/N\hat{p}$ is near one the deviance is approximately

$$2N\sum N\hat{p}(n/N\hat{p}-1)+(n/N\hat{p}-1)^2/2)=X^2,$$

because the sum $\sum N\hat{p}(n/N\hat{p}-1)=0$ as $\sum\hat{p}=1$.

7.5 Simple Examples of MLE

In parallel to our analysis of the multivariate Normal distribution we take a look at some simple examples and then derive more general conditions.

EXAMPLE 7.5.1 Fitting the saturated model in 2-dimensions, M=1.2. Consider an artificial two-way table and $N = 10$ observations on $X = (X_1, X_2)$:

$$(0,0),(0,1),(0,1),(1,0),(1,0),(1,0),(1,1),(1,1),(1,1),(1,1).$$

The observed table of counts, n, is

$n_{12}(x)$		x_2	0	1
x_1	0		1	2
	1		3	4

Consider the maximal log-linear model

$$\log f_{12}(x)=u_\phi(x)+u_1(x)+u_2(x)+u_{12}(x),$$

which, as both X_1 and X_2 are binary, can be simplified to

$$\log f_{12}(x)=u_\phi+u_1x_1+u_2x_2+u_{12}x_1x_2, \quad \text{for } x \text{ in } \{0,1\}^2.$$

The log-likelihood function, taken as a function of u, is the sum over the cells of the log-linear expansion weighted by the observed cell counts:

$$l(u;n)=10u_\phi+7u_1+6u_2+4u_{12}.$$

There are four parameters involved, but they are related by the constraint that $p(x)$ must sum to 1 :

$$e^{u_\phi}+e^{u_\phi+u_2}+e^{u_\phi+u_1}+e^{u_\phi+u_1+u_2+u_{12}}=1.$$

One could use this to express u_ϕ in terms of the other parameters but it is more symmetric to use the method of Lagrange multipliers and maximise

$$l(u;n)-\theta(\sum_x p(x)-1),$$

where θ is the Lagrange multiplier. Differentiating with respect to the four u-terms and the multiplier gives the likelihood equations

$$\begin{array}{rcr} \text{wrt } \theta & : & \sum p(x) - 1 = 0, \\ \text{wrt } u_\phi & : & 10 - \theta \sum p(x) = 0, \\ \text{wrt } u_1 & : & 7 - \theta \sum x_1 p(x) = 0, \\ \text{wrt } u_2 & : & 6 - \theta \sum x_2 p(x) = 0, \\ \text{wrt } u_{12} & : & 4 - \theta \sum x_1 x_2 p(x) = 0, \end{array}$$

ϒwhere each sum ranges over $x = (0,0), (0,1), (1,0), (1,1)$. The first two equations imply $\hat{\theta} = 10$; the next three then give

$$\hat{p}_1(1) = 0.7, \ \hat{p}_2(1) = 0.6 \ \text{ and } \ \hat{p}(1,1) = 0.4.$$

The other estimates for p can be obtained by subtraction, for instance

$$\hat{p}(1,0) = \hat{p}_1(1) - \hat{p}(1,1) = 0.7 - 0.4. = 0.3,$$

and consequently the maximum likelihood estimates of all the cell probabilities under the saturated model are just the observed proportions. The estimates of the u-terms can then be extracted by inversion of the log-linear expansion. They are

$$\hat{u}_\phi = -2.30, \ \hat{u}_1 = 1.10, \ \hat{u}_2 = 0.69, \quad \text{and} \quad \hat{u}_{12} = -0.41.$$

□

EXAMPLE 7.5.2 Fitting the independence model, $M = 1 + 2$. Now consider estimating the parameters from the same observed sample, but under the model of independence. As u_{12} is constrained to be zero, the likelihood equations are identical save that the equation obtained by differentiating with respect to u_{12} is missing from the set:

$$\begin{array}{rcr} \text{wrt } \theta & : & \sum p(x) - 1 = 0, \\ \text{wrt } u_\phi & : & 10 - \theta \sum p(x) = 0, \\ \text{wrt } u_1 & : & 7 - \theta \sum x_1 p(x) = 0, \\ \text{wrt } u_2 & : & 6 - \theta \sum x_2 p(x) = 0. \end{array}$$

We immediately get

$$\hat{\theta} = 10, \ \hat{p}_1(1) = 0.7 \ \text{ and } \ \hat{p}_2(1) = 0.6.$$

As $\sum p(x) = 1$, subtraction gives $\hat{p}_1(0) = 0.3$ and $\hat{p}_2(0) = 0.4$, so that the observed and fitted margins are identical:

observed

	x_2	0	1	total
x_1	0	0.1	0.2	0.3
	1	0.3	0.4	0.7
		0.4	0.6	1.0

fitted

	x_2	0	1	total
x_1	0			0.3
	1		?	0.7
		0.4	0.6	1.0

We are left to determine the internal cell probabilities. Under the independence model

$$p_{12}(1,1) = p_1(1)p_2(1)$$

so that the maximum likelihood estimate $\hat{p}(1,1) = (0.7)(0.6) = 0.42$. Subtraction from the marginal totals give estimates of the other internal cells as

$\hat{p}(x)$	x_2	0	1
x_1	0	0.12	0.18
	1	0.28	0.42

and estimated u-terms of

$$\hat{u_\phi} = -2.12,\ \hat{u}_1 = 0.85,\ \hat{u}_2 = 0.41, \quad \text{with } u_{12} = 0.00$$

The estimates of the non-zero u-terms differ considerably from the estimates in the saturated model. □

EXAMPLE 7.5.3 A three-way contingency table. The counts n_{123} in Table 7.5.1 are a fragment of a data set concerning breast cancer reported by Morrison *et al.* (1973). The classifying factors are: X_1 diagnostic center, X_2 nuclear grade, and X_3 survival. In addition to the two-way margin n_{23} we require the other two-way margins, n_{12} and n_{13}. There are a variety of interesting hypotheses but we consider only two: the graphical model of the conditional independence of survival and grade for each centre, $M = 1.2 + 1.3$; and the hierarchical model of no three-way interaction, $M = 1.2 + 1.3 + 2.3$. First consider $M = 1.2 + 1.3$. The conditional independence hypothesis is equivalent to setting the u-terms, u_{23} and u_{123}, to zero, so the corresponding log-linear expansion is

$$\log f_{123}(x) = u_\phi + u_1x_1 + u_2x_2 + u_3x_3 + u_{12}x_1x_2 + u_{13}x_1x_3.$$

Weighting this by the cell counts and adding gives the log-likelihood function

$$l(u;n) = 474u_\phi + 221u_1 + 261u_2 + 324u_3 + 102u_{12} + 153u_{13}.$$

Table 7.5.1: Three year survival of 474 breast cancer patients according to nuclear grade and diagnostic centre. From Morrison *et al.* (1973).

nuclear grade, X_2		malignant		benign		
	survival, X_3	died	survived	died	survived	total
centre, X_1	Boston	35	59	47	112	253
	Glamorgan	42	77	26	76	221
	total	77	136	73	188	474

	nuclear grade, X_2	
	malignant	benign
Boston	94	159
Glamorgan	119	102

	survival, X_3	
	died	survived
Boston	82	171
Glamorgan	68	153

Differentiating, and not forgetting the Lagrange multiplier,

$$\begin{aligned}
\text{wrt } \theta &: & \textstyle\sum p(x) - 1 &= 0, \\
\text{wrt } u_\phi &: & 474 - \theta \textstyle\sum p(x) &= 0, \\
\text{wrt } u_1 &: & 221 - \theta \textstyle\sum x_1 p(x) &= 0, \\
\text{wrt } u_2 &: & 261 - \theta \textstyle\sum x_2 p(x) &= 0, \\
\text{wrt } u_3 &: & 324 - \theta \textstyle\sum x_3 p(x) &= 0, \\
\text{wrt } u_{12} &: & 102 - \theta \textstyle\sum x_1 x_2 p(x) &= 0, \\
\text{wrt } u_{13} &: & 153 - \theta \textstyle\sum x_1 x_3 p(x) &= 0.
\end{aligned}$$

The three equations corresponding to the one-way interactions simplify to

$$N\hat{p}_1(1) = n_1(1),\ N\hat{p}_2(1) = n_2(1) \ \text{ and } \ N\hat{p}_3(1) = n_3(1)$$

and because the probabilities sum to 1 these also imply $N\hat{p}_1(0) = n_1(0)$, $N\hat{p}_2(0) = n_2(0)$ and $N\hat{p}_3(0) = n_3(0)$. Putting these together gives the expression

$$N\hat{p}_a(x_a) = n_a(x_a) \ \text{ for a=\{1\},\{2\},\{3\}},$$

affirming the equality between the observed counts and the fitted values in the one-way margins. The equations corresponding to the two-way interactions are $N\hat{p}_{12}(1,1) = n_{12}(1,1)$ and $N\hat{p}_{13}(1,1) = n_{13}(1,1)$. Using the one-way interactions gives the other cells in the two-way margins; in summary

$$N\hat{p}_a(x_a) = n_a(x_a) \ \text{ for a=\{1\},\{2\},\{3\},\{1,2\},\{1,3\}}.$$

The internal cell probabilities can now be derived from the marginals using the conditional independence:

$$\hat{p}(x) = \frac{\hat{p}_{12}(x_1, x_2)\hat{p}_{13}(x_1, x_3)}{\hat{p}_1(x_1)}.$$

The *fitted values* under the model 1.2 + 1.3 of the independence of survival and grade given centre are just N times the fitted cell probabilities, and are

	nuclear grade, X_2	malignant		benign	
	survival, X_3	died	survived	died	survived
Boston	fv(1.2+1.3)	30.47	63.53	51.53	107.47
	fv(1.2+1.3+2.3)	35.73	58.27	46.27	112.73
Glamorgan	fv(1.2+1.3)	36.62	82.38	31.38	70.62
	fv(1.2+1.3+2.3)	41.27	77.73	26.73	75.27

The table also includes the fitted values under the no three-way interaction model, $M = 1.2 + 1.3 + 2.3$. Its log-linear expansion differs from the conditional independence model by the additional term $u_{23}x_2x_3$. The required modification to the likelihood equations given above is the additional constraint that

$$\text{wrt } u_{23} : 188 - \theta \sum x_2 x_3 p(x) = 0,$$

which implies that fitted values satisfy

$$N\hat{p}_a(x_a) = n_a(x_a) \quad \text{for } a = \{1\}, \{2\}, \{3\}, \{1,2\}, \{1,3\} \text{ and for } a = \{2,3\}.$$

There is now a problem. This is not a model of conditional independence and there is no direct analytic expression for the internal cell probabilities. The fitted values have to be computed in some other manner. They are closer to the observed values than are the conditional independence fitted values, but this is only to be expected as the model has an extra free parameter.

These two models have deviances:

	model	dev	df
survival⫫grade|centre,	$1.2 + 1.3$	4.072	2
no three-way interaction,	$1.2 + 1.3 + 2.3$	0.088	1

where, for example,

$$\text{dev}\,(1.2 + 1.3) = 35\log\left(\frac{35}{30.47}\right) + 59\log\left(\frac{59}{63.53}\right) + \cdots + 76\log\left(\frac{76}{70.62}\right) = 4.072$$

This statistic is associated with 2 df because exactly two u-terms, u_{23} and u_{123}, are set to zero. The deviance for $1.2 + 1.3$ is computed from its fitted values in the same manner; it has 1 df since only u_{123} is set to zero.

Pearson's goodness of fit statistic for this survival data, are, for the independence model $1.2 + 3$, $X^2 = 19.080$ (while the dev $= 17.828$) and for the conditional independence model, $1.2+1.3$, $X^2 = 0.0836$ (while dev $= 0.0823$). □

All the variables are binary in the simple examples discussed here; the extension required to write down the appropriate derivatives for variables with more than two categories is straightforward and included as an exercise. Note that for a 2×3 table, the u-term $u_2(x_2)$ on $x_2 = 0, 1, 2$ has exactly 2 free parameters: $u_2(1)$ and $u_2(2)$; and so contributes two equations, rather than one, to the set of likelihood equations. More generally the term $u_{12}(x)$ contributes $(r_1 - 1)(r_2 - 1)$ degrees of freedom.

7.6 Estimates for Conditional Independence Models

The results of Chapter 4 on minimal divergence apply equally well to graphical log-linear as to graphical Gaussian models. We use the representation of the likelihood as a divergence to derive the likelihood equations for the maximum likelihood estimates and deviance.

Proposition 7.6.1 *The maximum likelihood estimator of a log-linear graphical model with model formulae M based on a random sample from the Multinomial distribution exists and satisfies the likelihood equations*

$$N\hat{p}_a^M = n_a$$

whenever the subset a of vertices in the graph form a clique.

Proof: The log-likelihood of p^M is given by $l(p^M) = l(\frac{1}{N}n) - NI(\frac{1}{N}n; p^M)$ which is maximised when the divergence is minimised. If the vertices in a form a clique then application of the lemma, Lemma 4.4.2, concerning arbitrary marginals in graphical models and used in the proof of Theorem 6.6.1, establishes that the marginal probabilities, p_a^M, are arbitrary. Hence choosing p^M to agree with $\frac{1}{N}n$ on this margin uniquely minimises the divergence and $N\hat{p}_a^M = n_a$. □

The theorem can be summarised as the requirement that in fitting a graphical model, 'observed = fitted' for every marginal table corresponding to a complete subgraph. In fact, it is easy to see that a version of this proposition applies more generally to fitting hierarchical log-linear models: 'observed = fitted' in every marginal table classified by X_a, for which the u-term, u_a, is arbitrary in in the log-linear expansion. One only has to check that the hierarchy condition ensures that the marginal distribution of X_a is arbitrary. Now consider certain special cases where there are direct estimates.

Conditional independence

Consider the graphical log-linear model, M, determined by the conditional independence of $X_b \perp\!\!\!\perp X_c | X_a$, of p, q, and r dimensions respectively.

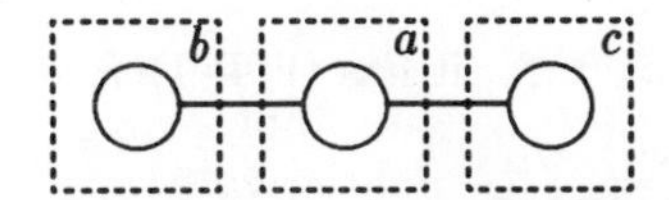

The diagram is meant to suggest that all within-box edges exist, as do those between a and b, and, a and c. There are exactly two cliques in the independence graph, $a \cup b$ and $a \cup c$, so the fitted table of probabilities satisfies the likelihood equations:

$$N\hat{p}^M_{ab} = n_{ab} \quad \text{and} \quad N\hat{p}^M_{ac} = n_{ac}.$$

and imply $N\hat{p}^M_a = n_a$ by summation. Because $X_b \perp\!\!\!\perp X_c | X_a$, the table of probabilities factorises and the parameters in the internal cells can be calculated from the margins

$$p^M_{abc} = \frac{p^M_{ab} p^M_{ac}}{p^M_a}; \text{ so that } N\hat{p}^M_{abc} = \frac{n_{ab} n_{ac}}{n_a}.$$

Substituting in $\text{dev}(M) = 2NI(\frac{1}{N}n, \hat{p}^M)$ gives an explicit formula for the deviance

Proposition 7.6.2 Deviance against conditional independence. *The deviance against $X_b \perp\!\!\!\perp X_c | X_a$ is*

$$\text{dev}(X_b \perp\!\!\!\perp X_c | X_a) = 2 \sum n_{abc} \log \frac{n_{abc} n_a}{n_{ab} n_{ac}},$$

where the sum is taken over all cells, x, of the table. Denoting the number of cells in the marginal distributions of X_a, X_b and X_c by r_a, r_b and r_c respectively, the number of degrees of freedom for the deviance is

$$\text{df}(X_b \perp\!\!\!\perp X_c | X_a) = r_a(r_b - 1)(r_c - 1).$$

When all the variables are binary this is $2^p(2^q - 1)(2^r - 1)$. There are two ways to derive this value. Firstly, this is a joint test of the independence of X_b and X_c at each value of X_a, and regarding X_b and X_c as single variables taking r_b and r_c values, respectively, gives the result immediately. Secondly, the calculation follows from the number of u-terms set to zero: all those u-terms containing at least one element from b and one from c.

An example of fitting a conditional independence model to a three-way table has been considered in Example 7.5.3. The model of independence is derived as a special case of conditional independence by setting $a = \phi$ so that $r_a = 1$. The fitted values for the internal cells are $n_b n_c / N$ and the deviance of the model is

$$\text{dev}\,(X_b \perp\!\!\!\perp X_c) = 2 \sum n_{bc} \log \frac{n_{bc} n_\phi}{n_b n_c}$$

where $n_{bc} = n_K$ and $n_\phi = N$; it has $(r_b - 1)(r_c - 1)$ degrees of freedom.

Edge exclusion deviances

The independence graph is defined in terms of pairwise conditional independences and there is no edge between two variables if those two are independent conditional upon the remaining variables. An edge exclusion deviance is the deviance for this hypothesis. Setting $b = \{i\}$, $c = \{j\}$ and $a = K \backslash \{i, j\}$ gives

Corollary 7.6.3 Edge exclusion deviance. *The deviance for testing the graphical model $X_i \perp\!\!\!\perp X_j | X_{K\backslash\{i,j\}}$ with one missing edge is*

$$\text{dev}\,(X_i \perp\!\!\!\perp X_j | X_{K\backslash\{i,j\}}) = 2 \sum n_K \log \frac{n_K n_{K\backslash\{i,j\}}}{n_{K\backslash i} n_{K\backslash j}}.$$

It has an asymptotic chi-squared distribution with $r_{K\backslash\{i,j\}}(r_i - 1)(r_j - 1)$ degrees of freedom.

No further simplification is possible here, unlike the graphical Gaussian model in which the edge exclusion deviance simplified to a function of the partial correlation coefficient; however there are more efficient ways of computing the triangular array of all edge exclusion deviances simultaneously.

EXAMPLE 7.6.1 Table 7.6.1 classifying psychiatric patients by their symptoms comes from Wermuth (1976b) and is discussed by Benedetti and Brown (1978). Denote the variables Validity by X_1, Solidity by X_2, Acute Depression by X_3, and Stability, with values (introvert, extrovert), by X_4. There

Table 7.6.1: A cross-classification of psychiatric patients. From Wermuth (1976b).

		Depression,3	yes		no	
Validity,1	Solidity,2	Stability,4	intro	extro	intro	extro
energetic	rigid		15	23	25	14
	hysteric		9	14	46	47
psychasthenic	rigid		30	22	22	8
	hysteric		32	16	27	12

are $\binom{4}{2} = 6$ pairwise conditional independences, and the corresponding edge exclusion deviances are

1	*			
2	4.78	*		
3	33.00	22.38	*	
4	12.87	3.39	7.64	*
	1	2	3	4

For instance, reported in the top left corner of this triangular array is the deviance of the model with formulae 1.3.4 + 2.3.4 in which $X_2 \perp\!\!\!\perp X_2 | (X_3, X_4)$. This is a joint test for independence in four two-way tables, the first of which is

		x_2 0	1
x_1	0	15	9
	1	30	32

The deviance is

$$\text{dev}\,(1.3.4 + 2.3.4) = 15\log\frac{15}{(45)(24)/(86)} + \cdots$$

$$+32\log\frac{32}{(62)(41)/(86)} + 23\log\frac{23}{(37)(45)/(75)} + \cdots + 12\log\frac{14}{(61)(22)/(81)}$$

and equals 4.78. Each statistic has 4 df and the 10% point of the chi-squared distribution on 4 df is 7.78. By comparison some of the exclusion deviances are large and some are small.

Discarding the two edges that are clearly not significant gives the graph

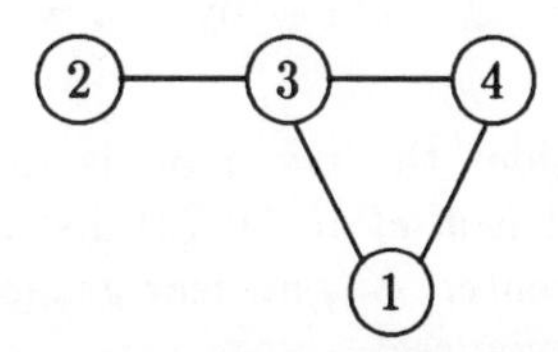

The independences in this independence graph can be summarised in a single statement: $X_2 \perp\!\!\!\perp (X_1, X_4) | X_3$, and the corresponding graphical model $M =$ 1.3.4 + 3.2 has direct estimates. The interpretation of the model is that Solidity is independent of Validity and Stability given the separating variable, Depression. The deviance is a joint test for independence in two 2×4 tables and so has $2(2-1)(4-1) = 6$ df. It has a value of 8.8792 which, on 6 df, suggests a good fit. Neither this deviance, nor its degrees of freedom, equals the sum of the corresponding deviances for excluding the edges separately. □

7.7 Partitioning the Deviance

The deviance, defined in terms of a maximised log-likelihood ratio test statistic and evaluated in terms of an information divergence, has been introduced for log-linear models in Section 7.4 of this chapter, and illustrated in Section 7.5. Here, by way of illustrative examples, we consider two distinct manners of partitioning the deviance against conditional independence into meaningful components. To contrast them, consider a 2^3 table and the independence of X_2 and X_3 given X_1, so that there are two degrees of freedom for the joint test.

(i) The first method is to partition the 2 df into two 1 df tests, one of which tests for $u_{123} = 0$, or equivalently, for the equality of conditional cprs $\log \text{cpr}(X_2, X_3 | X_1 = 0) = \log \text{cpr}(X_2, X_3 | X_1 = 1)$; and the other to testing $u_{23} = 0$, or equivalently that the partial log-cpr, $\log \text{cpr}(X_2, X_3 | X_1) = 0$. These two models, with formulae $M_1 = 1.2 + 1.3 + 2.3$ and $M_2 = 1.2 + 1.3$, form a chain in the lattice of hierarchical models in Figure 7.3.1 from 1.2.3 to 1.2 + 1.3.

(ii) The second method partitions the 2 df in the joint test for $X_2 \perp\!\!\!\perp X_3 | X_1$, according to the values of the conditioning variable X_1: into local tests for $X_2 \perp\!\!\!\perp X_3 | X_1 = 0$ and $X_2 \perp\!\!\!\perp X_3 | X_1 = 1$.

Both techniques can be traced back to Goodman (1968, 1970). There is no immediate parallel with graphical models for continuous observations because conditional independence is characterised by a single parameter in Gaussian models.

Partitioning along a chain

Consider the example:

EXAMPLE 7.7.1 Reconsider the two models that have been fitted to the breast cancer data of Morrison *et al.* (1973) discussed in Example 7.5.3. The variables are diagnostic center, X_1, nuclear grade, X_2 and survival, X_3, and the fitted models have deviances:

	model	dev	df
saturated	1.2.3	0.000	0
no three-way interaction	$M_1 = 1.2 + 1.3 + 2.3$	0.088	1
survival⫫grade|centre	$M_2 = 1.2 + 1.3$	4.072	2

The deviance of M_1 is a test that the two conditional cross-product ratios (cpr) between grade and survival, one for each centre, are the same. Model

M_2 is nested in M_1 and the difference in deviances, dev $(M_1) -$ dev (M_2), is a test that the common value of the cpr between grade and survival, the partial cpr, is zero. The analysis of deviance table and the corresponding chain of models is

deviance difference due to	dev	df
no three-way interaction	0.088	1
zero partial cpr $(X_2, X_3 \mid X_1)$	3.984	1
total $X_2 \perp\!\!\!\perp X_3 \mid X_1$	4.072	2

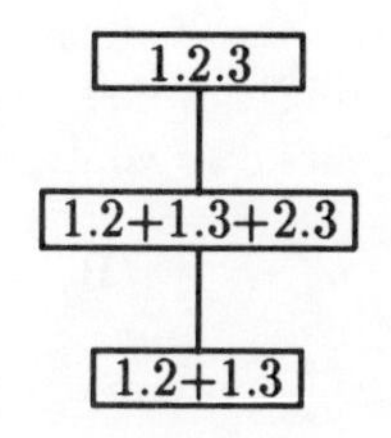

and deserves further comment: while the difference in the cprs is not significant and a common value is acceptable, this common value differs from zero. Partitioning the two degrees of freedom for testing conditional independence has revealed a significant association. If one's aim is in finding a parsimonious model to represent the data then $M_2 = 1.2 + 1.3$ is adequate; on the other hand if the object of the analysis is to examine the survival-grade interaction in the two diagnostic centres, then one has to conclude there is some (albeit small) evidence of interaction, an interaction which is the same in the two centres. □

EXAMPLE 7.7.2 Reconsider the Bishop *et al.* (1975) data comparing infant survival at two clinics, reproduced in Table 1.2.1. The variables clinic, care and survival are denoted by X_1, X_2 and X_3, respectively. Consider partitioning the deviance against the independence of survival from the other two factors.

For the independence model $M = 1.2+3$ in which survival, X_3, is independent of clinic and care, we compute the totals in the (X_1, X_2) and X_3 margins and calculate the fitted values from:

$$\text{fv}\,(1.2 + 3) = N\hat{p} = n_{12}n_3/N.$$

For the model $1.2 + 1.3$ of the independence of survival and care given clinic the fitted values are given from

$$\text{fv}\,(1.2 + 1.3) = N\hat{p} = n_{12}n_{13}/n_1,$$

and are:

fitted values			survival, X_3		
clinic, X_1	care, X_2	model	no	yes	total
A	less	obs	3.00	176.00	179
		fv(1.2+3)	6.51	172.49	
		fv(1.2+1.3)	2.63	176.36	
A	more	obs	4.00	293.00	297
		fv(1.2+3)	10.80	286.20	
		fv(1.2+1.3)	4.37	292.63	
B	less	obs	17.00	197.00	214
		fv(1.2+3)	7.78	206.22	
		fv(1.2+1.3)	17.01	197.99	
B	more	obs	2.00	23.00	25
		fv(1.2+3)	0.91	24.09	
		fv(1.2+1.3)	1.99	23.01	
		total	26.00	689.00	715

For example, in the leading cell (26)(179)/715 = 6.51 and (7)(179)/476 = 2.63. The deviances are twice the sum over cells of obs log(obs/fitted):

$$\begin{aligned} \text{dev}\,(1.2+3) &= 2\{3\log(3/6.51) + \cdots + 23\log(23/24.09)\} = 17.828, \\ \text{dev}\,(1.2+1.3) &= 2\{3\log(3/2.6) + \cdots + 23\log(23/23.0)\} = 0.082. \end{aligned}$$

Quite clearly the conditional independence model gives a much better fit than the complete independence model.

Consider partitioning the deviance between the saturated model 1.2.3 and the base model 1.2 + 3. There are two issues involved. Firstly, if possible, we should like to give meaning to the difference in deviances dev (1.2 + 3) − dev (1.2 + 1.3). As the model 1.2 + 3 is nested in 1.2 + 1.3, it is, of course, a maximised log-likelihood ratio test statistic. But also, recall from Proposition 2.2.6 that $X_3 \perp\!\!\!\perp (X_1, X_2)$, on the one hand, and that the conditional independence $X_2 \perp\!\!\!\perp X_3 | X_1$ and the marginal independence $X_1 \perp\!\!\!\perp X_3$, on the other, are equivalent; and furthermore, recall from Proposition 4.5.2 that the information proper is additive over these independences. The deviance difference between 1.2+3 and 1.2+1.3 is a test for the marginal independence $X_1 \perp\!\!\!\perp X_3$, and is clearly rejected.

Secondly, there is a symmetry in the roles played by variables X_1 and X_2. By interchanging these we may obtain another partition of the deviance of 1.2+3. Hence we fit the model 1.2+2.3 and display the results in the lattice

diagram

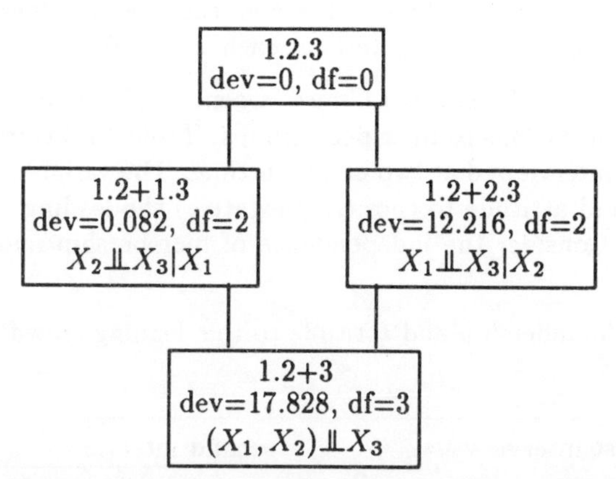

The analysis of deviance can be drawn up in two ways, corresponding to the two chains in this lattice from the maximal to the minimal model.

deviance due to	dev	df	deviance due to	dev	df
$X_2 \perp\!\!\!\perp X_3 \mid X_1$	0.08	2	$X_1 \perp\!\!\!\perp X_3 \mid X_2$	12.22	2
$X_1 \perp\!\!\!\perp X_3$	17.75	1	$X_2 \perp\!\!\!\perp X_3$	5.61	1
$(X_1, X_2) \perp\!\!\!\perp X_3$	17.83	3	$(X_1, X_2) \perp\!\!\!\perp X_3$	17.83	3

In standard analysis of variance, with a Normal errors model, a consequence of a balanced design is that the analysis of variance be invariant to the chain in the lattice. This is not the case here. It is not so much due to the lack of balance as to the manifest fact that there is no *a priori* reason to suppose that the deviance attributable to the conditional independence of X_2 and X_3 given X_1 should be the same as that attributable to their marginal independence. The deviance differences are estimating different features of the joint distribution. From the identity in information divergence discussed at the end of Section 4.5, it should no longer be surprising to find the equality of the deviance differences

$$17.75 - 12.22 = 5.53 = 5.61 - 0.08.$$

According to the block independence lemma $X_3 \perp\!\!\!\perp (X_1, X_2)$ is equivalent to the two conditional independence statements $X_2 \perp\!\!\!\perp X_3 | X_1$ and $X_1 \perp\!\!\!\perp X_3 | X_2$, but the information proper is not additive over these statements. Hence there is no corresponding deviance additivity. □

Partitioning by the conditioning variable: local tests

Another especially relevant instance of partitioning the deviance into additive components occurs with tests for conditional independence. The deviance against conditional independence is a joint or combined test of independence at each of the different values of the conditioning variables. Here we show how to break this up into local tests at each value. We start with an example.

EXAMPLE 7.7.3 The Leading Crowd. Coleman (1964) discusses a survey of schoolboy relationships to their peer-groups. Table 7.7.1 comes from a sample of 3398 boys interviewed at two points in time. The variables are membership of (yes/no) and attitude to (positive/negative) the leading crowd among the peer group. Consider the independence of membership and attitude at the

Table 7.7.1: Membership and attitude to the 'leading crowd'. From Coleman (1964).

first interview			second interview				
		mem2	yes	yes	no	no	
mem1	att1	att2	pos	neg	pos	neg	total
yes	pos		458	140	110	49	757
yes	neg		171	182	56	87	496
no	pos		184	75	531	281	1071
no	neg		85	97	338	554	1074
		total	898	494	1035	971	3398

second interview conditional on the variables taken at the first interview. The deviance against this hypothesis has a value of 14.83 on 4 df, which is significant at the 5% point, indicating some form of dependency. The 4 degrees of freedom come from four 1 df tests for independence, one for each of the four possible categories of the variables at the first interview. Knowledge that the overall deviance is large does not tell us much about the form of the interaction; for example, are the four categories homogeneous? and if not, does one group stand out? Partitioning allows us to explore this in more detail:

1st-interview				estimated
mem1	att1	dev	df	cpr
yes	pos	3.56	1	0.38
yes	neg	3.56	1	0.38
no	pos	2.87	1	0.26
no	neg	4.85	1	0.36
total		14.83	4	

Only the deviance in the last category reaches the 5% significance level, and one might decide to treat this group differently from the others. However since the conditional cross-product ratios are so similar, it would certainly appear better to fit a single common parameter, the partial cpr, to model this interaction between membership and attitude at the second interview. This leads to fitting models composed entirely of two-way interaction terms, discussed later in Chapter 9. This decomposition of the deviance has suggested more interesting parsimonious models. □

The theory of this decomposition is briefly outlined here. Partition K into $a \cup b \cup c$, condition upon X_a, and consider the deviance against the conditional independence of X_b and X_c. The order of summation in the deviance can be rearranged as a summation over x_{bc} for each value of x_a:

$$2\sum_{x} n_{abc} \log \frac{n_{abc}n_a}{n_{ab}n_{ac}} = \sum_{x_a} \left\{ 2\sum_{x_b, x_c} n_{abc} \log \frac{n_{abc}n_a}{n_{ab}n_{ac}} \right\}.$$

The summand within the curly brackets is the deviance against the independence of X_b and X_c at each value of X_a. Hence

$$\text{dev}\,(X_b \perp\!\!\!\perp X_c | X_a) = \sum_{x_a} \text{dev}\,(X_b \perp\!\!\!\perp X_c | X_a = x_a),$$

which establishes that the deviance is additive over the values taken by X_a.

Reparameterisation

There are other sensible ways to parameterise the log-linear expansion of the density function; the choice is dictated by the ease with which we can identify interesting submodels. In this analysis of conditional independence the following construction is useful, where, to keep the notation simple, it is supposed that X_1 and X_2 are binary random variables. The log-linear expansion for the marginal distribution of (X_1, X_2) is

$$\log f_{12}(x_1, x_2) = u_\phi + u_1 x_1 + u_2 x_2 + u_{12} x_1 x_2 \quad \text{for} \quad (x_1, x_2) \text{ in } \{0,1\}^2,$$

so that X_1 and X_2 are independent if and only if $u_{12} = 0$. In the conditional distribution, each of the four parameters depends on the value taken by X_a, so that the log-linear expansion for (X_1, X_2) conditional on $X_a = x_a$ is

$$\log f_{12|a}(x_1, x_2; x_a) = u_a + u_{\{1\}\cup a} x_1 + u_{\{2\}\cup a} x_2 + u_{\{1,2\}\cup a} x_1 x_2$$

for (x_1, x_2) in $\{0,1\}^2$. Multiplying by f_a can only affect the first u-term u_a. Written is this way the u-term, $u_{\{1,2\}\cup a}$, has the interpretation that, for each value of x_a, it is the conditional cross-product ratio between X_1 and X_2. Estimates of these parameters are reported in Example 7.7.3 above. This parameterisation emphasises that a test for the conditional independence of X_1 and X_2, is a test that the parameters $u_{\{1,2\}\cup a} = 0$, for all values of x_a and one degree of freedom is contributed by each. When a has two vertices, as in the last example, there are 4 degrees of freedom for the joint test.

7.8 Diagnostics for Log-linear Models

Diagnostic procedures are concerned to check the validity of the assumptions made in model fitting, and, if necessary, identify aberrations, and refit a modified model to obtain more reliable estimates. Unlike the study of diagnostics for models based on Normal distributions there are fewer available methods appropriate to the Multinomial distribution, see Haberman (1974) and Bishop *et al.*.

Residual analysis

A basic idea of diagnostic analysis is to compare the observed counts and fitted values in each cell of the contingency table. The Pearson residuals are defined as $(\text{obs} - \text{fv})/\sqrt{\text{fv}}$ for each cell in the table. The deviance residuals are defined as the square root of the absolute value of $\text{obs} \log \text{obs}/\text{fv}$ and signed according to whether the fitted values exceed the observed values or not. Inspection of either set of these statistics is a useful exercise in fitting any log-linear model to a contingency table. We make some remarks:

(i) In order to make comparative judgements about discrepancies between the observed and fitted values in individual cells they have to be standardised. A heuristic argument that can be made precise for the standardisation in the Pearson residual goes as follows: the Multinomial distribution can be derived as the distribution of a set of independent Poisson variables conditioned on their sum. The expected value is equal to the variance in the Poisson distribution, and the Normal approximation to the Poisson is good when the expected value is large. Consequently, as the fitted value converges to this expected value in large samples, the Pearson residual has a standard Normal distribution. The result follows for the deviance residuals as a corollary by using a Taylor series expansion.

(ii) Large residuals go to identify lack of model fit. There are at least two potential sources of inadequacy: the first is that there are not sufficient interactions in the model to account for the interactions in the data. For instance, the raw ratio of the observed values to the values fitted under the model of complete independence is often informative about missing interactions because the main effects have been eliminated. A second source is the possible failure of any log-linear model to fit the data. For instance, if the observation came from a mixture distribution composed of one type of individual that responded 0 on every variable, and another that responded independently on each, then the density function is

$$\log f(x) = \alpha p(x) + (1 - \alpha) q(x),$$

where $p(0) = 1$, $q(x)$ expresses independence, and α is the proportion of the zero-responders in the population. Only the saturated model can provide an

adequate fit, but a large residual in the all-zero cell from the independence model may indicate departures in this direction.
(iii) A great deal of care is needed in interpreting residuals due to their lack of independence. For instance, while there are 4 cells in a 2×2 table, when fitting the model of independence each standardised residual is based on an identical absolute difference between observed and fitted. There is only 1 degree of freedom to test model fit.
(iv) The parallel diagnostic procedure, in fitting graphical Gaussian models, to computing Pearson or deviance residuals is to compare observed and fitted correlation coefficients. This is equivalent to comparing fitted correlations from one model to the fitted values from the saturated model and the appropriate standardisation is determined by the Wishart distribution.

The parallel to comparing the observed and fitted partial correlations is to compare the observed and fitted u-terms distinguishing between those interactions that are and are not included in the model.

A parallel to multivariate Normal diagnostic procedures based on the original sample, is to assess the effect of deleting individual observations from the sample, which we turn to next.

Deleting observations from a contingency table

No statistical analysis should hang on the outcome of deleting a single observation from the sample, and in regression analysis, this consideration has generated a substantial literature devoted to methods for identifying outliers and measures of influence: see Cook and Weisberg (1982), or Atkinson (1985).

The term outlier, which in regression designates an observation that lies far away from the average, in the context of a contingency table becomes an observation which has a low probability of occurrence. This implies they are comparatively easy to identify, occurring in cells with a relatively small cell count. Our concern is that they might have an undue influence on the choice and fit of the model.

Most procedures for measuring the influence of an observation delete that observation from the sample, refit the model, and then compare the results from analysing the full sample with those from the analysis of a depleted sample. Comparing the values of the fitted parameters before and after deletion is a standard comparison, however, it will not work satisfactorily for many models for a contingency table. For example, consider estimating the parameters of the saturated model. In our parameterisation of the log-linear model, adjusting the last cell in the table, $(1,1,1,1)$ in a 2^4 table, affects only the value of the four-way interaction, while adjusting the first cell, $(0,0,0,0)$, affects every u-term.

A more natural measure of the influence of a deletion is to compare the edge exclusion deviances against the saturated model for the original and the perturbed table. This does not depend on the choice of parameterisation for the log-linear model, and while it does not give a unique measure of influence, it does reduce the diagnostic task to more manageable proportions.

EXAMPLE 7.8.1 Consider the attitude data from Stouffer and Toby (1951) in Table 7.8.1. Each respondent was presented with four situations, with

Table 7.8.1: Role conflict response of 216 respondents. From Stouffer and Toby (1951).

		X_3	0		1	
X_1	X_2	X_4	0	1	0	1
0	0		42	23	6	25
0	1		6	24	7	38
1	0		1	4	1	6
1	1		2	9	2	20

the response from each being classified as 0 denoting 'universalistic' or as 1 denoting 'particularistic'. There are several interesting ways to perturb this table and for illustrative puposes we choose the following: the two observations in the cells with a single count are deleted separately and together, and for the purpose of comparison one observation is deleted from cell (0, 0, 0, 0). The edge exclusion deviances from the saturated model are:

original table

1	*			
2	8.22	*		
3	2.29	10.29	*	
4	4.56	17.53	19.00	*
	1	2	3	4

perturbed cell (0,0,0,0)

1	*			
2	8.14	*		
3	2.26	10.07	*	
4	4.46	17.03	18.49	*
	1	2	3	4

perturbed cell (1,0,0,0)

1	*			
2	11.69	*		
3	4.76	11.47	*	
4	8.58	18.87	19.90	*
	1	2	3	4

perturbed cell (1,0,1,0)

1	*			
2	10.30	*		
3	1.00	11.47	*	
4	6.79	18.39	20.63	*
	1	2	3	4

	cells (1,0,0,0) and (1,0,1,0)			
1	*			
2	13.77 (4)	*		
3	0.74 (3)	10.29 (3)	*	
4	10.81 (4)	19.73 (4)	18.94 (3)	*
	1	2	3	4

Each deviance has 4 df in the original table but because of the zero count in the margin in the last perturbation, the degrees of freedom have to be adjusted in the manner described at the end of this section. The values are given in brackets and each test for the conditional independence of X_3 from another variable has lost a degree of freedom.

There are some fairly disquieting differences between the deviances for the table with two deleted observations: only the test for the edge $(2,3)$ remains unaltered, and some tests change quite substantially: the test for excluding the edge $(1,2)$ increases from 8.22 to 13.77. In particular the change in the test for $(1,4)$ from 4.56 to 10.81 clearly might lead to selecting a different model. The change in deviances is not all in one direction, for instance $(1,3)$ moves from 2.29 to 0.74. In conclusion, the edge exclusion deviances from this table are not invariant to deletion of entries in cells with low probabilities. This concurs with the belief that fitting models with large numbers of parameters requires large sample sizes. □

When a table has zero entries in marginal tables corresponding to a u-term in the model, it can be seen from Proposition 7.4.1 that the coefficient of the u-term in the log-likelihood function $l(u;n) = \sum_a \sum_{x_a} n_a(x_a)u_a(x_a)$ will be zero. Consequently, the likelihood is flat as the parameter varies and the parameter is not estimable. On the other hand the maximum value of the likelihood is usually well defined though usual methods of computing the value may not converge. The degrees of freedom for any test that sets the parameter to zero have to be adjusted.

7.9 Exercises

1: Discuss the possibility of constructing a two dimensional Bernoulli random vector with a table of probabilities p given by $p(0,0) = 0.1$, $p(0,1) = 0.2$, $p(1,0) = 0.3$ and $p(1,1) = 0.4$, by throwing two independent biased coins.

2: Consider the saturated log-linear model for a 2×2 table:

$$\log p_{12}(x) = u_\phi + u_1(x_1) + u_2(x_2) + u_{12}(x_1, x_2),$$

with constraints

$$\sum_{x_1} u_1(x_1) = 0,\ \sum_{x_2} u_2(x_2) = 0,\ \sum_{x_1} u_{12}(x_1, x_2) = 0 \text{ and } \sum_{x_2} u_{12}(x_1, x_2) = 0.$$

Show that $u_{12}(1,1) = -u_{12}(0,1) = -u_{12}(1,0) = u_{12}(0,0)$ and is given by $4 \log \text{cpr}(X_1, X_2)$.

3: Write out the log-linear expansion

$$\log f_{12}(x) = u_\phi + u_1(x_1) + u_2(x_2) + u_{12}(x_1, x_2)$$

explicitly for each cell of a 2×3 table, using the standard constraints on the u-terms.

4: Suppose that X_1, X_2, X_3, and X_4 are binary random variables. Find the independence graph if the log-linear expansion of p is
(a) $\log p = \text{const} + x_2 + x_3 + x_4 + x_1$;
(b) $\log p = \text{const} + x_2 + x_3 + x_4 + x_1 + x_2x_3 + x_2x_4 + x_3x_4 + x_2x_3x_4$;
(c) $\log p = \text{const} + x_2 + x_3 + x_4 + x_1 + x_2x_3 + x_3x_4 + x_4x_1 + x_1x_2$;
(d) $\log p = \text{const} + x_2x_3x_4x_1$.

5: The 2^4 table p classified by (x_1, x_2, x_3, x_4) is proportional to

216 504 24 56 54 126 6 14 36 4 144 16 144 16 576 64

when displayed in standard order. Find the log-linear expansion of p and use it to derive the pairwise conditional independences between X_1, X_2, X_3, X_4. Construct the independence graph.

6: Write down the log-linear expansion and the model formula when the associated independence graph of the graphical model is

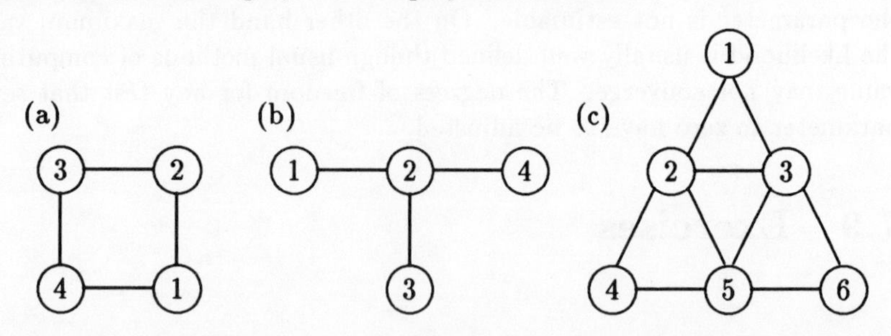

7: Find the lattice diagram of the equi-probability models in three dimensions.

8: Verify there are 167 models in the lattice of all hierarchical log-linear models in four dimensions, including those of equi-probability. More generally, the following table, for the number of distinct log-linear models for a contingency table with a given dimension, is based on Lauritzen (1982), see also Darroch *et al.* (1980).

	dimension			
	2	3	4	5
hierarchical	5	19	167	7,580
graphical	5	18	113	1,450
decomposable	5	18	110	1,233
equi-probability	4	8	16	32

9: Write down the likelihood equations for the two-way independence model appropriate to an $r \times c$ contingency table.

10: Find the maximum likelihood estimates for the non-hierarchical model $\log f_{123}(x_1, x_2, x_3) = u_\phi + u_{123}x_1x_2x_3$, with $x_i = 0, 1$ based on 36 observations:

	$x_2 = 0$		$x_2 = 1$	
x_3	0	1	0	1
$x_1 = 0$	1	2	3	4
$x_1 = 1$	5	6	7	8

11: Write down the likelihood equations for a 2^4 contingency table when fitting the graphical model with the chordless 4-cycle as independence graph.

12: Verify the fitted values for the model $1.2 + 1.3$ for the breast cancer data in Example 7.5.3. Verify that the estimates of the u-terms under two models are

	$1.2 + 1.3$	$1.2 + 1.3 + 2.3$
u_ϕ	3.4	3.6
u_1	0.18	0.14
u_2	0.53	0.26
u_3	0.73	0.49
u_{12}	−0.68	−0.69
u_{13}	0.08	0.14
u_{23}	–	0.40
u_{123}	–	–

13: Example 7.6.1, in analysing a cross-classification of psychiatric patients, fitted the model $M = 1.3.4 + 3.2$. Show that excluding each of the four edges from M leads to fitting the models

edge	model	edge exclusion dev diff	df
(1,3):	1.4+3.4+3.2	31.58	2
(1,4):	1.3+3.4+3.2	13.57	2
(2,3):	1.3.4+2	16.97	1
(3,4):	1.4+1.3+3.2	4.99	2

and that each of these has direct estimates. Verify the values of the deviance differences of these models to M. Note that the edge (3,4) may be dropped from M. The edge exclusion deviances are calculated from the differences between these deviances and the deviance of M. Note that these deviance differences are not the same as the original edge exclusion deviances: though the edge is fixed, its context is different.

14: One way to estimate the strength of an interaction in a fitted model is to calculate the edge exclusion deviances with the fitted values from the model M replacing the observed counts. Show that for the psychiatric patient data, with $M = 1.3.4 + 3.2$, these estimated edge strengths are

1	*			
2	0.00	*		
3	30.15	15.47	*	
4	13.58	0.00	4.78	*
	1	2	3	4

Compare these to the edge exclusion deviances given in the example.

15: The parameter estimates are one way to measure the strength of included interactions. Show that for the psychiatric patient data, with $M = 1.3.4 + 3.2$

	estimate	est s.e.
$\hat{u}_{23}$	0.8858	0.2174
$\hat{u}_{14}$	−0.9224	0.3334
$\hat{u}_{13}$	−1.320	0.3038
$\hat{u}_{34}$	−0.5847	0.3149
$\hat{u}_{134}$	0.1781	0.4604

16: Show that the exclusion deviances against conditional independence for the leading crowd data in Example 7.7.3 are

mem1	*			
att1	3.67	*		
mem2	988.80	15.70	*	
att2	4.00	262.40	14.83	*
	mem1	att1	mem2	att2

Each statistic has 4 df. Note the extremely strong association between membership at the two points in time, and between attitude at the two points, which entirely dwarf the other effects in the table.

17: The following detergent preference data comes from Ries and Smith (1964). Some 1008 people were given two brands of detergent, X and M, and subsequently asked 4 questions, concerning water softness (soft or not), previous use (yes or no), water temperature (high or low), brand preference (X or M).

	previous use, 3	yes		no	
	temperature, 4	high	low	high	low
softness, 1	preference, 2				
yes	X	19	57	29	63
	M	29	49	27	53
no	X	47	84	75	134
	M	90	107	53	92

Calculate the edge exclusion deviances and using a significance level of 5% construct the independence graph. Fit the corresponding graphical model.

18: Reconsider the analysis of the care clinic survival data discussed in Example 7.7.2 with clinic, care and survival denoted by X_1, X_2 and X_3. Show that the deviances of all the appropriate models for the conditional distribution of survival, X_3, given the other two variables, are

model	dev	df
1.2.3	0.00	0
1.2 + 1.3 + 2.3	2.29	1
1.2 + 1.3	105.18	2
1.2 + 2.3	53.440	2
1.2 + 3	151.02	3

Draw up the decompositions of the deviance along the chains of the appropriate lattice and attribute meaning to the components.

19: Anderson (1954) considers a two-interview panel study where at each interview the respondent was asked if he or she had seen an advertisement for a certain product and if he or she had bought that product.*

*Reproduced by permission from Lazarfeld, P.E. (Ed.) (1954) *Mathematical Thinking in the Social Sciences.* The Free Press.

		second interview			
	see, 3	yes		no	
first interview	buy, 4	yes	no	yes	no
see, 1	buy, 2				
yes	yes	83	8	35	7
	no	22	68	11	28
no	yes	25	10	95	15
	no	8	32	6	493

Calculate the edge exclusion deviances and partition the deviance for the conditional independence of see and buy at the second interview, according to the 4 groups at the time of the first interview.

Chapter 8

Model Selection

How may one select a graphical model? Even in moderate dimensions there are thousands of possible graphical models to choose from, many of them very similar; should one fit them all? if not, how should those fitted be chosen? how should the best one be determined? what are the properties of the fitted estimates in the chosen model? and so on.

Selecting a graphical model is an instance of the well known statistical problem of determining, on the basis of an observed sample, which parameters should be set to zero and which should be freely estimated. On the one hand, including extra parameters leads to a better fit to the the data; on the other, the fewer parameters there are the better, because they are more efficiently estimated, Altham (1984), and the resulting model is simpler. There have to be enough parameters in the model to represent the real effects present in the data structure while there should be few enough to guard against analysing noise induced by sampling variation. Including all parameters works in one direction, excluding them all in the other, and a principal aim of model selection is to reach a compromise and select one (or more) models to parsimoniously represent these effects.

The essential contribution of statistical modelling is the way in which it deals with noise. Under a given probability model for the observations, the known sampling distribution of the likelihood ratio sets a criterion for the effect of inherent random variation. The observed effect can then be compared to this standard, and a decision made as to whether the effect is real or apparent. If a particular effect can be explained by random variation then it can be discarded and the model simplified. An acceptable fit is one where the fitted model only differs from the observed data by an amount consistent with sampling variation.

Before proceeding, it should be pointed out that model selection is not essentially a computing problem, for, even when we are able to compute and examine the fit of all possible models, there is no unequivocal answer to the question of which is the best fitting model. This is because modelling has many diverse aims, for instance there may be particular interest in one parameter, or in one conditional independence, and this may lead to a different model than that selected, say, for empirical forecasting. None the less, methods that attempt to reach a solution to the selection problem without fitting every possible model are to be preferred, and at the end of a good selection procedure, a great deal can be inferred about models that have not been explicitly fitted.

After a rather general discussion of the issues involved the chapter presents a few practical techniques by way of examples.

8.1 Issues in Model Selection

A standard problem facing the statistical analyst is to select one, or perhaps a few, well fitting models from a potentially large number of possibilities; choosing a graphical model is another area where this issue is manifest. We briefly touch on some of the general statistical issues arising in model search and selection procedures, which are very similar to the well known problems of regression modelling, see Draper and Smith (1981).

The lattice of models

The universe of contending models often has a lattice structure and the regression problem of choosing the 'best' subset generates a Boolean lattice. Given a sample of N observations on a response random variable Y together with a vector of p explanatory regressor variables, the maximal model of the lattice is the linear regression

$$EY = \alpha + \beta_1 X_1 + \beta_2 X_2 + \ldots + \beta_p X_p,$$

with its associated directed independence graph. For example, with $p = 3$,

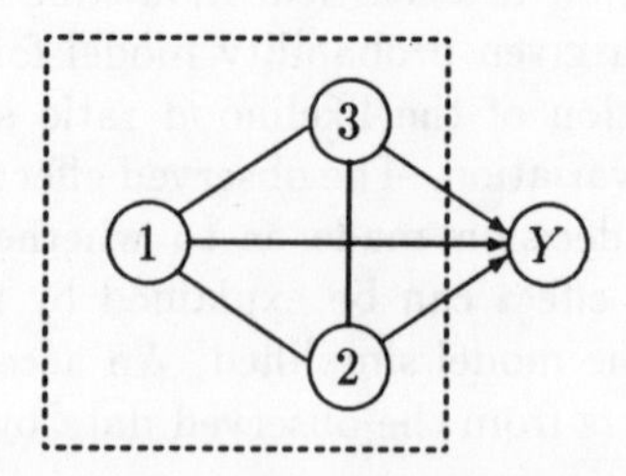

The elements of the lattice are the 2^p possible regression models generated by excluding edges from this graph, and using an obvious model formulae

notation, as in Section 7.3, the lattice diagram is

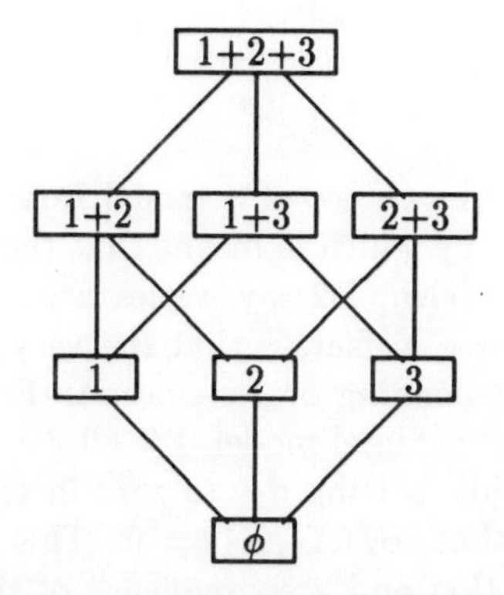

The minimal model is the model ϕ containing just the constant $EY = \mu$.

The lattice of graphical Gaussian models for a k-dimensional random vector X is generated by setting terms to zero in the expansion of the log density function

$$\log f_K(x_1, x_2, \ldots, x_k) = \text{const} + \sum_{i,j} d_{ij} x_i x_j.$$

The diagonal elements of the inverse variance matrix $D = \{d_{ij}\}$ are arbitrary, and the model selection problem is to decide which of the $\binom{k}{2}$ off diagonal elements can be set to zero. There are $2^{\binom{k}{2}}$ possible models, which when $k = 3$, can be displayed in the lattice

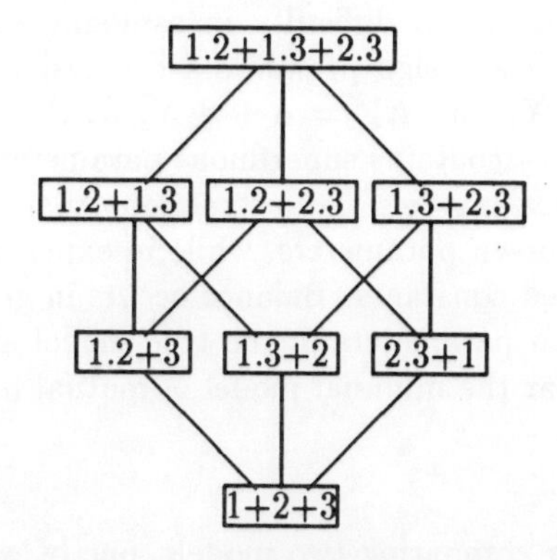

The graphical model selection problem is harder than the regression problem because it involves substantially more possibilities when $k > 3$; but both problems correspond to choosing one (or a few) elements from a Boolean lattice. The extended log-linear model selection problem from the lattice of

all possible hierarchical models is further complicated because the lattice is not Boolean, see Figure 7.3.1, *i.e.* it is not generated by a set of binary conditions.

Non-orthogonality

One problem besetting the choice of a model from the graphical Gaussian lattice is lack of balance, by which is meant that the increase in deviance by dropping the interaction term, 1.2 say, varies according to the model from which it is dropped. Worse in fact, is that the very meaning of the interaction term may change depending on its context. For instance, with $k = 3$, setting d_{12} to zero in the maximal model, $1.2 + 1.3 + 2.3$, is a hypothesis that $\text{cov}(X_1, X_2|X_3) = 0$, while setting d_{12} to zero in the context of the model $1.2 + 3$ is a hypothesis that $\text{cov}(X_1, X_2) = 0$. This is an important feature of graphical model selection and a consequence of the natural imbalance in multivariate observational studies.

Exactly the same phenomenon may occur in the regression lattice above: testing that $\beta_3 = 0$ in the context of the maximal model $1 + 2 + 3$ can be interpreted as a test for $\text{cov}(Y, X_3|X_1, X_2) = 0$ while in the context of the model $1 + 3$ it is a test that $\text{cov}(Y, X_3|X_2) = 0$. However, when the study is designed, and is partially balanced in the sense that

$$\text{cov}(X_2, X_3|X_1) = 0,$$

then the partial covariances $\text{cov}(Y, X_3|X_1, X_2)$ and $\text{cov}(Y, X_3|X_2)$ must be equal, and the tests are the same. Note that if all the regressors are orthogonal, corresponding to the mutual independence of X_1, X_2 and X_3 and complete balance, there is no difficulty in assessing each regression coefficient in isolation. Such a design prohibits a conclusion that while both the models $EY = \alpha + \beta_1 X_1$ and $EY = \alpha + \beta_2 X_2$ fit the data well, the model $EY = \alpha + \beta_1 X_1 + \beta_2 X_2$ contains superfluous parameters.

However, in graphical models the interactions that, if zero, may lead to orthogonality are unknown parameters, while in experimental design models they are pre-determined constants. Balance occurs in graphical models when the relevant interaction parameters in the true model are near zero, such as when this model is near the minimal model of mutual independence.

The deviance

The standard test for comparing two models, one of which is nested in the other, is based upon the generalised likelihood ratio test statistic. When one model is the saturated model, this is called the deviance and otherwise, for nested models, the deviance difference. In regression, when it is assumed that the scale parameter is known, deviance computations reduce to decompositions of the total sum of squares. As a statistic the deviance has certain advantages:

(i) it is invariant to different one to one parameterisations of the model;

(ii) it allows a unified approach to graphical model selection from either Gaussian, log-linear or mixed graphical models;

(iii) it can be interpreted as an information divergence, so that the relative sizes of the deviance give an instant comparison of two arbitrary models; and above all,

(iv) the deviance has a well-understood asymptotic sampling distribution.

The essential sampling results are that

(a) the difference in deviances between two nested models has a chi-squared distribution on the null hypothesis; and

(b) the deviance differences are independently distributed when they are components of a single chain of models passing from the minimal to the maximal model. This is intimately connected to the chi-squared partitions that permeate the analysis of variance.

These sampling results are asymptotic, and are proved as a corollary to the more general results on generalised likelihood ratio tests, see Cox and Hinkley (1974). In certain instances, where the sample size is relatively small and the number of variables in the model is large these asymptotic approximations are not adequate and their use is misleading. We discuss the use of exact procedures later. Note that the joint distribution of the edge exclusion deviances, which correspond to taking the first component of distinct chains in the lattice, is unfortunately not covered by this second result.

Sampling distribution of the selected model

The effects of model search and selection on the sampling distribution of the fitted model are not well understood. Even in the much simpler linear regression scenario, for example, it is not clear what distribution is followed by the regression coefficients in a model selected by a standard stepwise procedure. Since the selection procedure usually excludes terms with small estimated coefficients, the observed coefficients in the selected model are potentially biased away from zero, however the magnitude of such bias is unknown. Results from standard distribution theory are strictly applicable only in inference from a fixed model. There is a similar lack of information on the relative power of different procedures.

Multiple testing

Model selection consists of a sequence of comparisons between models. If each comparison is construed as a significance test and performed at the 5% level, is it possible to specify the size (type I error rate) of the overall procedure? Certain procedures can be constructed with this property of a given error rate, see for example, the simultaneous tests based on Bonferroni inequalities and discussed in Seber (1977); or, in context of contingency tables, a procedure due to Aitkin (1979), and used for selecting graphical log-linear models by Whittaker (1982). However, the calculation of the overall type I error rate is made under the assumption of the minimal model which, in the context of graphical models, is the model of mutual independence. This rather unrealistic assumption casts doubt on the value of the exercise. Of course, this is not to say that tests that take the problem of simultaneous inference seriously are without value: for example, if one of several additional terms is to be included, and the choice is to be based on the largest test statistic, then the significance level should pertain to the distribution of the maximum. For further discussion of this area, see Miller (1981) and Gabriel (1969).

8.2 Log-linear Model Selection

In this section we briefly summarise the contribution made by the concepts and theory of graphical models to understanding the difficulties of log-linear model selection. While the general statistical issues, for example, of multiple testing and of elucidating the distributional properties of the selected model, are no different from the corresponding regression problem, model selection is more difficult with discrete data than for continuous data because the lattice of all possible models is, for a given dimension k, both larger and has a more complicated structure, see Figure 7.3.1 for $k = 3$. The relative simplicity of graphical Gaussian models is due to the absence of interaction terms higher than two in the multivariate Normal density function. A variety of authors have suggested ways of tackling the selection procedure; we mention discussion by Goodman (1970, 1971), Bishop *et al.* (1975), Brown (1976), Wermuth (1976b), Benedetti and Brown (1978), Whittaker and Aitkin (1978), Sakamoto and Akaike (1978), Aitkin (1979), Whittaker (1982, 1984b), Edwards and Kreiner (1983), Havranek (1984), and Edwards and Havranek (1985, 1987), though this is far from comprehensive.

A list of the contributions of graphical modelling to understanding model selection includes the following.

(i) Associating an independence graph with each log-linear model and the concomitant discovery of a new incremental unit in stepwise model search procedures: the inclusion or exclusion of an edge from the independence graph.

(ii) The elucidation of a subset of log-linear models defined entirely by conditional independence constraints: the graphical log-linear models.

(iii) The characterisation of all log-linear models that have direct estimates, the decomposable models, by the triangulation property of their graphs.

(iv) Demonstration of a one to one correspondence of graphical log-linear with graphical Gaussian models, and finding other classes of models with such a property.

(v) An explanation of the equivalence of certain test statistics in inter-model comparisons by the reducbility properties of the independence graph.

Restricting model search to graphical models simplifies model selection in two ways: firstly, the overall magnitude of the problem is reduced, but secondly and more importantly, the resulting graphs focus the aim of the selection procedure.

BFH strategies

In their excellent book on log-linear modelling Bishop, Fienberg and Holland (1975, p.155ff) outline four strategies that can be used to select 'best' models:

1. standardised u-terms in the saturated model;

2. fitting models with all terms of uniform order;

3. examining the deviance components for models with direct estimates; and

4. fitting the partial association models.

Graphical concepts illuminate some of the relationships between these differing strategies, considered here in the context of the four dimensional lattice of all hierarchical log-linear models. The lattice is too complicated to display in its entirety, but some of its elements are

maximal	1.2.3.4
all three-way	1.2.3+1.2.4+1.3.4+2.3.4
all two-way	1.2+1.3+1.4+2.3+2.4+3.4
one pairwise independence	1.2.3+1.2.4, 1.2.3+1.3.4, ..., 1.3.4+2.3.4
mutual independence	1+2+3+4
conditional equi-probability	1.2.3, 1.2.4, ..., 2.3.4

The third and fourth BFH strategies are closely related to fitting graphical models. Fitting the partial association models is exactly equivalent to evaluating the edge exclusion deviances from the saturated model, the basic test of pairwise conditional independence. The third strategy, fitting the *decomposable* models, extends the fourth more systematically, and in Chapter 12 a proof is given that shows decomposable models are those purely graphical models with triangulated independence graphs. In fact all decomposable models can be fitted by fitting all conditional equi-probability models, Whittaker (1984b).

The first BFH strategy, inspecting the u-terms in the saturated model standardised by their standard errors, and calculated by fitting the model 1.2.3.4 at the head of the lattice, is suggested by Goodman (1970) for log-linear models. Since each parameter estimate is 'adjusted for' every other parameter estimate, any test for significance is a partial, not marginal, test. The procedure surveys the whole lattice in just one model fit, and in fact it is a standard way of choosing an initial base model. But as the sole selection stratagem, it raises certain concerns:

(i) it has to be shown that the selection is invariant to the particular parameterisation employed – that is, invariant to the constraints imposed to ensure the u-terms are identifiable and care must be taken to keep the model hierarchical;

(ii) the procedure requires standard errors for the parameters;

(iii) the procedure tests parameters separately and not jointly but a move in the lattice usually corresponds to setting more than one parameter to zero simultaneously;

(iv) there is no guarantee that the procedure will detect a conditional independence.

All j-way interaction search procedures

The second BFH strategy, fitting the models of uniform order, sequentially tests the hypotheses:

> the k-way interaction may be dropped from the model, all k-1-way interactions may be dropped from the model, all three-way interactions may be dropped from the model, all two-way interactions may be dropped from the model;

and is the odd-man-out. Though the procedure can lead to simplification in the sense that higher order interactions may be eliminated, it may well miss the fact that a two-way interaction, say 1.2, and all higher order interactions containing 1.2, may be simultaneously eliminated. There is a danger that this simplification procedure may not detect pairwise conditional independence and thus may not make use of the power of graphical models to explain and simplify complex contingency tables. Figuratively, if the full lattice of hierarchical log-linear models is enclosed in the dashed box:

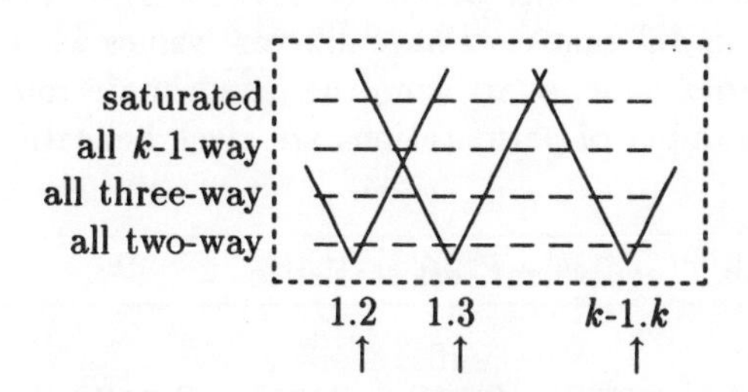

we oppose analyses based on decomposing the lattice by the horizontal rows and propose the analysis based on decomposing the lattice into vertical cones.

Reducibility and redundancy

To illustrate the type of redundant inter-model comparisons that may occur consider the next example.

EXAMPLE 8.2.1 ***Redundancy*** With $k = 4$, there are $2^6 = 64$ possible models in the graphical lattice. Consider the subset A of 8 models for which variable 1 is independent of the remaining 3 variables:

$$A = \{1 + 2.3 + 2.4 + 3.4, 1 + 2.3 + 2.4, 1 + 2.3 + 3.4, 1 + 2.4 + 3.4, \\ 1 + 2.3 + 4, 1 + 2.4 + 3, 1 + 2 + 3.4, 1 + 2 + 3 + 4\}$$

and the same set in which variables 1 and 2 interact: $B = \{1.2 + M; M \text{ in } A\}$. Now variable 2 is a separating set for each model in A and B, and each model factorises according to $f_{1234} = f_{12}f_{234}/f_2$. The diagram illustrates:

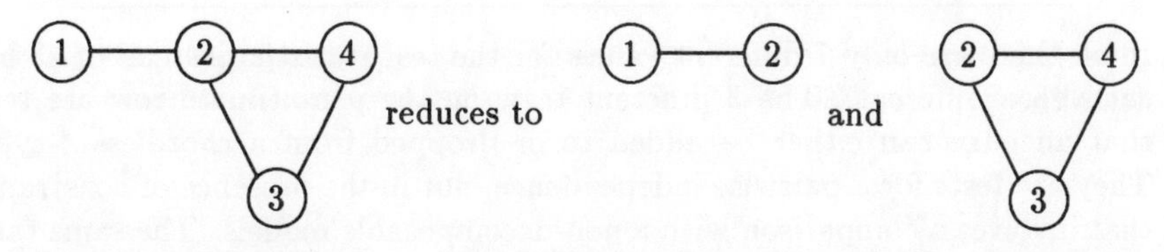

Consequently, any intra-A comparison is identical to the corresponding intra-B comparison and the difference in deviances between each model in A and the corresponding model in B is constant. The information in the deviances conveyed by fitting 16 models may be obtained by fitting exactly 9 models: for instance, the 8 in A together with $1.2 + 3 + 4$ from B. □

EXAMPLE 8.2.2 Equivalent tests. Consider the example of testing for no 1.2 interaction in a four-way contingency table. Suppose that the dotted line in

1o·······o2 indicates that the test compares 1o o2 with 1o——o2

and so corresponds to the deviance difference dev (1 + 2) − dev (1.2). In the following array, tests on different rows have different values while tests on the same row have the same value, apart from the penultimate row which has 3 different values; the number of permutations are given beneath. The graphs are labelled clockwise.

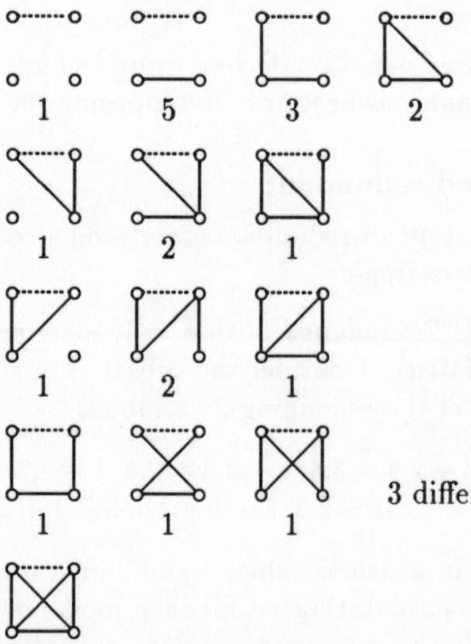

hypothesis	equivalent test statistics				
$X_1 \perp\!\!\!\perp X_2$	1	5	3	2	
$X_1 \perp\!\!\!\perp X_2 \mid X_3$	1	2	1		
$X_1 \perp\!\!\!\perp X_2 \mid X_4$	1	2	1		
	1	1	1		3 different tests
$X_1 \perp\!\!\!\perp X_2 \mid (X_3, X_4)$	1				

In all there are only 7 different values for the test statistic, instead of 23 had each been different. The 3 different tests on the penultimate row are tests that an edge can either be added to or dropped from a chordless 4-cycle. They are tests for a pairwise independence, but in the presence of constraints that involves a comparison with a non-decomposable models. The same table holds for graphical Gaussian as well as graphical log-linear models. □

Screening procedures

Certain model selection procedures, for instance the screening technique of Brown (1976), attempts to evaluate the interaction components needed in a model by comparing the values of the test in different contexts. For example, the 'marginal' test for the 1.2 interaction in a four-way table is

$$\text{dev}\,(1+2) - \text{dev}\,(1.2)$$

and is to be compared to the 'partial' test

$$\text{dev}\,(1.3 + 1.4 + 2.3 + 2.4 + 3.4) - \text{dev}\,(1.2 + 1.3 + 1.4 + 2.3 + 2.4 + 3.4).$$

If both tests are significant then the interaction 1.2 is included in the model and if both are not it is excluded, otherwise it remains undecided. By itself, this test for a two-way interaction term can be informative about an independence graph. However the 'marginal' test for the 1.2.3 interaction, $\text{dev}\,(1.2 + 1.3 + 2.3) - \text{dev}\,(1.2 + 1.3 + 2.3 + 1.2.3)$ and the 'partial' test $\text{dev}\,(1.2.4 + 1.3.4 + 2.3.4) - \text{dev}\,(1.2.3 + 1.2.4 + 1.3.4 + 2.3.4)$, is not informative, and may suppress the information in the two-way interactions. The meaning of these tests changes in different contexts, and the comparison is telling us as much about the context of the interaction as about the interaction itself.

8.3 Graphical Model Search Strategy

We review different manners of dissecting the lattice of possible graphical models.

Incremental search procedures: backwards or forwards?

Incremental search procedures start by fitting an initial base model and then make small, usually one step movements in the lattice to improve the fit. The class of models in the neighbourhood of the current model from which to choose the succeeding model, and a stopping rule, need to be determined. Well known instances are backward elimination and forward selection procedures which consider the class of models that differ from the current model by the exclusion or inclusion of a single interaction term. In backward elimination each contending model is nested in the current model and the deviance difference has a null chi-squared distribution; hence the procedure eliminates the least significant edge and a popular stopping rule is to continue to exclude edges until none has an observed P-value greater than 5%.

Forward inclusion proceeds in reverse, and such a procedure was suggested by Dempster (1972) in the original paper on covariance selection, while his student Wermuth (1976b) used a backward elimination procedure. There are many obvious variations, for example, stepwise procedures which alternate between a backward elimination step and a forward inclusion step. There is a myriad of initial starting points in the class: the maximal model for backward elimination, the minimal model for forward inclusion, the model determined by excluding all insignificant edges from the maximal model is often a good starting point for a stepwise procedure. There is little evidence to suggest that its choice affects the final model selected, but a lot to suggest that a wise choice gets to the final model faster. On the other hand, in practice it seems that selection procedures are quite sensitive to the choice of stopping rule. Two standard choices are the acceptability of the overall deviance of the model, and the acceptability of all the deviance differences between this model and its successor. The latter concentrates on the individual interactions, the former on the model as a whole.

The philosophy of a simple two step procedure

While in practice forward selection seems to do about as well as backwards elimination, there is a philosophical distinction to be made. A graphical model is a model for the joint distribution of the set of variables under study simplified by conditional independence constraints. backward elimination methods starting from the maximal model directly test these conditional independences. On the other hand, forward inclusion from the mutual independence model tests for marginal independences, which is a roundabout manner of attacking the problem.

This consideration suggests the following two step procedure:

1. An independence graph is defined by its pairwise conditional independence statements. At the first step test these directly by computing the P-values for all $\binom{k}{2}$ edge exclusion deviances from the saturated model. Drop the non-significant edges and jump to the model with the corresponding graph, say G^1.

2. The separation theorem established the global Markov property of an independence graph. Conduct a partial test for this property by computing the P-values of all edge inclusion deviances for the edges missing from G^1. Include all significant edges and jump to G^2.

EXAMPLE 8.3.1 If $k = 4$ the first step consists of comparing each graph with one missing edge

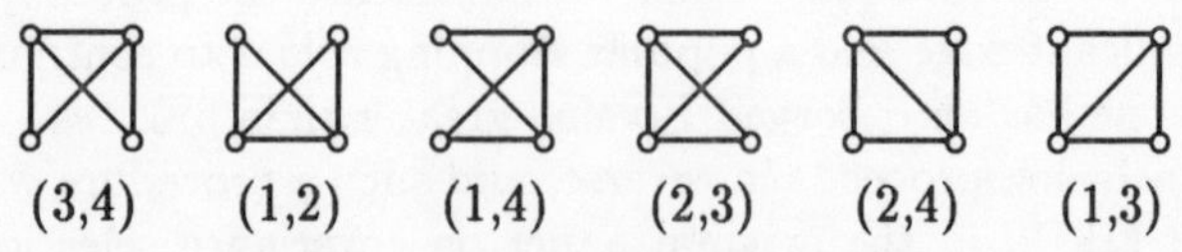

against the complete graph. Suppose that the P-values for edges (1,4), (2,4) and (1,3) are small while the others are large, then G^1 is

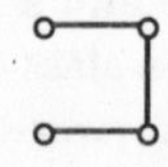

At the second step, retesting the edges (1,3) and (2,4) corresponds to testing the statements $1 \perp\!\!\!\perp 3 | 2$ and $2 \perp\!\!\!\perp 4 | 3$ both of which are conditional independence statements in marginal distributions deducible from the global Markov property. □

Global search procedures

The result of fitting all possible models, though computationally feasible even for quite large values of k, is by no means a trivial exercise, and is far too voluminous to digest; worse, it contains much redundant information, as seen in Examples 8.2.1 and 8.2.2.

Adequate models: To simplify the plethora of models, define a model M to be ***adequate*** if the deviance of M is sufficiently small. Of all the acceptable models, ones with the fewest parameters are of the greatest interest. Define a model M to be ***minimally adequate*** if it is adequate and if there is no model N nested in M that is also adequate. Rather than report the deviances of all models, Havranek (1984), Edwards and Havranek (1985, 1987) report the deviances for those models that are minimally adequate, and consider ways to evaluate these. The procedure is *fast* in the sense that relatively rapid evaluation is possible because if it is determined that a model, M say, is adequate, then all models containing M are also deemed adequate, and there is no need to fit them. This rule is in accordance with Gabriel's (1969) principle of coherence. Unlike the incremental search procedures, the procedure thus selects and rejects models on the basis of the overall deviance and not on deviance differences. The dual to the set of minimally adequate models, the maximally inadequate, is also a set of interesting models.

Restricting the lattice

It is sometimes possible to reduce the size of the underlying universe of potential models. One way, developed in detail later, is to partition the variables into response and explanatory variables. An idea of the reduction in size of the lattice is gained from the regression lattice in Section 8.1 above, which has 1 response and 3 explanatory variables, and contains 8 elements; the corresponding graphical lattice for $1 + 3 = 4$ response variables has 64 elements.

Another strategy is to restrict the class of graphical models in some way and one possibility is to consider only the decomposable models, those models which have direct estimates. This can reduce a task of fitting $2^{\binom{k}{2}}$ models to fitting just 2^k; an approach on these lines is considered by Whittaker (1984b). A difficulty encountered with stepwise search procedures through the decomposable models is that the lattice is not closed with respect to the insertion or deletion of an edge. To illustrate, if the current model has graph

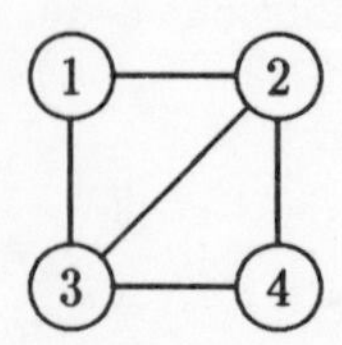

then it is decomposable and while removal of the edge (1,2) gives another decomposable model, removal of the edge (2,4) leads to the chordless 4-cycle which is not decomposable. Only those edges which do not form part of a proper separating set can be tested at any step of the procedure.

One restriction which is sometimes appropriate for log-linear models and discussed in Chapter 9, is to eliminate all models with interactions of higher order than two, so that the all two-way interaction model becomes the maximal element of the lattice replacing the saturated log-linear model. This lattice is closed under insertion and deletion of individual two-way interactions.

Final remarks

Experience confirms that carrying out the model selection procedure is a valuable exercise in data exploration: it gives information both on the relative strengths of associations and the aliasing of effects due to non-orthogonality. Even if the outcome does not lead to a simple model the procedure may suggest further models that are not members of the orignal lattice, for example, models with latent variables, or may suggest models with more specific structure than envisaged originally, for example, symmetry constraints. Model selection is affected by many of the other issues and decisions that are taken in the whole process of modelling, and the selection procedure can reveal the need for diagnostic checks and robust estimates.

8.4 Model Selection: a Continuous Example

The issues involved in model selection are well illustrated by the examples in the following two sections. The first study, Frets' heads, is an example of graphical Gaussian model selection and the analysis gives a detailed comparison of backward and forward procedures using stopping rules based on the overall deviance and based on deviance differences. The example is rather exacting because of the small sample size and the large but approximately equal correlations, but highlights some of the choices that have to be made, and illustrates the rewards of search procedures. The second study is an example of finding a suitable log-linear model to fit a six dimensional contingency table.

Even though the sample size is two orders of magnitude larger, and on the whole the variables are closer to independence, suggesting that the included effects are nearly orthogonal, there remain some difficulties.

EXAMPLE 8.4.1 Frets' Heads. Frets (1921) reported measurements of the head length and head breadth of the first and second adult sons in a sample of 25 families. The data are reproduced in Exercise 1. The correlation matrix

Table 8.4.1: Frets' Heads: correlation matrix of the head length and breadth of the first and second adult sons.

length of first, X_1	1.0			
breadth of first, X_2	.735	1.0		
length of second, X_3	.711	.693	1.0	
breadth of second, X_4	.704	.709	.839	1.0
	1	2	3	4

in Table 8.4.1 exhibits some fairly strong associations between the variables classifying the data. The symmetries in the variables make the structure easy to understand, but no attempt is made to incorporate these into the model.

Selection based on the edge exclusion deviances

The edge exclusion deviances from the saturated model with the complete independence graph are

1	*			
2	4.99	*		
3	1.27	0.44	*	
4	0.59	1.30	12.41	*
	1	2	3	4

Compared to the 5% point of chi-squared with 1 df only the interaction terms 1.2 and 3.4, between length and breadth for each son, are significant. Excluding the others gives the model 1.2 + 3.4 with independence graph

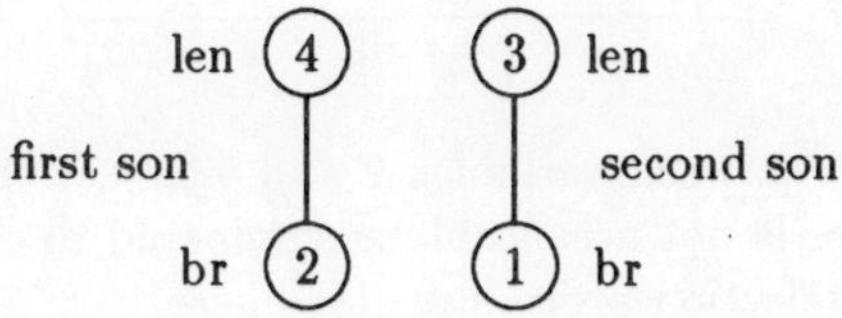

The graph breaks into two disconnected components and, by the global Markov property, has the implication that the measurements on the first son (X_1, X_2) are independent of those on the second son (X_3, X_4). However, this is not consistent with the magnitude of the inter-son correlations in the matrix above and we are loath to accept this conclusion. Fortunately, fitting this model directly gives a deviance of 24.38 on 4 df with an observed probability of 0.0002. Thus we have an example in which the initial edge exclusion deviances suggest a model with a rather poor fit.

Backward elimination with deviance difference stopping rule

The procedure starts with the saturated model and at each stage eliminates the smallest interaction. It stops when all of the exclusion deviance differences for the remaining interactions are significant.

1. The exclusion deviances of the six edges from the saturated model are calculated, as above. The interaction between variables 2 and 3 is the smallest, its exclusion deviance is not significant as compared to chi-squared on 1 df, so that the model $1.2 + 1.3 + 1.4 + 2.4 + 3.4$ becomes the new base model. (It has a deviance of 0.44 on 1 df with a P-value of 0.515.)

2. The five exclusion deviance differences from $1.2 + 1.3 + 1.4 + 2.4 + 3.4$ are

1	*			
2	6.25	*		
3	2.54	x	*	
4	0.31	4.29	15.41	*
	1	2	3	4

The 1.4 interaction is the smallest, the drop in deviance of 0.311 has a P-value of 0.584 and so is not significant. The model $1.2+1.3+2.4+3.4$ (with a deviance of 0.750 on 2 df) becomes the new base model.

3. The four exclusion deviances from $1.2 + 1.3 + 2.4 + 3.4$ are

1	*			
2	8.00	*		
3	6.20	x	*	
4	x	6.04	19.07	*
	1	2	3	4

The interaction between variables 2 and 4 is just the smaller, but the drop in deviance is not acceptable as compared to chi-squared on 1 df. The backward selection procedure stops here.

The chosen model is $1.2 + 1.3 + 2.4 + 3.4$ with the chordless 4-cycle as its independence graph, here adorned by the fitted partial correlations

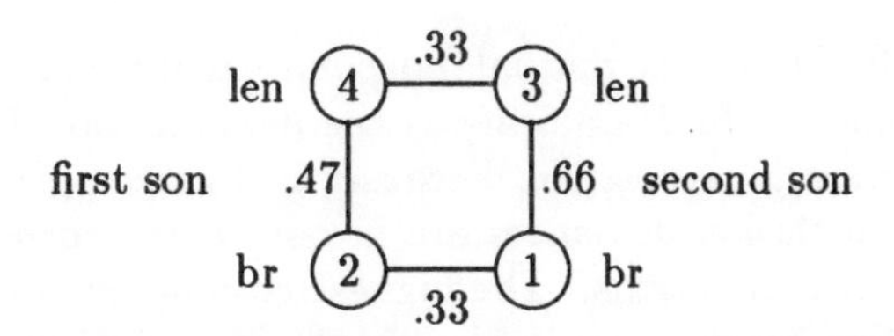

It has a deviance of 0.75 on 2 df indicating a very close fit of the model to data. It respects the symmetries of the study with like variables being associated with like.

Backward elimination with overall deviance stopping rule

At each stage the model with the overall smallest deviance is selected as the current model for the next iteration; and the procedure terminates when the model is not acceptable at the 5% level. Because the comparisons are nested within the current model, the model with the smallest overall deviance has the smallest edge exclusion deviance. Hence the execution proceeds as above until step 3:

1. The 2.4 interaction is the smallest; dropping this leads to the model $1.2 + 1.3 + 3.4$; it has a deviance of 6.789 on 3 df with an observed probability of .078; hence the procedure continues.

2. The three exclusion deviances from the model $1.2 + 1.3 + 3.4$ are

	1	2	3	4
1	*			
2	19.39	*		
3	17.59	x	*	
4	x	x	30.46	*

   However the smallest is the 1.3 interaction, with an overall deviance of 24.38 on 4 df and an observed probability of 0.0002. The procedure cannot continue and the accepted model is $1.2 + 1.3 + 3.4$.

Basing the stopping rule on the level of the overall deviance has not led to the same model as that attained by a deviance difference stopping rule.

Forward inclusion

Forward inclusion is the mirror image of backward elimination. At each stage the one step ahead edge inclusion deviance differences are calculated; the stopping rule may use either the overall deviance or the deviance difference. We shall use the difference and report the overall deviance in parentheses.

1. The procedure takes the mutual independence model 1+2+3+4 (which has a deviance of 74.23 on 6 df) as the minimal model. A short cut in fitting all models is possible, because, in this complete independence model, the inclusion deviances are in one to one correspondence with the empirical correlations. The highest correlation corresponds to the 3.4 interaction; its inclusion deviance is 30.46 on 1 df, which is significant and leads to the model $1 + 2 + 3.4$ (with a deviance of 43.77 on 5 df).

2. The five edge inclusion deviances are calculated by fitting the five models:

1	*			
2	19.39	*		
3	17.58	16.37	*	
4	17.10	17.43	x	*
	1	2	3	4

   The largest is 1.2, is significant and results in the model $1.2 + 3.4$ (with a deviance of 24.38 on 4 df).

3. The four edge inclusion deviances at this stage are

1	*			
2	x	*		
3	17.58	16.37	*	
4	17.10	17.43	x	*
	1	2	3	4

   Because of the relatively simple independence structure of each fitted models, these edge inclusion deviance differences are exactly those calculated at step 2. The significant term 1.3 is included and the new model is $1.2 + 1.3 + 3.4$ (a deviance of 6.79 on 3 df).

4. The three edge inclusion deviance difference at this stage are

1	*			
2	x	*		
3	x	3.44	*	
4	2.06	6.04	x	*
	1	2	3	4

which now differ substantially from those of the previous step. The interaction 2.4 is significant and its inclusion leads to the 4-cycle 1.2 + 1.3 + 2.4 + 3.4 (with a deviance of 0.75 on 2 df).

5. The two inclusion deviances from this model are

1	*			
2	x	*		
3	x	0.31	*	
4	0.16	x	x	*
	1	2	3	4

Neither of these two interactions is significant and so the procedure terminates with the chordless 4-cycle.

In this example, forward inclusion and backward elimination select the same model.

Minimally adequate models

Consider the following four models containing the 1.2, 3.4 and one other interaction; each has three degrees of freedom, and their deviances and P-values are:

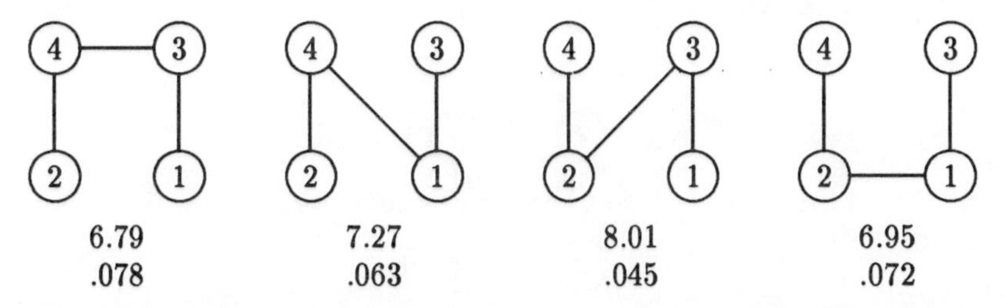

6.79	7.27	8.01	6.95
.078	.063	.045	.072

The three of these that are acceptable at the 5% level constitute the set of all minimally adequate models; because, as is easy to verify, deleting any edge from any one of the three leads to a model that is not acceptable at the 5% level (and thus not adequate). Any other adequate model has to contain at least one of these models and hence cannot itself be minimally adequate, this includes the 4-cycle model chosen by the incremental procedures above.

Conclusions

The example is interesting because it illustrates how different model selection procedures may lead to different models. It shows that the initial computation of edge exclusion deviances is not enough to determine the final model.

However appealing the philosophy of the simple two step procedure discussed at the beginning of Section 8.3, Frets' heads is an example where it does not converge: the first elimination step defining the independence graph leads to the model of independence between the two sons; the second step, testing the global Markov properties of this graph, asserts (step 3 of the forward inclusion procedure) that all the four excluded edges need to be reinstated in the model. This returns the procedure to the initial starting point.

The way in which given interactions change meaning according to their context in the model is illustrated by the way in which the exclusion deviances radically alter in the development of the selection procedures. It suggests that the initial phase of a forward inclusion procedure can be accelerated by exploiting the simple structure of the graphs. Perhaps surprisingly, it is seen that forward and backward procedures reach the same model: the real difference arises from choosing a stopping rule based on deviance differences or one based on overall deviances. Though the sample size is relatively small, and the values of the observed correlations are rather similar, the problem is not numerically ill-conditioned, and its lessons are widely applicable.

The effort made in following the progess of the model selection procedure is rewarded by the increase in understanding the inter-relationships manifest in the data. In this example, it suggests that other models may well explain the data more parsimoniously. Thus, a model based on latent variables representing, say, the genetic determinant of size, for each son, might have a graph

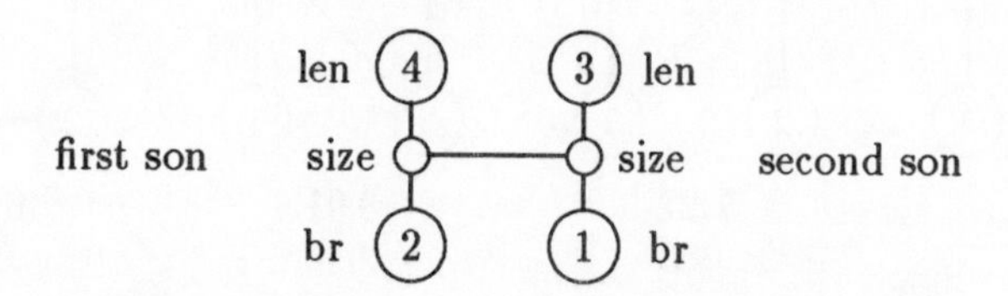

Exploiting symmetries in the values of the parameters can ensure that this model is estimable. The feature of this model is that it 'predicts' that the four models graphed in the above section on adequate models, with just one connection between the sons, together with the two chordless 4-cycles $1.2 + 1.3 + 2.4 + 3.4$ and $1.2 + 1.4 + 2.3 + 3.4$, should fit the observed correlation matrix reasonably well. The argument is that if the interactions are strong enough then the true conditional independence $X_1 \perp\!\!\!\perp X_4 | (\text{size}_1, \text{size}_2)$, for example, implies that $X_1 \perp\!\!\!\perp X_4 | (X_2, X_3)$ approximately. The conditioning variables (X_2, X_3) act as surrogates for the unobserved size variables. $\square$

8.5 Model Selection: a Discrete Example

We now give an example of graphical log-linear model selection for a multi-way contingency table.

EXAMPLE 8.5.1 Prognostic factors for coronary heart disease. This information in Table 8.5.1 was collected at the beginning of a 15 year follow up study of probable risk factors for coronary thrombosis, comprising data on all 1841 men employed in a Czechoslovakian car factory. The reference is Reinis *et al.* (1981), and our source is Edwards and Havranek (1985). The

Table 8.5.1: Prognostic factors in coronary heart disease. From Edwards and Havranek (1985).

				B	no		yes	
F	*E*	*D*	*C*	*A*	no	yes	no	yes
neg	<3	<140	no		44	40	112	67
			yes		129	145	12	23
		>140	no		35	12	80	33
			yes		109	67	7	9
	>3	<140	no		23	32	70	66
			yes		50	80	7	13
		>140	no		24	25	73	57
			yes		51	63	7	16
pos	<3	<140	no		5	7	21	9
			yes		9	17	1	4
		>140	no		4	3	11	8
			yes		14	17	5	2
	>3	<140	no		7	3	14	14
			yes		9	16	2	3
		>140	no		4	0	13	11
			yes		5	14	4	4

six variables that cross-classify the table are all binary and denote

- *A* smoking: yes,no
- *B* strenuous mental work: yes,no
- *C* strenuous physical work: yes,no
- *D* systolic blood pressure: <140,>140
- *E* ratio of beta and alpha lipoproteins: <3,>3
- *F* family anamnesis of coronary heart disease: yes,no

There is no reason to suppose that these six probable risk factors have a treatment-response structure and we shall treat them all on an equal footing. In our first analysis we adopt our two step procedure, first testing for pairwise conditional independence and then testing the global Markov property of the suggested model.

1. To begin, the $\binom{6}{2} = 15$ edge exclusion deviances from the saturated model are computed.

A	*					
B	22.65	*				
C	42.80	684.99	*			
D	28.72	12.23	14.81	*		
E	40.02	17.24	18.63	31.06	*	
F	21.31	22.79	22.15	18.35	18.32	*
	A	B	C	D	E	F

Each exclusion deviance is a test for independence at each of the 2^4 levels of the four remaining variables, hence there are 16 df associated with each chi-squared distribution, so that the critical value for a test at the 5% significance level is 26.30. The information in the exclusion deviances is immediately revealing: it is apparent that the $B \times C$ margin needs detailed inspection; its large value is due to the approximate dichotomous nature of the variables, presumably strenuous mental and physical work are approximate surrogates for blue and white collar workers in the factory. Otherwise, the magnitude of these dependences is small: when divided by the sample size each deviance is an estimate of the information divergence against the corresponding independence hypothesis, and as $N = 1841$, here, these values are very small.

Excluding the non-significant edges suggests the independence graph

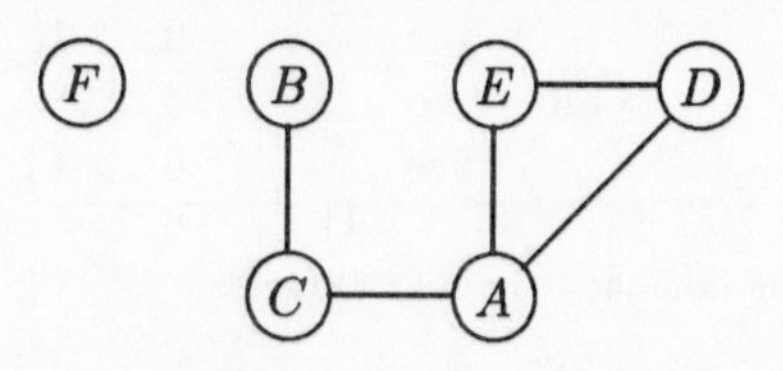

The complete independence of F from the other variables is a surprise, as a family history of coronary heart disease and systolic blood pressure are both thought to be partially hereditary. Otherwise the graph suggests a coherent ordering for the variables: history - behaviour - biology.

2. The overall deviance of this model is 83.75 on 51 df with an observed P-value of 0.0026, and indicating that the Markov properties of the graph

are not consistent with the data. The edge inclusion deviance differences (with df) are

	A	B	C	D	E	F
A	*					
B	6.0 (2)	*				
C	x	x	*			
D	x	0.2 (1)	0.8 (2)	*		
E	x	20.7 (1)	21.7 (2)	x	*	
F	1.1 (1)	4.7 (1)	0.2 (1)	1.1 (1)	3.0 (1)	*

Of the four interactions $A.B$, $B.E$, $C.E$ and $B.F$, that might be included in the model, $B.E$ and $C.E$ are large and the fact that they now appear significant suggests that the three variables B, C and E are non-orthogonal. This is not surprising in light of the strong interaction between B and C, and suggests that one or other of the $B.E$ and $C.E$ edges needs to be included.

3. Including all four interactions leads to the model with independence graph

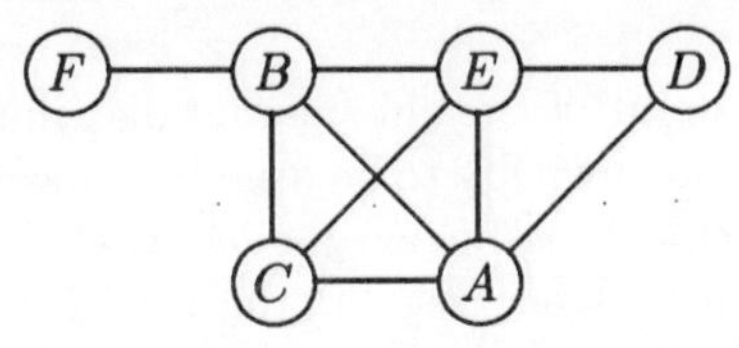

The overall deviance of this model, with formulae $B.F + A.B.C.E + A.D.E$, is a quite acceptable 44.59 on 42 df. The edge exclusion deviance differences from this model are

	A	B	C	D	E	F
A	*					
B	6.3 (4)	*				
C	26.1 (4)	668.0 (4)	*			
D	16.5 (2)	x	x	*		
E	28.5 (5)	6.8 (4)	7.5 (4)	18.3 (2)	*	
F	x	4.7 (1)	x	x	x	*

and, in line with the above discussion it is seen that three, $A.B$, $B.E$ and $E.C$ of the four interactions that have just been included, might now be dropped.

We shall stop the procedure here. The model $B.F+A.B.C.E+A.D.E$ fits well so that it is unlikely that any important effects have been omitted and is reducible to three components $B.F$, $A.B.C.E$ and $A.D.E$. The $B.F$ interaction, was too small to pick up in the first pass of the analysis. The ambiguity in the model selection procedure is confined to the $A.B.C.E$ component, which can be further analysed as a four-way table.

A coherent simultaneous confidence set

We shall use this example to illustrate a coherent simultaneous test procedure, along the lines discussed in Whittaker (1982) and Edwards and Havranek (1985). We consider evaluating a 5% confidence set of models.

The starting point is the likelihood ratio test for complete independence. There are $57 = 2^6 - 7$ parameters set to zero in going from the maximal saturated model to the minimal model of complete independence, represented by θ say, and 7 nuisance parameters, α say, representing main effects. The large sample chi-squared distribution of the likelihood ratio statistic is used to construct a joint confidence region for θ. It is that region for θ which twice the generalised log-likelihood ratio satisfies the inequality

$$2(l(\hat{\theta}, \hat{\alpha}) - l(\theta, \hat{\alpha})) \leq 75.62,$$

the 5% significance point of the chi-squared distribution of 57 df.

The region is far too unwieldy to be useful: at most we require a list of those graphical (or hierarchical log-linear) models that have θ parameter values in the interior of the confidence region, the set of all *acceptable* models, but better still, we need only report the models near the boundary of this region, the *minimally acceptable* models. Thus we wish to find all those models, M, for which dev$(M) \leq 75.62$, and for which there is no model N, nested in M, which is also in the interior of the region.

This procedure, based on comparing the overall deviance of a model to a *common critical value*, is *coherent* in the sense of Gabriel (1969), because, if M and N are two models with M nested in N then dev$(M) \geq$ dev(N), so that whenever M is acceptable then N is acceptable too, while if N is rejectable then M is rejectable too. Certain models are examined in the next table:

	deviance	df
$M = A + B + C + D + E + F$	843.96	57
$M + B.C$	157.99	56
$M + B.C + A.E$	140.58	55
$M + B.C + A.E + E.D$	127.78	54

	deviance	df
$N = A.D.E + A.C + F + B.C$	83.75	51
$N + B.E$	63.01	50
$N + C.E$	62.08	50
$N + A.B$	83.75	50

It can be seen from the table that the $B.C$ interaction must be included; so too must all the edges determined significant by their edge exclusion deviances at the first stage. After some trial and error it is found that the model $A.D.E + A.C + F + B.C$ is not acceptable, but that adding either $B.E$ or $C.E$ leads to minimally acceptable models; furthermore these are the only minimally acceptable models. □

8.6 Exercises

1: *Frets' heads.* The data from Frets (1921) consists of $N = 25$ observations on sons' head measurements; see also Mardia, Kent and Bibby (1979, p.121). They are

X_1	X_2	X_3	X_4	X_1	X_2	X_3	X_4	X_1	X_2	X_3	X_4
191	155	179	145	192	150	187	151	181	145	182	146
195	149	201	152	179	158	186	148	175	140	165	137
181	148	185	149	183	147	174	147	192	154	185	152
183	153	188	149	174	150	185	152	174	143	178	147
176	144	171	142	190	159	195	157	176	139	176	143
208	157	192	152	188	151	187	158	197	167	200	158
189	150	190	149	163	137	161	130	190	163	187	150
197	159	189	152	195	155	183	158				
188	152	197	159	186	153	173	148				

(a) Verify that the eigen values of the empirical correlation matrix for the Frets head data, are, in order of size: 3.20, 0.38, 0.27, 0.16; the smallest is 4% of the sum, so that the matrix is reasonably well conditioned.
(b) Verify that the fitted partial and marginal correlations for the model with formula $1.2 + 3.4 + 1.3 + 2.4$ are

	1	2	3	4
1	1.00			
2	.471	1.00		
3	.329	.000	1.00	
4	-.000	.325	.659	1.00

	1	2	3	4
1	1.00			
2	.735	1.00		
3	.711	.666	1.00	
4	.667	.709	.839	1.00

2: The following data comes from Lawley and Maxwell (1971), and has also been analysed by Knuiman (1978). The correlation matrix for $N = 220$ boys tested on 6 school subjects is

	Gae	Eng	his	ari	alg	geo
Gaelic	1.0					
English	.439	1.0				
history	.410	.351	1.0			
arithmetic	.288	.354	.164	1.0		
algebra	.329	.320	.190	.595	1.0	
geometry	.248	.329	.181	.470	.464	1.0

Find the exclusion deviances, and note that these divide the variables into two cognate groups. However, verify that the likelihood ratio test for the independence of these two groups is 52.54 on 9 df. For comparison, the deviance against the mutual independence of the six variables is 287.43 on 14 df.

3: *Fowl Bones.* Wright (1954) reports the following correlation matrix for six bone measurements on 276 white leghorn fowl. The measurements were taken on the skull (length and breadth), on the wings (humerus and ulna) and on the legs (femur and tibia).

	skl	skb	hum	uln	fem	tib
skull length	1.0					
skull breadth	.584	1.0				
humerus	.615	.576	1.0			
ulna	.601	.530	.940	1.0		
femur	.570	.526	.875	.877	1.0	
tibia	.600	.555	.878	.886	.924	1.0

Verify that the model

skl.skb + (skl+skb).hum + hum.uln + hum.fem + fem.tib + uln.tib

has a deviance of 17.65 on 8 df, and that the edge exclusion deviances from this model are

	skl	skb	hum	uln	fem	tib
skl	*					
skb	37.50	*				
hum	53.53	33.64	*			
uln	0	0	251.20	*		
fem	0	0	58.02	0	*	
tib	0	0	0	81.84	188.24	*

The corresponding graph for this model seems to have some respect for the anatomy of a chicken. Check that including the interaction between length and tibia reduces the deviance of the overall model to a more satisfactory, 9.96 on 7 df, but that the graph is not so readily interpretable.

4: Dudzinski and Arnold (1973) describe a comparative study of sheep and cattle grazing. Measured variables on $N = 18$ pasture samples are: 1 total dry matter, 2 green dry matter, 3 percent edible green, 4 green grass leaf, 5 dry grass leaf, 6 green clover, 7 dry clover, 8 stem, and 9 inert matter.

	1	2	3	4	5	6	7	8	9
1	1.00								
2	0.11	1.00							
3	0.04	0.86	1.00						
4	0.11	0.98	0.84	1.00					
5	0.42	-0.11	-0.33	-0.13	1.00				
6	0.11	0.76	0.80	0.64	-0.17	1.00			
7	0.22	-0.36	-0.57	-0.39	0.21	-0.24	1.00		
8	0.34	-0.48	-0.71	-0.48	0.39	-0.43	0.72	1.00	
9	-0.50	0.13	-0.11	0.12	-0.06	0.06	0.30	0.19	1.00

Investigate graphical models to fit this data.

5: *Cork Borings.* The weight of cork borings (centigrams) in four directions: north, east, south, and west, respectively, for 28 trees is taken from Rao (1948).

N	*E*	*S*	*W*	*N*	*E*	*S*	*W*	*N*	*E*	*S*	*W*
72	66	76	77	30	35	34	26	32	30	34	28
91	79	100	75	46	38	37	38	39	36	39	31
60	53	66	63	39	39	31	27	63	45	74	63
56	68	47	50	39	35	34	37	50	34	37	40
56	57	64	58	42	43	31	25	54	46	60	52
79	65	70	61	32	30	30	32	43	37	39	50
41	29	36	38	37	40	31	25	47	51	52	43
81	80	68	58	60	50	67	54	48	54	57	43
32	32	35	36	33	29	27	36				
78	55	67	60	35	37	48	39				

Investigate graphical models to fit this data.

6: Use a model selection procedure to find a model for the marginal $ABCE$ table of the coronary heart disease risk factor data: the table is

		C	no		yes	
A	B	E	< 3	> 3	< 3	> 3
no	no		88	58	261	115
	yes		224	170	25	20
yes	no		62	60	246	173
	yes		117	148	38	36

and the classifications are A smoking (yes,no), B strenuous mental work (yes,no), C strenuous physical work (yes,no), E ratio of beta and alpha lipoproteins ($< 3, > 3$).

7: The following table reported in Bartowiak (1987) is taken from the Wroclaw Coronary Heart Disease Prevention Study of $N = 6651$ men of working age, working in industrial plants in Wroclaw. After a first screening of the many variables four categorical variables of interest were selected; they are: A noise and vibration at place of work (neither noise nor vibaration/noise alone/noise and vibration); B smoking (never smoked/less than 10 per day or smoked for less than 2 years/ more than 10 and for longer than 2 years); C work in a hurry (no/yes); D physical effort involved in work (great/medium/small).

		C_1			C_2		
		B_1	B_2	B_3	B_1	B_2	B_3
A_1	D_1	10	0	26	31	7	100
	D_2	51	6	123	114	13	250
	D_3	166	17	171	325	39	485
A_2	D_1	25	1	75	107	14	309
	D_2	119	10	300	288	39	741
	D_3	44	7	85	130	18	286
A_3	D_1	33	4	105	207	21	531
	D_2	82	11	178	209	28	527
	D_3	12	4	31	46	5	85

Investigate graphical models to fit this data.

Chapter 9

Methods for Sparse Tables

A table may be sparse because the the size of the overall sample is small, or because the number of categories classifying the table is large, or because the number of dimensions is large. Increasing the number of variables in the cross-classification causes the number of cells to grow exponentially. In any of these cases, fitting log-linear graphical models faces certain problems, in particular the diffiulties posed by infinite parameter estimates that may arise if there are marginal tables with zero entries. The failure to satisfy the large sample assumptions may mean that the actual null distribution of a generalised likelihood ratio test statistic is far from the asymptotic chi-squared approximation, and misleading results in model selection procedures is a probable consequence. If there is not enough information to estimate certain parameters in the model sensibly, the question arises of how we can best extract the information that is there.

We consider two techniques for dealing with the problems posed by sparse tables: one is to restrict attention to two-way interaction models, the other is to condition on the margins of the table and perform exact tests.

Restricting models to the class of all two-way interaction models works because the sufficient statistic for any model in this class is a set of two-way marginal tables and, unlike the full table, these tables are not sparse. While we deplore techniques of contingency table analysis that inspect all two-way margins, as they do not admit the possibility of explaining interaction through other variables, but fitting a single model in which these tables form the sufficient statistic entirely overcomes this objection.

The chapter begins with a discussion of the partial cross-product ratio and then discusses how to interpret the parameters of the two-way interaction model in these terms. The Markov properties of any two-way interaction model are still characterised by its independence graph, so that the models are in direct one to one correspondence with purely graphical models. In fact, the two-way interaction model for binary variables has exactly the same number of interaction parameters as the graphical Gaussian model of the same dimension. A practical illustration is provided by the analysis of an original data set from a sample survey in Rochdale, where a crucial point of the analysis is concerned to see how the partial cross-product ratio changes as the partialling set varies.

The chapter then turns to a discussion of exact tests for conditional independence as an alternative method of analysing sparse tables, but which is reliable when the total sample size is small.

9.1 The Partial Cross-product Ratio

A basic measure of interaction between categorically valued random variables is the log cross-product ratio (log-cpr) introduced in Section 2.1. It is a measure with an exact parallel to the correlation coefficient, the standard measure of interaction between continuous random variables, and it is intimately related to the u-terms in the log-linear expansion of the density function. As with the correlation coefficient, but unlike u-terms in general, the log-cpr is essentially a pairwise measure and consequently is easily interpretable. Zero is the natural origin for the measurement scale and a feeling for the amount of interaction attached to a particular value of the log-cpr is given in Example 2.1.2 of Chapter 2.

The conditional log cross-product ratio, defined on the distribution of a pair of variables conditioned on the rest, provides an immediate generalisation to k-dimensional distributions. A notable feature of the multivariate Normal distribution is that the conditional correlation coefficient is a partial correlation in the sense that its value is invariant to the value of the conditioning values. Here we investigate the concept of a *partial* log cross-product ratio.

This section examines certain examples from this perspective before turning to more general considerations. The first is the clinic survival data where there is essentially one interaction of interest; the second, with two interactions, is the Bartlett plum root example, used to illustrate the equivalence of logistic and log-linear models; and the third is the leading crowd example, which illustrated deviance partitioning.

EXAMPLE 9.1.1 Care clinic survival data again. Reconsider the Bishop *et al.* (1975) data reproduced in Section 1.2 concerning the survival rate of infants attending two clinics and the amount of care received by the mother. The three variables under study are clinic X_1, care X_2, and survival X_3. We have seen, in Example 7.7.2 that while there is evidence of interaction in the (X_2, X_3) margin (a deviance of 5.61 on 1 df against independence) this pair of variables is nearly independent when conditioned on X_1, and the model

$1.2 + 1.3$ has a deviance of only 0.082 on 2 df. The conclusion reached is that marginalising over X_1 induces the interaction between X_2 and X_3.

We can repeat this analysis with more emphasis on estimation. By fitting models to the three-way table we may calculate the conditional cprs, and estimate a common value, the partial cpr; in the two-way table we calculate the marginal cpr. In summary

fitted model	est. interaction (standard errors)		
1.2.3	$\log \widehat{\text{cpr}}(X_2, X_3 \mid X_1 = 0)$	=	0.222 (0.47)
	$\log \widehat{\text{cpr}}(X_2, X_3 \mid X_1 = 1)$	=	−0.008 (0.77)
1.2 + 1.3 + 2.3	$\log \widehat{\text{cpr}}(X_2, X_3 \mid X_1)$	=	0.110 (0.78)
2.3	$\log \widehat{\text{cpr}}(X_2, X_3)$	=	1.038 (0.56)

These estimates entirely concur with the previous conclusions: firstly, the conditional interactions are approximately the same for each X_1 category, and are consistent with a common value of zero; and secondly, the marginal interaction differs considerably from zero. □

There are two, rather than one, interesting parameters in the next example.

EXAMPLE 9.1.2 Bartlett's plum root cuttings. This data was first used by Bartlett (1935) to test for no three-way interaction in contingency tables using maximum likelihood. Two hundred and forty plum root cuttings were

Table 9.1.1: Bartlett's plum root cuttings. From Bartlett (1935).

		survival		
length	time	dead	alive	total
long	at once	84	156	240
	in spring	156	84	240
short	at once	133	107	240
	in spring	209	31	240

observed for each of the 4 =2×2 combinations of length of cutting and time of planting. The three classifying variables: length, time and survival, are denoted by the binary variables X_1, X_2, and X_3 respectively, where survival of the cutting, X_3, is the response variable. The fitted log-linear model of no three-way interaction is

$$N \log \hat{f}_{1,2,3}(x) = 4.37 \\ + 0.56x_1 + 0.71x_2 - 0.32x_1x_2 + 0.71x_3 - 1.03x_1x_3 - 1.43x_2x_3,$$

which has the same parameters relating the response variable, survival, to the explanatory variables, as the equivalent logistic regression model.

The measures of the interaction between X_3 and X_1 and between X_3 and X_2 in all appropriate log-linear (or logistic) models are given in the table

interaction	
survival-length, 1.3	survival-time, 2.3
$\log \widehat{\text{cpr}}(X_1, X_3 \mid X_2 = 0) = -0.84$	$\log \widehat{\text{cpr}}(X_2, X_3 \mid X_1 = 0) = -1.24$
$\log \widehat{\text{cpr}}(X_1, X_3 \mid X_2 = 1) = -1.29$	$\log \widehat{\text{cpr}}(X_2, X_3 \mid X_1 = 1) = -1.69$
$\log \widehat{\text{cpr}}(X_1, X_3 \mid X_2) = -1.02$	$\log \widehat{\text{cpr}}(X_2, X_3 \mid X_1) = -1.43$
$\log \widehat{\text{cpr}}(X_1, X_3) = -0.91$	$\log \widehat{\text{cpr}}(X_2, X_3) = -1.37$

All four conditional measures are estimated in the model 1.2.3, the two partial measures are estimated in the model 1.2+1.3+2.3, the two marginal measures are estimated from the models 1.2 and 1.3, repsectively.

The difference between the two conditional cprs is 0.45, for both the survival-length and survival-time interactions, which is just the estimated coefficient of three-way interaction in 1.2.3. This emphasises the linear dependence between log cross-product ratios. The likelihood ratio test statistic for no three-way interaction has the value 2.29 on 1 df, so that the equality of these conditional cprs is acceptable, and the estimate of its common value (the partial log-cpr) is

$$\log \widehat{\text{cpr}}(X_1, X_3 \mid X_2) = -1.02.$$

The marginal and partial cpr both lie within the two conditional values; the estimate of the marginal coefficient is $\log \widehat{\text{cpr}}(X_1, X_3) = -0.91$, which is not very different in numerical value from the partial value. This is a reflection of the balanced nature of the experiment; Simpson's phenomenon is not working here. In this example inference about the effect of cutting length on survival made in the two-way margin are much the same as that made in the full three-way table.

The same is true of the effect of planting time on survival. The corresponding conditional measures for the interaction between X_2 and X_3 are: 1.24 when $X_1 = 0$ and 1.69 when $X_1 = 1$; the partial log-cpr is $\log \widehat{\text{cpr}}(X_2, X_3 \mid X_1) = -1.43$. Again the partial cpr lies between the two conditional values and is close to the marginal value, $\log \widehat{\text{cpr}}(X_2, X_3) = -1.37$. The essential postulate of this model is the additivity of the time and length effects: that is, the effect of cutting length is the same whatever the value of planting time, and vice versa, the effect of planting time is the same whatever the value of cutting length. □

Additivity of effects on the logistic scale is equivalent to no three-way interaction on the log-linear scale, which in turn is equivalent to the hypothesis that the conditional cprs, $\log \mathrm{cpr}\,(X_2, X_3|X_1 = 0)$ and $\log \mathrm{cpr}\,(X_2, X_3|X_1 = 1)$, are equal. As in Section 2.5, when they are equal, their common value is called the partial cross-product ratio and denoted by $\log \mathrm{cpr}\,(X_2, X_3|X_1)$. Similarly the $\log \mathrm{cpr}\,(X_1, X_3|X_2)$ is the partial cross-product ratio between X_1 and X_3 having adjusted for X_2.

EXAMPLE 9.1.3 The leading crowd again. Reconsider the four-way table of Example 7.7.3 wherein membership and attitude at time 1 are denoted by (X_1, X_2) and at time 2 by (X_3, X_4). Suppose interest lies in estimating the interaction between X_3 and X_4. By fitting the saturated model to the table we have seen that the four conditional cprs, $\log \mathrm{cpr}\,(X_3, X_4|X_1 = x_1, X_2 = x_2)$ for $x_1, x_2 = 0, 1$, are approximately equal. A test of their equality is to fit the model with formula 1.2.3 + 1.2.4 + 3.4. Its log-linear expansion has the structure

$$\log f_{1234}(x) = g_{123}(x) + h_{124}(x) + u_{34}x_3x_4,$$

and the parameter u_{34} estimates the partial cross-product ratio. The fitted model has a deviance of 0.34 on 3 df, and it is remarkable how well such a model fits. The estimated value $\hat{u}_{34} = \log \widehat{\mathrm{cpr}}\,(X_3, X_4|X_1, X_2) = 0.34$ and a test that this common partial interaction is zero is given by the deviance difference between this and the conditional independence model 1.2.3 + 1.2.4. The test is overwhelmingly rejected; while this interaction is small, it is not zero.

A comparison of the values of the conditional and of the partial cross-product ratio between X_3 and X_4 as the conditioning/partialling set is varied is presented here:

estimated 3.4 interaction (with standard errors)			
conditioning set	est. (s.e.)	partialling set	est. (s.e.)
$X_1 = 0, X_2 = 0$	0.38 (0.20)	X_1, X_2	0.34 (0.09)
$X_1 = 0, X_2 = 1$	0.38 (0.20)		
$X_1 = 1, X_2 = 1$	0.26 (0.16)		
$X_1 = 1, X_2 = 1$	0.36 (0.16)		
$X_1 = 0$	0.47 (0.13)	X_1	0.43 (0.08)
$X_1 = 1$	0.41 (0.11)		
$X_2 = 0$	0.43 (0.10)	X_2	0.42 (0.07)
$X_2 = 1$	0.40 (0.11)		
ϕ	0.53 (0.07)	ϕ	0.53 (0.07)

For instance, the conditional measure $\log \widehat{\mathrm{cpr}}\,(X_3, X_4|X_1 = 0, X_2 = 0) = 0.38$ while the partial measure is $\log \widehat{\mathrm{cpr}}\,(X_3, X_4|X_1, X_2) = 0.34$. The conditional cprs are summarised well by the partial cprs. A notable feature of this table is how the interaction weakens as more variables are allowed for. This is not necessarily the case, as evidenced by the plum root data above, but it occurs quite frequently in observational studies. Note also how the estimated standard errors of these estimates decrease when the dimension goes down and when a common value is assumed. □

A conclusion shared by these examples is how little information is lost by fitting models with a common value of the measures of conditional interaction.

Theory

Partition the k-dimensional random vector X into (X_i, X_j, X_a), where $a = \{1, 2, \ldots, k\} \backslash \{i, j\}$, so that X_a is a vector, the 'rest', while X_i and X_j are one dimensional random variables. If X_a takes the value x_a, the conditional cross-product ratio of X_i and X_j is defined as $\operatorname{cpr}(X_i, X_j | X_a = x_a)$ in the conditional density function of (X_i, X_j) given X_a. If this ratio is functionally independent of x_a then the *partial cross-product ratio* of X_i and X_j given X_a, $\operatorname{cpr}(X_i, X_j | X_a)$, is defined as its common value.

We consider those log-linear expansion which possess a partial cpr.

Proposition 9.1.1 *In the case where X_i and X_j are binary, a partial cross-product ratio is well-defined if and only if the the log-linear expansion of the density function is of the form*

$$\log f_{ija}(x) = g_{ia}(x) + h_{ja}(x) + u_{ij} x_i x_j,$$

where the functions g and h have the coordinate projection property, but are otherwise arbitrary.

Proof: The mixed derivative measure of conditional interaction, defined in Section 2.3, is

$$i_{ij|a} = \nabla_i \nabla_j \log f_{ij|a} == \log \frac{f_{ij|a}(0,0) f_{ij|a}(1,1)}{f_{ij|a}(0,1) f_{ij|a}(1,0)};$$

as x_i and x_j only take the values 0 and 1. But, by definition, this is the conditional cross-product ratio between X_i and X_j given $X_a = x_a$, and hence

$$i_{ij|a} = \log \operatorname{cpr}(X_i, X_j | X_a = x_a).$$

Now, by the log-linear expansion of the proposition, $\nabla_i \nabla_j \log f_{ija} = u_{ij}$ so that

$$i_{ij|a} = u_{ij}$$

and does not depend on x_a. Hence neither does $\log \operatorname{cpr}(X_i, X_j | X_a = x_a)$ and the conditional cprs are the same. Their common value defines the partial cpr.

To go the other way, the partial cpr is well defined if the mixed derivative of the conditional density is constant. That is, if $\nabla_i \nabla_j \log f_{ija} = u_{ij}$ is constant with respect to x_a. Integrating over x_i and x_j establishes that the log-linear expansion of f must be of the form stated. □

The same analysis extends to the case when X_i and X_j are categorical variables taking values in finite sets of the form $\{0, 1, \ldots, I-1\}$. The two-way interaction term $u_{ij}x_ix_j$, is replaced by the function $u_{ij}(x_i, x_j)$.

An extended definition of the partial cpr

If the conditional cross-product ratios are not equal the partial cross-product ratio is not defined. It seems a pity to waste such a concept and rather than limit its application it seems worthwhile to extend the definition to cover this eventuality. The following extension is reasonably natural and results in a definition for the partial cross-product ratio which is always estimable. Suppose that X has the density function f_K given by an arbitrary table of probabilities p_K. Consider the family of tables $\{q_K\}$ all of whom have a maximal log-linear expansion with a well-defined partial cpr, that is,

$$\{q_K;\ \log q_{ija}(x) = g_{ia}(x) + h_{ja}(x) + u_{ij}x_ix_j\}.$$

Now choose q such that it is nearest to p in the sense of minimal divergence $I(p;q)$, and define the partial log cpr $(X_i, X_j|X_a)$ with respect to p as the corresponding parameter of q. That the measure always exists and is unique follows from the corresponding properties of the divergence.

Proposition 9.1.2 *For a k-way table classified by $X = (X_1, X_2, \ldots, X_k)$ the partial cross-product ratio between X_i and X_j adjusted for X_a where $a = K\backslash\{i,j\}$ exists, is unique, and is identical to the common value of the conditional cross-product ratios in the special case when these are equal.*

The generalisation to variables that may take more than two levels just requires replacing the two-way interaction term $u_{ij}x_ix_j$ by $u_{ij}(x_i, x_j)$ in the log-linear expansion of q_K.

Varying the choice of g and h in the defining class of models for q_K leads to a wider choice of definition for the partial cross-product ratio. For instance, with $k = 4$ and $i = 3$, $j = 4$, the partial cross-product ratio is defined by minimising with respect to the terms in the log-linear expansion of q_K: $\log q_{1234}(x) = g_{123}(x) + h_{124}(x) + u_{34}x_3x_4$, where g_{123} and h_{124} are arbitrary. If these three-way interactions are simplified to two-way interactions in

$$\log q_{1234}(x) = u_{12}(x) + u_{13}(x) + u_{23}(x) + u_{14}(x) + u_{24}(x) + u_{34}x_3x_4.$$

The interaction parameter u_{34} is still interpretable as a partial cross-product ratio in this model because $\nabla_3\nabla_4 \log q_{1234} = u_{34}$. The model of all two-way interactions, considered in the next section, allows a partial cross-product ratio to be estimated simultaneously for each pair of variables, and by economising on parameters, leads to more efficient estimates. As an illustration of estimating the 3.4 interaction when the three-way interactions between the other variables are replaced by two-way interactions, consider the next example.

EXAMPLE 9.1.4 In the leading crowd data, it turns out that the parameter estimate of $\log \text{cpr}\,(X_3, X_4|X_1, X_2)$ equals 0.336, the same to 3 decimal places for each of the following models:

1.2.3+1.2.4+3.4
(dev=0.342, df=3)

(1+2).3+1.2.4+3.4
(dev=0.382, df=4)

1.2.3+(1+2).4+3.4
(dev=1.187, df=4)

(1+2).(3+4)+1.2+3.4
(dev=1.206, df=5)

generated by eliminating three-way interactions from the first. □

9.2 The All Two-way Interaction Model

An interesting class of models is constructed for a k-dimensional categorical random vector X with density function f_K, by insisting that all terms higher than the two-way interaction terms in the expansion of $\log f_K$ are zero:

$$\log f_K = u_\phi + \sum_i u_i + \sum_{i,j} u_{ij}.$$

The u-terms are functions of x with the coordinate projection property that, for example, $u_i(x) = u_i(x_i)$ is a function of x_i alone, and satisfy the now usual identifiability constraints that

$$u_i(x_i) = 0 \quad \text{and} \quad u_{ij}(x_i, x_j) = 0 \text{ whenever } x_i = 0 \text{ or } x_j = 0.$$

This, the *all two-way interaction* model, has the associated model formula $1.2 + 1.3 + \cdots + (k-1).k$. It has several interesting properties:

Proposition 9.2.1 *(a) each two-way interaction term, u_{ij}, is directly interpretable as a partial log cross-product ratio;*

(b) the criterion for pairwise conditional independence is that $u_{ij} = 0$; and consequently the non-zero two-way interactions determine the conditional independence graph of X;

(c) the conditional distribution for each variable given the rest is a logistic linear regression;

(d) the sufficient statistics are the set of two-way marginal tables; with increasing dimension the number of parameters goes up as $\binom{k}{2}$ and not as 2^k when each element of X is binary.

Proof: For (a), that each u_{ij} term is a log partial cpr comes by showing the mixed derivative $i_{ij|a} = u_{ij}$ and using $i_{ij|a} = \log \text{cpr}\,(X_i, X_j | X_a = x_a)$ as in Proposition 9.1.1 above.

For (b), the characterisation of pairwise conditional independence is a direct consequence of the factorisation criterion, or equivalently, can be viewed as a property of the mixed derivative measure of interaction.

For (c), consider X_i and put $a = K\backslash\{i\}$. The logistic regression formulation follows because

$$\log f_{i|a}(x_i; x_a)/f_{i|a}(0; x_a) = \log f_{ia}(x_i, x_a)/f_{ia}(0, x_a)$$

so that the marginal density of X_a cancels. Substitution and application of the identifiability constraints on $u_{ij}(x_i, x_j)$ gives

$$\log f_{i|a}(x_i; x_a)/f_{i|a}(0; x_a) = u_i(x_i) + \sum_{j \in a} u_{ij}(x_i, x_j).$$

For (d), from Proposition 7.4.1 the likelihood function is $l(p) = \sum_x n(x) \log p(x)$, where the sum is taken over the cells x of the table. Substituting the two-way interaction expansion of p gives

$$l(p) = \sum_x n(x)\{u_\phi(x) + \sum_i u_i(x) + \sum_{i,j} u_{ij}(x)\}.$$

Reversing the order of summation and using the coordinate projection property of the u-terms simplifies this log-likelihood to

$$l(p) = u_\phi n_\phi + \sum_i u_i n_i + \sum_{i,j} u_{ij} n_{ij},$$

where the table of counts n has the margins $n_\phi = \sum_x n(x)$, $n_i = \sum_{x_a} n(x)$, where $a = K\backslash\{i\}$ and $n_{ij} = \sum_{x_a} n(x)$, where $a = K\backslash\{i,j\}$. The set of two-way marginal tables of counts $\{n_{ij}\}$ is therefore the sufficient statistic. □

The first property, that all two-way interactions are interpretable as partial log-cprs, is an alternative formulation of the assumption that u-terms higher than the two-way interaction terms are zero, and gives some notion of its magnitude. For 2^k tables this property has the implication that each of the 2^{k-2} conditional cprs, $\text{cpr}\,(X_i, X_j | X_a = x_a)$, are equal, and equal to the partial cpr, and furthermore, this is true simultaneously for each of the $\binom{k}{2}$

pairs of variables (X_i, X_j). All this, of course, is a familiar assumption to those who use the multivariate Normal distribution as a probability model. Therein the conditional correlation coefficients between any pair of variables are the same for all values of the conditioning variables, and this is simultaneously true for all pairs of variables.

The characterisation of pairwise conditional independence establishes a direct parallel to the graphical Gaussian model, most especially for 2^k tables. In both the log-linear two-way interaction and the graphical Gaussian models there is a single free parameter that controls the independence relationship $X_i \perp\!\!\!\perp X_j |\text{rest}$: in the former, $u_{ij} = 0$, in the latter, $d_{ij} = 0$, where d_{ij} is an element of the inverse variance matrix. Conditional independence between any pair of variables is parameterised by a single scalar, the mixed derivative measure of partial interaction.

In high dimensions log-linear graphical models run into problems induced by small cell counts and large numbers of free parameters. There are several attendant consequences: the approximations to sampling distributions given by standard asymptotic theory are not valid, it is not possible to construct efficient estimates of the parameters, and tests for pairwise conditional independence lack power.

By contrast, in two-way interaction models, pairwise conditional independence is determined by a single degree of freedom, and the resultant tests are more powerful than the tests with 2^{k-2} degree of freedom based on testing that all conditional log cross-product ratios are zero. With a reduced set of parameters, the asymptotic approximations emanating from maximum likelihood considerations are better, on both the null and the alternative distributions.

Furthermore, as the maximal model contains a manageable number of parameters we may hope to estimate them all efficiently. This is fundamentally important for model selection; we can only hope to test for pairwise conditional independence if we have enough information to estimate all the relevant parameters in the maximal model. It is not unusual for the parameters of the fully saturated model to be so poorly estimated, and the chi-squared approximation to the deviance to be so poor, that the values of the edge exclusion deviances are entirely meaningless. But the *difference* in deviance between the all two-way and any reduced two-way interaction model is a generalised likelihood ratio test between two efficiently estimated models, and has a distribution adequately approximated by the chi-squared approximation. For further discussion see Fienberg (1979) and Kreiner (1987).

Against the positive picture painted by these attributes of the class of all two-way interaction models has to be set the one negative feature of closure. The family of density functions is not closed under marginalisation, so that, if, for instance, $X = (X_1, X_2, X_3, X_4)$ has a density belonging to the class of two-way interaction models with formula $1.2 + 1.3 + 1.4 + 2.3 + 2.4 + 3.4$ then f_{123} has the model formula 1.2.3 and not the formula $1.2 + 1.3 + 2.3$.

9.3 A Case Study: Rochdale

In this section we discuss an application of the two-way interaction model to the exploratory analysis of a survey of households in Rochdale, conducted to elicit information about the factors affecting the pattern of economic life and how these have changed in recent years. The survey was performed under an initiative sponsored by the ESRC entitled 'Social Change and Economic Life in Rochdale', see Davies (1987) and Penn (1988), for more details; the data are available for further analysis on request.[1] The survey design is reasonably straightforward. All members of the sample come from the Rochdale urban region and are chosen by the following method. An interview is made if, firstly, an address is selected by choosing a name at random from the electoral register, and secondly the respondent is selected (again at random) from the individuals between 20 and 60 living at that address.

For the following example we take a particular cluster of eight variables relating to women's economic activity and husband's unemployment. Dichotomising variables where necessary, gives a 2^8 cross-classification of the $N = 665$ households:

- A wife economically active : no,yes
- B age of wife > 38 : no,yes
- C husband unemployed : no,yes
- D child ≤ 4 : no,yes
- E wife's education, O level+ : no,yes
- F husband's education, O level+ : no,yes
- G Asian origin : no,yes
- H other household member working : no,yes.

This is still a large number of dimensions and the resulting eight-way table, Table 9.3.1, has 256 cells. It is written in standard order with H varying fastest, then G, and A varying the slowest. Of the 256 cells, 165 are empty, and 217 have 3 or less observations; while on the other hand, there are a few cells that contain some 30, 40 and even 50 observations: while this a rather sparse table it also appears to be imbalanced. Such a cross-classification is not untypical of data arising from observational studies as in social survey analysis. In spite of these apparent limitations we show that graphical modelling methods give a sensible and robust interpretation.

The all two-way interaction model

The real cost of the survey design that leads to such a sparse table is that it is

[1] Reproduced by permission from R.B. Davies.

Table 9.3.1: Women's economic activity: an eight-way table.

5	0	2	1	5	1	0	0	4	1	0	0	6	0	2	0
8	0	11	0	13	0	1	0	3	0	1	0	26	0	1	0
5	0	2	0	0	0	0	0	0	0	0	0	0	0	1	0
4	0	8	2	6	0	1	0	1	0	1	0	0	0	1	0
17	10	1	1	16	7	0	0	0	2	0	0	10	6	0	0
1	0	2	0	0	0	0	0	1	0	0	0	0	0	0	0
4	7	3	1	1	1	2	0	1	0	0	0	1	0	0	0
0	0	3	0	0	0	0	0	0	0	0	0	0	0	0	0
18	3	2	0	23	4	0	0	22	2	0	0	57	3	0	0
5	1	0	0	11	0	1	0	11	0	0	0	29	2	1	1
3	0	0	0	4	0	0	0	1	0	0	0	0	0	0	0
1	1	0	0	0	0	0	0	0	0	0	0	0	0	0	0
41	25	0	1	37	26	0	0	15	10	0	0	43	22	0	0
0	0	0	0	2	0	0	0	0	0	0	0	3	0	0	0
2	4	0	0	2	1	0	0	0	1	0	0	2	1	0	0
0	0	0	0	0	0	0	0	0	0	0	0	0	0	0	0

impossible to detect many high order interactions, and one should hesitate to fit the saturated log-linear model that includes terms to represent all possible interactions. However we may fit the all two-way interactions model, because the sufficient statistics are the two-way marginal tables and the entries in these are quite respectable. This model has just 37 free parameters, 28 of which are two-way interactions, and for which there is a sample of 665 independent observations available. In regression analysis such a ratio of parameters to observations in commonplace.

Each variable is binary, so the deviance difference obtained by setting each interaction parameter to zero gives a 1 df test for conditional independence. Refitting the model 28 times is somewhat tedious, and an approximate but faster technique is possible if standard errors of the parameter estimates are available. In this case, as (asymptotically) the squared ratio of the estimate to its standard error is chi-squared distributed with 1 df under the null hypothesis, all 28 tests can be calculated in one fit of the the all two-way interaction model. A significant value for a particular interaction indicates that the edge should be retained in the independence graph.

The deviance of the fitted all two-way interaction model

$$A.B + A.C + \cdots + G.H$$

is 144.56 with 219 df. The fact this is not significant should not be taken to imply that higher order interactions are zero, because as discussed above, there is no reason to believe these could be efficiently estimated. The absence of higher order interaction is largely a matter of faith.

The maximum likelihood estimates of the two-way interaction terms are given in Table 9.3.2. The standard errors of these estimates are reasonably

Table 9.3.2: Parameter estimates from the two-way interaction model.

inter.	est.	s.e.	(est./se.)^2	inter.	est.	s.e.	(est./se.)^2
$A.B$	-0.38	0.24	2.50	$C.E$	-1.25	0.36	11.92
$A.C$	-1.33	0.30	19.87	$C.F$	-0.71	0.29	6.18
$A.D$	-1.49	0.25	34.12	$C.G$	1.09	0.37	8.78
$A.E$	0.63	0.21	9.23	$C.H$	0.28	0.34	0.67
$A.F$	0.09	0.20	0.20	$D.E$	0.24	0.25	0.93
$A.G$	-2.17	0.47	21.30	$D.F$	0.38	0.25	2.37
$A.H$	0.22	0.25	0.82	$D.G$	1.32	0.43	9.32
$B.C$	-0.17	0.34	0.24	$D.H$	-1.08	0.45	5.80
$B.D$	-2.97	0.34	77.23	$E.F$	1.03	0.18	32.96
$B.E$	-0.78	0.20	14.53	$E.G$	-0.82	0.43	3.66
$B.F$	0.01	0.20	0.00	$E.H$	-0.18	0.22	0.61
$B.G$	-0.56	0.46	1.50	$F.G$	-1.39	0.38	13.26
$B.H$	1.74	0.27	41.04	$F.H$	-0.24	0.21	1.21
$C.D$	0.01	0.37	0.00	$G.H$	-0.06	0.50	0.01

small, in line with the overall sample size.

The independence graph

We have remarked already that choosing the independence graph is equivalent to the model selection problem and here the choice is to be made from a possible 2^{28} graphs. Here, we adopt the quick model selection method of selecting interactions for which the square of the standardised parameter estimate exceeds 3.84. There are just 14 such terms and the reduced model that contains these interactions is

$$A.(C+D+E+G)+B.(D+E+H)+C.(E+F+G)+D.(G+H)+F.(E+G).$$

It has a deviance of 160.63 on 233 df. By itself, this is just the generalised likelihood ratio test statistic against the fully saturated model, and because of sparsity is a fairly meaningless figure. However, the difference in deviance between the all two-way and this reduced two-way interaction model is a generalised likelihood ratio test between two efficiently estimated models, and hence has an adequate chi-squared approximation. The increase in deviance, $160.63 - 144.56 = 16.17$, is not significant on 14 df and suggests that the reduced model that excludes 14 of the 28 possible two-way interaction terms provides a good representation of the data.

The conditional independence graph associated with the reduced model, with and without edge strengths, is fortunately planar, and systematises the information about the non-zero interactions rather succinctly:

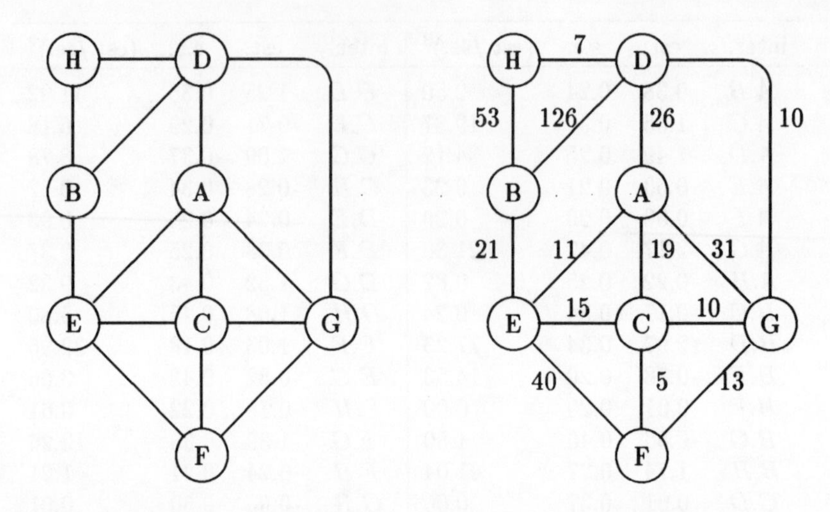

Maximum likelihood estimates of the strength of each interaction are estimated by calculating the conditional independence edge exclusion deviances from the eight-way table of *fitted* values. The edge strengths are rounded down so as not to overload the diagram.

On the whole the graph makes sociological sense, connecting variables such as wife's and huband's education, E and F, which one might expect to be strongly related. The strongest interaction is the edge between wife's age and a child under 4, (B, D); the next is between wife's age and the presence of another working member of the family, (B, H). There is also a strong interaction on the edge (A, G) indicating a different level of wife's economic activity in the Asian and non-Asian communities.

As each variable is binary, each interaction is parameterised by a single scalar u_{ij}, interpretable as a partial log-cpr. These parameters, given in the previous table, could be used to indicate the strength of each edge, rather than the estimated information divergences used here, and have the advantage of indicating direction as well as strength. However, the technique does not easily generalise to variables with more than two levels.

This graph gives a good starting point for any further model selection procedure. All of the standardised interaction estimates of the included terms in this reduced model are significant and so no further term can be dropped. Incorporating each of the deleted interactions now involves only 14 model fits, but does not affect the selected model.

Wife's economic activity and husband's unemployment

But it is the missing edges in the graph that provide the necessary simplification of the data set to guide the further analysis of the interaction between the wife's economic activity and her husband's unemployment. It is well known in Britain that wives whose husbands are unemployed are less likely to work than those whose husbands are employed. Evidence of this has come from several studies and is summarised in Davies (1987). The relevant two-way margin from the eight-way table is

		husband's unemployment, C no	yes	total
wife's A	yes	421	23	444
activity,	no	165	56	221
	total	586	79	

with a reasonably strong interaction, $\log \text{cpr}(A, C) = -1.83$.

One possibility is that this relationship is due to the disincentive effect of the social security system, with increased earnings resulting in lower benefits. Alternatively, it has been argued that this relationship is explainable by other characteristics of the household that simultaneously effect the employment status of husband and wife. We can discount this last point from our analysis: the interaction between husband's unemployment and wife's economic activity cannot be wholly explained by the other factors; for if it could, variables A and C would have a zero partial interaction, and would not be adjacent in the graph.

The graph contains more information about the nature of this interaction. From the analysis of collapsibility in Section 4.6, the mixed derivative of interaction on the edge (A, C) can only depend on variables E and G. Consequently it is only these variables that affect the magnitude of the $A.C$ interaction. Varying the partialling set affects this interaction in the following way:

partialling set a	estimated $\log \text{cpr}(A, C \mid a)$	(s.e.)	model
ϕ	-1.83	(.26)	$A.C$
E	-1.65	(.27)	$A.C + (A + C).E$
G	-1.47	(.28)	$A.C + (A + C).G$
E, G	-1.33	(.29)	$A.C + (A + C).(E + G) + E.G$
E, G, D	-1.33	(.30)	$A.C + (A + C + D).(E + G) + E.G$

The marginal (A, C) interaction is -1.83; this interaction weakens from -1.83 to -1.65 by adjusting for education, E; from -1.83 to -1.47 by adjusting for

Asian origin, G; and to -1.33 by adjusting for both. The term $E.G$ has to be included in the model because, as can be seen from the interaction graph, when the table is collapsed over the variable F, an interaction between E and G is liable to be induced. The last row is included to confirm that adjusting for a further variable, such as D, has no effect.

One way to summarise the effect of adjusting is to construct, by means of the iterative proportional fitting algorithm, a two-way table with the same margins as the margins of the observed (A, C) table above, but with a log cross-product ratio of -1.33, rather than -1.83.

		husband's unemployment, C		
		no	yes	total
wife's A	yes	413.4	30.6	444
activity,	no	172.6	48.4	221
	total	586.0	79.0	

and though the entries are nearer independence, the move is not enormous.

A submodel for a single response

It is for the social scientist to decide if the fitted model with its associated independence graph makes substantive sense; it certainly provides a clear representation of the data and suggests which hypotheses are worthy of further examination. As another example of how the graph guides analysis we consider the problem of determining the factors which influence variable A, wife's economic activity. The graph identifies the determinants of wife's economic activity as the boundary set of the vertex A. They are: husband's unemployment C, a child under 4 D, the wife's educational level E, and membership of the Asian community G. The other variables: age, husband's education and other working persons in the household, may be neglected and further analysis proceeds by collapsing the eight-way table onto this five dimensional margin.

The deviance of the all two-way interaction model for the 5 variables (A, C, D, E, G) is

$$\text{dev}\,(A.(C + D + E + G) + C.(D + E + G) + D.(E + G) + E.G) = 13.95$$

on 16 df. It appears to suggest a good fit, but as this five-way margin is still sparse we cannot conclude higher order interactions are negligible. To serve as a bench mark, note there is a deviance difference of 152.26 on 4 df between this model and the model in which A is completely independent of the other variables: $\text{dev}\,(A + C.(D + E + G) + D.(E + G) + E.G) = 166.21$ on 20 df. This indicates the substantial variation in A that is explained by the other factors.

All interaction estimates in the model are (naturally) non-zero, and the interaction graph of this subset is complete. The two-way interaction estimates involving the response factor A, with standard errors in brackets, are

husband's unemployment	$A.C$	-1.33 (.30)
child under 4	$A.D$	-1.32 (.21)
wife's education	$A.E$	0.69 (.20)
Asian	$A.G$	-2.17 (.47)

A straightforward interpretation summarises this information in the estimated logistic regression equation

$$\log \frac{\text{Prob}(A=1|C,D,E,G)}{\text{Prob}(A=0|C,D,E,G)} = \text{const} \underset{(.30)}{-\ 1.33C} \underset{(.21)}{-\ 1.32D} \underset{(.20)}{+\ 0.69E} \underset{(.47)}{-\ 2.17G}$$

Had this equation been the main motivation for the analysis of this data then the analyst might prefer to condition on the explanatory variables, and take the conditional distribution of A, to be Binomial. Such an analysis, which gives practically identical numerical results, can only be performed in the context of log-linear modelling if the margin of the explanatory variables contains no zeros.

9.4 Exact Conditional Tests

The edge exclusion deviances from the fully saturated model are an important tool for model selection, and we need reliable estimates of their observed significance levels, the P-values. However, high-dimensional tables tend to be sparse and in consequence the asymptotic approximation to their sampling distributions is often very poor. In the previous sections of this chapter, fitting the all two-way interaction model has been suggested as a way to cope with this problem, but there are other methods of dealing with sparse tables and we consider one based on evaluating the exact sampling distribution of the cell counts.

The Multinomial distribution

A basic tool in all forms of statistical inference is the sampling distribution of the sufficient statistic. For the log-linear models of Chapter 7 the sufficient statistic was derived in Proposition 7.4.1: in sampling from the cross-classified Multinomial of size 1, the sufficient statistic for the parameters of the model (the table of probabilities, p, or the u-terms $\{u\}$) is the cell count n that records the total number of observations from the sample in each cell of the table. Hence, we require to evaluate the sampling distribution of

$$N_K(x) = \sum_{l=1}^{n_\phi} \delta(x, X^l).$$

where $X^1, X^2, \ldots, X^{n_\phi}$ is a sample of size n_ϕ taken on the random vector X. The answer turns out to be the Multinomial distribution of size n_ϕ, and consequently this distribution plays the parallel role to the Wishart distribution in the analysis of graphical Gaussian models. It is the cross-classified Multinomial distribution of size that 1 corresponds to the multivariate Normal distribution.

An aside on notation: we would like to adopt the standard convention that uses capital letters to denote random variables, but unfortunately N has already been used to denote the sample size. We may avoid confusion by (i) always making the full set of variables classifying the table explicit and so writing N_K for the random table of counts with observed value n_K; and by (ii) denoting the sample size by n_ϕ rather than by N throughout this section.

The sampling distribution of N_K is calculated by evaluating the joint probability that $N_K(x) = n_K(x)$, for all cells x in the table and for each possible configuration n_K. Here is a derivation. Suppose we choose a listing for the cells $x(1), x(2), \ldots, x(m)$ where $m = \prod_{i=1}^k r_i$ is the number of cells in the table; for instance, in a 2×3 table we might choose $x(1) = (0,0), x(2) = (1,0), \ldots, x(6) = (1,2)$. Now the event $N_K(x) = n_K(x)$ for all x occurs if the sample is so ordered that: the first $n_K(x(1))$ of the sample fall in cell $x(1)$, and the second $n_K(x(2))$ of the sample fall in cell $x(2)$, and ... and the last $n_K(x(m))$ of the sample fall in cell $x(m)$. This event has probability

$$\prod_{i=1}^{m} p_K(x(i))^{n_K(x(i))} = \prod_x p_K(x)^{n_K(x)}.$$

Now note firstly, that any ordering of the sample has the same probability, and secondly, that there are exactly $\frac{n_\phi!}{\prod_x n_K(x)!}$ permissible orderings. We have proved

Proposition 9.4.1 *The sampling distribution of the observed cell counts from a sample n_ϕ independent and identically distributed Multinomial random vectors of size 1 is the Multinomial distribution of size n_ϕ:*

$$P(N_K(x) = n_K(x) \text{ for all cells } x) = \frac{n_\phi!}{\prod_x n_K(x)!} \prod_x p_K(x)^{n_K(x)}.$$

To evaluate the sampling distribution of the edge exclusion deviance under the null hypothesis of pairwise conditional independence, $X_b \perp\!\!\!\perp X_c | X_a$ we need to calculate the probability that N_K falls in the set A:

$$A = \{n_K : 2\sum_x n_K \log \frac{n_K n_a}{n_{ab} n_{ac}} > t\}.$$

for any value t. Under the Multinomial distribution, the probability of A is

$$P(A) = \sum_{n_K \in A} \frac{n_\phi!}{\prod_x n_K(x)!} \prod_x p(x)^{n_K(x)}.$$

This probability, $P(A)$, is a function of the parameters p, and though p satisfies $X_b \perp\!\!\!\perp X_c | X_a$ that is, $p_{abc} = p_{ab} p_{ac} / p_a$, these marginal probabilities are still unknown; consequently this expression is impossible to use numerically.

Conditioning

Is this the only relevant sampling distribution? If we are interested in testing the independence, $X_b \perp\!\!\!\perp X_c$, the counts in the margins, n_b and n_c, by themselves hold no information about the independence relationship, and it is the conditional distribution of N_{bc} given the margins $N_b = n_b$ and $N_c = n_c$, that is of interest. It turns out that this conditional distribution has no nuisance parameters, and so allows the numerical calculation of exact P-values in repeated sampling of tables with fixed margins. The conditioning argument is due to Fisher (1935) who illustrated it with his famous but fictitious tea-tasting experiment.

Conditioning plays a dual role in statistical inference:

1. conditioning may be used to eliminate nuisance parameters from a sampling distribution of interest; and
2. it may be argued that the most relevant distribution for inference is the conditional distribution.

The argument that the pattern of interaction in the observed table should be assessed against all other possible tables with the same margins, has some force.

Eliminating the nuisance parameters in the derivation of the conditional distribution for the simple case of a 2×2 table is the next task.

Proposition 9.4.2 Conditioning in a 2×2 table. *The conditional distribution of the cell counts N_{12} given the margins n_1 and n_2 is free of nuisance parameters.*

Proof: From the Multinomial distribution, the joint density function of N_{12} is the probability that $\{N_{12}(x) = n_{12}(x)$ for all cells $x\}$:

$$f_{N_{12}}(n_{12}) = \frac{n_\phi!}{\prod_x n_{12}(x)!} \prod_x p_{12}(x)^{n_{12}(x)}$$

where x ranges over the four values $(0,0)$, $(0,1)$, $(1,0)$, $(1,1)$. The expression simplifies: first write the Multinomial coefficient as $g(n)$, so

$$f_{N_{12}}(n_{12}) = g(n_{12})\exp\{\sum_x n_{12}(x)\log p_{12}(x)\}.$$

Next, recall the log-linear expansion for a 2×2 table is $u_\phi + u_1x_1 + u_2x_2 + u_{12}x_1x_2$, so substituting and summing over x gives

$$f_{N_{12}}(n_{12}) = g(n_{12})\ \exp\{u_\phi n_\phi + u_1n_1(1) + u_2n_2(1) + u_{12}n_{12}(1,1)\}.$$

We have evaluated $f_{N_{12}}(n_{12})$, the joint probability density function of $N_{12}(0,0)$, $N_{12}(0,1)$, $N_{12}(1,0)$, $N_{12}(1,1)$. As the variables are discrete and the transformation to N_ϕ, $N_1(1)$, $N_2(1)$, $N_{12}(1,1)$ is one to one, it is also the probability density function for this set of statistics. Consequently the conditional density of N_{12} is $f_{N_{12}|n_1,n_2,n_\phi} = f_{N_{12}}/\sum_{n_{12}} f_{N_{12}}$, where the sum is taken over the possible values for $n_{12}(1,1)$ when the margins are fixed. All u-terms other than the highest, u_{12}, disappear from the conditional density function leaving

$$f_{N_{12}|n_1,n_2,n_\phi} = g(n_{12})/\sum g(n_{12})\exp\{u_{12}n_{12}(1,1)\},$$

an expression involving factorials and a polynomial-type term in the interaction parameter. □

Under the hypothesis of independence, $X_1 \perp\!\!\!\perp X_2$, this conditional density is just a ratio of factorials. An alternative, more general argument that shows the conditional density is invariant to the nuisance parameters is to note from $f_{N_{12}}(n_{12}) = g(n_{12})\exp\{u_\phi n_\phi + u_1n_1(1) + u_2n_2(1) + u_{12}n_{12}(1,1)\}$ that there are exactly 4 scalar parameters (u_ϕ, u_1, u_2, u_{12}) and 4 sufficient statistics (n_ϕ, $n_1(1)$, $n_2(1)$, $n_{12}(1,1)$). When u_{12} is known, n_ϕ, n_1 and n_2 are sufficient for the nuisance parameters u_ϕ, u_1, and u_2. By sufficiency, the conditional distribution of N_{12} given n_ϕ, n_1 and n_2 does not depend on the values of u_ϕ, u_1 and u_2, and these nuisance parameters are eliminated. This leaves the exact form of the conditional density to be made explicit.

The Hypergeometric distribution

We wish to find the distribution of the quantities found in the 2×2 table of counts:

	$x_2 = 0$	$x_2 = 1$	total
$x_1 = 0$	–	–	–
$x_1 = 1$	–	$N_{12}(1,1)$	$n_1(1)$
total	–	$n_2(1)$	n_ϕ

Proposition 9.4.3 *Under independence the joint probability that $N_{12}(1,1) = n_{12}(1,1)$ conditioned on $n_1(1)$, $n_2(1)$ and n_ϕ fixed is the (central) Hypergeometric distribution*

$$\frac{\prod_{x_1} n_1(x_1)! \prod_{x_2} n_2(x_2)!}{n_\phi! \prod_{x_1,x_2} n_{12}(x_1,x_2)!}$$

valid for $n_{12} \geq 0$.

Proof: The expression can be derived from the last proposition by substituting for the factorial terms in $g(n_{12})$. Alternatively, since the distribution is invariant to p_K it can be derived from an equi-probable Multinomial joint distribution using a permutation argument. □

If the scalar random variable $N_{12}(1,1)$ has the 2×2 Hypergeometric distribution given here, we write $N_{12}(1,1) \sim H(n_1(1), n_2(1), n_\phi)$. The important point is that this probability distribution is known exactly.

EXAMPLE 9.4.1 Consider the 2×2 table of counts with margins

	$x_2 = 0$	$x_2 = 1$	total
$x_1 = 0$			4
$x_1 = 1$			14
total	3	15	18

based on a sample of size $n_\phi = 18$. The only permissible values of $n_{12}(1,1)$ are 11, 12, 13 and 14 leading to tables

	0	1
0	0	4
1	3	11

	0	1
0	1	3
1	2	12

	0	1
0	2	2
1	1	13

	0	1
0	3	1
1	0	14

which have probabilities 0.446, 0.446, 0.103 and 0.005 respectively. For instance, the probability of the first table is

$$\frac{4!\ 14!\ 3!\ 15!}{18!\ 0!\ 4!\ 3!\ 11!} = 0.446.$$

This example can be used to illustrate how big the difference in the P-values calculated from the asymptotic chi-squared approximation and those calculated from the exact sampling distribution of the deviance can be. The deviance statistic, 2dev $= \sum$ obs log(obs/fitted), and Pearson's statistic

$X^2 = \sum$ (obs-fitted)^2/fitted take the values

$n_{12}(1,1)$	dev	P-value asymp.	P-value exact	X^2	P-value asymp.	P-value exact
11	1.67	0.196	0.554	1.03	0.310	0.554
12	0.24	0.624	1.000	0.26	0.610	1.000
13	3.47	0.063	0.108	4.11	0.043	0.108
14	11.72	0.001	0.005	12.58	0.000	0.005

On the basis of this information there is no reason to prefer either one of the deviance and Pearson's statistic. Both underestimate the exact P-value, asserting the observed result is more significant than it actually is, and in this instance, by so much as to render these asymptotic approximations practically useless. □

The theoretical derivation of the exact conditional distribution of the internal cell counts N_{12} given the margins, n_1 and n_2, in a $r_1 \times r_2$ table is a fairly straightforward generalisation of the above argument for the 2×2 table. As can be seen from Exercise 8, in a 2×3 table the number of different tables with the same margins grows rapidly, and evaluating the probability is a non-trivial exercise. With even fairly small sample sizes it can become an impossible computational task.

Simulated inference

However, all is not lost and we take up an idea suggested by Agresti *et al.* (1979) and pursued by Kreiner (1987). Suppose the statistic T takes an observed value of t. If the exact distribution of T is free of nuisance parameters we may estimate the P-value of observing t by simulation. We may

(i) draw a random table from the right distribution, which has the same margins as the original table;

(ii) evaluate the statistic T on the random table; and

(iii) estimate the P-value as the proportion of random tables for which T is at least as great as t.

A neat feature of this procedure is how it directly mimics the conventional explanation of this sampling distribution.

The size of the simulation procedure can be determined in advance by a preset tolerance on a confidence interval for the estimated P-value. Thus if we wish to estimate $\theta = P(T \geq t)$, from a simulation sample of size M, then from Binomial considerations the estimate $\hat{\theta}$ has a variance of $\theta(1-\theta)/M$. By the Normal approximation to the Binomial distribution, the half-width of the asymptotic 95% confidence interval is $1.96\{\theta(1-\theta)/M\}^{1/2}$. For example, if θ is in the neighbourhood of 0.1 and $M = 100$, this half-width is approximately 0.059, a reasonable result for a not too onerous task. However, if we insist that no interval should have a half-width in excess of 0.01, whatever θ, then M must be at least $\theta(1-\theta)(1.96/0.01)^2$ which at worst is 9,800.

More difficult is to know how to choose a random table, with the correct row and column margins. We describe an idea due to Patefield (1981) in the context of a simple example.

EXAMPLE 9.4.2 Consider generating random 2×3 tables with given margins:

	$x_2 = 0$	$x_2 = 1$	$x_2 = 2$	total
$x_1 = 0$	–	–	–	–
$x_1 = 1$	–	$N_{12}(1,1)$	$N_{12}(1,2)$	4
total	–	3	5	18

Generating a table is equivalent to choosing $N_{12}(1,1)$ and $N_{12}(1,2)$ from the 'right' distribution. Consider ignoring $N_{12}(1,2)$ and generating $N_{12}(1,1)$ first. Combining the columns, $x_2 = 0$ and $x_2 = 2$, shows that $N_{12}(1,1)$ is the internal cell in a 2×2 table with fixed margins; and as combining columns does not affect row and column independence, $N_{12}(1,1) \sim H(4,3,18)$.

Suppose that we know how to generate an observation from any 2×2 Hypergeometric distribution, and we observe a value of $n_{12}(1,1)$. The next part of the procedure is to generate $N_{12}(1,2)$ from its conditional distribution, conditioning on the event $N_{12}(1,1) = n_{12}(1,1)$. But this too has a Hypergeometric distribution. The values in column $x_2 = 1$ are now determined, and if the column is excluded from the table it leaves a 2×2 table:

	$x_2 = 0$	$\{x_2 = 1\}$	$x_2 = 2$	total
$x_1 = 0$	–	$\{-\}$	–	–
$x_1 = 1$	–	$\{n_{12}(1,1)\}$	$N_{12}(1,2)$	$4 - n_{12}(1,1)$
total	–	$\{3\}$	5	$18 - 3$

So $N_{12}(1,2) \sim H(4 - n_{12}(1,1), 5, 18-3)$, and we only need to simulate from another 2×2 Hypergeometric distribution. □

Thus if we know how to generate random 2×2 tables with given margins we can extend the technique to generate 2×3 tables. It is left as an exercise for the reader to extend the method to $r_1 \times r_2$ tables.

EXAMPLE 9.4.3 Consider estimating the exact P-value for testing the independence of X_1 and X_2 if we observe

	$x_2 = 0$	$x_2 = 1$	$x_2 = 2$	total
$x_1 = 0$	–	–	–	–
$x_1 = 1$	–	2	1	4
total	–	3	5	18

where the other values are determined by subtraction. The observed values of the deviance and Pearson's statistic are 3.75 and 4.30, on 2 df, and with asymptotic P-values of 0.1538 and 0.1164 respectively. In 5 simulation experiments each with 1000 replications the estimated exact probabilities of observing Pearson's statistic greater than 4.30 are 0.132, 0.141, 0.146, 0.168, and 0.177. This shows that quite large simulation experiments need to be performed for decimal point accuracy (though a ball park figure will often suffice). In this example, the simulation suggests that the true P-value is nearer 0.15 than the asymptotic figure of 0.116. □

9.5 Exact Tests for Graphical Models

Before we can apply these techniques of exact conditional testing to graphical model selection, we need to to be able to test for conditional independence in k-way tables. We follow Kreiner (1987). In fact it is enough to think of three-way tables, because by redefining the variables all statements of conditional independence of the form $X_b \perp\!\!\!\perp X_c | X_a$, where the subsets a, b and c partition K, can be recast as $X_2 \perp\!\!\!\perp X_3 | X_1$.

Conditional independence

The basic idea of the extension is quite straightforward: the sufficient statistics for the Multinomial log-linear model determined by the independence $X_b \perp\!\!\!\perp X_c | X_a$ are the marginal tables n_{ab} and n_{ac}. All nuisance parameters in the Multinomial distribution of N_K are eliminated by taking the distribution of N_{abc} conditioned on n_{ab} and n_{ac}. To evaluate this conditional distribution, firstly, we break up the full k-way table into slices corresponding to each value, $x_a = 0, 1, \ldots, r_a - 1$, of the conditioning variable and then argue that slices are mutually independent given the count in the marginal table n_a.

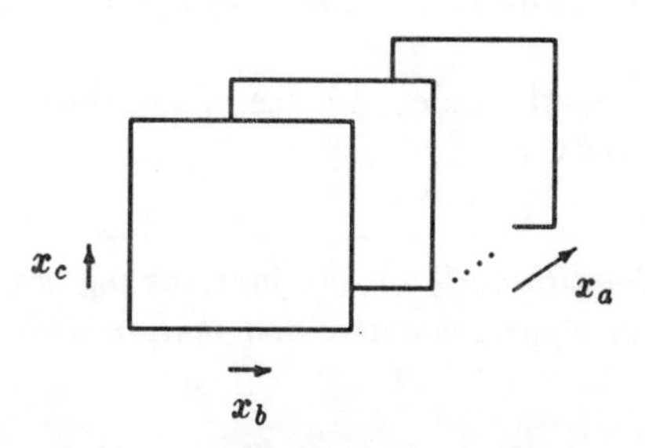

This means that within each slice the exact distribution of the table, conditioned on its margins, is the Hypergeometric distribution we met in the last section; and overall, the joint conditional distribution is the product Hypergeometric. Hence, within each slice we can execute an exact conditional test, a test for local independence at $X_a = x_a$; or we may consider an overall test statistic for conditional independence at all values of X_a, and evaluate its exact P-value, by summing product Hypergeometric probabilities over the relevant region.

We need a little extra notation in order to be able to talk about a slice of the table. The full table, $n_K = \{n_K(x)\}$, is cross-classified by the k variables in $K = \{1, 2, \ldots, k\}$ which is partitioned into subsets, $K = a \cup d$, containing p and q variables respectively. Even though n is a count rather than a probability, $n_{d|a}$ is a fairly suggestive notation and so we precisely define the *slice*, $n_{d|a}(x_d; x_a)$ for given x_a, as the q-dimensional table classified by the variables in d obtained by extracting

$$n_{d|a}(x_d; x_a) = n_{ad}(x_a, x_d)$$

from the full table of counts.

Proposition 9.5.1 *If the fully classified table of counts N_K has the cross-classified Multinomial distribution of size n_ϕ, then*

1. *the marginal distribution of any slice, $N_{d|a}$, of the table is Multinomial $M(n_a(x_a), p_{d|a})$ with density*

$$\frac{n_a(x_a)!}{\prod_{x_d} n_{d|a}(x_d)!} \prod_{x_d} p_{d|a}(x_d)^{n_{d|a}(x_d)},$$

for which $n_{d|a}(x_d; x_a) \geq 0,$; and

2. *conditionally on fixing the marginal table $N_a = n_a$ the r_a slices $\{N_{d|a}\}$ are mutually independent.*

The simple proof, which starts by writing down the full Multinomial distribution, is left as an exercise to the reader. (It is especially easy for those who know about the conditional Poisson representation of the Multinomial distribution.)

To be able to discuss the independence of variables within the slice of the table, partition d into $b \cup c$.

Corollary 9.5.2 *Conditionally on the margins n_{ab} and n_{ac} the distribution of N_{abc} is the product Hypergeometric distribution with density function*

$$\prod_{x_a} \frac{\prod_{x_b} n_{b|a}(x_b)! \prod_{x_c} n_{c|a}(x_c)!}{n_a(x_a)! \prod_{x_{bc}} n_{bc|a}(x_{bc})!}.$$

Note that by taking a to be empty, $b = \{1\}$ and $c = \{2\}$, we can recover the formula for the Hypergeometric distribution given above. This representation allows an easy simulation exercise: the full random table with margins n_{ab} and n_{ac} is generated by independently generating a random table with margins $n_{b|a}$ and $n_{c|a}$ for each slice $x_a = 0, 1, \ldots, r_a - 1$.

Tests in graphical models

Exact conditional tests of conditional independence have several immediate applications in graphical modelling; firstly, they may give exact P-values for edge exclusion deviances from the fully saturated model; secondly, they may test the independence implications of the separation theorem, instances of the global Markov property; and thirdly, in the special case where the graphical model of interest is defined entirely in terms of a single conditional independence, they may be applied to evaluate the exact P-value for the overall fit of the model.

EXAMPLE 9.5.1 Stouffer-Toby continued. Consider the Stouffer-Toby data from Example 7.8.1 again, used to illustrate the possible effects of perturbing the table by deleting observations from cells with small cell counts. We might expect the asymptotic approximations to fail in this case. Here we compare the exact and asymptotic P-values for the observed value of the deviance, computed in the orginal table and in the table perturbed by deleting the two observations in singleton cells. The exact values are estimated from simulating 1000 random tables for each hypothesis.

	original table				perturbed table			
			P-values				P-values	
hypothesis	dev	df	asymp.	exact	dev	df	asymp.	exact
$1 \perp\!\!\!\perp 2 \mid \{3,4\}$	8.22	4	.084	.115	13.8	4	.008	.012
$1 \perp\!\!\!\perp 3 \mid \{2,4\}$	2.29	4	.682	.720	0.7	3	.864	.870
$2 \perp\!\!\!\perp 3 \mid \{1,4\}$	10.29	4	.036	.066	10.3	3	.016	.028
$1 \perp\!\!\!\perp 4 \mid \{2,3\}$	4.56	4	.336	.419	10.8	4	.029	.054
$2 \perp\!\!\!\perp 4 \mid \{1,3\}$	17.53	4	.002	.002	19.7	4	.001	.000
$3 \perp\!\!\!\perp 4 \mid \{1,2\}$	19.00	4	.001	.002	18.9	3	.000	.002

The array of asymptotic and exact P-values are similar for both tables, though rigid adherence to a 5% significance level would change the status of the edge (2,3) in the original table. In this instance, the simulations suggest that the asymptotic values slightly over estimate the significance of the effect in both the original and the pertrubed tables. The effect of perturbation remains enormous, even when an exact analysis is conducted, and changes the selected model from formula $1 + 2.3.4$ to formula $1.2.4 + 2.3.4$.

For didactic purposes, consider the original table and the model $1.2 + 2.3.4$ defined by the two pairwise conditional independences $1 \perp\!\!\!\perp 3 \mid \{2,4\}$ and $1 \perp\!\!\!\perp 4 \mid \{2,3\}$. It has the graph

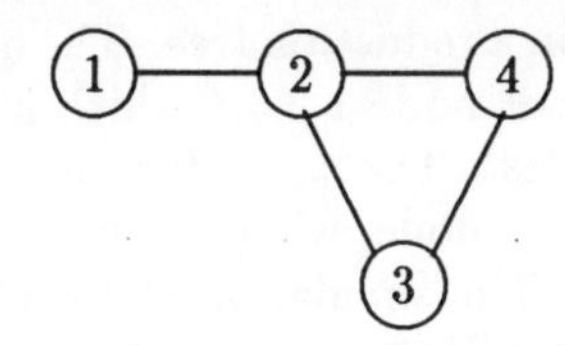

A consequence of the separation theorem is that $1 \perp\!\!\!\perp 3 \mid 2$ and $1 \perp\!\!\!\perp 4 \mid 2$, so that if the model fits, these hypotheses should also have acceptably high P-values. Both hypotheses have 2 df. The corresponding deviance statistics and P-values are 3.5 (asymp=0.178, exact=0.202) and 5.7 (asymp.=0.057, exact=0.069) respectively. □

Decomposable models

The extension of exact conditional testing to evaluating the P-value for a fitted decomposable model is straightforward but requires further effort in the simulation. To see what is involved consider application of these techniques to estimate the P-value of a goodness of fit statistic T in fitting the model $1.2 + 2.3 + 3.4$ to a four-way contingency table, in which the pairwise independences cannot be structured into a single independence statement. The independence graph is

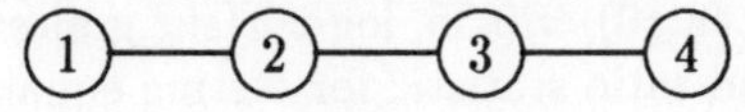

The idea is to reduce the model into its components, discussed fully in Chapter 12, which because the model is decomposable, are complete, and to lay out the components as nodes on a tree, the ***reduction tree***. The model formula is reducible to components 1.2 and 2.3.4 and 2.3.4 is further reducible to 2.3 and 3.4, so corresponding to this sequence of reductions the tree is

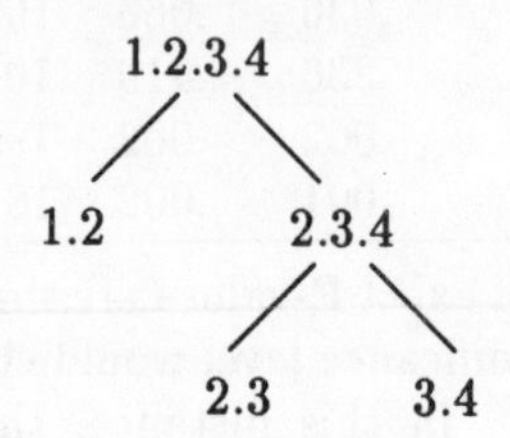

The tree maps a sequence of reductions, so that emanating from a node there are either two branches, a fork, or no branches, in which case the node is a terminal node corresponding to a complete component. Each fork corresponds to a conditional independence statement, for example, the fork at the node 2.3.4 corresponds to $2 \perp\!\!\!\perp 4 | 3$. The reduction tree is not unique, and for example, another can be formed by using vertex 3 as the initial separating set.

The simulation exercise needed to conduct an exact conditional test for the overall model operates on a reduction tree. The object is to generate a random four-way table at the root node 1.2.3.4 which has fixed margins specified by the list of terminal nodes. The procedure is to climb the tree, from these terminal nodes to the root node, where at each fork a table is simulated with the prescribed margins. The simulation at each fork corresponds to selecting a table from the product Hypergeometric distribution as explained above. In this example, the first simulation is to create at three-way table for the variables 2, 3 and 4, with given 2.3 and 3.4 margins. Then regarding this three-way table as fixed the four-way table at the root node is simulated from given 1.2 and 2.3.4 margins, and the value of the statistic T calculated. In consequence each simulated value of the statistic requires simulating two random tables.

9.6 Exercises

1: Recall the breast cancer data reported by Morrison *et al.* (1973) and discussed in Example 7.5.3. Show that

(a) the conditional log-cpr's between survival and nuclear grade are
log cpr(sur,gra|cen=0) = 0.35, log cpr(sur,gra|cen=1) = 0.35+0.12 = 0.47;

(b) the likelihood ratio statistic for testing equality is 0.088 on 1 df.

(c) the estimate of the partial log cpr(sur,gra|cen)= 0.40, while the marginal coefficient is log cpr(sur,gra) = 0.38.

2: The following data on smoking and cancer was reported by Doll and Hill (1950).

		smoker	
sex	group	yes	no
male	cancer	647	2
	control	622	27
female	cancer	41	19
	control	28	32

Estimate the conditional log-cpr between smoking and cancer for both males and females. Test if they are equal. Give the common estimate. Is this the same as the marginal estimate?

3: Show that in a 2^3 table of probabilities p for (X_1, X_2, X_3) that the partial cross-product ratio, defined as the value of u_{12} in the expansion

$$\log q_{123} = u_{12}x_1x_2 + u_{13}(x) + u_{23}(x)$$

for which q minimises $I(p;q)$, always lies between $\log \text{cpr}\,(X_1, X_2|X_3 = 0)$ and $\log \text{cpr}\,(X_1, X_2|X_3 = 1)$.

4: The following example is taken from Whittemore (1978). Consider the two $3 \times 2 \times 2$ tables classified by x_1, x_2 and x_3 respectively:

	x_1	0		1		2	
x_2	x_3	0	1	0	1	0	1
0		4	2	2	1	1	4
1		2	1	4	2	1	4

	x_1	0		1		2	
x_2	x_3	0	1	0	1	0	1
0		75	24	25	8	20	16
1		20	16	60	48	16	32

For each table, show that the conditional $\text{cpr}\,(X_2, X_3|X_1 = i)$ is the same whatever $i = 0, 1, 2$; and is the same as the $\text{cpr}\,(X_2, X_3)$ for the marginal table classified by X_2 and X_3.

5: Generate numerically a density function for $X = (X_1, X_2, X_3, X_4)$ from a particular log-linear expansion in the class of two-way interaction models. Show numerically by marginalisation that the expansion of the density of X_{123} does not necessarily have a zero three-way interaction term.

6: In a three-way table of Multinomial counts N_{123} classified by variables X_1, X_2, X_3, does $X_2 \perp\!\!\!\perp X_3 | X_1$ imply $N_{12} \perp\!\!\!\perp N_{13} | N_1$?

7: Compare the exact and asymptotic P-values for the deviance against independence calculated from observing

(a)

n	0	1
0	0	4
1	3	5

and (b)

n	0	1
0	0	8
1	6	10

8: Find analytically the exact conditional probability of the 2×3 table:

	$x_2 = 0$	$x_2 = 1$	$x_2 = 2$
$x_1 = 0$	1	4	2
$x_1 = 1$	2	2	3

given the margins, under the hypothesis of independent row and column effects. Find how many possible tables exist with these margins.

9: Explain how one may generate a random $r \times c$ table with given margins from a sequence of 2×2 Hypergeometric random variables.

10: Consider testing for independence between rows and columns in the table

	$x_2 = 0$	$x_2 = 1$	$x_2 = 2$	total
$x_1 = 0$	10	1	6	17
$x_1 = 1$	3	5	0	8
$x_1 = 2$	5	0	1	6
total	18	6	7	31

based on a sample of 31 observations. Find the asymptotic P-value against independence based on the deviance and compare it to a simulated exact value.

11: The following cross-classification of diabetic patients is taken from Plackett (1974), (see also Fienberg, 1977). There are three variables: A age at onset (<45,≥45), I dependence on insulin injections (yes,no), and F family history (yes,no); and the observed table of counts is

A	<45		≥45	
F I	yes	no	yes	no
yes	6	6	16	8
no	1	36	2	48

Compare the asymptotic P-values for Pearson's statistic with the simulated exact P-values, for the three possible pairwise conditional independences and comment.

12: Justify theoretically the derivation of the product Hypergeometric distribution outlined in Section 9.4, and prove Proposition 9.5.1.

13: Show that the product Hypergeometric distribution in Corollary 9.5.2 can be expressed as

$$\prod_{x_a} \frac{\prod_{x_b} n_{ab}(x_a, x_b)! \prod_{x_c} n_{ac}(x_a, x_c)!}{n_a(x_a)! \prod_{x_{bc}} n_{abc}(x_a, x_{bc})!}.$$

14: Write out the exact distribution of the four-way table N_{1234} conditioned on the margins n_{12}, n_{23} and n_{34}, under the hypothesis of the model with formula $1.2 + 2.3 + 3.4$.

Chapter 10

Regression and Graphical Chain Models

It often happens that the set of variables under study naturally partitions into two groups, (X, Y), where the q variables in Y can be considered *response* variables and the p variables in X, explanatory variables or *covariates.* For example, in a study of the effect of nutrient on growth rates, measures of growth might be classified as response variables and measures of nutrient as covariates. As the adjectives suggest, the response vector Y and the explanatory vector X do not enter into the model on the same footing, and this asymmetry has to be incorporated into the model; 'treatment-response', 'exogenous-endogonous' and 'independent-dependent' are three other synonyms for this partition. Though at first it may appear that such a partition adds another layer of complication to an already complex subject, in fact, the practitioner will rapidly become convinced that this additional structure overwhelmingly simplifies the statistical analysis and interpretation of graphical models. From the outset we categorically assert that the partitioning into explanatory and response variables is part of the initial model specification, and is *not* a conclusion to be drawn from any subsequent data analysis.

We consider models based on a partition of the variables into a chain of m ordered blocks, $X_K = (X_{b_1}, X_{b_2}, \ldots, X_{b_m})$. The probabilistic theory of chain graphs, defined by pairwise independences between variables conditioned on all other variables in the current block and in preceding blocks, represented by a mix of directed and undirected edges, and summarised by the Markov properties of the associated moral graph, was developed in Chapter 3. This chapter adds the distributional models needed for fitting and testing these chain models to data. The case of just two blocks is especially interesting: firstly, it corresponds to a multivariate regression in which concern is to model the distribution of the response variables conditionally on the covariates; and secondly, the analysis of the chain of m blocks devolves to the separate analysis of $m-1$ two block structures. The univariate linear regression model consists of two blocks, with b_1 containing p explanatory variables and b_2 containing the single response variable.

Summary of chapter: In the first section we discuss the general issue arising from partitioning variables into recursive blocks. In particular, we discuss possible implications for relevant independence hypotheses and the nature of the independence graph and changes needed in statistical inference procedures. The second section studies the relationship between those tools developed for the analysis of the all-response model and those needed for the regression model. In the following sections we report case studies involving two and several blocks of continuous variables, and then turn to some more theoretical relationships with multiple and multivariate Normal linear regression. The following sections treat categorical variables in a similar manner. Logit regression is the name given to log-linear models for conditional probabilities with discrete covariates (with continuous covariates these are termed logistic regression models); and we discuss how to fit these models individually, and how to combine them into a chain model for a recursive system of variables. We give two case studies for five-way tables. Before embarking, we briefly touch upon two issues that relate to regression ideas.

Cause and effect: In scientific studies it may be natural to suppose that the relationship between X and Y is causal, with X identified as the cause and Y as the effect. But, the mapping between cause-effect and explanatory-response variables may be inverted so that the cause corresponds to the response variable rather than to the explanatory variable. A classical example is provide by retrospective investigations of the effect of smoking on the occurrence of lung cancer where the previous smoking histories of individuals in a group of cancer patients are compared to those in a control group. The role of the explanatory variable is played by a variable that flags whether the patient belongs to the cancer group or the control group, and the response variable is a smoking indicator. There is no interest in modelling the relative sizes of the cancer and control groups as this is determined by the original survey design. But current medical belief would have it that smoking causes cancer rather than the converse: the manner of data collection may seriously affect the status of variables in causal studies.

Alternative model specifications: Enlarging the compass of the text to consider those models with a general response-covariate structure is akin to opening Pandora's box, and we make no attempt to review these models, see though Blalock (1971), Goldberger and Duncan (1973), and Joreskog (1981). We would point out that certain models specified in terms of a system of structural equations lead to ambiguous formulations. For instance

EXAMPLE 10.0.1 A causal model. Suppose that the Normally distributed random vector $Z = (Z_1, Z_2, Z_3, Z_4)$ has mean zero and variance the identity matrix. The observed random variables $X = (X_1, X_2, X_3, X_4)$ are generated by a mechanism such that, given X_1 and X_2, the response X_3 depends on X_1 and X_4 but not X_2, and the response X_4 depends on X_2 and X_3 but not X_1, as described by the following equations:

$$X_3 = \beta_{31}X_1 + \beta_{34}X_4 + Z_3 \quad \text{and} \quad X_4 = \beta_{42}X_2 + \beta_{43}X_3 + Z_4;$$

and further, X_1 and X_2 are supposed independent. The graph describing these equations is the chain graph (a), from which the moral graph (b) associated with the joint distribution of X, can be deduced:

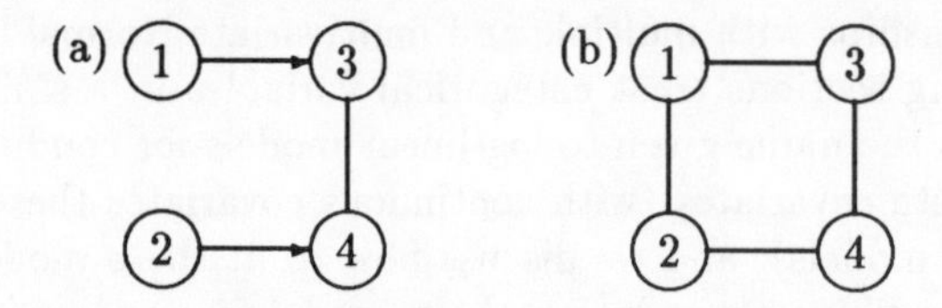

it is a four-cycle. Now, how does the supposition that X_1 and X_2 are generated by

$$X_1 = Z_1 \quad \text{and} \quad X_2 = Z_2,$$

affect the analysis? Seemingly not at all, for this would seem to be an alternative way to specify the independence of X_1 and X_2, so that these two equations, together with the two equations for X_3 and X_4 above, should be an identical specification of the probability mechanism.

However, this view is superficial. The independence graph of X is determined by zeros in the inverse variance of X. The four equations can be written in the form $AX = Z$ where the matrix A is given by

$$A = \begin{pmatrix} 1 & 0 & 0 & 0 \\ 0 & 1 & 0 & 0 \\ -\beta_{31} & 0 & 1 & -\beta_{34} \\ 0 & -\beta_{42} & -\beta_{43} & 1 \end{pmatrix}.$$

As $\text{var}(X) = \text{var}(A^{-1}Z) = A^{-1}IA^{-T}$ taking inverses gives

$$\text{var}(X)^{-1} = A^T A = \begin{pmatrix} 1+{\beta_{31}}^2 & 0 & -\beta_{31} & \beta_{31}\beta_{34} \\ 0 & 1+{\beta_{42}}^2 & \beta_{42}\beta_{43} & -\beta_{42} \\ -\beta_{31} & \beta_{42}\beta_{43} & 1+{\beta_{43}}^2 & -\beta_{34}-\beta_{43} \\ \beta_{31}\beta_{34} & -\beta_{42} & -\beta_{34}-\beta_{43} & {\beta_{34}}^2+1 \end{pmatrix}.$$

The corresponding independence graph is

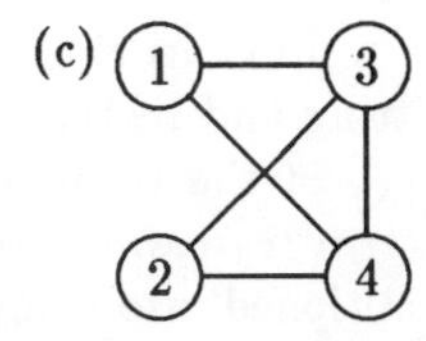

There is only one missing edge, that between X_1 and X_2, so that these variables are conditionally (and in fact marginally) independent. Setting $\beta_{31} = 0$ eliminates both the $(1,3)$ and the $(1,4)$ edges; and similarly $\beta_{42} = 0$ eliminates edges $(2,4)$ and $(2,3)$. The edge between X_3 and X_4 vanishes if the parameters satisfy $\beta_{34} + \beta_{43} = 0$, which leads to a four-cycle, but not the one given in the graph (b) above.

The two graphs, (b) and (c), are not consistent specifications for the joint distribution. □

The structural equation formulation may generate confusion for several reasons:

1. While formulating equations focuses on interactions present in the model, it does not always make clear the conditioning sets for those interactions missing from the model.

2. It is not always clear whether a parameter is free and to be estimated, or redundant and to be derived from the fitted parameters, or constrained, for instance, to be zero.

3. With models for continuous variables, there is a variety of parameters available, for instance, in a choice of regression or correlation coefficients. Furthermore, as these are not mixed derivative measures, they differ according to whether they apply to one of the conditional distributions or to the joint distribution. The situation is simpler with categorical variables because the standard measures, the u-terms and cross-product ratios are derived as mixed derivative measures.

4. Furthermore some models, in particular, the multivariate regression model, take measures from marginal distributions rather than measures from conditional distributions. In consequence, a potential difficulty faced in the analysis of recursive systems is that while several different parameterisations of a single model may be equivalent, the submodels generated by setting a subset of parameters to zero are not equivalent.

For these reasons, we specify all regression models in terms of conditional independence and discuss parameterisations, if at all, as a subsidiary question.

10.1 Conditional Probability Models

The all-response model discussed in previous chapters is modified to incorporate covariates by formulating, fitting and testing models for the conditional probability density function, $f_{Y|X}$, of the response variables, Y, given the covariates, X. We shall use the term *regression model* to denote a model for $f_{Y|X}$. When the variables are partitioned into several blocks, the joint model is built from the product of models for each term in the factorisation

$$f_K = f_{b_1} f_{b_2|b_1} \cdots f_{b_m|b_{m-1}\ldots b_1}.$$

The case of just two blocks is generic, because each factor in this product is a conditional probability.

We take the given partition as axiomatic together with the associated supposition that there is no interest whatsoever in *modelling* the conditional distribution of X given Y, and models in which Y is part of the conditioning set for elements of X are specifically prohibited. If, later,there is interest in predicting X from Y, it can be done with the fitted models via Bayes theorem: $f_{X|Y} = f_{Y|X} f_X / f_Y$.

We shall see that taking the conditional density function as the basic object of study (i) modifies the structure of permissible independence graphs but does not affect the interpretation of the graphs; (ii) separates inference procedures of estimation and testing into inference for each response-covariate model separately; and, (iii) sometimes, permits joint inference procedures to solve conditional procedures.

Independence in the conditional distribution

The variables are partitioned into a vector $Y = (Y_1, Y_2, \ldots, Y_q)$ of response variables and a vector $X = (X_1, X_2, \ldots, X_p)$ of covariates. A graphical model with covariates is a specification of the conditional density function $f_{Y|X}$ that incorporates a specified subset of the $\binom{q}{2}$ *intra-response* pairwise conditional independence statements:

$$Y_i \perp\!\!\!\perp Y_j \,|\, (Y_a, X), \quad \text{where } a = \{1, 2, \ldots, q\} \backslash \{i, j\},$$

and of the pq *response-covariate* pairwise conditional independence statements:

$$Y_i \perp\!\!\!\perp X_j \,|\, (Y_a, X_b), \quad \text{where } a = \{1, 2, \ldots, q\} \backslash \{i\} \text{ and } b = \{1, 2, \ldots, p\} \backslash \{j\}.$$

Proposition 10.1.1 *The conditional independence graph for the conditional distribution of Y given X is identical to that for the joint distribution of X and Y, in which the subgraph corresponding to X is complete. The graph has the global Markov property with respect to the conditional distribution of Y given X.*

Proof: As $f(y|x) = f(x,y)/f(x)$ the factorisation properties of $f(y|x)$ and $f(x,y)$ with respect to intra-response and response-covariate independences are identical. Now apply the separation theorem. □

The independence graph is identical to the graph constructed from the joint density of X and Y defined by the same conditional independence statements now regarded as a subset of all $\binom{p+q}{2}$ possible relationships.

The division of the $p+q$ variables into two groups substantially reduces the magnitude of the model selection problem. For if all the variables are response variables then there are $\binom{p+q}{2}$ independences to examine; while if only q are response variables then the number of independences is reduced to $\binom{q}{2}$ intra-response independences and pq response-covariate independences and

$$\binom{p+q}{2} > \binom{q}{2} + pq, \text{ whenever } p > 1.$$

If the covariates are also modelled in a separate exercise, leading to the examination of $\binom{p}{2}$ intra-covariate independences, there is still a subtantial reduction for while the number of independences is the same:

$$\binom{p+q}{2} = \binom{q}{2} + pq + \binom{p}{2},$$

the way in which they combine is different and the number of models under consideration falls from $2^{\binom{p+q}{2}}$ to the sum of those for f_X alone added to those for $f_{Y|X}$:

$$2^{\binom{q}{2}+pq} + 2^{\binom{p}{2}}.$$

For example, with $q = 2$ and $p = 3$, it falls $2^{10} = 1024$ to $2^{1+2.3} + 2^3 = 136$. The lattice of all possible response-covariate models is then only a fraction of the lattice of all graphical models.

This lattice has obvious maximal and minimal elements, but also two other pivotal models, one obtained by excluding all response-covariate edges from the saturated model and the other obtained by excluding all intra-response edges, as portrayed in Example 10.1.1. Note all edges that join a response node to a covariate node are directed, and all edges that join two responses are undirected, in accordance with previous convention. These two models are interesting in their own right: the former asserts that the vector of response variables is independent of the explanatory variables, while the latter postulates that the response variables are mutually independent given the explanatory variables. Both are sensible choices for the initial base model in a model selection procedure.

EXAMPLE 10.1.1 With $p = 2 = q$, the pivotal elements of the lattice are

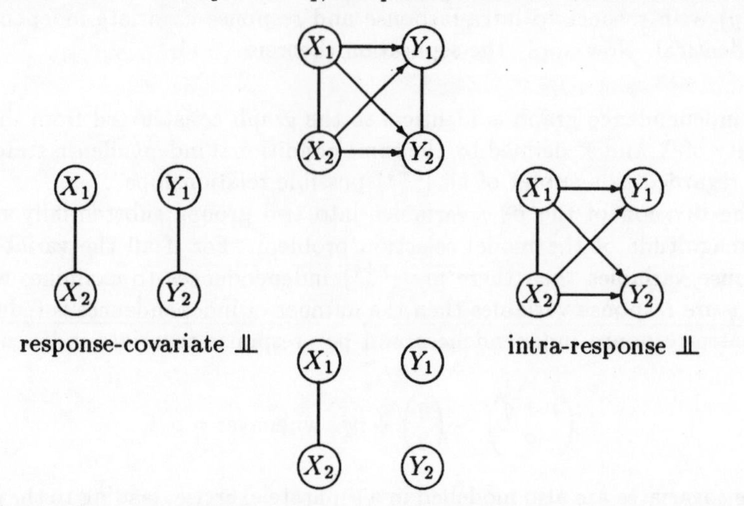

□

The combined chain model is composed of separate models for each block of responses given their covariates. The chain graph, with its mixture of directed and undirected edges flagging the varying conditioning sets for the elements in the system, is a natural way to describe the independence structure of the overall system. As the subgraph of the covariates is complete in each separate independence graph, there is no loss of information in suppressing these edges in a combined graph. So, with this understanding, the *chain graph* of the combined model introduced in Chapter 3 (which has nodes for the variables in all blocks, and directed edges between response and covariate, and undirected edges between response and response) provides a simultaneous description of the independence structure of each part of the combined model. The associated moral graph, in which parents are 'married' and directions dropped, gives a representation of the independence structure of the joint distribution which allows interpretation according to the global Markov property.

Inference in the conditional distribution

We turn to a discussion of inference for the model of the conditional distribution. In principle, we should like the standard program of writing down the likelihood function, finding maximum likelihood estimates, and conducting generalised likelihood ratio tests for model assessment to go through without alteration. We discuss a general condition which makes it permissible to do this for each separate response-covariate model.

Suppose that (i) the parameters of the combined model can be partitioned into θ and γ such that

$$f_{XY}(x, y) = f_{Y|X}(y; x, \theta) f_X(x; \gamma);$$

and that (ii) θ and γ are *variation-independent*, in the sense that the permissible values of one parameter are not affected by the value of the other; equivalently, the parameters take values in a product set. We say we have an instance of *parametric separability*.

We assume the observations are generated by independent random sampling so that the log-likelihood function for the combined model is

$$\sum_l \log f_{Y|X}(y^l; x^l, \theta) + \sum_l \log f_X(x^l; \gamma);$$

just the sum of the individual log-likelihood functions. As the parameters are variation-independent, the mechanics of maximisation, and the construction of likelihood ratio test statistics and intervals, can be entirely separated. Maximum likelihood estimates and their standard errors in one component of the model are the same whatever the model or estimates for the other component.

This condition of parametric separability is strong enough to guarantee separability for all schools of inference. For the Bayesian who selects a prior distribution for the parameters in which $\theta \perp\!\!\!\perp \gamma$, the condition guarantees posterior independence of the form $\theta \perp\!\!\!\perp \gamma | \text{data}$. Consequently, inference about θ cannot affect inference about γ.

For the repeated sampler: first, given the observed covariates, the sampling distribution of the responses depends on θ but not γ; and second, the marginal distribution of the covariates depends on γ but not θ. Another way to express parametric separability is to say that the covariates are *ancillary* for θ and *sufficient* for γ, or equivalently, that we have a *cut* in the sense of Barndorff-Nielsen (1978).

EXAMPLE 10.1.2 Consider a bivariate Normal example where the distribution of (X_1, X_2) is specified by $X_2|X_1 \sim N(\mu_2 + \beta x_1, \sigma^2)$ and $X_1 \sim N(\mu_1, \tau^2)$, then the parameters

$$\theta = (\mu_2, \beta, \sigma^2) \quad \text{and} \quad \gamma = (\mu_1, \tau^2)$$

are variation-independent. Inference about θ and γ is separable, and in making inference about the regression coefficient, β, the observed sample values of X_1 are regarded as fixed, while they are taken as random in making inference about the mean μ_1. □

Parametric separability not only splits up the original problem into tasks with known solution, but also shows how the final results can be reassembled to provide estimates for the combined model. The maximum likelihood estimate of (θ, γ) is $(\hat{\theta}, \hat{\gamma})$. By parametric separability, the log-likelihood function is a sum of separate log-likelihood functions and, as the parameters are variation-independent, the joint information matrix of the parameters is block diagonal. By large sample considerations of maximum likelihood estimation,

$\hat{\theta}$ and $\hat{\gamma}$ are independent. The maximum likelihood estimate of the joint density function is the product $\hat{f}_{XY} = \hat{f}_{Y|X}\hat{f}_X$ where $\hat{f}$ indicates that maximum likelihood estimates are substituted for the parameters. Furthermore, the deviance of the combined model is the sum of the deviances of the two component models. Not only do we know how to to combine the separate independence graphs into a single chain graph, but we also know how to combine the separate statistical analyses into a single set of parameters and fitted values. The same argument applies when there are several blocks to put together.

And this is not all: model selection for one component of the model does not affect the selection procedure for another; diagnostic checking procedures separate, and in particular, residuals may be computed and assessed separately for each part of the model.

Deviance decompositions

In regression the concept of 'explained variation' is often used to interpret results. The 'total to explain' is usually the total sum of squares of the response variable, and a guide to model performance is the proportion of this total variation attributable to fitting various sub-models. The total variation, when divided by the sample size, estimates the variance of the response unadjusted for any explanatory variables. The analogue in conditional independence analysis is the deviance against the model of mutual independence, which, when divided by the sample size estimates the corresponding information divergence. We have given an example of this decomposition in the analysis of the pit-prop data in Example 6.8.1, in which all variables are responses.

When a recursive block structure exists, because of the ordering between the variables, the decomposition is often more meaningful. We use an obvious generalisation of the model formulae introduced in Section 7.3.

Proposition 10.1.2 Block decomposition of independence. *The total information against independence can be partitioned into (i) the information against the independence of the explanatory variables and (ii) the information against the independence of the response variables from the explanatory variables and from each other. That is,*

$$\begin{aligned}\mathrm{dev}\,(X_1 + X_2 + \cdots + X_p + Y_1 + \cdots + Y_q) = \\ \mathrm{dev}\,(X_1 + X_2 \cdots + X_p) + \mathrm{dev}\,([X_1.X_2 \ldots X_p] + Y_1 + \cdots + Y_q).\end{aligned}$$

Proof: The expressions developed in Section 4.5 on the additivity of information divergence, together with the representation of the deviance as a divergence in Sections 6.5 and 7.4, provide a proof. □

Whether there are further sensible or unique decompositions of the terms on the right depends on the context. We give an illustration.

EXAMPLE 10.1.3 Deviance decomposition on the cork data. The correlation matrix for the 28 observations on a cork tree introduced in Exercise 5 of Chapter 8, is

N	1.000			
E	0.885	1.000		
S	0.905	0.826	1.000	
W	0.883	0.769	0.923	1.000
	N	E	S	W

For illustrative purposes we concern ourselves with three-way margin of the variables N, W and S, and consider decomposing the deviance against the complete independence of these three variables, $\text{dev}\,(N+W+S) = 103.74$ on 3 df. We shall need the following information on deviances against conditional and marginal independence, obtained by excluding edges from the saturated model in the three-way margin and the relevant two-way margins:

	3-way margin		2-way margin	
edge	dev diff	df	dev diff	df
$N.S$	7.92	1	47.78	1
$W.N$	2.54	1	42.41	1
$W.S$	13.55	1	53.42	1

First consider the case in which (N, S) form the first block and W the second, so that $p = 2$ and $q = 1$. The deviance decomposition in the proposition is

$$\text{dev}\,(N + S + W) = \text{dev}\,(N + S) + \text{dev}\,(N.S + W), \quad 103.74 = 47.78 + 55.96.$$

The total information in the data against independence comes in two roughly equal halves. The deviance $\text{dev}\,(N.S + W) = 55.96$ for the response model could be explained by including an edge between W and S, accounting for $\text{dev}\,(S + W) = 53.41$ and leaving a residual of $\text{dev}\,(N.S + W.S) = 2.54$; or by including an edge between W and N: $\text{dev}\,(N.S + W.N) + \text{dev}\,(N + W) = 13.55 + 42.40$; or by including both edges.

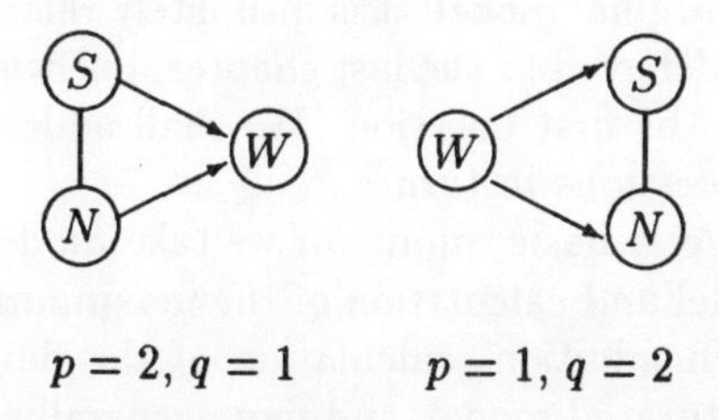

$p = 2, q = 1$ $\qquad$ $p = 1, q = 2$

Alternatively, suppose that W is the first block and (N, S) is the second so that $p = 1$ and $q = 2$. The block decomposition of independence in the proposition is uninformative, as in $\text{dev}(W + N + S) = \text{dev}(W) + \text{dev}(W + N + S)$ the first term is always zero. The total deviance is identical to the deviance for the response model.

Further analysis of this latter case is possible, but is essentially no different from a decomposition for an all-response model. Of the more interesting decompositions of the response deviance, two are defined by the pivotal models in Example 10.1.1, one excluding response-covariate edges, the other excluding intra-response edges. For the pivotal model in which there is no intra-response edge between N and S, the decomposition of the response deviance is $\text{dev}(N + W + S) = \text{dev}(N.W + S.W) + \text{dev}(W + N) + \text{dev}(W + S)$, or numerically, $103.74 = 7.92 + 42.40 + 53.41$. Corresponding to the pivotal model in which there is no response-covariate edge between W and (N, S), the decomposition is $\text{dev}(W+N+S) = \text{dev}(W+N.S)+\text{dev}(N+S) = 47.78+55.96$, numerically identical to our first example, but, due to the different block structure, with a different interpretation. □

10.2 Fitting Regression Models in the Joint Distribution

There are two related questions:

1. Can the regression model for $f_{Y|X}$ be fitted by fitting models for f_{XY}?

2. Can a given model for f_{XY} be thought of as model for $f_{Y|X} f_X$?

Our specific interest here is the first question: can analysis of the all-response model can be adapted to analyse a regression model? If so, we may re-use those algorithms and statistical procedures to fit chain models without extra work. As there is no such thing as a free lunch, the general answer must be no, except in some special cases, which fortunately includes the Multinomial and multivariate Normal distributions. The second question is concerned with whether a given all-response model can be interpreted as a combined regression model and marginal model; it is intimately related to collapsibility of distributions, a topic deferred to the last chapter, but whose solution provides part of the answer to the first question. We shall address some of the issues raised by these two questions in turn:

1. Model fitting: At a basic minimum we take model fitting to mean the specification of a model and calculation of the maximum likelihood estimates of the parameters. In addition, calculation of the deviance for the model requires fitting the saturated model, and more generally, model selection procedure may entail fitting all regression models consistent with the partition into response variables and covariates.

2. Equivalent independence hypotheses: We have seen above that there is a one to one correspondence between pairwise independence statements made in the joint distribution and in the distribution of the responses given the covariates, and it is necessary to fit models which respect exactly this set of independence relations. If the response-covariate interaction is arbitrary then the model for the joint distribution must have a complete subgraph for the covariates. The next example illustrates.

EXAMPLE 10.2.1 Consider the three following chain independence graphs for the distribution of (X_1, X_2, X_3, X_4) partitioned into two blocks, where (X_1, X_2) is the covariate for the response (X_3, X_4).

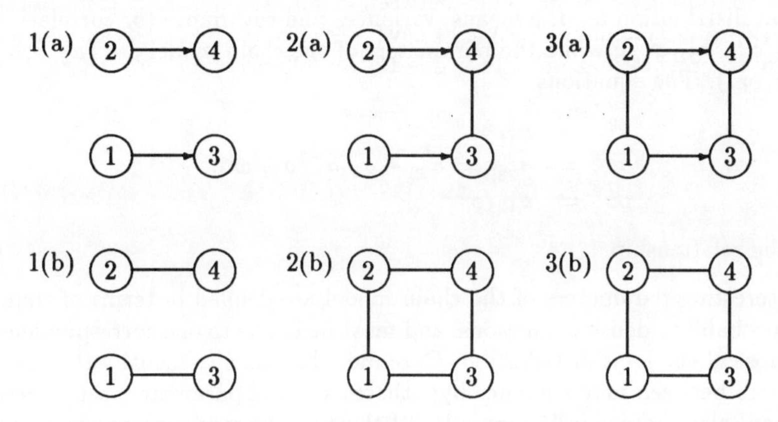

The undirected versions are shown in the lower three graphs. The independence graph of the joint distribution is determined from the the moral graph. In the second example the moral graph differs from the undirected graph (vertices 1 and 2 are connected in the moral graph) while in the first and third they are the same. In the third example the covariates have a complete subgraph and the moral graph coincides with the undirected version whatever the regression model. □

3. Conditional inference: If tests for an independence hypothesis are the same in f_{XY} as in $f_{Y|X}$, they are also the same in

$$f_{Y|X} h_X$$

where h is an arbitrary marginal density function for X. For equivalence of tests, we shall require the repeated sampling properties of tests to be invariant to h, which, for instance, is satisfied in making inference conditional on the observed covariates, or invariant to their distribution.

4. Closure of families of distributions: Inference in the joint distribution must be simple for the strategy to convey advantage, and this may be satisfied if we can choose the marginal density function h above, so that the joint distribution falls in a known family of distributions. We have seen in Chapters 6 and 7 that both of our two archetypical distributions, the multivariate Normal and the cross-classified Multinomial distributions, are closed under marginalisation and lead to related families of distributions under conditioning. (The mixed case proves more difficult, essentially because of lack of closure.)

5. *Parametrics:* Even though we may specify models entirely in terms of independence statements, we cannot avoid parametric families of distributions for model fitting. Modelling the response-covariate structure with a conditional distribution uses different parameters than the joint distribution.

EXAMPLE 10.2.2 For instance, in Example 10.1.2 above, the parameters of the joint distribution are the means, variances and covariance (or correlation) $(\mu_1, \mu_2, \sigma_{11}^2, \sigma_{22}^2, \sigma_{12})$; while the parameters of the chain model are (μ_2, β, σ^2) and (μ_1, τ^2). The equations

$$\begin{aligned}\tau^2 &= \sigma_{11}^2 \\ \sigma^2 &= \sigma_{22|1}^2 = \sigma_{22}^2 - \sigma_{12}\sigma_{11}^{-2}\sigma_{12} \text{ and} \\ \beta &= \sigma_{12}/\sigma_{11}^2.\end{aligned}$$

show the relationship. □

The interesting parameters of the chain model are defined in terms of conditional probability density functions, and must be in one to one correspondence with those of the joint distribution. Care must be taken to identify the correspondence between independence hypotheses and the parameters set to zero; as we have already seen in Example 10.0.1 there can be confusion as to whether parameters determine marginal or conditional independence. The choice of parameterisation may lead to differences, for example, the choice of correlation or regression coefficients in continuous models is important, for while a regression coefficient is a component of a mean, a correlation coefficient is a component of a variance.

6. *Commutativity of marginalisation and maximisation:* We wish to maximise the log-likelihood function for the regression model $\sum_l \log f_{Y|X}(y^l; x^l, \theta)$ by multiplying by a suitable marginal density $f_X(x; \gamma)$, and then maximising the log-likelihood from the resulting joint distribution

$$\sum_l \log f_{XY}(x^l, y^l; \psi),$$

where $\psi = \psi(\theta, \gamma)$. The parameters, (θ, γ) are transformed to the parameter, ψ, of the joint distribution. A necessary condition for these two procedures to reach the same estimates is that

$$\int_y f_{XY}(x, y; \hat{\psi})dy = f_X(x; \hat{\gamma});$$

or equivalently, that the operations of maximisation and marginalisation are commutative: the left hand side, which marginalises over the fitted joint density function, has to equal the right hand side, which is the maximised value over the marginal density function. If this condition is satisfied the estimated conditional density function may be extracted from

$$f_{Y|X}(y; x, \hat{\theta}) = f_{XY}(x, y; \hat{\psi})/f_X(x; \hat{\psi}).$$

In the analysis of the Normal and the Multinomial distributions, a condition that generally ensures commutativity is that the joint model preserves the observed marginal distribution of the covariates.

EXAMPLE 10.2.3 Two-way interaction models. Consider models for 4 categorical variables partitioned into two blocks $b_1 = \{1, 2, 3\}$ and $b_2 = \{4\}$. Can the all two-way interaction model $1.2 + 1.3 + 2.3 + 4.(1 + 2 + 3)$ for the joint distribution of the 4 variables be used to fit a response-covariate model for the response X_4? The subgraph of the covariates in the independence graph is complete. However, the maximum likelihood equations which require that observed and fitted margins are the same on the interactions in the model, only ensure that the margins (X_1, X_2), (X_1, X_3), and (X_2, X_3) are the same. Hence the full margin for the covariates is not preserved in the fitted model.

In general, this class of models cannot be used to fit suitable response-covariate models, though it may serve as a good approximation to $1.2.3 + 4.(1 + 2 + 3)$ which does have this property. Note that fitting the joint model $1.2 + 1.3 + 2.3 + 4.(1 + 2)$ can be used because it is enough to preserve the margin (X_1, X_2) for fitting this model. □

10.3 A Case Study: Noctuid Moth Trappings

This example was used by Dempster (1972) to illustrate covariance selection, a progentor of the graphical Gaussian model, and as such it deserves to be better known. The data had been introduced in a paper of Cochran (1938) concerning simple algorithms of recomputing coefficients when a new variable is added to the regression. We use it here to illustrate the features of fitting models to both the explanatory variables and to the response variable.

The response variable is the number of moths caught in a light trap in one night, transformed by adding 1 and taking logarithms; the covariates are five measures of weather conditions: minimum night temperature, previous day's maximum temperature, average wind speed during night, amount of rain during night, and lastly, percentage of starlight obscured by clouds. The sample correlation matrix is given in Table 10.3.1 below, but unfortunately the original observations are not published; in fact Cochran intimates that they come from a rather complicated design with an effective sample size of

Table 10.3.1: The correlation matrix of the noctuid moth trappings data. From Cochran (1938).

min	1.00					
max	0.40	1.00				
wind	0.37	0.02	1.00			
rain	0.18	-0.09	0.05	1.00		
cloud	-0.46	0.02	-0.13	-0.47	1.00	
moth	0.29	0.22	-0.24	0.11	-0.37	1.00
variance	14.03	14.54	2.07	17.11	7.87	3.55
	min	max	wind	rain	cloud	moth

$N = 72$. An interest of the study is to determine how the number of moths trapped is affected by climatic conditions.

The correlation matrix indicates that the numbers of moths trapped increases with temperature, decreases with wind and cloud, but somewhat surprisingly, increases with rain. There is a high correlation between maximum and minimum temperature together with some rather complicated relationships between the meteorological variables. The sign of the correlation coefficient between cloud and rain is also a little surprising.

Our analysis, unlike Dempster's, partitions the variables into response and covariates; the moth trapping measure is the univariate response variable, Y, and the metereological variables constitute the explanatory vector X. We first find a model for the covariates, then for the response, and finally consider a combined chain model.

The covariates

Consider the inter-relationships between the five metereological variables. Whether or not the joint distribution of these is multivariate Normal, the analysis sheds light on their interaction structure, relevant to the purpose of explaining the response variable. We first calculate the edge exclusion deviances, $\text{dev}\,(X_i \perp\!\!\!\perp X_j | \text{rest})$, for eliminating each of the $\binom{5}{2}=10$ edges from the graph:

min	*				
max	18.58	*			
wind	11.16	1.88	*		
rain	0.01	0.37	0.01	*	
cloud	17.30	3.08	0.37	13.71	*
$R^2(X_i;\text{rest})$	46.8	23.4	16.0	22.4	39.5
	min	max	wind	rain	cloud

The multiple correlation coefficients are highest for cloud cover and minimum temperature; our first stab at an independence graph to describe the weather conditions is

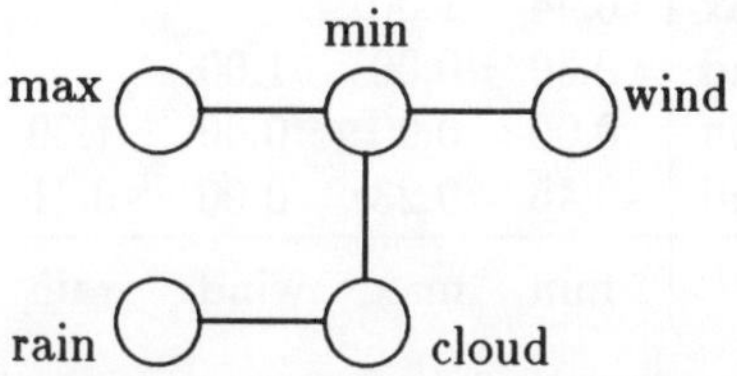

A feature of the graph is the relatively central role it assigns to the minimum temperature variable: it separates the other variables into three groups as shown, with rain being separated from min-temp by the cloud measure. While the overall deviance of the model is 7.20 on 6 df, well within the limits of sampling variation, the edge inclusion deviances are

min	*				
max	x	*			
wind	x	1.51	*		
rain	0.19	2.33	0.07	*	
cloud	x	4.62	0.14	x	*
	min	max	wind	rain	cloud

and an edge between max-temp and cloud has to be included.

The selected explanatory model: The model has the formula max.min.cloud + rain.cloud + min.wind, needing five interaction parameters. The edge inclusion deviances from this model are 0.19, 1.51, 0.54, 0.07, and 0.14; and the edge exclusion deviances are 16.92, 10.53, 21.99, 17.72, and 4.63; indicating that no further adjustment need take place. It has a deviance of 2.57 on 5 df with an observed P-value of 0.768, which should be compared to a total deviance against mutual independence of 65.15 on 10 df. The overwhelming proportion of interaction is accounted for by this model.

The selected model has the independence graph given in Figure 10.3.1.

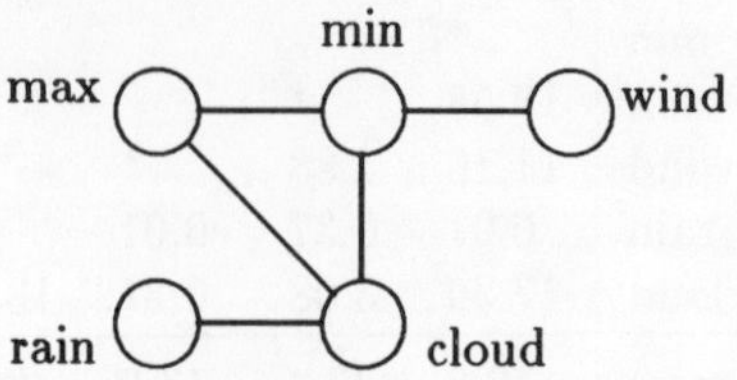

Figure 10.3.1: Independence graph for the meteorological variables.

The fitted partial correlations are

	min	max	wind	rain	cloud
min	1.00				
max	0.44	1.00			
wind	0.30	0.00	1.00		
rain	0.00	0.00	0.00	1.00	
cloud	-0.45	0.23	0.00	-0.41	1.00

with zeros on missing edges, and agreeing well with the empirical coefficients.

It is more difficult to disentangle the separate effects of the covariates on the response variable when they interact. Thus the effects of the variables in cliques: in particular (max, min, cloud), (rain, cloud) and (min, wind), may become confounded.

Response model selection

We now include the dependent variable in the analysis and look at the edge exclusion deviances corresponding to a response-treatment connection

	min	max	wind	rain	cloud	moth
multiple corr. coeff.	(48.8)	(25.0)	(27.7)	(22.7)	(45.0)	30.2
edge excl.devs. -moth	2.77	1.52	10.85	0.24	6.80	*

The multiple correlation coefficient for moth trappings is 30.2% so that a linear regression model can account for no more than this percentage of the variation. Only the edge exclusion deviances with wind and with cloud are greater than 3.84, the 5% point of the chi-squared distribution with 1 df.

The deviance of this graphical model is 8.23, which on 3 df is greater than 7.81, the 5% point of the chi-squared distribution. The edge inclusion deviances of the variables not in the regression model are

min	max	wind	rain	cloud
6.39	5.22	x	0.44	x

and suggest that both measures of temperature need to be included. The values should be compared with the values of the edge exclusion deviances from the saturated model, (2.77, 1.52, x, 0.24, x), suggesting that neither measure needs to be included. We have an apparent contradiction. The interaction between the covariates noticed above, has a direct effect on the selection process for the response variable. Clearly, one, but not both, of maximum temperature and minimum temperature have to be included in the regression equation, see the exercises.

The selected response model: We choose to include the minimum temperature: partly because it corresponds to the temperature measured at the same time as when the moths were trapped, unlike the maximum temperature; partly because it has a slightly higher deviance than that of the maximum temperature; but mainly because of its more central position in the independence graph of the covariates. The deviance of this model with explanatory variables (min,wind,cloud), is 1.85 which on 2 df constitutes a perfectly adequate fit and the value of R^2 rises to 28.4%, almost the maximum possible. The independence graph in Figure 10.3.2 consists of the complete subgraph on the five meteorological variables with wind, cloud and min-temp connected to moth trappings:

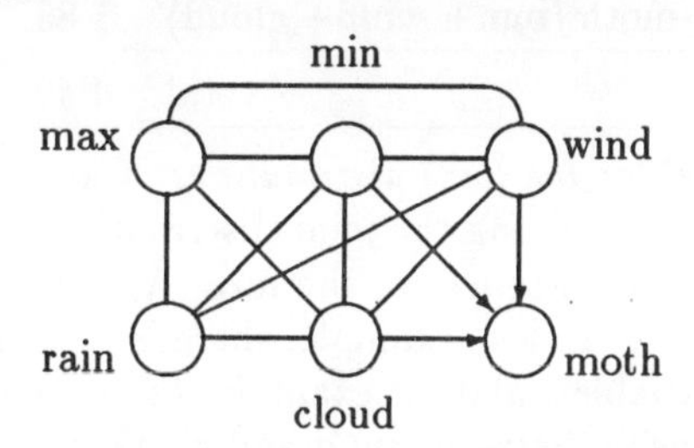

Figure 10.3.2: Independence graph for the selected response model.

It asserts that moth$\perp\!\!\!\perp$(max, rain)|(min, wind, cloud).

The combined chain graph model

These two models can be combined in a single chain graph model based on two blocks, one consisting of the explanatory variables, X, and the other of the response variable Y. The chain graph is given in Figure 10.3.3. Each independence statement in the two separate models is retained in this chain graph; for example, ignoring the second block reveals the graph of the covariates alone and wind$\perp\!\!\!\perp$cloud|(min-temp) alone. Moralising the chain graph by joining 'unmarried' parents and dropping directions, gives the independence graph for (X, Y), which in this instance, requires a single additional edge between wind and cloud.

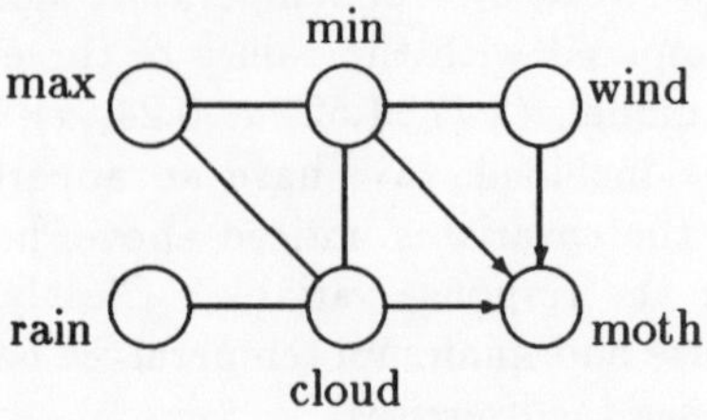

Figure 10.3.3: Selected chain graph.

The overall fit of the model can be summarised in the following analysis of deviance table. The total deviance to explain is the deviance against the complete independence of the six variables, some 91.05 on 15 df, and the combined model explains all but 4.42 units.

	residual		total	
model	dev	df	dev	df
max.min.cloud + rain.cloud + min.wind	2.57	5	65.15	10
(1.2.3.4.5)+ moth.(min+ wind+ cloud)	1.85	2	25.90	5
total	4.42	7	91.05	15

Dempster's model for the joint distribution: Can our analysis of this data be supplemented by examining the joint distribution of (X, Y)? At most this can reveal further independences of the form $X_i \perp\!\!\!\perp X_j \,|\, (Y, \text{rest})$ which need Y in the conditioning set. In this example, there is a pronounced asymmetry in the status of the variables, and for example, the notion that moth trappings mediate the relationship between wind and cloud is difficult to interpret.

On the other hand, if one can sustain the argument that these variables are all imperfect measures of local climate, then they should be treated on the same footing, and Dempster (1972) models the joint distribution of Y and X. We contrast the results. Using a forward selection procedure he arrived at the model with independence graph

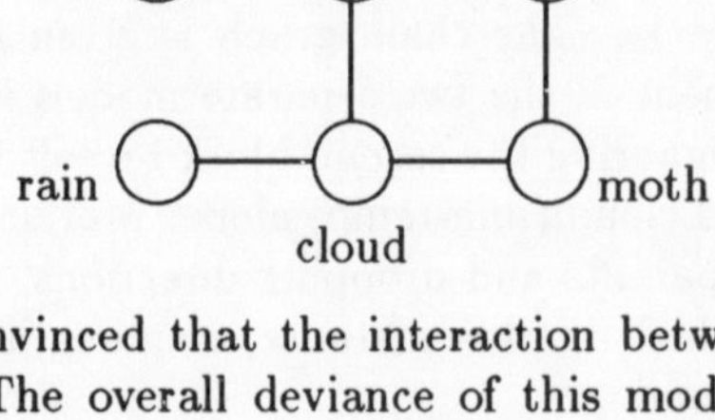

though he was not convinced that the interaction between wind and moth should be included. The overall deviance of this model is 15.66 on 9 df, corresponding to an observed P-value of 0.074, which is the first encountered acceptable P-value, and terminates Dempster's procedure. But, to make the results comparable with those above, model selection should be based on deviance differences rather than overall deviances.

If this is done, we see that two extra edges: max-cloud, and min-moth, need including; and the joint independence graph is identical to the chain graph in which directions are dropped. Though we are led to very similar conclusions, we remark that firstly, this joint model has the interpretation that wind$\perp\!\!\!\perp$cloud$|$(min, moth); and secondly, the joint model does not imply wind$\perp\!\!\!\perp$cloud$|$(min), entailed by the chain model, as integrating the joint distribution over moth, leads to an edge between wind and cloud. Thus, the chain model and the joint model are distinct, with neither model a subset of the other.

10.4 Several blocks: a case study

We now consider an example in which the variables are partitioned into blocks and recursively ordered. The data was reported by Kerchoff (1974) in a study of ambition and attainment, and is also analysed by Joreskog and Sorbom (1981). Our treatment is in accordance with the model selection techniques proposed in a previous chapter, but we short-circuit some of the details. The variables are measures of

> X_1: father's education, X_2: father's occupation, X_3: number of siblings, X_4: intelligence, X_5: grades, X_6: educational expectation, and X_7: occupational aspiration,

observed from a sample of size $N = 767$ twelth grade males. All are positive

Table 10.4.1: A correlation matrix relating ambition and attainment. From Kerchoff (1974).

	X_1	X_2	X_3	X_4	X_5	X_6	X_7
X_1	1.000						
X_2	0.611	1.000					
X_3	−0.108	−0.152	1.000				
X_4	0.250	0.277	−0.100	1.000			
X_5	0.248	0.294	−0.105	0.572	1.000		
X_6	0.410	0.446	−0.213	0.489	0.597	1.000	
X_7	0.331	0.303	−0.153	0.335	0.478	0.651	1.000

(a sign change accounts for the negative interaction to the number of siblings in the family) but vary quite substantially. The interest of the analysis is to relate the individual's background to their expectations and attainment while at school. To this end we suppose that the variables are partitioned into four blocks:

> $b_1 = \{1,2\}$: father's education and occupation; $b_2 = \{3\}$: number of siblings; $b_3 = \{4,5\}$: intelligence and grades; $b_4 = \{6,7\}$: educational expectation and occupational aspiration.

Our aim is to construct a regression model for the conditional distribution of each block given all its preceding blocks. Each block is analysed separately, and model selection operates in exactly the same manner as for the response-treatment model discussed in Section 10.1 above. The edge exclusion deviances corresponding to intra- or inter- response interaction from models for the four blocks are

	1	2	3	4	5	6	7
Block 1							
1	*						
2	358.43	*					
Block 2							
3	0.28	9.21	*				
Block 3							
4	3.59	3.85	0.73	*			
5	1.20	10.01	1.02	250.62	*		
Block 4							
6	7.08	25.48	11.57	24.30	67.20	*	
7	8.18	1.35	0.46	1.35	18.31	189.56	*

These values testify to the extremely strong connections between intelligence and grades, 4.5; between father's education and occupation, 1.2; between grades and educational expectation, 5.6; and between expectation and occupational aspiration, 6.7. On the basis of these edge exclusion deviances, we entertain response treatment models for each block with formulae

$$\begin{aligned} b_1 &: \; () + 1.2 \\ b_2 &: \; (1.2) + 2.3 \\ b_3 &: \; (1.2.3) + 2.5 + 4.5 \\ b_4 &: \; (1.2.3.4.5) + 5.6 + (2+3+4+5).7 + 6.7 \end{aligned}$$

An obvious shorthand is used for the covariate part of the model formula.

The residual deviances from fitting each of these component models is given in the next table, together with the respective 'total deviance to explain'. The latter is calculated by fitting the model in which any variable in the response block is independent of all other variables, both in the same block and in preceding blocks. For example, the total deviance to explain associated with block 3 is dev((1.2.3) + 4 + 5) = 401.86 on 7 df, and the formula comes by eliminating all interactions in the response component of the model.

model	residual dev	df	total dev	df
1.2 for b_1	0.00	0	358.44	1
(1.2)+2.3 for b_2	0.29	1	18.22	2
(1.2.3)+2.5+4.5 for b_3	28.57	5	401.86	7
(1.2.3.4.5)+5.6+(2+3+4+5).7+6.7 for b_4	30.71	5	955.88	11
total 1+2+3+4+5+6+7	59.57	11	1734.39	21

The model formula

$$/1.2/(1.2) + 2.3/(1.2.3) + 2.5 + 4.5/(1.2.3.4.5) + 5.6 + (2 + 3 + 4 + 5).7 + 6.7/$$

denotes the combined model, and has a deviance of almost 60 on 11 df. Though significant, inspection of this table shows that the vast majority of the variation in the data is accounted for by the 10 parameters of this model. Its chain graph is

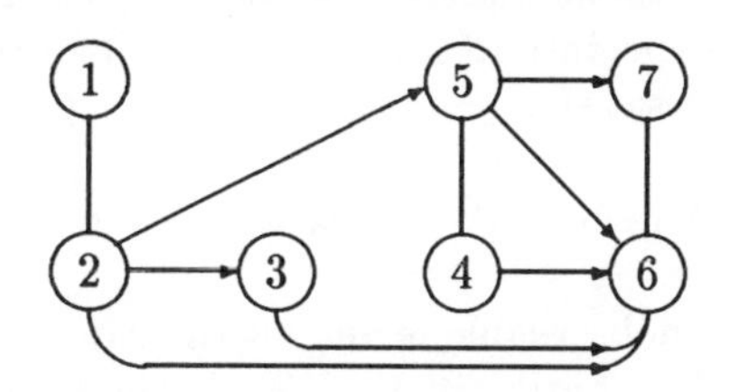

This result coincides well with intuition. For instance, in this recursive model it turns out that grades, X_5, are affected by intelligence alone, and influence educational aspiration. In fact, educational expectation has a connection to all but one variable included in the study; but perhaps this is not so surprising in view of the avowed aims of the study.

The associated moral graph, which needs edges between all parents, has formula

$$1.2 + 2.3 + 2.4 + 2.5 + 2.7 + 3.4 + 3.5 + 3.7 + 4.5 + 4.7 + 5.7 + 4.6 + 6.7.$$

In this instance, this graph is not particularly informative about the pattern of interaction, and, for example, conceals the independence between number of siblings and intelligence, 3 and 4.

10.5 Regression Models for Continuous Variables

The techniques of graphical modelling are closely related to the widely used and well understood techniques of regression analysis; we shall see that graphical models shed light on aspects of regression, and conversely. The treatment here presupposes that the reader has a familiarity with the standard regression set-up as described, for instance, in Seber (1977) or Draper and Smith (1981).

We first consider the multiple linear regression model with p continuous explanatory variables, X, and with just one continuous response variable, Y.

The conditional Normal regression model

The centrepiece of classical regression theory is the analysis of the model which relates the observed values, y and x, by the linear equation $Y = \alpha + \beta x + e$, where e is unobservable, where Y and e are random, with $Ee = 0$ and $\text{var}\,(e) = \sigma^2$; and where α, β and σ^2 are unknown parameters to be estimated. The model extends to several explanatory variables by replacing the term βx by the bilinear form $\beta^T x$ in which x and β are now p-dimensional. This formulation stresses the additivity of systematic and error components.

However as this model has no discrete counterpart, we rewrite the scenario in terms of conditional expectations and variances. The conditional density of Y given X is $f_{Y|X}$ and satisfies

$$E_{Y|X}(Y) = \alpha + \beta^T x, \quad \text{var}_{Y|X}(Y) = \sigma^2.$$

There is no mention of unobservable error; all the random terms enter into the model through the distribution of Y. We further assume the Normality of the conditional distribution of Y given $X = x$, so that

$$Y|x \sim N(\alpha + \beta^T x, \sigma^2).$$

The conditional expectation $E_{Y|X}(Y)$ regarded as a function of $X = x$ is the *regression function*, and a property of the multivariate Normal distribution is that all its regression functions are linear. By generalising $E_{Y|X}(Y) = \alpha + \beta^T x$ to $g\{E_{Y|X}(Y)\} = \alpha + \beta^T x$ for a suitable function g and relaxing the assumption that $\text{var}_{Y|X}(Y)$ is constant with respect to x, the class of generalised linear models proposed by Nelder and Wedderburn (1972) can be derived, see McCullagh and Nelder (1983).

The following proposition points out that the variable selection problem of regession analysis is identical to that of determining which of the edges joining the response to the covariates should be included in the graph.

Proposition 10.5.1 *In the conditional Normal linear regression framework, the hypothesis that the i-th regression coefficient is zero is equivalent to the hypothesis that Y and the X_i are conditionally independent given the remaining variables in the model.*

Proof: We may formulate the conditional independence restriction as the condition that $f_{Y|X}(y; x)$ is invariant to x_i. In the conditional Normal model, the conditional distribution

$$f_{Y|X}(y; x) = n(y; \alpha + \beta^T x, \sigma^2);$$

is invariant with respect to x_i if and only if the term $\alpha + \beta^T x$, does not depend on x_i. By inspection of the density function n, this is possible if and only if $\beta_i = 0$. □

EXAMPLE 10.5.1 Corresponding to the independence graph of Y with explanatory variables (X_1, X_2, X_3)

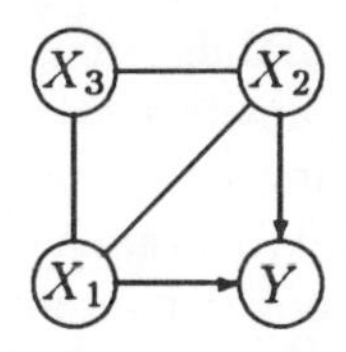

is the regression equation $E_{Y|X}(Y) = \alpha + \beta_1 X_1 + \beta_2 X_2$, and the coefficient $\beta_3 = 0$. □

Different parameterisations for a three variable model

Regression coefficients are a feature of interactions between variables, important for prediction; and are not, however, measures of interaction strength. An inkling of the difficulties that arise by defining models in terms of parameters rather than in terms of conditional independence can be gained by analysing a system with just three variables. Consider a tri-variate Normal vector with vertex set $K = \{1, 2, 3\}$, and the possible partitions into blocks corresponding to the decomposition of the joint density function f_{123} into (i)

f_{123}, (ii) $f_1 f_{23|1}$, and (iii) $f_1 f_{2|1} f_{3|12}$, respectively.

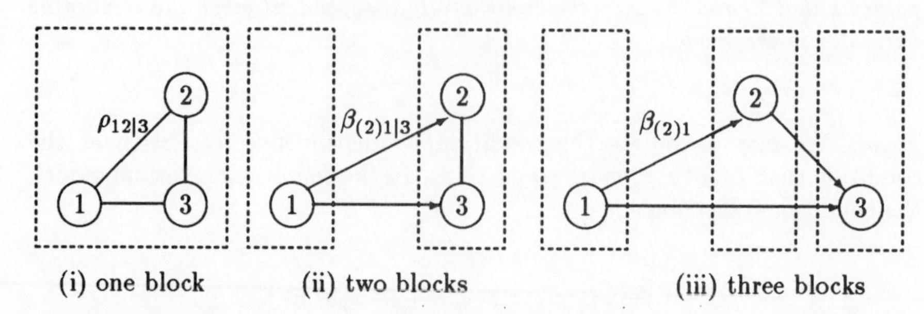

(i) one block (ii) two blocks (iii) three blocks

The natural parameter to associate with the interaction between variables X_1 and X_2 is different in each case. For

(i) the single block, it is the partial correlation coefficient between X_1 and X_2 adjusted for X_3, $\rho_{12|3}$;

(ii) two blocks, it is the partial regression coefficient of X_2 on X_1 having adjusted for X_3, $\beta_{(2)1|3}$;

(iii) three blocks, it is the marginal regression coefficient of X_2 on X_1 ignoring X_3, $\beta_{(2)1}$.

Because the multivariate Normal distribution is closed under marginalisation and conditioning, these parameters are direct functions of the variance $V = \text{var}(X_{123})$, and can be computed by expressing V as (i) $\text{var}(X_{123})$, (ii) $\text{var}(X_1)$ and $\text{var}(X_2, X_3|X_1)$, and (iii) $\text{var}(X_1)$, $\text{var}(X_2|X_1)$, $\text{var}(X_3|X_1, X_2)$.

These measures may or may not remain invariant to imposing conditional independence constraints. For instance, if V satisfies $X_3 \perp\!\!\!\perp X_1|X_2$, so that there is no edge between 1 and 3 in any of the three diagrams, then variable 3 does not mediate the relationship between variables 1 and 2. Consequently $\rho_{13|2} = 0$ and $\beta_{(3)1|2} = 0$.

These constraints affect the value of the interaction parameter in the first diagram (i) because $\rho_{12|3} \neq \rho_{12}$; and in the second (ii) because $\beta_{(2)1|3} \neq \beta_{(2)1}$ even if $X_3 \perp\!\!\!\perp X_1|X_2$.

It is only in the recursive block structure exemplified by the third diagram that the parameter measuring the interaction between 1 and 2 remains the same, and in fact this partition has the remarkable property: removing any edge between two variables will not affect the measure of interaction between any other pair of variables.

Multivariate regression

We extend the discussion to q response variables, in the framework of the multivariate regression model; and show that though analyses can be put into direct correspondence, the usual hypotheses of multivariate regression are not those of the graphical modeller: while the latter examines the joint distribution of the response variables, the regression modeller focuses on the marginal distribution of each response.

The multivariate Normal regression model is that the response vector Y can be expressed as

$$Y = \alpha + BX + e$$

where B is a $q \times p$ matrix of regression coefficients and e is an unobservable q-dimensional Normal vector with mean zero and variance-covariance matrix V, distributed independently of X; for example, see Rao (1973) or Seber (1984). When $q = 1$ this is just the multiple linear regression model. The covariate X may be either fixed or random; but if X is random we take it to be Normally distributed random with variance var (X) and, without loss of generality, set $EX = 0$.

The parameters of the model, B and V, can be interpreted in the following way: firstly, the conditional distribution of Y given X is $Y|X \sim N(\alpha + BX, V)$, so

$$E_{Y|X}(Y) = \alpha + BX \quad \text{and} \quad \text{var}_{Y|X}(Y) = V;$$

$\alpha + BX$ is a conditional expectation and V is a conditional variance. Furthermore, in the joint distribution of X and Y

$$B = \text{cov}(Y, X)\text{var}(X)^{-1} \text{ and } V = \text{var}(Y) - \text{cov}(Y, X)\text{var}(X)^{-1}\text{cov}(X, Y);$$

so that BX is the linear least squares predictor of Y from X and V is the partial variance of Y given X. As we have pointed out before, the conditional and the partial expectations are identical in the multivariate Normal distribution.

The difference to carrying out q separate multiple regression analyses is that firstly, multivariate regression gives estimates of the covariance structure of the response variables, V, and secondly, its more general framework allows tests for linear hypotheses of the form H: $DBC = 0$, where C and D are specified. For example, if $p = 1$ and $q = 2$, the hypothesis that the coefficients of the different response variables are equal corresponds to $(1\ {-1})B = 0$. The focus of interest rests in B and usually V is a nuisance parameter, and while the maximum likelihood estimates of B are the same as given by separate analyses for each response variable, an estimate of V is needed for tests of these hypotheses.

The regression method for the analysis of the dependence of Y on X tests hypotheses of the form $d_i^T Bc_j = 0$, where d_i and c_j are indicator vectors that isolate the element of B connecting the response variable Y_i to the covariate X_j. There are pq such parametric hypotheses and their relationship to the conditional independence statements is given in the next proposition.

Proposition 10.5.2 Equivalent regression and independence hypotheses. *In the multivariate Normal linear regression model the following parametric hypotheses and conditional independence statements are equivalent:*

(a) $H_1 : d_i^T V^{-1} c_j = 0$ *and intra-response independence* $Y_i \perp\!\!\!\perp Y_j | (Y_a, X)$, *where a denotes the remaining Y's;*

(b) $H_2 : d_i^T V^{-1} B c_j = 0$ *and the response-covariate independence statement* $Y_i \perp\!\!\!\perp X_j | (Y_a, X_b)$, *where a denotes the remaining Y's and b the remaining X's ;*

(c) $H_3 : d_i^T B c_j = 0$ *and the conditional independence in the marginal distribution of* Y_i *and* X: $Y_i \perp\!\!\!\perp X_j | X_a$, *where a denotes the remaining X's.*

Proof: The proof of (a) is the characterisation of pairwise conditional independence by a zero element in the inverse variance, and noting that $V = \text{var}_{Y|X}(Y)$ is the variance of Y given X.

For (b), let $V_X = \text{var}(X)$ and $C = \text{cov}(Y, X)$. The variance covariance matrix of (X, Y) is

$$\begin{pmatrix} V_X & C^T \\ C & V + BV_X B^T \end{pmatrix}$$

and, by the inverse variance lemma in Section 5.7, has inverse

$$\begin{pmatrix} V_X^{-1} + B^T V^{-1} B & -B^T V^{-1} \\ -V^{-1} B & V^{-1} \end{pmatrix}.$$

Application of the lemma to the off diagonal element gives the result.

For (c), note that

$$d_i^T B = d_i^T \text{cov}(Y, X) \text{var}(X)^{-1} = \text{cov}(Y_i, X) \text{var}(X)^{-1}.$$

so that $d_i^T B c_j = 0$ is a hypothesis about the marginal distribution of Y_i and X. □

In summary, tests for intra-response conditional independence correspond to zeros in V^{-1} and tests for response-covariate independence statements correspond to zeros in $V^{-1}B$. Tests for zeros in the elements of B do not correspond to assertions about the independence graph of Y given X but correspond to separate assertions about the independence graphs of each Y_i given X. The usual hypotheses of multivariate regression do not coincide with the objectives of the graphical modeller.

Of course when $q = 1$ there is no difference and in this case of multiple regression, a graph constructed on the basis of tests for zero regression coefficients can be interpreted as an independence graph.

EXAMPLE 10.5.2 Suppose that a graph is constructed on the basis of testing coefficients in the regression model. For instance, with $q = 1$ and $p = 2$ and the model $EY = \beta_1 X_1 + \beta_2 X_2$, the test for $\beta_1 = 0$ and the test for $Y \perp\!\!\!\perp X_1 | X_2$ are equivalent. In this case the resulting graph can be interpreted as an independence graph. On the other hand if $q = 2$ and $p = 1$ with the model $EY_1 = \beta_1 X$, $EY_2 = \beta_2 X$ the test for $\beta_1 = 0$ corresponds to the test for $Y_1 \perp\!\!\!\perp X$ and not to the test for $Y_1 \perp\!\!\!\perp X | Y_2$. In consequence, if it is accepted that β_1 is zero then, unless $\beta_2 = 0$ or $Y_1 \perp\!\!\!\perp Y_2 | X$, the resulting regression graph

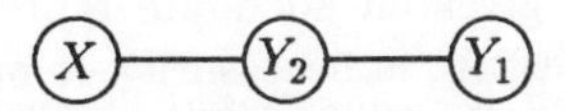

is *not* an independence graph and is seriously misleading in that it suggests that variable Y_2 mediates the relationship between X and Y_1. □

EXAMPLE 10.5.3 Social status and participation. Hodge and Treiman (1968) studied the relationship between social status and participation. For a sample of 530 women they report data on

X_1 : income, X_2 : occupation, and X_3 : education,
Y_1 : church attendance, Y_2 : memberships, and Y_3 : friends' seen.

The objective of the analysis is to determine how the explanatory variables X affect the response variables in Y. All variables are expressed in standardised form and the sample correlation matrix of the variables is recorded in Table 10.5.1. Analysis of the marginal distribution of X reveals no simple structure. A model fitting exercise suggests the graphical model defined by

Table 10.5.1: Correlation matrix for social status and participation. From Hodge and Treiman (1968).

X_1	1.00					
X_2	0.30	1.00				
X_3	0.31	0.34	1.00			
Y_1	0.10	0.16	0.16	1.00		
Y_2	0.28	0.19	0.32	0.36	1.00	
Y_3	0.18	0.14	0.23	0.21	0.26	1.00
	X_1	X_2	X_3	Y_1	Y_2	Y_3

the independence graph

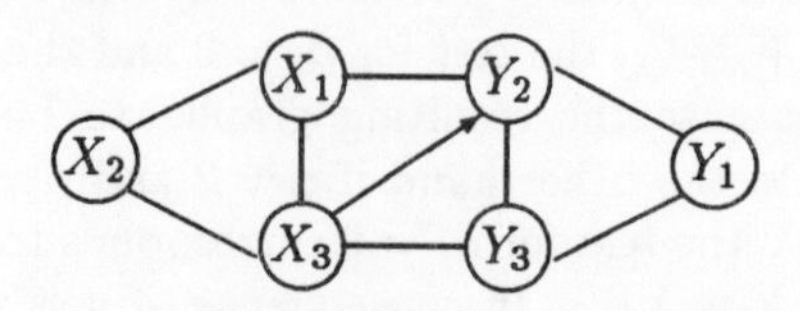

The deviance of the model is 9.43 on 6 df, and the observed P-value of 0.15 suggests that the model gives an adequate representation of the data. The subgraph of the covariates, X, is necessarily complete, but the responses, Y, are all interconnected as well. There are two remarks to make: firstly, church attendance, Y_1, is separated from the covariates by the two other response variables, and so 'explained' by them; and secondly, occupation, X_2, plays no effective explanatory role. The relationships within this set of variables are dominated by the triangle composed of income, education and memberships, and the model has the implication that

$$(Y_1, Y_3) \perp\!\!\!\perp (X_1, X_2) | (X_3, Y_2).$$

The conclusions from a regression analysis are, to some extent, different. The fitted equations, with t-values in brackets, are

$$\begin{array}{lllllll}
\widehat{EY_1} = \text{const} & + & 0.03X_1 & + & 0.11X_2 & + & 0.11X_3 \\
 & & (0.73) & & (2.31) & & (2.37) \\
\widehat{EY_2} = \text{const} & + & 0.19X_1 & + & 0.05X_2 & + & 0.25X_3 \\
 & & (4.46) & & (1.08) & & (5.64) \\
\widehat{EY_3} = \text{const} & + & 0.11X_1 & + & 0.04X_2 & + & 0.18X_3 \\
 & & (2.40) & & (0.90) & & (3.87)
\end{array}$$

and the fitted variance-covariance matrix for these standardised variables is

	Y_1	Y_2	Y_3
Y_1	0.96		
Y_2	0.29	0.86	
Y_3	0.16	0.17	0.93

The sample size is large enough to compare the t-statistics against the standard Normal distribution and, in contradistinction to above, we conclude that the Y_1 depends on X_2 and X_3, and that Y_3 depends on X_1. That Y_3 may depend on X_1 as well as X_3 can be anticipated from the independence graph by integrating or collapsing out Y_1 and Y_2. The difference is a consequence of the different foci of interest: regression models on the marginal distributions and graphical models on the joint distribution.

However, the conclusion that Y_1 may depend on X_2 is in direct contradiction to the conclusion, $Y_1 \perp\!\!\!\perp X_2 | (X_1, X_3)$, embodied in the fitted graphical model. This conflict suggests that the overall goodness of fit measured by the deviance may mask the interaction between Y_1 and X_2, and in fact, further analysis revealed the deviance of the independence $Y_1 \perp\!\!\!\perp X_2 | (X_1, X_3, Y_2, Y_3)$, to be 3.99 on 1 df, which is not acceptable compared to the chi-squared distribution. When $q > 1$ testing for an association between a reponse variable and an explanatory variable differs according to the framework chosen by the analyst. □

10.6 Regression: Some Sampling Results

We consider two well known results of sampling distribution theory for multiple regression models (q=1) in the light of graphical models.

The variance of parameter estimates

Under conditional Normal theory, with X fixed, the sample variance-covariance matrix, $S = S_{(X,Y),(X,Y)}$ holds the sufficient statistics for the unknown parameters β and σ^2. It is well known, see Seber (1977) or Draper and Smith (1981), that the maximum likelihood estimator, $\hat{\beta}$, of the regression coefficient is

$$\hat{\beta} = S_{XX}{}^{-1} S_{XY},$$

and that this estimate is Normally distributed as

$$\hat{\beta} \sim N(\beta, \frac{\sigma^2}{N} S_{XX}^{-1}).$$

To help apply the results from the chapters on covariance and inverse variance, denote this sample variance-covariance matrix, S, by the variance operator $\text{var}_N(.)$, so that

$$\text{var}_N(X,Y) = \begin{pmatrix} S_{XX} & S_{XY} \\ S_{YX} & S_{YY} \end{pmatrix}.$$

This is certainly a permissible nomenclature as S is symmetric and positive definite. The regression coefficients then satisfy the likelihood equations $\text{var}_N(X)\hat{\beta} = \text{cov}_N(X,Y)$, which exactly correspond to the normal equations for determining the linear least squares prediction coefficients. The sampling variance of the coefficients,

$$\text{var}(\hat{\beta}) = \frac{\sigma^2}{N} \text{var}_N(X)^{-1},$$

is the inverse of the empirical variance of the covariates; so we see firstly, that the variance of the regression estimator, $\hat{\beta}_i$, is directly proportional to the partial variance of X_i given the rest,

$$\text{var}\,(\hat{\beta}_i) = \frac{\sigma^2}{N}\text{var}_N(X_i|\text{rest})^{-1},$$

and secondly, that the correlation of $\hat{\beta}_i$ and $\hat{\beta}_j$ is the negative of the partial correlation between X_i and X_j given the rest:

$$\text{corr}\,(\hat{\beta}_i, \hat{\beta}_j) = -\text{corr}_N(X_i, X_j|\text{rest}).$$

The independence graph of the covariates illuminates the structure of the variance-covariance matrix for the regression coefficients: if the covariate graph has no edges, then the study is completely balanced and the regression estimates are mutually independent; if a pair of vertices are strongly connected then the corresponding regression estimates are highly correlated, and an example of this situation occurs in the analysis of the noctuid moth data in Section 10.3. More generally, the overall structure of the covariates independence graph indicates some of the directions in which the observed study departs from an ideal notion of balance.

Exact F-tests for the deviance

The standard test procedure for excluding one or more variables from the regression is based on the F-distribution. Suppose that the covariates are partitioned into (X_a, X_b) of dimension p_a and p_b respectively. The regression is

$$E_{Y|a\cup b}(Y) = \alpha + \beta_{a|b}^T X_a + \beta_{b|a}^T X_b \quad \text{and} \quad \text{var}_{Y|a\cup b}(Y) = \sigma^2.$$

Consider the hypothesis that $\beta_{b|a} = 0$, which corresponds to simultaneously deleting a set of inter-response edges from the independence graph. The appropriate analysis of variance table for testing that these specified regression coefficients are zero is

variation due to	df	sumsq	F-ratio
X_b adjusted for X_a	p_b	ss (1)	$F = \frac{\text{ss}\,(1)/p_b}{\text{ss}\,(3)/(N-1-p_a-p_b)}$
X_a ignoring X_b	p_a	ss (2)	
residual (adj X_a, X_b)	$N-1-p_a-p_b$	ss (3)	
total	$N-1$	ss	

where the sums of squares are computed from the sample variance matrix as

$$\begin{aligned} \text{ss} &= NS_{YY}, \qquad \text{ss}\,(2) = NS_{Ya}S_{aa}^{-1}S_{aY} \qquad \text{and} \\ \text{ss}\,(1,2) &= NS_{Y,a\cup b}S_{a\cup b,a\cup b}^{-1}S_{a\cup b,Y}; \end{aligned}$$

which in our alternative notation we may write as

$$\begin{aligned}
\text{ss} &= N\text{var}_N(Y), \qquad \text{ss}(2) = N\text{var}_N(\hat{Y}(X_a)) \quad \text{and} \\
\text{ss}(1,2) &= N\text{var}_N(\hat{Y}(X_a, X_b)).
\end{aligned}$$

The remaining sums of squares are given by subtraction

$$\begin{aligned}
\text{ss}(1) &= \text{ss}(1,2) - \text{ss}(2) = N\{\text{var}_N(Y|X_a) - \text{var}_N(Y|X_a, X_b)\} \quad \text{and} \\
\text{ss}(3) &= \text{ss} - \text{ss}(1,2) = N\text{var}_N(Y|X_a, X_b).
\end{aligned}$$

The significance test for the hypothesis compares the observed value of F with the tabulated percentage point of the F-distribution on $(p_b, N-1-p_a-p_b)$ degrees of freedom. An explicit expression relating the F-statistic and the deviance is given in the next proposition.

Proposition 10.6.1 *The relationship between the deviance for testing the independence of Y and X_b conditional on X_a and the F-statistic for testing that a subset of the regression coefficients is zero is*

$$\text{dev}(Y \perp\!\!\!\perp X_b | X_a) = N\log\{1 + \frac{p_b F}{(N-1-p_a-p_b)}\}$$

where F is given in the above analysis of variance table and the deviance is given by Proposition 6.7.1.

Proof: First, express the sums of squares defining the F-statistic in terms of partial variances, to give

$$F = \frac{(N-1-p_a-p_b)}{p_b}\{\frac{\text{var}_N(Y|X_a)}{\text{var}_N(Y|X_a, X_b)} - 1\}.$$

Now, the deviance against conditional independence, Proposition 6.7.1, is

$$\text{dev}(Y \perp\!\!\!\perp X_b | X_a) = -N\log\frac{\det \text{var}_N(X_a, X_b, Y)\det \text{var}_N(X_a)}{\det \text{var}_N(X_a, Y)\det \text{var}_N(X_a, X_b)},$$

which simplifies on using Schur's identity to

$$\text{dev}(Y \perp\!\!\!\perp X_b | X_a) = -N\log \det \text{var}_N(Y|X_a, X_b)/\det \text{var}_N(Y|X_a).$$

As Y is one dimensional the determinants are scalars, and substituting for the ratio of partial variances in terms of F delivers the identity. □

This equation shows the close relationship between the two test statistics; there are several remarks to make.

(i) Under the assumptions of the conditional Normal linear regression model, when Y is random and the covariates are fixed, the F-statistic is *exactly* F-distributed, see Seber (1977) for a proof. But this distribution is the same whatever the values of the explanatory variables; hence the F-statistic has the same F-distribution when the covariates are random. Thus a simple transformation of the deviance can be used to derive an exact test for conditional independence.

(ii) It is known that when N is large the statistic $p_b F$ is approximately chi-squared with p_b df. Now because

$$\log(1+\theta) = \theta + O(\theta^2), \text{ for } \theta \to 0,$$

it follows that

$$\text{dev}\,(Y \perp\!\!\!\perp X_b | X_a) = p_b F + O(N^{-1}).$$

This is a direct proof that tests based on the deviance and on the F-statistic are asymptotically equivalent. They have the same asymptotic chi-squared distribution with p_b degrees of freedom under the null hypothesis. In exact regression theory the residual sum of squares goes to estimate the nuisance scale parameter σ^2. Thus the effect of using the asymptotic approximation is equivalent to assuming that this parameter is known without error.

(iii) The first term in the Taylor series for the logarithmic function gives the following approximation to the deviance

$$N p_b F / (N - 1 - p_a - p_b).$$

As the exact moments of the F-distribution are known this gives a way to calculate *Bartlett correction factors*, see Porteous (1985ab).

(iv) The edge exclusion deviances from the saturated model are obtained by taking X_b as well as Y, to be one dimensional, and the statistics are compared against the chi-squared distribution with 1 df. The familiar identity, $F = t^2$, relating the F-statistic with 1 df for the numerator to Student's t-statistic, gives the corresponding identity

$$\text{dev}\,(Y \perp\!\!\!\perp X_b | X_a) = N \log\{1 + \frac{t^2}{(N - 2 - p_a)}\}$$

and proves the asymptotic equivalence of the exclusion deviance and the t-test for the hypothesis that just one regression coefficient is zero. This relationship shows how graphical modelling is used to select a regression equation. The first step in the model selection procedure is to compute, in one pass, all the edge exclusion deviances from $\text{var}_N(X, Y)^{-1}$; and then to compare the entries in the row corresponding to the edges between Y and the explanatory variables with the chi-squared distribution. This procedure is (asymptotically) the same as the first step of the familiar backward elimination procedure of multiple regression.

10.7 Logistic Regression

The issues in modelling categorical variables are, if anything, rather simpler than their continuous counterpart: there is no nuisance scale parameter to worry about; the standard parameterisation in u-terms is a mixed derivative measure and remains invariant to conditioning; the relationship between the logistic regression model and the log-linear model is straightforward, and is not confused by the development of marginal models as in multivariate regression; there is very little exact sampling theory available, which may not be a good thing, but at least keeps the exposition simple.

A well-known class of conditional probability models for categorical response variables are the logistic regression models. Certain authors make a distinction between logistic models with a continuous covariate and logit models with a categorical covariate, paralleling the difference between linear regression and ANOVA models. We rather blur this and allow the term logistic regression to denote a mixture of continuous and categorical covariates; linear logistic regression refers to linearity in the continuous covariates, but in a wider sense refers to the absence of interactions between two or more covariates and the response. We do not attempt to give a full discussion, but only indicate the close relationship between certain log-linear and logistic regression models. We begin with an example.

EXAMPLE 10.7.1 We return to the plum root stock data of Bartlett (1935) discussed in Example 9.1.2. The data was collected from a planned experiment to determine how the length of the cutting and the time it is planted affect its chances of survival. Assuming that the survival of any one cutting is independent of any other leads to a Binomial distribution for the number surviving, so that if θ is the probability of survival, total number alive $\sim$ Bino$(240, \theta)$, for each of the 4 treatment combinations. A natural way to model the relationship between survival and the two explanatory variables is by the linear model

$$\log \frac{\theta}{(1-\theta)} = v_\phi + v_1 x_1 + v_2 x_2, \qquad x_1 = 0, 1;\ x_2 = 0, 1;$$

for the logit of the probability, θ. An essential postulate of this model is the additivity of the time and length effects: that is, the effect of cutting length is the same whatever the value of planting time, and vice versa, the effect of planting time is the same whatever the value of cutting length.

This is an example of the ***logistic multiple regression*** model, which gains its name because the inverse of the logit function $\log\{\theta/(1-\theta)\} = t$ is the logistic function $\theta = e^t/(1+e^t)$. As the logit function may take all values in $(-\infty, +\infty)$ when $0 < \theta < 1$, all real values of the coefficients v_ϕ, v_1, v_2 and of the covariates are permissible. In particular, this formulation is valid if x_1 and x_2 are continuous variables; hence it is a regression type of model. When x_1 or x_2 are categorical taking more than 2 values, the coordinate projection notation for the right hand side naturally generalises to $v_\phi + v_1(x_1) + v_2(x_2)$, with the constraints $v_1(0) = 0 = v_2(0)$.

We show that this logistic regression model is equivalent to a particular log-linear model. First note that θ is the conditional probability of Y:

$$\theta = P(Y = 1|X_1 = x_1, X_2 = x_2) = \frac{\exp\{v_\phi + v_1x_1 + v_2x_2\}}{1 + \exp\{v_\phi + v_1x_1 + v_2x_2\}},$$

by inverting the logit function as above. Consequently

$$P(Y = 0|x_1, x_2) = 1 - \theta = \frac{1}{1 + \exp\{v_\phi + v_1x_1 + v_2x_2\}}.$$

Putting these two expressions together gives the single formula

$$P(Y = y|x_1, x_2) = \frac{\exp\{(v_\phi + v_1x_1 + v_2x_2)y\}}{1 + \exp\{v_\phi + v_1x_1 + v_2x_2\}} \quad \text{for } y = 0, 1,$$

and for $x_1 = 0, 1$ and $x_2 = 0, 1$. The denominator, though a somewhat complex function of x_1 and x_2, is not a function of y, and so this expression for $P(Y = y|x_1, x_2)$ is of the form

$$P(Y = y|x_1, x_2) = \exp\{v_\phi y + v_1x_1y + v_2x_2y + g_{12}(x_1, x_2)\},$$

for $y = 0, 1$. The log conditional density function is

$$\log f_{Y|12}(y; x_1, x_2) = v_\phi y + v_1x_1y + v_2x_2y + g_{12}(x_1, x_2).$$

Multiplying by the marginal distribution of (X_1, X_2) only affects the term g_{12}, so that the joint distribution is

$$\log f_{12Y}(x_1, x_2, y) = u_{12}(x) + v_\phi y + v_1x_1y + v_2x_2y,$$

where $u_{12} = g_{12} + \log f_{12}$.

This is a log-linear model for the joint distribution of variables classifying a three-way table. It has two distinguishing features which generalise to higher dimensions:

(i) the response-explanatory interactions are two-way interactions: the direct consequence of the additivity of the x_1 and x_2 affects on the logit scale;

(ii) the order of the intra-covariate interaction u_{12} is equal to the number of explanatory variables: in consequence the fitted values for the full table preserve the marginal table of the explanatory variables.

The parameters v_1 and v_2 can be recovered by fitting either the logistic regression model directly, which in this example has only 3 parameters, or the log-linear model with 7 parameters. For the plum root stock data the fitted log-linear model is

$$\begin{aligned} N\log \hat{f}_{12Y}(x_1, x_2, y) &= 4.37 \\ &+ 0.56x_1 + 0.71x_2 - 0.32x_1x_2 + 0.71y - 1.03x_1y - 1.43x_2y. \end{aligned}$$

The deviance of this model, 2.29 on 1 df, is a test for no three-way interaction between x_1, x_2 and Y, and is acceptable. The signs of the x_1y and x_2y coefficients indicate that survival increases with long cuttings and with cuttings planted at once. Including the fitted u-term $\hat{u}_{12} = 4.37 + 0.56x_1 + 0.71x_2 - 0.32x_1x_2$ preserves the (X_1, X_2) margin, so that the coefficients of the x_1, x_2 and x_1x_2 terms are uninteresting in interpreting the response.

The fitted logistic regression model

$$N\log \frac{\hat{f}_{Y=1|x_1,x_2}}{\hat{f}_{Y=0|x_1,x_2}} = 0.71 - 1.03x_1 - 1.43x_2,$$

highlights the interesting parameters. Note that there is no term involving y in the right hand side. The model has the same deviance as the log-linear model above, which provides a test for the additivity of the effects of x_1 and x_2 on Y. □

Any chain model based on a partition $b_1, b_2, \ldots, b_m$ can be analysed by combining separate analyses of $m - 1$ regression models for the response and its covariates and an all-response model for b_1. The cross-classified Multinomial distribution is closed under marginalisation and conditioning: both marginal distributions and conditional distributions are cross-classified Multinomial with appropriate parameters, tables of probabilities become tables of marginal and conditional probabilities respectively. Each regression model can be analysed as a joint model containing an interaction term high enough to reproduce the observed marginal table of covariates.

10.8 Two Case Studies with Categorical Variables

We give two examples of fitting graphical chain models to analyse a five-way table: the first has two blocks and a single response variable, and is of special interest because it gives an instance when the fitted block recursive and all-response models are the same; the second example has three blocks.

EXAMPLE 10.8.1 Danish homework survey. The data are taken from Edwards and Kreiner (1983) who analysed an investigation conducted at the Institute for Social Research, Copenhagen; data was collected in 1978-1979, for several purposes, one of which was to estimate the amount of tax evasion in the building industry. A sample of employed men in the age group 18-67 were asked if, in the preceding year, they had done work to the home, that formerly they would have paid a craftsmen to do. The reply to that question, recorded in Table 10.8.1, is the response variable. There are four covariates

Table 10.8.1: A five-way table from a Danish homework survey. From Edwards and Kreiner (1983).

		B	apartment			house		
C	D	E	A age			A age		
work	tenure	resp	<30	31-	45+	<30	31-	45+
skill	rent	Yes	18	15	6	34	10	2
		no	15	13	9	28	4	6
	own	yes	5	3	1	56	56	35
		no	1	1	1	12	21	8
unskilled	rent	yes	17	10	15	29	3	7
		no	34	17	19	44	13	16
	own	yes	2	0	3	23	52	49
		no	3	2	0	9	31	51
office	rent	yes	30	23	21	22	13	11
		no	25	19	40	25	16	12
	own	yes	8	5	1	54	191	102
		no	4	2	2	19	76	61

and one response variable:

> A age : <30, 31-45, 45+; B type of accommodation : apartment, house; C mode of tenure : rent, own; D work of respondent : skilled, unskilled, office; E response to question : yes, no.

The covariate model: Consider first a graphical model for the explanatory variables. The edge exclusion deviances from the saturated model are

A	*			
B	57.2 (12)	*		
C	174.1 (12)	555.9 (9)	*	
D	86.8 (16)	14.8 (12)	32.7 (12)	*
	A	B	C	D

with degrees of freedom in brackets. There appears to be just one conditional independence, the graph is

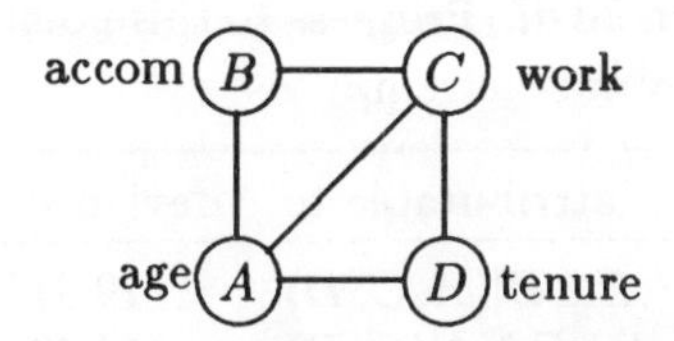

with the interpretation that type of accommodation, B, and type of work, D, are independent conditioned on the other two variables. The model $(B+D).A.C$ fits. The large exclusion deviances indicate that the $B\times C$ and maybe the $A\times C$ marginal tables need inspection. The $B\times C$ margin is

B C	rent	own
aptmnt.	346	44
house	295	906

There are relatively few respondents in the owned apartment category. (Intriguingly Kreiner points out that the Danish word for 'apartment' originally meant a 'home that is rented'.) It is clear that observed sampling design for the response variable E is not balanced.

The response model: The minimal model of complete independence expresses the hypothesis that the proportion of respondents who answered yes is the same for all 36 categories, and has a deviance (the 'deviance to explain') of 158.9 on 35 df. The log-linear formula for this independence model is $A.B.C.D+E$, which contains the term $A.B.C.D$ for the 4 covariates. That the response model always contains such a term is a direct consequence of the fact that no independences relating two explanatory variables are permitted in the response model. The edge exclusion deviances are taken from the saturated model in 5 dimensions:

	A	B	C	D
dev	38.77	19.81	73.71	52.98
df	24	18	18	24

The $E.B$ interaction is the first to be excluded, the $E.A$ interaction is on the borderline of significance, but the other edges have to be retained. The new edge exclusion deviances after eliminating B are

	A	B	C	D
dev	19.90	x	87.97	35.44
df	12	x	9	12

and as the $E.A$ edge has an observed P-value of 0.069 this edge can also be eliminated. Nothing else can be excluded and the response model finally selected is $A.B.C.D{+}E.C.D$, so that response$\perp\!\!\!\perp$(age,type of accom)|(tenure,work), and has a deviance of 39.7 on 30 df. Progress to this point can be summarised in an analysis of deviance table:

attributable to	deviance	df
B adj.for A: $E\perp\!\!\!\perp B\|(A,C,D)$	19.81	18
A ignoring B: $E\perp\!\!\!\perp A\|(C,D)$	19.90	12
sub-total: $E\perp\!\!\!\perp(A,B)\|(C,D)$	39.7	30
total attr. to real effects	119.2	5
total	158.9	35

For some purposes of investigation this will be sufficient. The important explanatory variables are (C, D) and the table can be reduced to the corresponding three-way margin and the 'explained' deviance can be further partitioned.

The combined model: The block recursive model incorporating the two sets of conditional independences has the graph

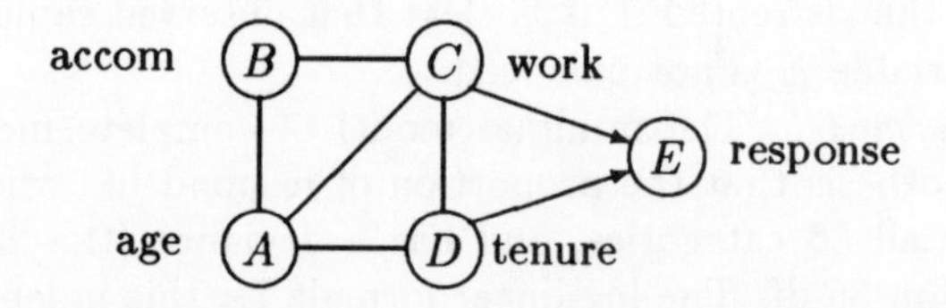

The deviance of the chain model $/A.B.C + A.C.D/(A.B.C.D) + E.C.D/$ is $14.8 + 39.7 = 54.53$ on $12 + 30 = 42$ df with an observed P-value of 0.093. The parents of the response variable are adjacent in this graph, hence the associated moral graph requires no extra edges and the independence structure of the chain model and the all-response model are identical. In particular, the fitted values for the chain model $n_\phi \hat{f}_{E|ABCD}\hat{f}_{ABCD}$ simplify to

$$n_\phi \frac{n_{CDE}}{n_{CD}} \frac{n_{ABC}n_{ACD}}{n_{AC}},$$

and are identical to those calculated under the all-response model $(B + D).A.C + E.C.D$. The all-response model is reducible to three components summarised by the marginal tables $A\times B\times C$, $A\times C\times D$, and $E\times C\times D$. Further analysis of the table can now concentrate on these margins without fear of missing important associations nor analysing interactions that can be explained as indirect relationships. □

*EXAMPLE 10.8.2 **College plans.*** Sewell and Shah (1968) studied the relationship between a student's intention to attend college, parental encouragement and three other factors, social class, sex and a measure of intelligence. The data derives from 10,318 Wisconsin high school seniors, and has been analysed by Fienberg (1977) and Cox and Snell (1981). The names and levels taken by the variables are

> X_1 sex: male, female; X_2 social class: low, lower middle, upper middle, high; X_3 intelligence: low, lower middle, upper middle, high; X_4 parental encouragement: low, high; X_5 college plans: yes, no.

In this standard order, with variable 5 varying most quickly, the counts in the 128 cells are given in Table 10.8.2. The principal questions of interest are

Table 10.8.2: College plans in standard order. From Sewell and Shah (1968).

4	349	13	64	9	207	33	72	12	126	38	54	10	67	49	43
2	232	27	84	7	201	64	95	12	115	93	92	17	79	119	59
8	166	47	91	6	120	74	110	17	92	148	100	6	42	198	73
4	48	39	57	5	47	123	90	9	41	224	65	8	17	414	54
5	454	9	44	5	312	14	47	8	216	20	35	13	96	28	24
11	285	29	61	19	236	47	88	12	164	62	85	15	113	72	50
7	163	36	72	13	193	75	90	12	174	91	100	20	81	142	77
6	50	36	58	5	70	110	76	12	48	230	81	13	49	360	98

Reproduced by permission from the University of Chicago Press.

how the three background variables affect parental encouragement, and how all four variables affect college plans, so we analyse this data as a recursive system consisting of three blocks, $b_1 = \{1,2,3\}$, $b_2 = \{4\}$, and $b_3 = \{5\}$.

An analysis of the three variables in the first block, based on the edge exclusion deviances in the three-way marginal table, quickly shows that the model $1+2.3$ in which sex is independent of both social class and intelligence, is the only model possible. The deviance of the model is 20.25 on 15 df, while the deviance to explain is 864.85 on 24 df.

The total deviance to explain in the system is the deviance against the complete independence of the five variables, 8203.45 on 118 df; its large magnitude is not so much a reflection of the magnitude of dependence in the table but of the large sample used. While the subsequent analysis is thus able to detect some very small interactions, we wish to give a broader picture, and so we sometimes neglect relatively small interactions even if significant.

An omnibus test of linearity in the logistic regression is the deviance of the fitted linear logistic regression containing all explanatory variables. In the second block with 3 explanatory variables and one response, X_4, the model 1.2.3 + (1 + 2 + 3).4 has a deviance of 55.82 on 24 df; and, in the third block with 4 explanatory variables, and one response, X_5, the model 1.2.3.4+ (1+2+3+4).5 has a deviance of 73.82 on 55 df. Thus we feel fairly confident about the ability of linear regression to represent the data faithfully. A more detailed investigation can be made by comparing the edge exclusion deviances corresponding to an interaction with the response variable, from two possible maximal models: the saturated model and the linear logistic model:

edge and exclusion	deviances	df	from models	
	1.2.3.4		1.2.3 + (1 + 2 + 3).4	
4.1	209.69	16	174.76	1
4.2	1550.02	24	1498.87	3
4.3	615.24	24	572.46	3
	1.2.3.4.5		1.2.3.4 + (1 + 2 + 3 + 4).5	
5.1	65.56	32	15.68	1
5.2	394.54	48	336.55	3
5.3	594.05	48	527.51	3
5.4	1661.79	32	1648.27	1

The two columns are very similar. In relation to the other values, the 1.5 interaction deviance difference is small and may be neglected in explaining X_5, but all three terms need to be included in the model for X_4. The corresponding chain graph is

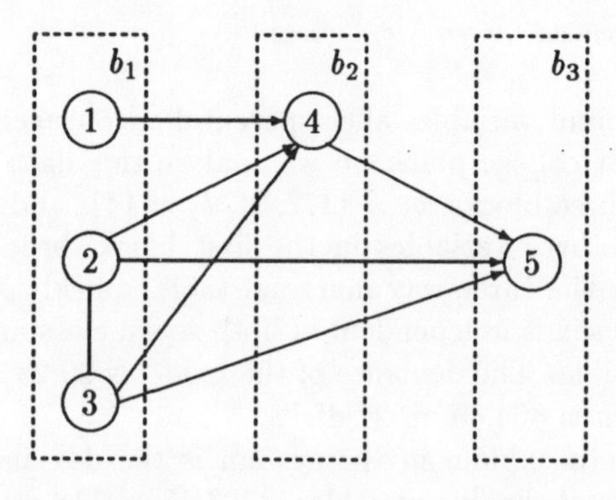

The overall fit of the model may be summarised in the analysis of deviance table

	residual		total	
model	dev	df	dev	df
$()+1+2.3$	20.25	15	864.85	24
$(1.2.3)+(1+2+3).4$	55.82	24	2841.09	31
$(1.2.3.4)+(2+3+4).5$	89.50	56	4497.51	63
total	166.57	95	8203.45	118

The vast majority of the variation is accounted for.

Linearised two-way interaction models: We summarise the fitted model by reporting parameter estimates to attach to each edge. The estimated value of the two-way interaction parameter, $u_{34}(x_3, x_4)$, measuring the interaction between parental encouragement and intelligence is (0.54, 0.96 1.52). The moderate values of these estimates are entirely consistent with our remark that the large deviances are attributable to large sample sizes and not to large interactions. The estimate is composed of the log cross-product ratios in the three 2×2 tables of parental encouragement against intelligence, that contrast each of the three higher levels of intelligence to the lowest in turn. It increases approximately linearly by about 0.5 for every step in intelligence.

The linear logistic model can be further simplified by exploiting this additional linearity and fitting the model with a single parameter for the two-way interaction:

$$u_{34}(x_3, x_4) = u_{34}x_4x_3, \qquad \text{for } x_4 = 0,1 \text{ and } x_2 = 0,1,2,3.$$

If the same is done for social class, $u_4(x_2, x_4) = u_{24}x_4x_2$, the difference in deviance between full linear logistic and the linearised version is 34.4 on 4 df which, though significant, is not unduly large.

The estimated model is

$$\log\frac{f_{4|123}(1)}{f_{4|123}(0)} = \underset{\substack{(.046)\\ \text{sex}}}{-0.60x_1} + \underset{\substack{(.022)\\ \text{class}}}{0.80x_2} + \underset{\substack{(.021)\\ \text{iq}}}{0.50x_3}$$

This data stongly suggests that sex effect on the frequency of parental encouragement is of the same order of magnitude as taking one step down either the ladder of social class or of intelligence.

The similarly linearised version of the linear logistic model for college plans (X_5), increased the deviance by a small 17.8 on 4 df, and yielded the following fitted model:

$$\log\frac{f_{5|1234}(1)}{f_{5|1234}(0)} = \underset{\text{sex}}{0.0x_1} + \underset{\substack{(.03)\\ \text{class}}}{-0.46x_2} + \underset{\substack{(.03)\\ \text{iq}}}{-0.58x_3} + \underset{\substack{(.07)\\ \text{par enc}}}{-2.46x_4}$$

Judging by the edge exclusion deviances the term due to sex is unnecessary and is set to zero, though since sex affects parental encouragement, it has an indirect effect on college plans. The effect of positive parental encouragement on college plans is approximately equal to that of moving from the lowest to the highest intelligence group, as $(3)(-0.58) \approx -2.46$. □

10.9 Exercises

1: Express the parameters ρ_{12}, $\beta_{2|1}$, $\rho_{12|3}$ and $\beta_{2|1;3}$ in terms of the elements of the variance matrix $V = \text{var}(X_1, X_2, X_3)$. Show that, in general, $\rho_{12|3} \neq \rho_{12}$ and $\beta_{2|1;3} \neq \beta_{2|1}$ even if $X_3 \perp\!\!\!\perp X_1 | X_2$.

2: Recall the social status and participation data given in Example 10.5.3.
(a) Find the multiple correlation coefficients for the three response variables (Y_1, Y_2, Y_3).
(b) Calculate the deviances of (i) the minimal model of complete independence of the Y's from each other and from X; (ii) the model defined by $Y \perp\!\!\!\perp X$; and (iii) the model determined by the mutual independence of the Y's conditional on X.
(c) Verify that the edge exclusion deviances for a response-explanatory model are

Y_1	0.89	3.99	0.05	*		
Y_2	17.14	0.08	21.09	50.41	*	
Y_3	2.77	0.21	7.48	7.81	10.75	*
	X_1	X_2	X_3	Y_1	Y_2	Y_3

and that the deviance of the model obtained by dropping all edges less than 4 is 9.43.
(d) Verify that the edge inclusion deviances from this model are

Y_1	0.03	5.23	0.41	*		
Y_2	x	1.18	x	x	*	
Y_3	3.06	1.57	x	x	x	*
	X_1	X_2	X_3	Y_1	Y_2	Y_3

(e) Verify that the deviance of the model including an $X_2.Y_1$ interaction is reduced to 4.20 on 5 df and find the fitted partial correlations.

3: For Cochran's moth data, given in Table 10.3.1,
(a) Compute the regression equation of moth on (wind, cloud). Show that this model has an R^2 of 21.8%. This should be compared to an R^2 of 30.2%

based on the full set of explanatory variables, which suggests that something important has been omitted.
(b) Evaluate the forward inclusion deviances for the remaining variables (min, max, rain).
(c) Draw up the appropriate analysis of variance table and calculate the F-statistics to test for the significance of including wind and cloud.
(d) Draw up the ANOVA table with entries for the additional contribution of mintemp; and for the subsequent contribution of maxtemp and rain together.
(e) Reanalyse the moth data to show that together with wind and cloud, one, but not both, of min and max temperature needs to be in the regression model for moth trappings.

4: Finish the model selection procedure on the Kerchoff data set in Table 10.4.1. Show that the inter-response edge exclusion deviances from the fitted response model are 133.40, 27.25, 18.16, 35.44, 8.47, 6.11, and note how close they are to the corresponding values calculated from the saturated model.

5: The following correlation matrix due to Hodapp based on a sample size of 98 is taken from Wermuth (1988). The variables are f: systolic blood pressure, e: diastolic blood pressure, d: anxiety at work, c: anger at work, b: weight, and a: age.

f	1.000					
e	0.738	1.000				
d	0.034	-0.059	1.000			
c	-0.012	-0.042	0.135	1.000		
b	0.351	0.371	-0.101	0.211	1.000	
a	0.270	0.139	-0.058	0.283	0.390	1.000
means	128.31	85.46	8.23	4.38	0.42	32.74
sds	13.47	11.38	3.43	2.91	0.04	11.67
	f	e	d	c	b	a

It is thought that blood pressure is a function of age and weight, but that stress is also a direct influence. Find a well fitting chain graph model.

6: For the analysis of the Danish homework contingency table,
(a) Verify analytically and numerically that the fitted values under the chain model $/A.B.C + A.C.D/(A.B.C.D) + E.C.D/$ and the all-response model $(B + D).A.C + E.C.D$ are identical.
(b) Compute the standardised residuals, $(n - n_\phi \hat{p})/\sqrt{n_\phi \hat{p}}$, for the all-response model $(B + D).A.C + E.C.D$ and comment on the shape of the histogram.

7: The marginal three-way table $E \times C \times D$ from the Danish homework survey is

		C=1			C=2		
	D	1	2	3	1	2	3
E	1	85	81	120	156	129	361
	2	75	143	137	44	96	164

Draw up the analysis of deviance table taking E as the response variable.

8: Consider the all two-way interaction models for the four-way margin 1234 and five-way margin 12345 in the analysis of the college plans data, Example 10.8.2. Verify that these models have deviances of 47.23 on 33 df and 138.72 on 88 df respectively. Verify that the edge exclusion deviances and dfs for the response variable 4 from the two-way interaction model for the four-way table, and variable 5 from its two-way interaction model, are

4.1	112.59	1	5.1	15.20	1
4.2	735.22	3	5.2	336.56	3
4.3	2188.65	3	5.3	531.44	3
			5.4	1652.86	1

respectively.

9: For the multivariate regression model, show that the maximum likelihood estimators of B and V are

$$\hat{B} = \text{cov}_N(Y, X)\text{var}_N(X)^{-1} \quad \text{and} \quad \hat{V} = \text{var}_N(Y|X)$$

where $\text{var}_N(X, Y)$ is the sample variance-covariance matrix. Note that the regression coefficient estimators are identical to the ordinary least squares estimates obtained separately from each of the q multiple regression equations.

10: The data from Dyke and Patterson (1952) has one response, cancer knowledge, and four explanatory factors. See also Goodman (1970).

		B				B^c			
		D		D^c		D		D^c	
		E	E^c	E	E^c	E	E^c	E	E^c
A	C	23	8	8	4	27	18	7	6
	C^c	102	67	35	59	201	177	75	156
A^c	C	1	3	4	3	3	8	2	10
	C^c	16	16	13	50	67	83	84	393

where the variables denote A read newspapers : yes, no; B listen to radio : yes, no; C attend lectures : yes, no; D solid reading : yes, no; E response, good knowledge of cancer : yes, no. Analyse these data.

Chapter 11

Models for Mixed Variables

We fall short of giving a complete treatment of graphical models for mixed discrete and continuous variables. Even though one of the principal lessons to be drawn from graphical modelling is the importance of jointly modelling the variables, a full treatment of the subject requires a text of its own, and probably written in a rather different way as the subject matter is more difficult than the material in either of the two pure cases.

There are two necessary extensions of the machinery already developed. The first is to extend the graph theory to incorporate marked vertices corresponding to the two types of variables. Fortunately, the discussion of conditional independence given in Chapter 3 is not affected by this extension and the Markov properties of independence graphs remain unaltered. The second extension is to specify a flexible family of distributions, rich enough to encompass the two types of variables and simple enough to deal with tractably. Substantial progress in this direction has been made by Lauritzen and Wermuth (1984, 1989), Wermuth and Lauritzen (1990), who introduce the *conditional Gaussian* family of distributions, which has the necessary property of containing the special cases of the Normal and Multinomial distributions. The graphical models discussed in this chapter are all of this form.

The full CG distribution is defined by the following two part asymmetric construction: the marginal distribution of the discrete variables is Multinomial and, conditional on these variables, the continuous variables have a multivariate Normal distribution. The CG distribution contains both the purely discrete and continuous distributions as special cases.

An important feature of this construction is that the sufficient statistics for the saturated model are the cell counts and, *within* each cell, the mean vector and the variance matrix of the continuous variables. (In fact this remark can be used as a theoretical starting point to generate the family of CG distributions.) Given a graph, CG graphical models are defined as those families of CG distributions that satisfy the conditional independence statements implicit in the associated independence graph.

The problem that arises with the CG distribution for mixed variables is that, unlike either of the pure cases, the distribution is not closed under marginalisation. That is, even if the full vector of variables has a CG distribution, it may be that the distribution of a given subset is not CG. One must either face up to the difficulties associated with mixture distributions, which are notoriously problematic as they fall outside the exponential family of distributions, or put restrictions on permissible graphs, which corresponds to assuming a very strong notion of decomposability, see Leimer (1989).

An extended family of models: the *hierarchical interaction* models, Edwards (1990), are generated from the CG distribution by setting terms to zero in the hierarchical interaction expansion of the parameters. The construction is a straightforward generalisation of our construction of log-linear expansions for purely discrete variables, and one which greatly enhances the applicability of the CG distribution.

Another major extension is to the family of *CG regression* models, which widens the class of models for mixed variables appropriate for applications with a covariate structure. These are constructed from the joint CG distribution of the full set of mixed variables by taking those distributions formed by conditioning on any subset of the variables. Of course, this is fairly trivial if one chooses to condition exactly on the set of discrete variables, in which case the procedure just returns the multivariate Normal distribution with mean and variance dependent on the categories, which was a part of the initial joint CG distribution. However, if one chooses to condition on the continuous variables, then the procedure generates a family of multivariate logistic models.

Related methods: Graphical modelling is intimately related to many classical statistical procedures, and the CG distribution is a rich enough family of distributions to contain many examples. Best known is multiple linear regression with $q = 1$ and p continuous covariates where choosing an independence graph is reformulation of the variable selection problem. When all the variables are categorical then a subset of ordinary log-linear models for a cross-classified contingency table is another example. CG regression models with a continuous response and categorical explanatory models correspond to fixed effects analysis of variance models, while the logistic regression model has a categorical response and continuous explanatory variables.

Summary of chapter: In the first section we introduce the CG distribution, its parameters and derive its Markov properties. Before writing down the likelihood function we discuss concise model formulae to represent families of CG distributions corresponding to hierarchical mixed interaction models. After making some general remarks about graphical modelling with mixed variables we illustrate the notions with the analysis (in MIM) of four case studies. We have chosen specific case studies so as to be able to fit any CG regression models in the joint CG distribution. We have omitted a derivation of the likelihood equations and the deviance for these mixed models, and also a discussion of the CG regression models.

11.1 The CG Distribution

We need slight changes in notation necessary to distinguish the two types of variables. The full vector of random variables X becomes $X = (I, Y)$ where I is the p-dimensional vector of discrete variables and Y is the q-dimensional vector of continuous variables. The full set of k vertices K is correspondingly partitioned into $K = \Delta \cup \Gamma$. A typical value taken by (I, Y) is denoted (i, y) and a marginal vector of (I, Y) as (I_a, Y_b) where $a \subseteq \Delta$ and $b \subseteq \Gamma$.

The conditional Gaussian distribution is defined by the following construction: suppose that the vector of p categorical variables, I, has the cross-classified Multinomial distribution with parameters $\{p(i)\}$, the table of probabilities, and that the conditional distribution of the q continuous variables, Y given $I = i$, is multivariate Normal with expectation $E(Y|I = i) = \mu(i)$ and variance $\mathrm{var}\,(Y|I = i) = V(i)$. The joint distribution is obtained by multiplying these functions together according to $f_{IY}(i, y) = f_I(i) f_{Y|I}(y; i)$ and is called the CG distribution because of this conditional Gaussian construction.

Definition. The $p+q$-dimensional random vector $X = (I, Y)$ has the *CG distribution* if and only if its density function is of the form

$$f_{IY}(i, y) = p(i)\,(2\pi)^{-q/2} \det(V(i))^{-1/2} \exp\{-\frac{1}{2}[y - \mu(i), V(i)^{-1}(y - \mu(i))]\}$$

for $y \in R^q$ and i taking values in the p-fold Cartesian product of the index sets for each categorical variable. The family of all CG distributions with a given dimension is $CG(p, q)$ and we write $f \in CG(p, q)$ to denote a member. The CG distribution is indexed by the *moment parameters* $\{p(i), \mu(i), V(i)\}$.
□

Consider specialising to the case of one categorical and one continuous variable.

EXAMPLE 11.1.1 The $CG(1,1)$ family of distributions. With $p = 1$ and $q = 1$, the CG distribution of (I, Y) is the joint distribution

$$f_{IY}(i, y) = \exp\{u_\phi + u(i)\} \frac{1}{\sqrt{2\pi}\sigma(i)} \exp\{-\frac{1}{2}(\frac{y - \mu(i)}{\sigma(i)})^2\},$$

where the support of y is the whole real line, and the support of i is a finite set of integers. In general, both the mean and variance of Y depend on i. Taking logarithms gives

$$\log f_{IY}(i,y) = \text{const} + u(i) - \log\sigma(i) - \frac{1}{2}\frac{\mu(i)^2}{\sigma(i)^2} + \frac{\mu(i)y}{\sigma(i)^2} - \frac{1}{2}\frac{y^2}{\sigma(i)^2},$$

which, by appropriately identifying coefficients, we may write as

$$\log f_{IY}(i,y) = \lambda_\phi + \lambda(i) + (\eta_\phi + \eta(i))y - \frac{1}{2}(\psi_\phi + \psi(i))y^2.$$

There are three distinct components to this log density function: the discrete part indexed by the λ terms; the linear part, indexed by the η terms; and the quadratic part indexed by the ψ terms. The expression is made identifiable by setting $\lambda(0) = 0$, $\eta(0) = 0$ and $\psi(0) = 0$. It is made to integrate to unity by insisting that $\psi_\phi + \psi(i) > 0$ for all i, and suitably choosing the norming constant λ_ϕ. The parameters λ, η and ψ, are canonical parameters, which is a name taken from the exponential family of distributions, see Barndorff-Nielsen (1978).

We make two remarks. Firstly, direct application of the factorisation criterion gives a parametric condition for independence: if (I, Y) has a CG distribution then $Y \perp\!\!\!\perp I$ if and only if $\eta(i) = 0$ and $\psi(i) = 0$ for all i.

Secondly, the one-way analysis of variance model in which the number of units allocated to each treatment is determined at random is a special case of the CG(1,1) model. The standard ANOVA model supposes that $E(Y|I = i) = \mu(i)$ and $\text{var}(Y|I) = \sigma^2$ for all i. To ensure variance homogeneity the canonical parameter $\psi(i) = 0$ and the parameter $\psi_\phi = \sigma^{-2}$. Consequently by identifying the coefficients of the linear interaction, $\eta_\phi + \eta(i) = \mu(i)/\sigma^2$. The hypothesis of no treatment effect is equivalent to the condition that $\eta(i) = 0$ for all i. □

Closure

The marginal distributions of the CG distribution are not necessarily themselves members of the CG family, so the family is not closed under marginalisation, and consequently it may not be closed under conditioning. To examine this in a little more detail, consider the $CG(1,1)$ family. The family $CG(1,0)$ is the family of all Multinomial distributions for a single discrete variable (the support of I in $CG(1,0)$ is assumed the same as in $CG(1,1)$) and $CG(0,1)$ denotes the family of all univariate Normal distributions. Now, the marginal distribution of I comes by integrating out Y from the joint density function, $f_{IY} \in CG(1,1)$, and, trivially from the definition of the CG density, implies that $f_I \in CG(1,0)$. On the other hand, the marginal distribution of Y, $f_Y(y) = \sum_i f_{IY}(i,y)$, is a mixture of Normal distributions with mixing weights $p(i)$, and in general is not itself Normal, so that $f_Y \notin CG(0,1)$. It is well known that mixture distributions pose awkward estimation problems.

The CG family of distributions is not the only way to combine probability models for discrete and continuous variables and other constructions are possible.

EXAMPLE 11.1.2 The logistic-Normal distribution is not CG. For instance, we might suppose that the joint distribution of (I,Y) comes by supposing Y to have a marginal Normal distribution and I given Y to have a logistic regression, with Y taking the role of explanatory variable. For simplicity suppose that I is a binary variable, and $p = P(I = 1|y)$ is the probability that I takes the value 1. It is supposed that this probability is linearly related to the explanatory variable on the scale determined by the logit function, so that

$$\log \frac{p}{1-p} = \alpha + \beta y.$$

Inverting this expression gives the conditional density function of I as

$$f_{I|Y}(i;y) = \frac{\exp i(\alpha+\beta y)}{1+\exp(\alpha+\beta y)}, \quad \text{for} \quad i = 0,1.$$

Suppose further, that Y has the standard Normal distribution, $Y \sim N(0,1)$, then the the joint density function of (I,Y) is

$$f_{IY}(i,y) = \frac{\exp i(\alpha+\beta y)}{1+\exp(\alpha+\beta y)} \frac{1}{\sqrt{2\pi}} \exp\{-\frac{1}{2}y^2\}.$$

On collecting terms together

$$\log f_{IY}(i,y) = \text{const} + \alpha i + \beta i y - \frac{1}{2}y^2 - \log\{1+\exp(\alpha+\beta y)\}.$$

This logistic-Normal density is of a different functional form to the CG density above. It is not in the linear exponential family of distributions (Cox and Hinkley 1974) because the last term on the right cannot be expressed as a product $g(\alpha,\beta)t(y)$ where g and t are two fixed functions of their explicit arguments.

There is no *a priori* reason why the CG family of distributions should be preferred, and in practice the appropriate choice of density function will depend on the specific modelling context. However, we only consider CG models in this text. □

CG Parameterisations

There is a variety of ways to parameterise the CG distribution: most intuitive is the moments parameterisation incorporated in the definition, but equally important is the reparameterisation in exponential family form. Write f_{IY} above as

$$f_{IY}(i,y) = \exp\{\alpha(i) + [\beta(i), y] - \frac{1}{2}[y, D(i)y]\},$$

where $D(i)$ is symmetric and positive definite. The *canonical parameters* $\alpha(i)$, $\beta(i)$ and $D(i)$, are in one to one correspondence with the moment parameters: it is easy to verify that

$$D(i) = V(i)^{-1} \quad \text{and} \quad \beta(i) = V(i)^{-1}\mu(i)$$

and the expression for $\alpha(i)$ is given in the exercises.

The elements of $D(i)$ are familiar as elements of an inverse variance matrix and determine the interaction between two continuous variables for each cell i. But calculating the mixed derivative measure of interaction shows that $D(i)$ may also be an ingredient of discrete-continuous and of discrete-discrete interaction. The other two sets of canonical parameters are more difficult to interpret. The β parameters which relate the continuous variables to the discrete ones do not have the intuitive simplicity of means or regression coefficients. Though note that $[\beta(i), y] = [\mu(i), V(i)^{-1}y]$ is a weighted inner product between y and the mean vector $\mu(i)$.

In this canonical form, the density is fairly amenable to application of the factorisation criterion for conditional independence. Testing for the pairwise independence of two continuous variables is easiest because the criterion is confined to a single component of the density function, the quadratic term, though as this may depend on the value i, the situation is more complicated than in the corresponding pure case. However, it is not possible to index the interaction between individual elements of the vector I in this parameterisation, and, in parallel to our treatment of the purely discrete case, we need a hierarchical interaction expansion.

Recall from Section 2.4 for a two dimensional categorical variable, $I = (I_1, I_2)$, that $\log f_{12}(i_1, i_2) = \log p(i_1, i_2)$ is reparameterised to $\log f_{12}(i_1, i_2) = u_\phi + u_1(i_1) + u_2(i_2) + u_{12}(i_1, i_2)$ where each u-term is a coordinate projection function, for instance, $u_2(i) = u_2(i_2)$, and there are enough constraints on the u-terms to ensure identifiability. This isolates the interaction between I_1 and I_2 measured by the parameter u_{12} from the main effects measured by (the one-way interactions) u_1 and u_2.

We do the same for the canonical parameters: we endow each set with a hierarchical interaction expansion, so that

$$\alpha(i) = \sum_{a \subseteq \Delta} \lambda_a(i), \quad \beta(i) = \sum_{a \subseteq \Delta} \eta_a(i) \quad \text{and} \quad D(i) = \sum_{a \subseteq \Delta} \psi_a(i),$$

where the sum is taken over all subsets of the set of discrete indices Δ. Because $\beta(i)$ is a vector of dimension q then so is $\eta_a(i)$ for each a, and similarly $\psi_a(i)$ is a matrix of order $q \times q$.

We may now re-express f_{IY} above as

$$f_{IY}(i,y) = \exp\{\sum_{a \subseteq \Delta} \lambda_a(i) + \sum_{a \subseteq \Delta} [\eta_a(i), y] - \frac{1}{2}\sum_{a \subseteq \Delta} [y, \psi_a(i)y]\}.$$

We use the same constraints adopted for the u-term expansion in log-linear expansion of the Multinomial density function, so that whenever one element of i is zero, say the j-th so that $i_j = 0$, then $\lambda_a = 0$ for all $j \in a$. This convention extends to the parameters η and ψ. In summary: whenever $j \in a$ and $i_j = 0$, then

$$\begin{aligned} \lambda_a(i_a) = 0, \quad \eta_a^r(i_a) &= 0 \quad \text{for all } r \in \Gamma, \quad \text{and} \\ \psi_a^{rs}(i_a) &= 0 \quad \text{for all } r,s \in \Gamma. \end{aligned}$$

EXAMPLE 11.1.3 $CG(1,1)$. In this instance the η and ψ terms are scalars, and

$$\log f(i_1, y_1) = \lambda_\phi + \lambda_1(i) + \{\eta_\phi + \eta_1(i)\}y_1 - \frac{1}{2}\{\psi_\phi + \psi_1(i)\}y_1^2.$$

If I_1 is binary, there are exactly 6 parameters: λ_ϕ, $\lambda_1(1)$, η_ϕ, $\eta_1(1)$, ψ_ϕ, and $\psi_1(1)$, though λ_ϕ is determined by the requirement that the density integrate to unity. They can be interpreted in terms of expected values: in particular, note that

$$\psi_\phi = \frac{1}{\text{var}\,(Y|I=0)} \quad \text{and} \quad \psi_1 = \frac{1}{\text{var}\,(Y|I=1)} - \frac{1}{\text{var}\,(Y|I=0)}$$

so that a test for $\psi_1 = 0$ is a test for equality of inverse variances, and, in this one dimensional example, also for equality of variances themselves. □

EXAMPLE 11.1.4 $CG(2,2)$. When $p = 2$ and $q = 2$ the density is

$$\begin{aligned} \log f(i_1, i_2, y_1, y_2) &= \lambda_\phi + \lambda_1(i_1) + \lambda_2(i_2) + \lambda_{12}(i_1, i_2) \\ &+ [\eta_\phi + \eta_1(i_1) + \eta_2(i_2) + \eta_{12}(i_1, i_2), y] \\ &- \tfrac{1}{2}[y, \{\psi_\phi + \psi_1(i_1) + \psi_2(i_2) + \psi_{12}(i_1, i_2)\}y], \end{aligned}$$

where $y = (y_1, y_2)$, and the η-terms are vectors and the ψ-terms are matrices. While this expansion appears more complicated than in the $\{\alpha, \beta, D\}$ set-up, this parameterisation permits a flexible specification of hypotheses: for instance if the variance of Y does not depend on I_2 then the matrices ψ_2 and ψ_{12} must both be zero. We shall return to this example again as it conveniently exemplifies many of the distinguishing features of the mixed case. □

Markov properties

The Markov properties of the family of CG distributions follow directly from the earlier material developed in Chapter 3. It is easy to check that the density function is always positive and to elicit a condition for pairwise independence given the remaining variables.

Proposition 11.1.1 Pairwise conditional independences in the CG distributions. *If the random vector* (I, Y) *has the CG distribution given by*

$$f_{IY}(i, y) = \exp\{\sum_{a \subseteq \Delta} \lambda_a(i) + \sum_{a \subseteq \Delta} [\eta_a(i), y] - \frac{1}{2} \sum_{a \subseteq \Delta} [y, \psi_a(i) y]\}$$

then

1. $I_j \perp\!\!\!\perp I_k | (I_{\Delta \setminus \{j,k\}}, Y_\Gamma)$ *if and only if* $\lambda_a = 0$, $\eta_a = 0$ *and* $\psi_a = 0$ *whenever* $j, k \in a \subseteq \Delta$;

2. $I_j \perp\!\!\!\perp Y_r | (I_{\Delta \setminus \{j\}}, Y_{\Gamma \setminus \{r\}})$ *if and only if* $\eta_a^r = 0$ *and* $\psi_a^{rs} = 0$ *whenever* $j \in a \subseteq \Delta$ *and for all* $s \in \Gamma$; *and*

3. $Y_r \perp\!\!\!\perp Y_s | (I_\Delta, Y_{\Gamma \setminus \{r,s\}})$ *if and only if* $\psi_a^{rs} = 0$ *for all* $a \subseteq \Delta$.

The proof is a direct application of the factorisation criterion for conditional independence, Proposition 2.2.3. We use the convention that ψ^{rs} is the (r, s)-th element of the matrix ψ, so that superscripts refer to continuous variables and subscripts to discrete variables.

EXAMPLE 11.1.5 CG(2,2) continued. Suppose I_1 and I_2 are binary. When $p = 2$ and $q = 2$ the possible pairwise conditional independences are summarised by

1. $I_1 \perp\!\!\!\perp I_2 | Y_{12}$ if and only if $\lambda_{12} = 0$, $\eta_{12} = 0$ and $\psi_{12} = 0$. There are $1 + 2 + 3 = 6$ parameters set to zero as η is a vector and ψ a matrix.

2. $I_1 \perp\!\!\!\perp Y_1 | (I_2, Y_2)$ if and only if $\eta_1^1 = 0$, $\eta_{12}^1 = 0$, and $\psi_{12}^{11} = 0$, $\psi_{12}^{12} = 0$, $\psi_1^{11} = 0$, $\psi_1^{12} = 0$. There are 6 parameters set to zero.

3. $Y_1 \perp\!\!\!\perp Y_2 | I_{12}$ if and only if $\psi_\phi^{12} = 0$, $\psi_1^{12} = 0$, $\psi_2^{12} = 0$, $\psi_{12}^{12} = 0$. There are 4 parameters set to zero.

□

11.2 Interaction Model Formulae

It is unnecessarily complicated to refer to a model either by writing out the free parameters in full, or conversely, by writing out those set to zero. To simply denote a model we adopt the following syntax, which generalises the model formulae for discrete variables introduced earlier in Chapter 7 and extended in Chapter 10.

To distinguish the discrete and continuous components of interaction we use the symbols $I_1, I_2, \ldots, I_p$ and $Y_1, Y_2, \ldots, Y_q$.

EXAMPLE 11.2.1 Consider the saturated model in $CG(2,2)$

$$\begin{aligned}\log f(i_1, i_2, y_1, y_2) &= \lambda_\phi + \lambda_1(i_1) + \lambda_2(i_2) + \lambda_{12}(i_1, i_2) \\ &+ [\eta_\phi + \eta_1(i_1) + \eta_2(i_2) + \eta_{12}(i_1, i_2), y] \\ &- \tfrac{1}{2}[y, \{\psi_\phi + \psi_1(i_1) + \psi_2(i_2) + \psi_{12}(i_1, i_2)\}y].\end{aligned}$$

We wish to write this model as something like

$$\begin{aligned}&(1 + I_1 + I_2 + I_1.I_2) + \\ &\quad (1 + I_1 + I_2 + I_1.I_2).(Y_1 + Y_2) + (1 + I_1 + I_2 + I_1.I_2).(Y_1^2 + Y_1.Y_2 + Y_2^2)\end{aligned}$$

where the interaction operator '.' and the addition operator '+' combine terms, and the parentheses are used to simplify expressions. The linear and quadratic parts of the continuous components are distinguished by the square.

As we restrict attention to hierarchical models we need only mention the highest order interactions explicitly. Eliminating the lower order terms in the discrete variables simplifies this formula to

$$I_1.I_2 + I_1.I_2.(Y_1 + Y_2) + I_1.I_2.(Y_1^2 + Y_1.Y_2 + Y_2^2)$$

and doing the same for the continuous variable simplifies it further to $I_1.I_2.(Y_1^2 + Y_1.Y_2 + Y_2^2)$. Additional simplification suggests we write this saturated model in $CG(2,2)$ as

$$I_1.I_2.(Y_1 + Y_2)^2.$$

It is the missing interactions in the expression that go to define the model. For instance, the model, M, in which the continuous variables have constant variance and are independent for all levels of the discrete variables, while their means depend additively on the discrete variables, may be described as

$$M = (I_1 + I_2).(Y_1 + Y_2) + Y_1^2 + Y_2^2.$$

The absence of a quadratic term involving I_1 or I_2 implies variance homogeneity, the missing $Y_1.Y_2$ interaction term implies the independence $Y_1 \perp\!\!\!\perp Y_2 | (I_1, I_2)$, and the missing $I_1.I_2.(Y_1 + Y_2)$ from the linear interaction component implies additivity, as in the usual analysis of variance model. □

Model syntax

We summarise the rules for this model syntax. There are discrete symbols

$I_1, I_2, \ldots, I_p$ and continuous symbols $Y_1, Y_2, \ldots, Y_q$. There is a null term 1, or, if this is misleading, 1_ϕ. There are two binary operations '.' and '+' which obey

$$I + I = I, \quad Y + Y = Y, \quad I.I = I, \quad Y.Y = Y^2, \quad \text{and}$$
$$\text{if} \quad T = 1, I, Y \text{ or } Y^2 \quad \text{then} \quad T + 1 = T.1 = T;$$

and the operations are commutative, associative and distributive in an obvious manner. Parentheses simplify the expressions and are expanded using the distributive rules. There is one further condition, there must be a Y^2 term in the model formulae corresponding to every continuous variable in the density function. Interaction terms of the form $Y.Y.Y$ cannot occur.

We can construct the independence graph directly from the model formula.

EXAMPLE 11.2.2 A CG 4-cycle. Consider the model defined by applying the statements $I_1 \perp\!\!\!\perp I_2 | (Y_1, Y_2)$ and$Y_1 \perp\!\!\!\perp Y_2 | (I_1, I_2)$ to the saturated $CG(2,2)$ density function.

The following construction shows how to calculate its model formula.

1. Expand the saturated model $I_1.I_2.(Y_1 + Y_2)^2$ by adding 1 to each I and Y to get

$$(1 + I_1 + I_2 + I_1.I_2) + (1 + I_1 + I_2 + I_1.I_2).(Y_1 + Y_2) \\ + (1 + I_1 + I_2 + I_1.I_2).(Y_1^2 + Y_1.Y_2 + Y_2^2).$$

2. Eliminate any term including $I_1.I_2$ and $Y_1.Y_2$ interactions to give

$$(1 + I_1 + I_2) + (1 + I_1 + I_2).(Y_1 + Y_2) + (1 + I_1 + I_2).(Y_1^2 + Y_2^2).$$

3. Finally, eliminate redundancies to give

$$(I_1 + I_2).(Y_1^2 + Y_2^2).$$

Its graph is constructed by inserting an edge for each interaction, where we use the convention that open circles represent continuous variables and filled circles, or dots, represent discrete variables:

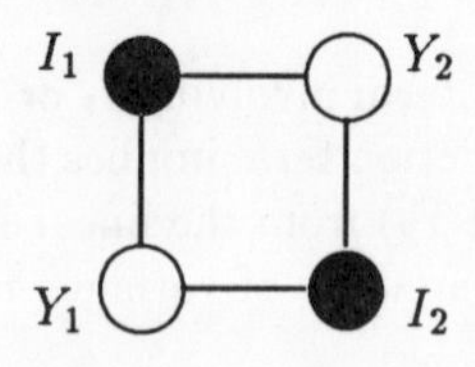

the chordless 4-cycle. If I_1 and I_2 are binary random variables then the parameters that have been set to zero in the saturated CG model are: $\lambda_{12} = 0$, the vector $\eta_{12} = 0$ and the matrix $\psi_{12} = 0$; there are $(1+2+3) = 6$ such parameters. To account for the independence between Y_1 and Y_2 the matrix $\psi_a^{12} = 0$ for all $a \subseteq \Delta$; there are 4 parameters involved. It appears that 10 parameters are set to zero, however one parameter is counted twice, the off diagonal element in the matrix ψ_{12}^{12} is set to zero under both independence statements; hence only 9 are set to zero. □

Other implementations of model formulae syntax are possible and the one used in the graphical modelling package MIM, see Appendix A and Edwards (1987), is stricter. In fact there are two differences. The first is cosmetic: rather than writing the saturated model in $CG(2,2)$ as $I_1.I_2.(Y_1+Y_2)^2$ it is written as $I_1.I_2/I_1.I_2.Y_1+I_1.I_2.Y_2/I_1.I_2.Y_1.Y_2$, making the discrete, linear and quadratic interaction components explicit and separating them by '/'. For instance, the saturated model in $CG(0,3)$ is $/Y_1+Y_2+Y_3/Y_1.Y_2.Y_3$ rather than $(Y_1+Y_2+Y_3)^2$ and in $CG(1,2)$ the model $I.Y_1^2+Y_1.Y_2+I.Y_2+Y_2^2$ becomes $I/I.Y_1+I.Y_2/I.Y_1+Y_1.Y_2$. (We use the symbol '/' in Chapter 10 to separate formulae for distinct blocks in a regression or chain model.) Secondly, the product rule, $I.I = I$, is adopted for continuous variables as well as discrete ones, so that $Y.Y = Y$. Consequently the presence of Y_1 in the quadratic component implies that the corresponding y_1^2 appears in the density function; and the presence of $Y_1.Y_2$ in the quadratic component implies that the corresponding y_1^2, y_2^2, and y_1y_2, appear in the density function. The quadratic interaction terms $Y_1.Y_2.Y_3$ and $Y_1.Y_2+Y_1.Y_3+Y_2.Y_3$ are synonymous. An effect of this is to prohibit models of the form $I.Y_1^2+I.Y_1.Y_2+Y_2^2$.

11.3 The Likelihood Function

In this section we write down the likelihood function for models that have a joint CG distribution. In principle we only need to put the results for the two pure cases together, and the essential result obtained therein: that the likelihood equations for hierarchical models preserve those margins for parameters that are free to vary in the model, is still valid.

However, the representation of the log-likelihood function in terms of information divergence does not extend so easily because the lack of closure under marginalisation in the CG family means that one cannot simply derive the equations by equating marginal distributions. Had we space to give a complete derivation, and tackle questions of existence and uniqueness, it would be based on the properties of the linear exponential family of models (Barndorff-Nielsen, 1978, Andersen, 1980, Glonek, Darroch and Speed, 1988) of which the CG family is a special example. Therefore we do not give a general treatment. We also exclude the question of when direct estimates exist, as this depends on the notion of decomposability for mixed models, see Leimer (1989), Frydenberg and Lauritzen (1989).

Recall that $x = (i, y)$ is an observation on a $p+q$-dimensional partitioned random vector $X = (I, Y)$ where i indexes the cell of the p-way table cross-classifed by I, and y is q-dimensional continuous observation. Consider a sample of N observations $x^l = (i^l, y^l)$ for $l = 1, 2, \ldots, N$. As in the discrete case, $n(i)$ denotes the number of observations falling in i-th cell; and we use the same notational convention for extracting statistics on margins as before, so that $n_a(i_a)$ is the count in i_a-th cell of the margin classified by I_a. Consequently $n_\phi = N$.

The q-dimensional vector $\bar{y}(i) = n(i)^{-1} \sum_l y^l$ denotes the sample mean in the i-th cell; $S(i)$ denotes the sample variance matrix of order $q \times q$ in the i-th cell (with divisor $n(i)$); $S_{bb}(i)$ is the sample variance matrix of the marginal vector Y_b in the i-th cell. Note that $S_{bb}(i_a)$ is not defined.

We shall also need the raw sums and sums of squares of the continuous observations:

$$t(i) = \sum_l y^l \quad \text{and} \quad T(i) = \sum_l y^l (y^l)^T,$$

where both sums are taken over the observations in the i-th cell. The marginal totals are $t_a(i_a) = \sum t(i)$ and $T_{aa}(i_a) = \sum T(i)$, where these sums are taken over all cells i which have i_a as the margin.

Writing θ for the vector of parameters, the log-likelihood function is formed from

$$l(\theta) = l(\theta; x^1, x^2, \ldots, x^N) = \sum_l \log f(x^l; \theta);$$

where $x = (i, y)$, and from Section 11.1 above

$$\log f_{IY}(i, y) = \sum_a \lambda_a(i) + \sum_a [\eta_a(i), y] - \frac{1}{2} \sum_a [y, \psi_a(i) y],$$

where the sum is taken over $a \subseteq \Delta$. In evaluating this expression for the likelihood, the only hard part is to verify the identities

$$\begin{aligned} \textstyle\sum_l \sum_a \lambda_a(i) &= \textstyle\sum_a \sum_{i_a} \lambda_a(i_a) n_a(i_a) \\ \textstyle\sum_l \sum_a [\eta_a(i), y] &= \textstyle\sum_a \sum_{i_a} [\eta_a(i_a), t_a(i_a)] \\ \textstyle\sum_l \sum_a [y, \psi_a(i) y] &= \textstyle\sum_a \sum_{i_a} \operatorname{tr}\{\psi_a(i_a) T_{aa}(i_a)\}. \end{aligned}$$

These hold because firstly, the expression for each hierarchical expansion is derived in the same manner as the log-likelihood function for the u-terms in the log-linear model in Section 7.4; and secondly, the trick of using the trace to simplify the quadratic term is the same as employed in deriving the log-likelihood function for the Gaussian model in Section 6.5. Putting it all together, we have

Proposition 11.3.1 The likelihood function for the CG distribution. *The log-likelihood function for the parameters of the $CG(p, q)$ distribution based*

on a random sample of N observations is

$$l(\lambda, \eta, \psi) = \sum \lambda_a(i_a)n_a(i_a) + \sum [\eta_a(i_a), t_a(i_a)] + \sum \text{tr}\, \{\psi_a(i_a)T_{aa}(i_a)\}.$$

where the sums are taken over each cell i_a of each margin a.

The sample count, sum and sum of squares for each cell constitute a set of sufficient statistics. The likelihood function is of the form $l(\theta) = \sum_j \theta_j t_j$ where the θ's are parameters and the t's are statistics; but it is constrained by the condition that probabilities must sum to one. It is maximised by using Lagrange multipliers to enforce the constraint, differentiating, and setting the derivatives to zero. The linearity of the log-likelihood function in the parameters raises the hope that the second derivatives are constant, and thus lead to unique global maxima. The resulting equations can be solved using a modified iterative proportional scaling algorithm see Frydenberg and Edwards (1989).

EXAMPLE 11.3.1 $CG(1,1)$. When there are just two groups, so that $p = 1$ and $I = I_1$ is binary, and there is a single continuous measurement, so that $q = 1$, there are 6 parameters, corresponding to the two probabilities, two means and two variances. To make the connection between the subscripts on the parameters and the discrete variable more explicit we write I_1 rather than just I. The log-density function is

$$\log f_{I_1,Y}(i, y) = \lambda_\phi + \lambda_1(i) + [\eta_\phi + \eta_1(i), y] - \frac{1}{2}[y, \{\psi_\phi + \psi_1(i)\}y],$$

for $i = 0, 1$. The log-likelihood function is

$$l(\lambda_\phi, \lambda_1(1), \eta_\phi, \eta_1(1), \psi_\phi, \psi_1(1)) = \\ \lambda_\phi n_\phi + \lambda_1(1)n_1(1) + \eta_\phi t_\phi + \eta_1(1)t_1(1) + \psi_\phi T_\phi + \psi_1(1)T_{11}(1);$$

where $n_1(1)$ is the number of observations for which $I_1 = 1$, and where $t_1(1)$ and $T_{11}(1)$ are the sum and sum of squares of those y's for which $I_1 = 1$. The parameter λ_ϕ acts as a normalising constant. An extension to where I_1 takes three levels is given in the exercises. □

11.4 Issues in Modelling

At a formal level the basic ideas of model fitting for CG distributions: evaluation of the likelihood function, maximisation, fitted values, computation of the deviance, model selection and diagnostic analysis are straightforward extensions of the material developed for the graphical log-linear and Gaussian models. Here we outline notions of modelling for mixed variables,

1. *Conditional independence.* First and foremost the analysis is guided by an independence graph to represent the associations in the data, and the notion of conditional independence gives a common framework to examine interaction between discrete and continuous variables.

2. *The deviance.* While conditional independence is the common currency of interaction, the common unit of goodness of fit is the deviance. The likelihood ratio naturally determines a common objective function for categorical and continuous variables alike. The very simplest example is the one way analysis of variance model with random cell counts, and formula $I.Y + Y^2$. The log-likelihood is

$$\sum_i n(i) \log p(i) - \frac{n_\phi}{2} \log \sigma^2 + \frac{1}{2\sigma^2} \sum_{ij} (y_{ij} - \mu_i)^2,$$

an expression formed from the sum of two likelihoods corresponding to the categorical and the continuous model respectively.

Secondly, consider a combined test for the independence of two discrete variables and the independence of two continuous variables specified by $(I_1+I_2)+(Y_1^2+Y_2^2)$ against the model $(I_1.I_2)+(Y_1+Y_2)^2$. The deviance attributable to the hypothesis is

$$\text{dev} = 2 \sum n_{12} \log \frac{n_{12} n_\phi}{n_1 n_2} - n_\phi \log (1 - r_{12}^2)$$

where r_{12} is the sample correlation coefficient. The two separate test statistics are combined in a specific manner, and searching for a good fitting overall model can involve trading between these components.

Decompositions of the deviance that generalise the classical analysis of variance table often give a concise summary of the source of variation in a data set.

3. *Regression structures.* Important simplifications accrue if one may assume a regression structure, where certain variables are taken to be covariates as discussed in Chapter 10. The lack of symmetry between discrete and continuous variables that exists in the model may be reflected in the subject context, and while it may not always be true that the discrete variables are explanatory, in most studies some form of block structure is necessary to model the data.

The machinery for dealing with the general chain graph model requires an extension of the CG family to CG regression models. Unlike the two pure cases, these models may be quite complex due to possible lack of closure of the corresponding conditional distribution, and specifically it may not be possible to numerically fit the conditional model by reinterpreting the fit of a joint model, a trick heavily utilised in the Chapter 10.

4. *Hierarchical interaction models.* The hierarchical interaction expansions used to derive log-linear models in Chapter 7 play an important part in the linear and quadratic interaction terms of the CG distribution. A parsimonious choice of the quadratic term leads to the idea of modelling variance heterogeneity, an idea which goes beyond the graphical modelling ideas pertinent to modelling pure systems. For example, in the $CG(1,1)$ family, it may be relevant to ask how the variance of Y depends on the level of I, and if it is permissible to assume homogeneity. This model is intermediate between, to one extreme, independence in which the whole distribution of Y is invariant to I and, to the other, complete dependence in both the mean and the variance.

   Parsimonious interaction modelling generalises the analysis of the two-way interaction log-linear model discussed in Chapter 9 and has a correspondingly richer structure that includes, for example, models of variance homogeneity and models with parallel regression lines. If variance homogeneity is acceptable, then the analyst is able to model the linear interaction $I_1.I_2 \ldots I_p.(Y_1 + Y_2 + \cdots + Y_q)$ and the quadratic interaction $(Y_1 + Y_2 + \cdots + Y_q)^2$ entirely separately.

5. *Alternative probability models.* There may be more than one representation of the study as a probability model, and in the final case study, dealing with some defence expenditure data we give an example in which it may be permissible to analyse the data either as a mixed or as a purely discrete variable system. The first case study provides an example where a single graphical model does not fully represent the data.

11.5 Case Studies

In this section we present four case studies illustrating different aspects of graphical modelling with mixed variables.

EXAMPLE 11.5.1 Fisher's iris data. The first example provides an instance where the result of the analysis is to describe the structure of the different groups defined by the discrete variable, rather than to provide a single overall model. The data are reproduced in several texts, one of which is Kendall and Stuart (1961). There are three different species of iris denoted here by $I = 1$, $I = 2$ and $I = 3$, respectively, and four measurements are taken from each: petal length X_1, petal width X_2, sepal length X_3, and sepal width X_4. The mean vectors and variance matrices for a sample of 50 observations from each species are given in Table 11.5.1.

Table 11.5.1: Means and variance matrices for Fisher's iris data.

species 1				
1	0.12176			
2	0.09723	0.14082		
3	0.01603	0.01146	0.02956	
4	0.01012	0.00911	0.00595	0.01088
mean	5.006	3.428	1.462	0.246
	1	2	3	4
species 2				
1	0.26110			
2	0.08348	0.09650		
3	0.17924	0.08100	0.21640	
4	0.05466	0.04038	0.07164	0.03832
mean	5.936	2.770	4.260	1.326
	1	2	3	4
species 3				
1	0.39626			
2	0.09189	0.10192		
3	0.29722	0.06995	0.29850	
4	0.04811	0.04668	0.04785	0.07392
mean	6.588	2.974	5.552	2.026
	1	2	3	4

First steps: As a first step in the analysis the following models are fitted:

	model formula	dev	df
mutual independence	$I + Y_1^2 + Y_2^2 + Y_3^2 + Y_4^2$	1434.91	34
independence given I	$I.(Y_1^2 + Y_2^2 + Y_3^2 + Y_4^2)$	275.37	18
independence from I	$I + (Y_1 + Y_2 + Y_3 + Y_4)^2$	712.67	28
variance homogeneity	$I.(Y_1 + Y_2 + Y_3 + Y_4) + (Y_1 + Y_2 + Y_3 + Y_4)^2$	149.67	20

The deviances of these four models give some feeling for the overall orders of magnitude involved. There is a substantial amount of interaction, both between the discrete and continuous variables and between the continuous variables within the categories of the discrete variable. Clearly the model of variance homogeneity is not acceptable and one purpose of the analysis is to explore the heterogeneity more systematically.

Consider the edge exclusion deviances for the edges between the continuous variables, obtained by deleting in turn the six interactions from the model $I.(Y_1 + Y_2 + Y_3 + Y_4)^2$; and these overall figures partitioned for each species separately:

edge	edge exclusion deviances overall	df	within species I=1	I=2	I=3
(Y_1, Y_2)	44.38	3	36.88	3.74	3.76
(Y_1, Y_3)	87.22	3	1.40	25.20	60.62
(Y_1, Y_4)	3.94	3	0.57	2.58	0.79
(Y_2, Y_2)	0.97	3	0.10	0.58	0.29
(Y_2, Y_4)	25.94	3	0.13	12.47	13.34
(Y_3, Y_4)	32.95	3	4.07	27.23	1.64

The overall edge exclusion deviance is partitioned into the sum of three separate deviances, one for each species, e.g. $44.38 = 36.88 + 3.74 + 3.76$. If we are guided by the deviances in the overall column then these test statistics suggest that the chordless 4-cycle

$$I.(Y_1^2 + Y_2^2 + Y_3^2 + Y_4^2 + Y_1.Y_3 + Y_3.Y_4 + Y_4.Y_2 + Y_2.Y_1)$$

might fit, and in fact, this model has a deviance of 4.26 on 6 df. Its graph makes the 'right' connections, the lengths are adjacent to the lengths, breadths to the breadths, petals to petals and sepals to sepals:

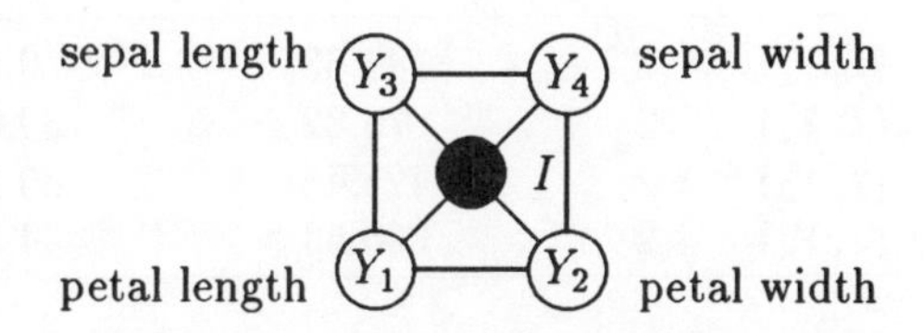

Distinct models: On the other hand, the differences between species indicate that a single independence graph for the continuous variables will not do justice to the data. Any single graph will have too many edges and in fact each species can be fitted by a simple but different model, given by the three

graphs

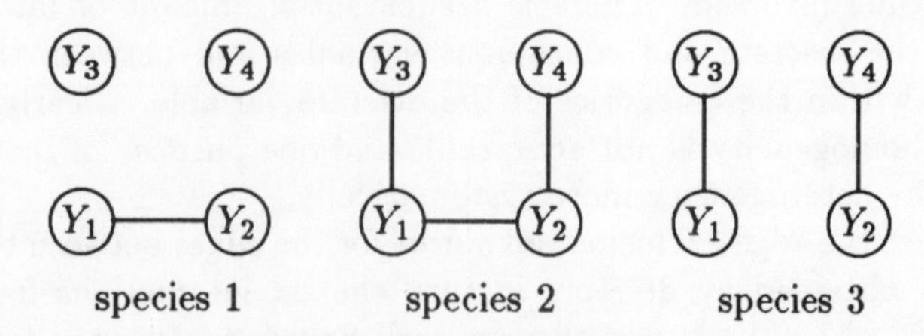

Even though species 2 and 3 are the most similar, the same model does not fit both. Such a discussion parallels that of local independence in partitioning the deviance for certain log-linear models in Section 7.7.

Differences in mean structure: We see that the variance matrices of the four measurements are different in the three species. In this context it is hard to give unequivocal descriptions of how the means of the variables differ with the species. To illustrate, if we compare the edge exclusion deviances for the differences between species for each of the four variables, in the context of (a) a standard univariate analysis of variance, and (b) a homogeneous variance model, we reach different conclusions as to the magnitude of the interaction. For (a), we may calculate the deviance attributable to a linear interaction between the species, I, and a continuous measure, Y_1 say, by computing the deviance difference between the two models

$$I.(Y_1+Y_2+Y_3+Y_4)+(Y_1^2+Y_2^2+Y_3^2+Y_4^2) \text{ and } I.(Y_2+Y_3+Y_4)+(Y_1^2+Y_2^2+Y_3^2+Y_4^2).$$

While for (b), which allows for correlation between the continuous measures, the attributable deviance is the difference between

$$I.(Y_1+Y_2+Y_3+Y_4)+(Y_1+Y_2+Y_3+Y_4)^2 \text{ and } I.(Y_2+Y_3+Y_4)+(Y_1+Y_2+Y_3+Y_4)^2.$$

The results are summarised in:

edge	df	(a) univariate	(b) var homog
(I, Y_1)	2	144.63	9.52
(I, Y_2)	2	76.82	39.89
(I, Y_3)	2	425.48	60.23
(I, Y_4)	2	396.53	44.56

Not only are the magnitudes rather different, even the rank orderings determined by columns (a) and (b) are different.

In conclusion: The analysis of the iris data illustrates that data sets cannot necessarily be adequately described by a single model. The next stage in the analysis of the iris data might well be to return to the original data and check

Table 11.5.2: Smoking habits and personality traits of 384 college students. From Wermuth and Lauritzen (1990).

$A \setminus B$		neither smoked			one smoked			both smoked		
		X	Y	m	X	Y	m	X	Y	m
smoker	X	46.61			32.40			29.10		
	Y	20.94	48.58		10.95	28.53		21.35	46.55	
	m	19.45	25.27	11	23.05	24.60	20	22.00	25.80	60
quit	X	10.16			34.06			35.99		
	Y	15.18	32.89		28.88	61.72		21.43	39.26	
	m	19.17	23.58	12	22.05	25.33	21	22.26	25.74	43
never	X	19.82			28.52			27.32		
	Y	9.85	26.46		13.53	30.35		11.11	31.30	
	m	17.81	19.53	47	19.56	20.69	91	19.86	21.48	79

out the differing conclusions concerning the interaction structure of the four variables in each of the three species.

□

EXAMPLE 11.5.2 College smoking data. The next example is taken from Wermuth and Lauritzen (1990) who report an analysis of survey data from Spielberger on smoking habits and personality traits of 384 college students. There are four variables:

> A students' smoking status (smoker, quit smoking, never smoked);
> B parents' smoking habits (neither smoked, one parent smoked, both smoked); X trait anxiety; and Y trait anger.

The trait measures are continuous psychological variables, and the question of interest is how a student's smoking behaviour, which is discrete, interacts with these measures. It is presumed that parents' smoking behaviour may be an influence on the other three variables, but not conversely, so that we are interested in the chain model based on two blocks: $\{B\}$ and $\{A, X, Y\}$. As the conditioning block is only one dimensional, this regression model may be fitted in the joint distribution.

The sufficient statistics for all graphical and hierarchical mixed interaction models are a subset of the mean vectors, variance matrices and cell counts, for each cell in the classification determined by the discrete variables. They are given in Table 11.5.2, where m denotes the mean, and the entry in the (m, m) cell of the table is the cell count. At one extreme, there is the saturated model in which all parameters are free to vary, and at the other, the model of complete independence which indicates the total amount of variation there is to explain. It is generally the case that there are substantial numbers of parameters to index each interaction in mixed models; so to improve the efficiency of the search, our strategy is to choose a parsimonious hierarchical model from $CG(2,2)$ as a maximal model, and then to compute edge exclusion deviances from this choice.

An exploratory analysis: There are two obvious choices for a parsimonious maximal model: firstly, the model of variance homogeneity, and secondly, variance homogeneity together with additive effects of the discrete variables on the linear interaction parameters. An alternative characterisation of this latter model is as a model that only contains two-way interactions.

	model formula	dev	df	P-value
saturated model				
	$M_1 = A.B.(X+Y)^2$	0	0	
variance homogeneity				
	$A.B+(A.B).(X+Y)+(X+Y)^2$	31.75	24	0.133
additive effects				
	$M_2 = A.B+(A+B).(X+Y)+(X+Y)^2$	34.82	32	0.345
mutual independence				
	$A+B+X^2+Y^2$	251.78	45	

Both are acceptable, and we choose the one with least parameters. The edge exclusion deviances from the saturated model and this model of additive effects are

	M_1			M_2		
edge	dev	df	P-value	dev	df	P-value
(B,A)	41.41	24	0.015	17.88	4	0.001
(X,A)	21.08	18	0.276	0.71	2	0.702
(X,B)	28.09	18	0.061	7.15	2	0.028
(Y,A)	50.97	18	0.000	28.80	2	0.000
(Y,B)	17.56	18	0.485	1.51	2	0.470
(Y,X)	114.87	9	0.000	103.96	1	0.000

where $M_1 = A.B.(X+Y)^2$ and $M_2 = A.B+(A+B).(X+Y)+(X+Y)^2$. The pattern of exclusion deviances from these two maximal models tells a similar story, and the difference in the deviances between columns is consistent with noise. But reducing the degrees of freedom associated with each edge changes the observed P-value, and indicates that the $X.B$ interaction may be real. The objective of fitting parsimonious models is illustrated here: the genuine interaction has been isolated into specific degrees of freedom.

A fitted model: This analysis suggests a model with formula $A.B+(A.Y+B.X)+(X+Y)^2$ and chain independence graph

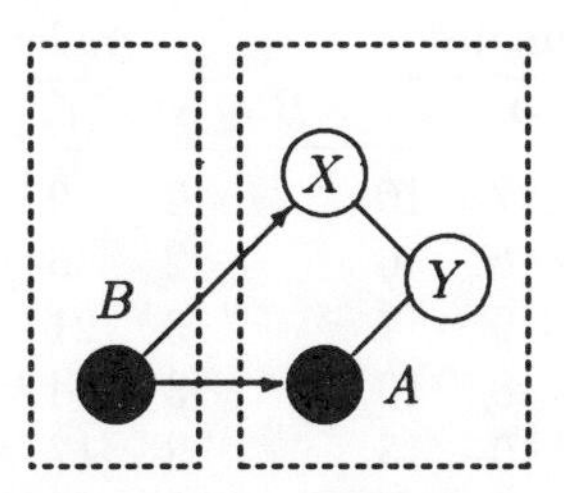

The deviance of this model is 27.64 on 32 df with a P-value of 0.688, and affords a good fit to the data. The computations are based on a sample size of 384 students; regarding the exclusion deviance differences above as measures of edge strength, it is seen that the interaction between the two psychological variables X and Y is strong, and those between A and B, and A and Y are moderate. The fitted counts, means and correlations are

$A\setminus B$		neither smoked			one smoked			both smoked		
		X	Y	m	X	Y	m	X	Y	m
smoker	X	1			1			1		
	Y	0.478	1		0.542	1		0.532	1	
	m	19.29	23.95	11	22.03	25.66	20	22.07	25.69	60
quit	X	1			1			1		
	Y	0.486	1		0.551	1		0.540	1	
	m	19.23	23.78	12	21.98	25.55	21	22.02	25.58	43
never	X	1			1			1		
	Y	0.420	1		0.476	1		0.467	1	
	m	17.83	19.96	47	19.80	20.91	91	19.94	20.97	79

The correlation coefficients between X and Y are fairly similar in each cell of the table. The mean levels of both X and Y change with A and B, in an additive fashion. Further analysis might investigate if the levels 'smoker' and 'quit smoking' of A may be combined, or the levels 'both' and 'one' of B. □

EXAMPLE 11.5.3 A clinical trial on rats. For this example, we take some data analysed by Edwards (1987, 1990) and previously reported by Morrison (1976) and Mardia *et al.* (1979). The randomised trial measured weight loss of male and female rats under three drug treatments. There are four variables, two are discrete: A, sex (male, female); B, treatment group (1, 2, 3); and two are continuous: X and Y, weight loss in the first and the second week of the

Table 11.5.3: A clinical trial on rats. From Morrison (1976).

males			females			males			females		
B	X	Y	B	X	Y	B	X	Y	B	X	Y
1	5	6	1	7	10	2	9	12	2	7	6
1	5	4	1	6	6	2	6	8	2	6	9
1	9	9	1	9	7	3	21	15	3	16	12
1	7	6	1	8	10	3	14	11	3	14	9
2	7	6	2	10	13	3	17	12	3	14	8
2	7	7	2	8	7	3	12	10	3	10	5

trial, respectively. The distribution of rats to sex and treatment group is fixed by the experimenter and models must respect this by conditioning on A and B. This is then a CG regression model, but since the conditioning variables are discrete it can be estimated in the joint CG distribution.

One might conjecture that X is independent of A given B, $X \perp\!\!\!\perp A|B$, indicating that weight loss does not depend on the sex of the rat, and, because of the temporal relationship, that Y is independent of both A and B given X, $Y \perp\!\!\!\perp (A, B)|X$, indicating that any drug effect in the second week is mediated by the response in the first week. The graph of this model is

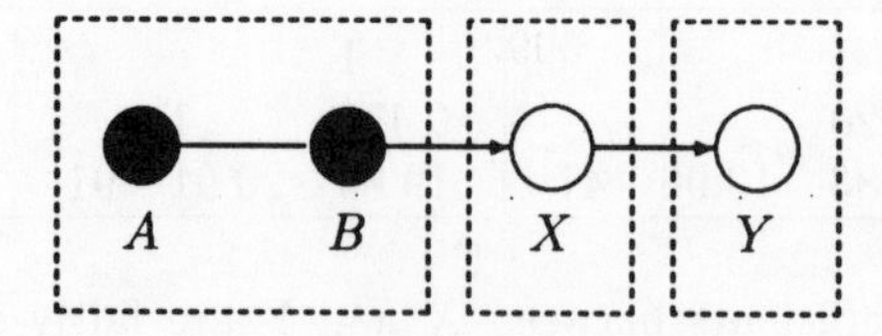

There is a line joining variables A and B because they are in the same block and their interaction is determined *a priori* by the experimental design.

A decomposition of a research hypothesis: An informative breakdown of this joint hypothesis is given in the next table. Clearly it is not possible to assume that the regression line of Y on X has common intercepts for each level of B, but furthermore the low P-value of the hypothesis $Y \perp\!\!\!\perp A|(B, X)$ in this table suggests that there may be sex effects that are not accounted for.

attributable deviance	dev	df	P-value
equal residual variances of Y on X for B	1.11	2	0.580
parallel regression lines for each B	1.45	2	0.512
common intercepts for each B	10.32	2	0.006
subtotal $Y \perp\!\!\!\perp B \mid X$	12.88	6	0.045
$X \perp\!\!\!\perp A \mid B$	4.37	6	0.629
$Y \perp\!\!\!\perp A \mid (B, X)$	19.86	9	0.019
Total $Y \perp\!\!\!\perp (A, B) \mid X$ & $X \perp\!\!\!\perp A \mid B$	37.11	21	

□

EXAMPLE 11.5.4 U.K. defence expenditure 1966-85. Table 11.5.4 relates to U.K. defence expenditure over a 20 year period, where the expenditure is counted at outrun prices, and is classified according to the category of expenditure C, where $C = 1$ denotes personnel expenditure, $C = 2$ denotes equipment expenditure, and $C = 3$ denotes other expenditure. How does the

Table 11.5.4: U.K. defence expenditure 1966-85.

year	C=1	C=2	C=3	total	year	C=1	C=2	C=3	total
66	856	926	390	2172	76	2864	2138	1156	6158
67	909	891	405	2205	77	3021	2565	1201	6787
68	997	852	422	2271	78	3293	2984	1178	7455
69	1030	822	415	2266	79	3912	3640	1625	9178
70	1182	836	485	2503	80	4556	4885	1741	11182
71	1399	952	477	2828	81	5058	5638	1910	12607
72	1552	1023	517	3092	82	5455	6297	2659	14412
73	1694	1153	637	3484	83	5726	6939	2822	15487
74	2026	1302	836	4164	84	5983	7838	3302	17122
75	2530	1792	1024	5346	85	6379	8193	3370	17943

pattern of spending change over the years? First, expenditure is logged and plotted against year in Figure 11.5.1. The plot reveals the interaction pattern quite clearly, personnel and other expenditure are increasing at a similar rate, but slower than expenditure on equipment.

A CG model: We consider models with the following basic structure: expenditure X is a continuous response variable; year Y is continuous with values in the range 1 to 20; and expenditure category C, is discrete with three levels; Y and C are both covariates. The point of interest is how expenditure varies with year and category, which we investigate in the form of regression

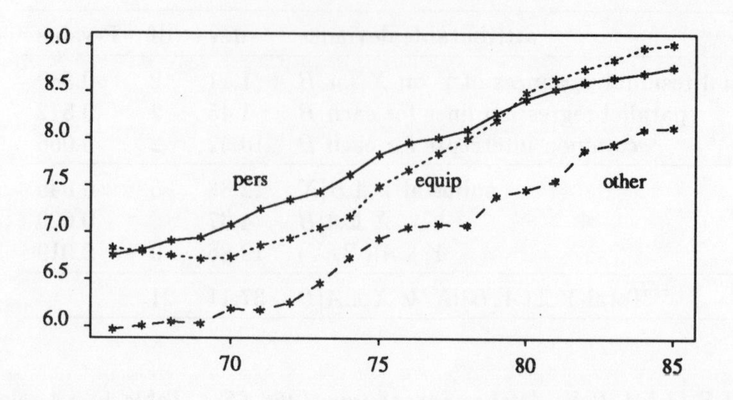

Figure 11.5.1: U.K. defence expenditure by category, 1966-85.

models for X on Y and C. These are based on the premise that given year and category, observations on expenditure are independent: thus in all there are 60 independent observations, and the sufficient statistics for the saturated model are the means and variances for each category:

$C = 1$

Y	33.250	
X	3.896	0.462
m	10.500	7.798
	Y	X

$C = 2$

Y	33.250	
X	4.697	0.703
m	10.500	7.687
	Y	X

$C = 3$

Y	33.250	
X	4.208	0.547
m	10.500	6.922
	Y	X

We wish to investigate the graphical structure

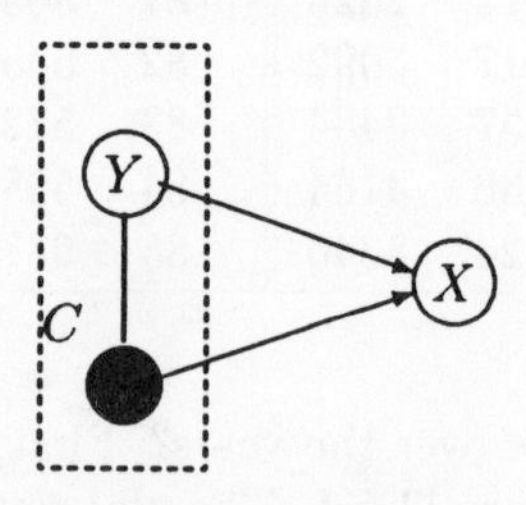

There is an edge between year and category as we have no interest in modelling this interaction, though because of the balanced nature of the data - 20 observations on each category and 3 observations on each year - its value is zero. We review a sequence of models, starting with the simplest, that give some insight into the structure of the data.

A sequence of mixed models: The model in which expenditure neither depends on year nor on category, $X \perp\!\!\!\perp (Y, C)$, has the formula $Y^2.C + X^2$

and a deviance of 235.03 on 7 df. The model of a common level and a common increase in expenditure over years is specified by $X \perp\!\!\!\perp C | Y$, with formula $Y^2.C + X^2 + X.Y$, and the deviance is 149.83 on 6 df. This model is characterised by X having identical regression lines on Y for each of the three levels of C, and is clearly untenable.

A more interesting model is one in which the different categories of expenditure have different intercepts with year but a common slope. The model formula $Y^2.C + X^2 + X.Y + X.C$ gives a deviance of 27.85 on 4 df. The value of the common regression coefficient is $\hat{\beta} = 0.128$ and the estimate of the common residual variance is $\hat{\sigma} = 0.025$. The model does not fit the data well, as is obvious from the plots of expenditure against year.

We fit a model in which each category is allowed a different regression coefficient on year, but with a common variance about the regression line. The regression estimates of this model are $\hat{\beta}_1 = 0.117$, $\hat{\beta}_2 = 0.141$, $\hat{\beta}_3 = 0.127$ and $\hat{\sigma} = 0.022$. The rate of increase for personnel and other expenditure, $C = 1$ and $C = 3$, are similar, while equipment expenditure rises at a somewhat faster rate.

The deviance of this model can be computed indirectly in several ways: most clearly, as a model of homogeneous partial variances, which means that the variance of X should be the same within each of the categories of expenditures, having adjusted for the effect of year. The residual sum of squares of X in each of the three categories is 0.106, 0.783 and 0.287, respectively, and the deviance for equality of variances is 18.57 on 2 df (it is assumed that the sample size is 20, to preserve deviance additivity). Thus a common value is not acceptable.

No single model fitted here entirely represents the data but we can conclude that the difference between personnel and other expenditure can be ascribed to chance, while the equipment expenditure has a relatively larger regression coefficient on year and a larger variance.

A continuous Gaussian model: An alternative form of analysis is to assume that the expenditure in each year is an observation on a trivariate Normal random vector. As before, the variable year is regarded as a continuous covariate, and the observed correlation and partial correlation matrices are

	Y	X_1	X_2	X_3
Y	1			
X_1	0.994	1		
X_2	0.972	0.975	1	
X_3	0.987	0.986	0.986	1

	Y	X_1	X_2	X_3
Y	1			
X_1	0.794	1		
X_2	−0.197	0.204	1	
X_3	0.384	0.073	0.686	1

respectively. Based on this sample of 20 observations the deviance 'to explain' from the model of complete independence ($Y^2 + X_1^2 + X_2^2 + X_3^2$) is 237.19 on 6 df. Further model fitting reveals two important interactions: (Y, X_1) and

(X_2, X_3), and that either the model $(Y+X_1)^2+(X_2+X_3)^2+Y.X_3$ (dev 2.78, 3 df) or the model $(Y+X_1)^2+(X_2+X_3)^2+X_1.X_3$ (dev 3.51, 3 df) fit well. They have graphs

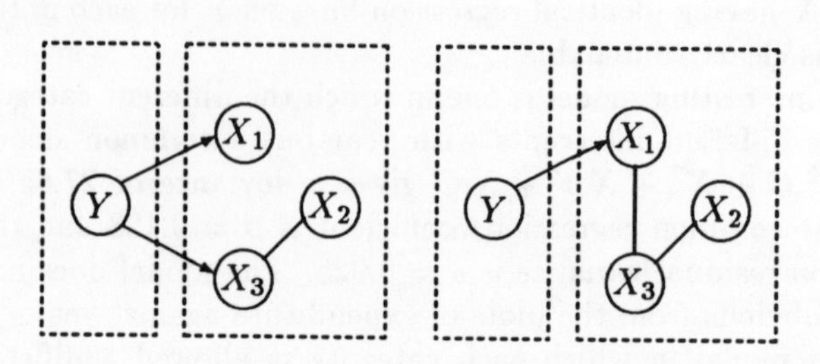

respectively.

A contrast: These two analyses are presented above to illustrate the effect of making different assumptions about the process generating the data.

The first analysis, in which the category of expenditure is treated as a discrete variable, makes the assumption that there are 60 independent observations as is in a straightforward univariate analysis of variance. This rules out dependence between successive years and dependence between expenditure categories. For instance, if working within the constraint of a fixed budget, one might expect the correlation to be negative. The second analyis allows for such a dependence and views the data as comprising 20 independent observations.

The foci of the two analyses are different. While both take year as fixed, in the mixed model the essential question of interest is the relationship between expenditure and year and how this varies with the category of expenditure. In the continuous model interest hinges on the conditional independence structure of the three expenditure variables. □

11.6 Exercises

1: A mixture of Normal distributions. Let Z be a Bernoulli random variable taking the values 0 and 1 with equal probabilities. Let the conditional density of (X, Y) in R^2 be bivariate Normal, $N(0, V)$, where if $Z = 0$, V is the 2×2 identity matrix and if $Z = 1$, V is $\begin{pmatrix} 1 & r \\ r & 1 \end{pmatrix}$. Find the conditional density of X given $Z = 0$ and given $Z = 1$, and find the marginal distribution of X. Find the conditional density of $T = X + Y$ given $Z = 0$ and given $Z = 1$. Show that while $f_X(x) = n(x; 0, 1)$, $f_T(t) = n(t; 0, 2)/2 + n(t; 0, 2 + 2r)/2$. Conclude that $f(t)$ is not $n(t; a, b)$ for any a or b.

2: Write out the density function of the graphical model in $CG(2,2)$ defined by $I_1 \perp\!\!\!\perp (Y_1, Y_2)|I_2$. If the discrete variables are both binary, how many parameters are set to zero?

3: Write out the density function in terms of the (λ, η, ψ) parameters for the model $Y_1^2 + I.Y_1.Y_2 + Y_2^2$. Show that the variances of Y_1 and Y_2 depend on I.

4: Give the model formulae for the the graphical models in $CG(2,2)$ with discrete variables A and B, and continuous variables, X and Y, and defined by the independences (i) $X \perp\!\!\!\perp A|B$, (ii) $Y \perp\!\!\!\perp A|(B, X)$, (iii) $Y \perp\!\!\!\perp B|X$, (iv) $Y \perp\!\!\!\perp (A, B)|X$ and $X \perp\!\!\!\perp A|B$.

5: Interpret the model Y^2 for (I, Y) in $CG(1,1)$ in terms of equi-probability. Take I to be a binary random variable.

6: Express the canonical parameter α of the full CG density function in terms of the moments parameter $\{p, \mu, V\}$.

7: Interpret the model $I_1.I_2 + (I_1 + I_2).Y^2$ in terms of linearities in the inverse variance of Y.

8: If (I, Y) has a CG distribution show that the unconditional mean and variance of Y are

$$E(Y) = \sum_i p(i)\mu(i) \quad \text{and} \quad \text{var}(Y) = \sum_i p(i)V(i) + \sum_i p(i)(\mu(i) - EY)^2.$$

9: Extend the $CG(1,1)$ log-likelihood function of Section 11.3.1 to the case where the discrete variable takes three levels rather than two.

10: Calculate the observed and expected Fisher information matrices for the parameters in the $CG(1,1)$ density and in the logistic-Normal density discussed in Section 11.1.

11: Show that the two models $(I_1 + I_3.I_4).I_2$ for $CG(4,0)$ and $(Y_1 + Y_2)^2 + (Y_2 + Y_3 + Y_4)^2$ for $CG(0,4)$ have the same independence graph if the type of vertex is discounted.

12: Consider the four graphical models in $CG(2,2)$ defined by applying or not the independences $I_1 \perp\!\!\!\perp Y_1|\text{rest}$ and $I_2 \perp\!\!\!\perp Y_2|\text{rest}$. Find the model formula for each. If I_1 and I_2 are binary random variables how many parameters, not including constraints, are set to zero.

13: *Miners' wheeze.* This data set relates to breathlessness and wheeze in a national sample of coal miners, see Plackett (1974) [1] for more details.

[1] Reproduced by permission from Plackett, R.L. (1974) *The Analysis of Categorical Data.* O.U.P.

Breathlessness	Yes		No		Total
Wheeze	Yes	No	Yes	No	
Age					
20-	9	7	95	1841	1952
25-	23	9	105	1654	1791
30-	54	19	177	1863	2113
35-	121	48	257	2357	2783
40-	169	54	273	1778	2274
45-	269	88	324	1712	2393
50-	404	117	245	1324	2090
55-	406	152	225	967	1750
60-	372	106	132	526	1136
Totals	1827	600	1833	14022	18282

Each age group is indexed by lower limit, e.g. 35-= 35-39 and the observed response is a 2×2 classification of breathlessness and wheeze. Due to the grouping on the age variable, methods have to be somewhat approximate, and a possible form of analysis is to regard the table as a three-way cross-classification, in which the covariate age is an explanatory variable and breathlessness and wheeze are response variables.

Show that if the model for the uncondensed data is a CG regression then the interaction between the covariate and the responses should be at most quadratic in the covariate. Fit an appropriate model.

14: Recall the rats data from Example 11.5.3. Verify that the edge exclusion deviance differences against the three choices of maximal model: $M_1 = A.B.(X+Y)^2$, $M_2 = A.B.(X+Y)+(X+Y)^2$ and $M_3 = A.B+(A+B).(X+Y)+(X+Y)^2$; are

	M_1			M_2			M_3		
edge	dev	df	P-value	dev	df	P-value	dev	df	P-value
BA	41.41	24	0.015	20.95	12	0.051	17.88	4	0.001
XA	21.08	18	0.276	3.05	6	0.803	0.71	2	0.702
XB	28.09	18	0.061	9.49	6	0.148	7.15	2	0.028
YA	50.97	18	0.000	30.65	6	0.000	28.80	2	0.000
YB	17.56	18	0.485	3.36	6	0.762	1.51	2	0.470
YX	114.87	9	0.000	105.09	1	0.000	103.96	1	0.000

15: One interesting investigation is the relationship between the u-term expansion of $\log f_I = \log p$ in the marginal distribution of I and the λ-terms in the expansion of $\log f_{IY}$. It is easy to see that in general they have different meanings, characterised as the difference between measures of marginal and partial interaction, and have to be distinguished notationally. Though, of course, if $m(i)$ and $V(i)$ do not vary with i, so that $I \perp\!\!\!\perp Y$, then, apart from u_ϕ and λ_ϕ, the u-terms and the λ-terms are identical.

16: Consider the deviance for the combined test of independence given in Section 11.3, item 2. Show that if the marginal distributions of each of the four variables are fixed, then the discrete component of N^{-1}dev converges to a quantity that is bounded above by $-\log \min_{ij}(p_i, p_j)$ whatever the interaction between I_1 and I_2. On the hand, show that the continuous component is unbounded and may dominate the expression for the deviance.

Chapter 12

Decompositions and Decomposability

In this last chapter we return to some of the theoretical issues raised in Chapter 3.

A notable success of the theory of graphical models was the identification of those log-linear models which possess analytic maximum likelihood estimates: decomposable models. For instance, it turns out that the model based on the class of density functions for which the independence graph is a Markov chain, as in the example below, has direct estimates, while that based on the chordless 4-cycle does not. Why is this? Decomposable density functions admit predictors and maximum likelihood estimates that can be computed without numerical iteration, which allows their properties to be studied analytically. In fact decomposable models possess many other interesting properties, and one can argue that a substantial amount, if not all, of the information about the interaction structure of a k-dimensional random vector can be extracted solely by fitting the decomposable models.

A Markov chain has the striking property that its joint distribution can be simply elucidated by conditioning on the immediate past. For example, a finite fragment of the independence graph of a simple Markov chain is

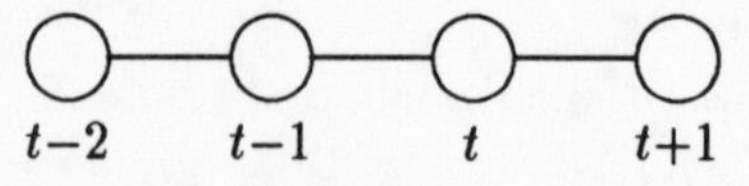

and the joint density of these variables factorises according to

$$f_{t-2,t-1,t,t+1} = f_{t+1|t} f_{t|t-1} f_{t-1|t-2} f_{t-2}.$$

Not only is there a backwards factorisation starting from the right in the graph, but also a forwards factorisation, based on conditioning on the immediate future, and starting from the left. In contrast starting from the middle does not result in an entire factorisation of the density function.

The existence of such a factorisation is interesting for several diverse reasons: (a) restricting the conditioning set simplifies the interpretation of the model, in particular the interaction structure; (b) it permits models to be specified in terms of conditional probabilities rather than joint probabilities; (c) the inference procedures based on the likelihood function for models that admit entire factorisations are relatively simple; (d) the high dimensional joint distribution, or observed data set, can be reduced into lower dimensional components, an important practical feature; and (e) such a factorisation is a way to relax the strict assumption of mutual independence in order to establish central limit type theorems. These remarks are extended to arbitrary finite graphs by developing a theory of decompositions. One part of the information carried by an independence graph is whether the distribution corresponding to the graph can be factorised, or decomposed, into simpler marginal distributions. Families of fully factorisable density functions with a given independence graph are known as decomposable models. The characteristic property of their graphs is triangulation and we see that the class of triangulated graphs is a natural generalisation of the graph for the Markov chain above. The theory also provides a more general criterion for collapsibility that determines if collapsing a high dimensional distribution over certain variables could lead to misleading interpretations between the remaining variables.

We regard the treatment given here as an introduction to more advanced material available in the research literature. In particular, we return to the theory for unmarked and undirected graphs, even though extensions of the notions of decompositions and decomposability exist for mixed recursive models. For simplicity, the discussion is generally phrased in terms of factorisation properties of a given density function, but it is clear from the examples and the propositions that these are properties of classes of density functions, which to be specific, correspond to log-linear and graphical Gaussian models.

12.1 Factorisation

The independence graph dramatically clarifies the interaction structure of a set of variables $X_K = (X_1, X_2, \ldots, X_k)$. The separation theorem has shown the correctness of the direct intuitive interpretation of this information: that non-adjacent variables are independent conditional only on a separating set. The graph highlights sets of variables which interact together and may suggest ways in which the probability model may be factorised. It is this latter aspect that is of interest here.

EXAMPLE 12.1.1 Take $k = 4$ and consider a general class of density functions for $X = (X_1, X_2, X_3, X_4)$, of the form

$$\log f_{1234}(x) = h_{12}(x) + h_{234}(x),$$

where the functions h are coordinate projection functions, that is, $h_{12}(x)$ is only a function of (x_1, x_2) and $h_{234}(x)$ is a function only of (x_2, x_3, x_4). The h functions are called ***interactions*** or ***interaction terms.*** It is assumed that these functions are well behaved enough to ensure that $f_{1234}(x)$ is a probability density function, and x takes values in the Cartesian product of the four sets containing the permissible values for each coordinate x_i. The factorisation criterion implies that $1 \perp\!\!\!\perp 3 | \{2, 4\}$ and $1 \perp\!\!\!\perp 4 | \{2, 3\}$ and consequently the independence graph is

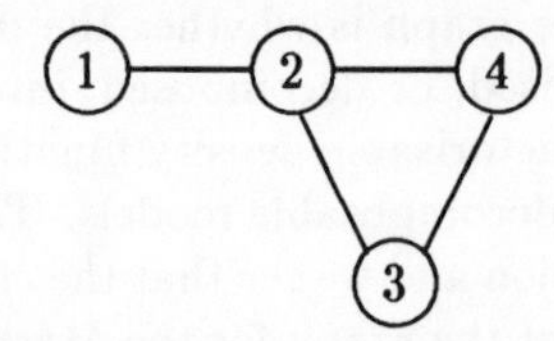

Now consider factorising the density into the product of marginal distributions. The marginal density of (X_1, X_2), obtained by integrating $f_{1234}(x)$ over (x_3, x_4), is

$$\begin{aligned} f_{12}(x_1, x_2) &= \exp(h_{12}(x_1, x_2)) \int \exp(h_{234}(x_2, x_3, x_4)) dx_3 dx_4, \\ &= \exp(h_{12}(x_1, x_2) + g_2(x_2)), \end{aligned}$$

or more succinctly as $f_{12} = \exp(h_{12} + g_2)$, where g_2 is a function only of x_2. A similar integration gives

$$f_{234} = \exp(g_2' + h_{234}),$$

where g_2' is a function only of x_2, as well. Now by integrating this marginal density over variables 3 and 4, the density of X_2 is

$$f_2 = \exp(g_2' + g_2).$$

Finally, substituting for the interaction terms h_{12} and h_{234} in the interaction expansion gives the factorisation of the joint density in terms of the marginal densities

$$f_{1234} = f_{12} f_{234} / f_2.$$

As this derivation shows, the exact nature of the interaction terms is not important. To get the factorisation all that matters is how the interaction expansion is structured. □

There are several points to make here. Firstly, the factors in the factorisation are not arbitrary functions but are marginal probability density functions; secondly, it is of interest to know if the factorisation is unique, or whether it depends on the order in which the pairwise independences are applied; and thirdly, how can we characterise graphs for which the factorisation is *full* in the sense that an independence statement corresponding to every pair of non-adjacent vertices is used in the factorisation of the density function? Consider the next example.

EXAMPLE 12.1.2 Suppose that (X_1, X_2, X_3, X_4) has the independence graph

There are three pairs of non-adjacent vertices: (1,3), (1,4), (2,4), corresponding to the three pairwise conditional independence statements $1 \perp\!\!\!\perp 3 | \{2,4\}$, $1 \perp\!\!\!\perp 4 | \{2,3\}$, and $2 \perp\!\!\!\perp 4 | \{1,3\}$. To simplify the joint density, f_{1234}, apply these, in the following order: firstly, application of

$$(1,3) \quad : \quad 1 \perp\!\!\!\perp 3 | \{2,4\} \text{ gives } f_{1234} = f_{124} f_{234} / f_{24}.$$

Now, consider the pair (1,4). The independence $1 \perp\!\!\!\perp 4 | \{2,3\}$ cannot be applied directly to simplify f_{124} because vertex 3 is required in the conditioning set. However, the global Markov property of the independence graph implies $1 \perp\!\!\!\perp 4 | 2$, which can be used:

$$(1,4) \quad : \quad 1 \perp\!\!\!\perp 4 | 2 \text{ gives } f_{1234} = f_{12} f_{24} f_{234} / f_2 f_{24} = f_{12} f_{234} / f_2.$$

Finally, the non-adjacent pair (2,4), needs

$$(2,4) \quad : \quad 2 \perp\!\!\!\perp 4 | 3 \text{ to give } f_{1234} = f_{12} f_{234} / f_2 = f_{12} f_{23} f_{34} / f_2 f_3.$$

An independence statement corresponding to every pair of non-adjacent vertices has been used, and so the density is fully factorised. It is a simple matter to check that it is unique: each of the 3!=6 orderings of the non-adjacent pairs leads to the same answer.

A more direct derivation, discussed later, is to apply the independence statements inherent in the graph, in the form of the local Markov property, to the recursive factorisation

$$f_{1234} = f_{1|234} f_{2|34} f_{3|4} f_4.$$

For instance, $1 \perp\!\!\!\perp \{3,4\} | 2$ implies $f_{1|234} = f_{1|2}$. Continuing gives $f_{1234} = f_{1|2} f_{2|3} f_{3|4} f_4$, which can then be expressed as the product of marginal densities. □

The next example shows there are graphs for which the density factors into the product of certain marginals but not sufficiently to exhaust the list of conditional independences.

EXAMPLE 12.1.3 Consider the chordless 4-cycle

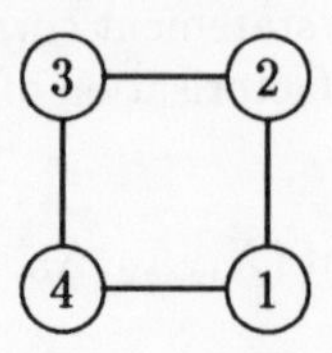

There are just two pairs of non-adjacent vertices, $(1,3)$ and $(2,4)$, corresponding to the conditional independence statements $1 \perp\!\!\!\perp 3 | \{2,4\}$ and $2 \perp\!\!\!\perp 4 | \{1,3\}$. Now application of

$$(1,3) \quad : \quad 1 \perp\!\!\!\perp 3 | \{2,4\} \text{ gives } f_{1234} = f_{124} f_{234} / f_{24}.$$

But it is not possible to use the non-adjacency of the pair $(2,4)$ to simplify the right hand side, because none of the density functions containing $\{2,4\}$ also contains a separating set for 2 and 4. Thus this factorisation into marginal densities is not full.

If we start with the other independence statement we get the factorisation

$$(2,4) \quad : \quad 2 \perp\!\!\!\perp 4 | \{1,3\} \text{ gives } f_{1234} = f_{123} f_{134} / f_{13},$$

and reach the same conclusion concerning further simplification. In consequence, this density neither factorises uniquely nor fully into the product of marginal density functions.

Note of course, that any member of the class of densities

$$\log f_{1234}(x) = h_{12}(x) + h_{23}(x) + h_{34}(x) + h_{14}(x)$$

has the chordless 4-cycle as its independence graph, and while this does not have a full factorisation into marginal density functions, there is of course, a factorisation into coordinate projection functions. □

All four dimensional independence graphs are fully factorisable apart from the chordless 4-cycle.

EXAMPLE 12.1.4 The following are examples of independence graphs for five dimensional density functions that are fully factorisable:

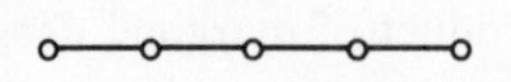

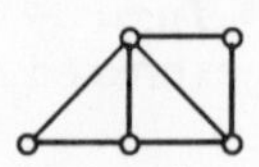

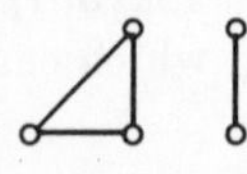

None of the following graphs is fully factorisable:

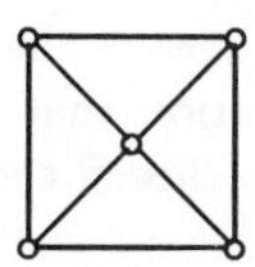

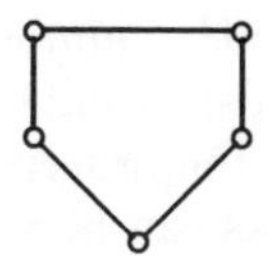

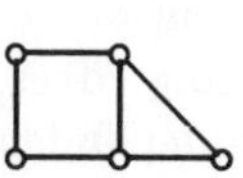

□

Simplifying the recursive factorisation

In the previous examples, every step in the factorisation is obtained by applying a pairwise conditional independence statement, inherent in the definition of the independence graph. There is a symmetry between the two vertices in each pair, and a non-symmetric approach is to use the independences generated by the equivalent local Markov representation of the independence graph. Suppose that we are given a specified order for the vertices: $1, 2, \ldots, k$, and consider the *recursive factorisation* identity

$$f_{12\ldots k} = f_{1|2\ldots k} f_{2|3\ldots k} \cdots f_{k-1|k} f_k,$$

satisfied by any joint probability density function. In applying this factorisation it is possible for vertex 1 to be conditionally independent of $k-1$ other vertices, vertex 2 to be conditionally independent of $k-2$ others, and so on, and furthermore

$$k-1+k-2+\cdots+3+2+1 = \binom{k}{2},$$

which is the maximum possible number of conditional independence statements.

EXAMPLE 12.1.5 Consider the simple Markov chain

determined by 6 pairwise conditional independences. Read the graph from left to right, according to the local Markov property, and apply the independences to simplify the recursive factorisation identity:

$$f_{12345} = f_{1|2345} f_{2|345} f_{3|45} f_{4|5} f_5.$$

The first independence is $1 \perp\!\!\!\perp \{3,4,5\}|2$, and implies $f_{1|2345} = f_{1|2}$. Repeated application gives

$$f_{12345} = f_{1|2} f_{2|3} f_{3|4} f_{4|5} f_5.$$

Now exactly three pairwise independences are used to simplify $f_{1|2345}$ to $f_{1|2}$, two to simplify the $f_{2|345}$ to $f_{2|3}$ and one to simplify $f_{3|45}$ to $f_{3|4}$. In all $3+2+1=6$ and this is equal to the number of missing edges in the graph. Note that the order $5,4,3,2,1$ also results in a simplification that utilises all the missing edges, but that if the ordering $3,2,4,1,5$ is chosen then only $2+1=3$ independence statements can be applied:

$$f_{12345} = f_{3|2415}f_{2|415}f_{4|15}f_{1|5}f_5 = f_{3|24}f_{2|41}f_{4|15}f_{1|5}f_5.$$

Not all orderings of the vertices will achieve a full factorisation. □

EXAMPLE 12.1.6 Consider the chordless four cycle

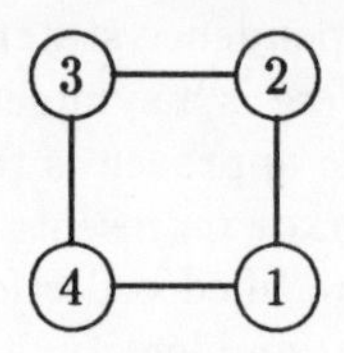

again. As one can check, whatever the ordering of the vertices, it is only possible to apply one of the two possible independences to any factorisation. □

An extension problem

One might speculate about the converse procedure to factorisation: when is it the case that the product of marginal density functions is a joint density function?

EXAMPLE 12.1.7 Suppose that we are given the joint density function, f_{1234}, of the four dimensional random vector $X = (X_1, X_2, X_3, X_4)$, and that we calculate from this the two dimensional margins: f_{12}, f_{23}, f_{34} and f_{14} together with all one dimensional margins. Consider the functions

$$g_{1234} = f_{12}f_{23}f_{34}/f_2f_3 \text{ and } h_{1234} = f_{12}f_{23}f_{34}f_{14}/f_1f_2f_3f_4.$$

Is either g or h a proper joint density function for X? A partial answer is that we can obtain g by application of conditional independence statements in the graph of Example 12.1.2 to f_{1234}, so that g is alright. This works because that graph is fully factorisable. However, the independence graph corresponding to h is the chordless 4-cycle, and as we saw in Example 12.1.3, this does not lead to h when factorised. Thus we suspect that h is not a proper density function, in general. However, certain examples exist when h is a density function, for instance, the so-called perfect tables discussed by Darroch (1976) and Whittemore (1978), but this is a too deep a matter to pursue further here. □

12.2 Partial Factorisations: Decompositions

While a density function may not admit a full factorisation a ***unique*** partial reduction is sometimes possible.

EXAMPLE 12.2.1 In the independence graph

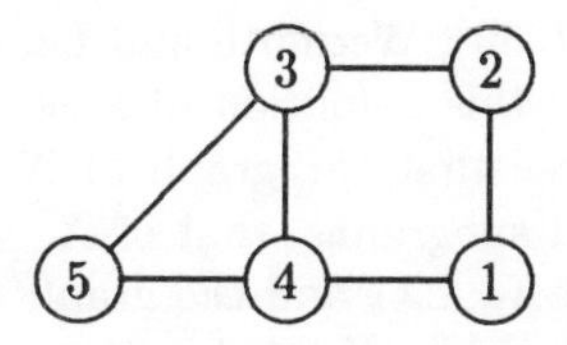

$5 \perp\!\!\!\perp 1|\{2,3,4\}$ and $5 \perp\!\!\!\perp 2|\{1,3,4\}$ so that the joint density admits the factorisation

$$f_{12345} = f_{1234} f_{345} / f_{34}.$$

Now using either of these two pairwise independence statements does not prevent, nor is prevented by, the application of any other independence, which is in contra-distinction to the application of either the independence between X_2 and X_4 or that between X_1 and X_3, as discussed in Example 12.1.3 above. Hence, application of $5 \perp\!\!\!\perp 1|\{2,3,4\}$, for instance, leads to a unique factorisation, while $3 \perp\!\!\!\perp 1|\{2,4,5\}$ does not. □

This lack of interference is formalised by the notion of a decomposition.

Definition. There ***exists a decomposition*** of the random vector X, or equivalently, X is ***reducible***, if and only if there exists a partition of X into (X_a, X_b, X_c) such that

(i) $X_b \perp\!\!\!\perp X_c | X_a$ and neither b nor c empty; and

(ii) the subgraph on a, in the independence graph of X, is complete.

If so, the ***components*** of X are $X_{ab} = (X_a, X_b)$ and $X_{ac} = (X_a, X_c)$. If such a decomposition does not exist then X is said to be ***irreducible***. □

Equivalently, we say that the density function, f_K, is reducible to f_{ab} and f_{ac} and that the independence graph G is reducible to subgraphs on $a \cup b$ and $a \cup c$. We see later in this section, that the independence graph of each component is the corresponding induced subgraph. The essence of this definition is the requirement that the separating set a be complete. In Example 12.2.1 above, the conditioning set for the independence $5 \perp\!\!\!\perp \{1,2\}|\{3,4\}$ is complete, while that for the independence between X_2 and X_4 contains $\{1,3\}$ and is not complete.

The concept of a decomposition is implicit in the construction of the fitted values for certain log-linear models, by Andersen (1974) and Sundberg

(1975); it is also implicit in the definition of decomposability proposed by Haberman (1974), with which it must not be confused; and it is made explicit by Lauritzen (1982); see also, and Leimer (1989). Historically, the term decomposable has come to define the class of independence graphs with full factorisations discussed above. Nor should a decomposition be confused with the notion of a 'reducible ordering', which is equivalent to the term 'decomposable', see Wermuth (1980), Wermuth and Lauritzen (1983).

Further motivation for this definition of a decomposition comes from the following remark: suppose that the graph of $X = (X_a, X_b)$ separates into two entirely disconnected subgraphs, that of X_a and that of X_b. The global Markov property implies $X_a \perp\!\!\!\perp X_b$ and the graph thereby concentrates attention on the interactions within X_a and within X_b. Expressed in the form $f_{b|a} = f_b$, the independence $a \perp\!\!\!\perp b$ asserts that inference concerning the vector X_b need not be conditioned on the value of X_a, and so statements about X_b are global rather than local. The analysis of a $p+q$-dimensional object is reduced to the examination of one p-dimensional and one q-dimensional object, a much simpler task. The tautology $f_{ab} = f_a f_{b|a}$ is replaced by the factorisation $f_{ab} = f_a f_b$. More generally when $K = a \cup b \cup c$, if $b \perp\!\!\!\perp c | a$, the joint density factorises into $f_K = f_{ab} f_{ac} / f_a$ and the $p+q+r$-dimensional graph is reduced to the analysis of one $p+q$- and one $p+r$-dimensional graph. The independence $f_{b|a \cup c} = f_{b|a}$ asserts that inference concerning the vector X_b need not be conditioned on X_c. The requirement that the subgraph on a is complete implies that there is no further independence constraint on the elements of X_a, so that this factorisation contains all the information about the joint distribution of (X_a, X_b, X_c).

EXAMPLE 12.2.2 Take $k = 5$ again and suppose that the joint density has the structure $\log f_{1234} = h_{123} + h_{134} + h_{45}$, with the independence graph

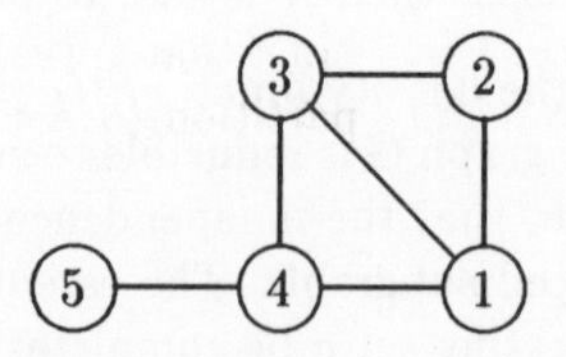

Then X is reducible to (X_{123}, X_{1345}), because in the partition $c = \{2\}$, $b = \{4,5\}$ and $a = \{1,3\}$, neither b nor c is empty and the subgraph on a is complete.

Consider the $a \cup b = \{1,3,4,5\}$ margin. The subgraph on $a \cup b$ is

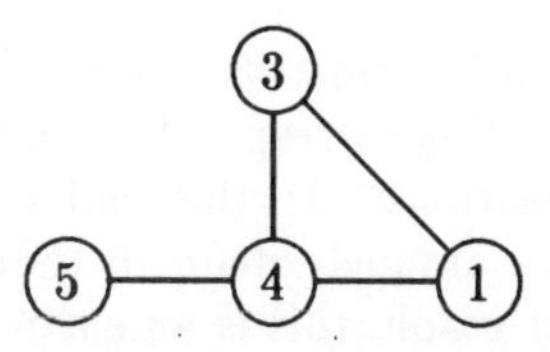

But application of the separation theorem to the full independence graph reveals the marginal conditional independences

$$5 \perp\!\!\!\perp 1 \mid \{3,4\} \text{ and } 5 \perp\!\!\!\perp 3 \mid \{1,4\}.$$

Consequently the independence graph of X_{1345} is identical to that of the induced $\{1,3,4,5\}$ subgraph from the full graph.

This component is itself reducible to (X_{134}, X_{45}), neither of whose components are further reducible. Thus the irreducible components of X_{12345} are X_{123}, X_{134} and X_{45}, and the density function factorises into

$$f_{1234} = f_{123} f_{134} f_{45} / f_{13} f_4.$$

In fact, it is easy to check that this can be achieved as a full factorisation of the joint density. □

In this last example we easily computed the independence graph of a component of a reduction of X, and found the independence graph to be the same as the induced subgraph. Recall that the induced subgraph for a subset of vertices is the graph with edges given by retaining just those edges which have both end-points in the subset. Is this true generally: if X is reducible to a component X_d is the independence graph of X_d the same as the induced subgraph on d?

Proposition 12.2.1 *If X has the independence graph G and is reducible to components X_d and X_e then the independence graph of X_d is no larger than the induced subgraph of G on d, in the sense that every missing edge is preserved.*

Proof: Take any two vertices i and j in d that are non-adjacent with respect to G. As X is reducible there exists a partition $\{a,b,c\}$ of K where a is complete in G and separates b and c. We may take $d = a \cup b$. As a is complete all its vertices are adjacent, and so i and j cannot both be members of a. Rewrite the pairwise independence $i \perp\!\!\!\perp j \mid K \setminus \{i,j\}$ as $i \perp\!\!\!\perp j \mid \{a \cup b \cup c\} \setminus \{i,j\}$, and note by construction that

$$(a \cup b) \setminus \{i,j\} \text{ separates } i \text{ and } j \text{ in } G.$$

By the global Markov property $i \perp\!\!\!\perp j \mid (a \cup b) \setminus \{i,j\}$ and c is no longer in the conditioning set. The edge (i,j) missing from G is still missing in the independence graph on d. □

12.3 Irreducible Components

We show that any joint density function has a unique partial factorisation in terms of the marginal densities corresponding to the irreducible components of a sequence of decompositions. To this end we must first show that the irreducible components are uniquely defined; though as the components are determined from the given graph, this is an entirely graph-theoretic matter.

EXAMPLE 12.3.1 Let us look at Example 12.1.1 again. The choice of subsets $a = \{2\}$, $b = \{1\}$ and $c = \{3,4\}$ gives a decomposition, for neither b nor c is empty, nor are any elements of c adjacent to those of b, and (trivially) a is complete. The independence graph decomposes from

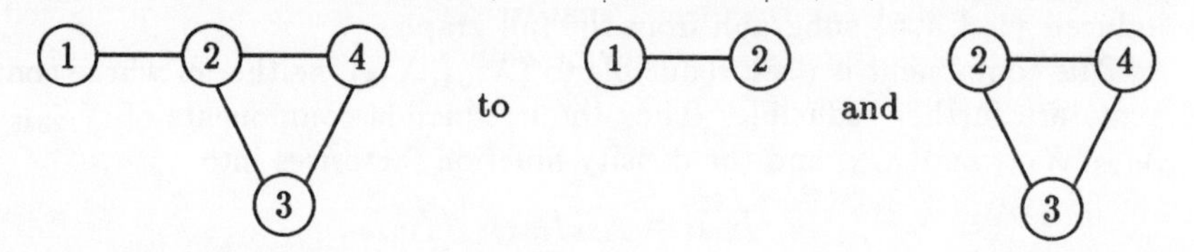

The components of this decomposition are (X_1, X_2) and (X_2, X_3, X_4) and neither component is further reducible.

But the choice of partition is not unique; for instance the partition $a = \{2,3\}$, $b = \{1\}$ and $c = \{4\}$ also works. The independence graph reduces from

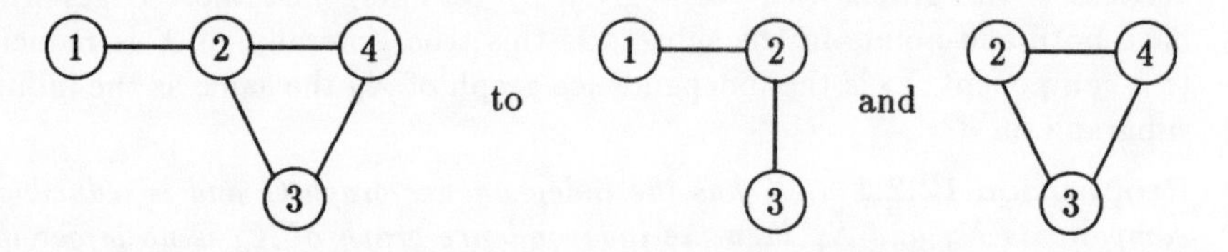

and the component $(X_a, X_b) = (X_1, X_2, X_3)$ is further reducible. If this is reduced as well then the irreducible components are

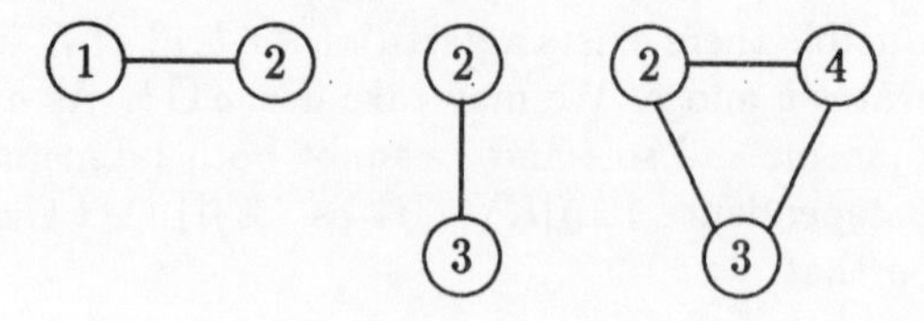

However (X_2, X_3) is a proper subset of the component (X_2, X_3, X_4). □

To avoid this possible non-uniqueness we take the maximal irreducible components.

Definition. The random vectors $X_{d_1}, X_{d_2}, \ldots, X_{d_m}$ are the *maximal irreducible components* of X if and only if

(i) each vector X_{d_i} is an irreducible component of X;

(ii) no subset, d_i, is a proper subset of any other, d_j; and

(iii) $d_1 \cup d_2 \cup \ldots \cup d_m = K$.

□

The requirement (ii), that the subsets are *pairwise incomparable*, ensures that they are maximal and eliminates any 'nesting' as exemplified above. The intersection of any pair is either empty or has a complete subgraph. Now it is conceivable that the set of maximal irreducible components produced by one sequence of decompositions may be different, even entirely different, from those produced by another. To see that this is not the case we first prove that irreducible components are preserved under decomposition.

Lemma 12.3.1 *An irreducible component of a random vector X deriving from one sequence of decompositions is always preserved in any other sequence of decompositions of X.*

Proof: We have to show that any given irreducible component of X, X_d, with $d \subseteq K$, is always entirely contained in at least one component resulting from any other arbitrary decomposition. If X is irreducible there is nothing to prove. If X is reducible there exists a partition of K into $\{a, b, c\}$ with b and c non-empty and non-adjacent and such that the subgraph of a is complete. The components of the decomposition have coordinates $a \cup b$ and $a \cup c$. Consider the intersection of d with this partition: $\{a \cap d, b \cap d, c \cap d\}$. The subgraph on $a \cap d$ is complete because a is complete, and the vertices of $b \cap d$ and $c \cap d$ are non-adjacent. Hence if both $b \cap d$ and $c \cap d$ are non-empty, X_d is reducible. But X_d is irreducible by assumption, so that at least one of these subsets must be empty. Suppose it is $c \cap d$. Then $d \subseteq a \cup b$, and so X_d is preserved in the arbitrary decomposition. □

Given the components X_d and X_e, say, of a decomposition of X, or of a component of X, then it is possible to deduce the partition that was used: the separating set is just the intersection, $d \cap e$, so that the partition must be $\{d \cap e,\ d \backslash (d \cap e),\ e \backslash (d \cap e)\}$. Furthermore the subgraph on a is complete. We can now prove the main result.

Proposition 12.3.2 Irreducible component factorisation. *The maximal irreducible components of X corresponding to the subsets $\{d_1, d_2, \ldots, d_m\}$ are unique and the density function of X, f_K, factorises uniquely into*

$$f_K = f_{d_1} f_{d_2} \cdots f_{d_m} / g$$

where the function g is a product of marginal density functions, $g = \prod f_a$, in which each subset a is an intersection of irreducible components, and is complete.

Proof: If X is irreducible then the assertion is trivial. Otherwise if X is reducible then it can be reduced to the components (X_a, X_b) and (X_b, X_c), say, and the density of X factorises into

$$f_K = f_{ab} f_{ac} / f_a.$$

Both terms in the numerator are density functions of components of X and the denominator corresponds to a complete subgraph. Continuing to reduce these components leads to the factorisation of the form asserted.

To show that this factorisation is unique suppose that $\{d_1, d_2, \ldots, d_m\}$ and $\{e_1, e_2, \ldots, e_n\}$ are two expansions into maximal irreducible components resulting from two different sequences of decompositions of K. Now d_1 is irreducible, so by the previous proposition, it is preserved in the decomposition of K to $\{e_1, e_2, \ldots, e_n\}$. Hence $d_1 \subseteq e_i$ for some i; but as the maximal irreducible components cannot nest, $d_1 = e_i$ for some i. Thus, for all subsets d_j in the first decomposition, there exists the same subset in the second decomposition. The converse holds by symmetry, so that it follows that the two factorisations are one and the same.

Finally to show that the function g is determined uniquely, suppose g_d and g_e are the functions emanating from two different sequences of decompositions into irreducible components. Then

$$f_{d_1} f_{d_2} \cdots f_{d_m} / g_d = f_K = f_{e_1} f_{e_2} \cdots f_{e_m} / g_e,$$

but as the numerators are identical, so are the denominators. □

EXAMPLE 12.3.2 The irreducible components of any density with an interaction structure of the form

$$\log f_{123456} = h_{12} + h_{23} + h_{34} + h_{14} + h_{125} + h_{16}$$

are $\{\{1, 2, 3, 4\}, \{1, 2, 5\}, \{1, 6\}\}$ and the density function factorises into $f_{123456} = f_{1234} f_{125} f_{16} / f_{12} f_1$. Note this is not a product of marginal density functions on cliques. □

Irreducible graphs in low dimensions

The irreducible graphs occupy a special place within the class of all graphs: as any other graph may be composed from these they have a status similar to that enjoyed by the primes among the integers.

Table 12.3.1: Irreducible graphical models in four and five dimensions.

dimension	complete graph	complete permutations	incomplete graph	incomplete permutations
4	ϕ	1		3
		4		
		6		
	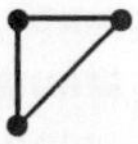	4		
	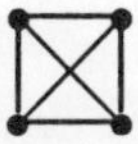	1		
5	ϕ	1		15
		5		10
		10	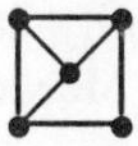	30
		10	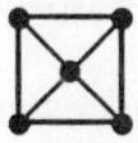	15
	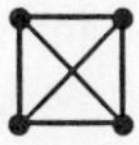	5		12
		1		

Irreducible components are either complete or incomplete. The complete components are the building blocks of the decomposable models, and are simple to interpet, factorise and fit. The incomplete irreducible components are rather more complicated, as instanced by the discussion about factorising the chordless 4-cycle, and Table 12.3.1 gives an idea of the varieties in which they arrive.

In k dimensions there are 2^k complete irreducible components but it seems hard to find an analytic expression for the number of incomplete irreducible components, though it is evident that this number increases rapidly as the dimension increases.

12.4 Decomposability

One of the outstanding successes of the theory of graphical models is the identification of those log-linear models for contingency tables which have direct maximum likelihood estimates: they are competely characterised as that subset of the graphical log-linear models with triangulated independence graphs. Decomposable models are those probability models with the following properties.

- The models are *multiplicative* in the sense that every density function in the model fully factorises into the product of marginal density functions. Such a factorisation is unique and entirely describes the properties of the model.

- The joint density function factorises into the product of marginal density functions on cliques; equivalently, the irreducible components are complete.

- The models are *recursive* in the sense that their vertices can be ordered so as to fully simplify the recursive factorisation of the joint density function. This is a Markov chain type property.

- The models have triangulated independence graphs.

- Maximum likelihood estimates of the model are directly calculable.

It is not surprising that models with so many pleasant properties have been defined in many different ways, and any one of these equivalent properties could be used to define the term 'decomposable'. However, we would not want to base a definition on the last, as to make it precise requires further assumptions about sampling models. Also, because our essential interest lies in probability modelling we exclude triangulation, which is an entirely graph-theoretic property. Properties couched in terms of vertex orderings relate to directed graphs, and are also eliminated.

The natural definition of decomposable is the multiplicative property, but this is comparatively more difficult to work with, so we adopt the following.

The classical definition of a decomposable model is given by Haberman (1974) and here it translates to: a model is decomposable if either (i) it is complete or (ii) it is reducible to two decomposable components. This is a recursive definition which, because our graphs are finite, is exactly equivalent to the assertion that a model is decomposable if and only if it has complete irreducible components.

Definition. A k-dimensional random vector X, or its density function, is *decomposable* if and only if there exists a sequence of decompositions to complete irreducible components. □

Decomposable random vectors have independence graphs consisting entirely of complete subgraphs. Hence the maximal irreducible components of a decomposable model are the cliques of the graph. Conversely, if any of the irreducible components of X are incomplete, for instance, such as one listed in Table 12.3.1, of which the simplest is a chordless 4-cycle, then X is not decomposable. We have seen that there are $2^{k(k-1)/2}$ distinct possible independence graphs for a k-dimensional random vector. Even for moderate k this is quite a large number, for example, if $k = 4$, there are 64 models and if $k = 6$ there are 32,768. By comparison there are just 2^k complete irreducible components, which generate all decomposable models.

The direct way to determine whether a given graph is the independence graph of a decomposable density function is to reduce it to its irreducible components, and check that each is complete. A sometimes faster way of showing that it is not decomposable is to find a chordless cycle of length greater than 3. The procedure works because of the the triangulation property satisfied by decomposable models.

Decomposable models and triangulated graphs

An undirected graph is *triangulated* if and only if the only chordless cycle in the graph contains no more than three vertices. Conversely, if one can find a chordless cycle with four or more vertices then the graph is not triangulated.

EXAMPLE 12.4.1 The graph (a) is not triangulated because of the existence of the chordless 4-cycle 1, 2, 3, 4, 1. (Imagine that vertex 5 is lifted out from the page.) The graphs (b) and (c) are triangulated.

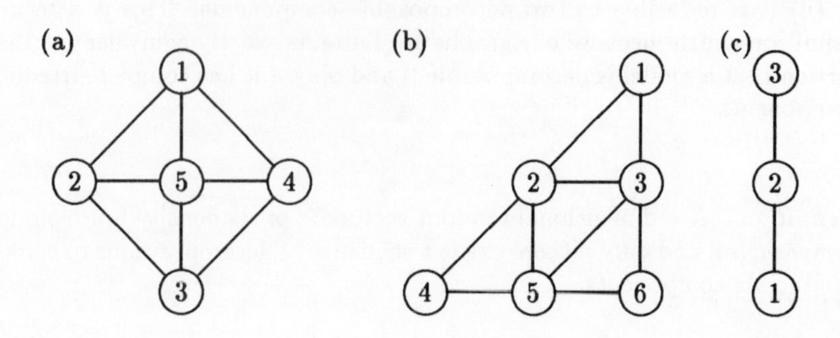

□

Triangulated graphs constitute an interesting set of graphs that has been well studied and has generated a large literature. They are also known under the names rigid circuit and chordal graphs, and form a subset of the so-called perfect graphs, see Rose (1970) and Golumbic (1980) for discussion and further references. The characterisation of decomposable models as those that have triangulated graphs was first recognised by Lauritzen, Speed and Vijayan (1984) in a 1978 preprint.

We need a lemma, very similar to Lemma 12.3.1 whose proof is left as an exercise to the reader.

Lemma 12.4.1 Inheritability. *All chordless cycles, including triangles, of a graph are preserved in a component under any decomposition of the graph.*

So triangulation is inheritable under decomposition.

Proposition 12.4.2 The triangulation theorem. *The random vector X (or its density function) is decomposable if and only if its independence graph G is triangulated.*

Proof: We suppose that G is connected, for if not, we may consider the connected subgraphs of G individually.

The first part is easy: suppose there exists a chordless p-cycle in G, with $p>3$. By the inheritability of chordless p-cycles under decomposition, at least one of the irreducible components of G must also contain the cycle. Consequently G is not the graph of a decomposable random vector.

To prove the converse, we show that if G is triangulated then it is either complete (and so decomposable), or not complete, but reducible.

So, suppose that G has no chordless cycle other than 3-cycles. First, if G itself is complete then it is irreducible and so is decomposable. If it is incomplete then we must show it is reducible for otherwise it would not be decomposable. The following construction of a decomposition is the hard part of the proof.

Take two vertices that are not adjacent in the graph of G and label them 1 and 2. There is said to be a path from 1 to 2 if there exist distinct vertices $q_1, q_2, \ldots, q_m$ with $m \geq 1$ such that the edges $(1, q_1), (q_1, q_2), \ldots, (q_m, 2)$ exist in the edge set of G. This is a ***short path*** if no proper subset of the path is also a path from 1 to 2. Hence only vertices that are adjacent in the list $1, q_1, q_2, \ldots, q_m, 2$ are adjacent in short paths and in particular the vertex 1 is not adjacent to q_j for all $j = 2, \ldots, m$.

We show that if $q_1, q_2, \ldots, q_m$ and $r_1, r_2, \ldots, r_n$ are both short paths from 1 to 2 with q_1 and r_1 distinct then q_1 and r_1 are adjacent. Suppose otherwise, as in the diagram

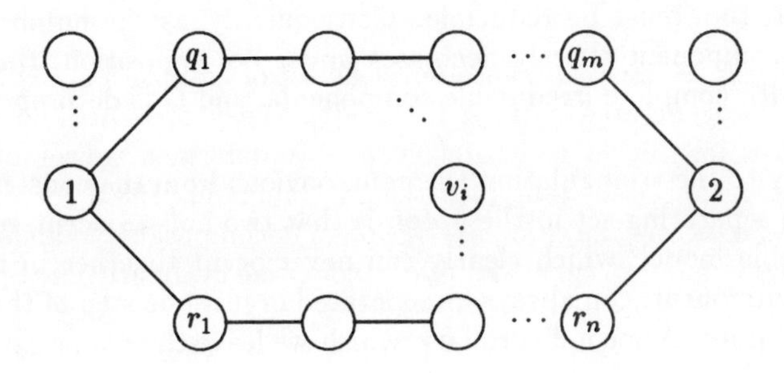

There is one short path from q_1 to r_1 via 1. Let $q_1, v_1, v_2, \ldots, v_p, r_1$ be another short path connecting q_1 to r_1, derived by shortening the path

$$q_1, q_2, \ldots, q_m, 2, r_n, \ldots, r_2, r_1$$

where necessary. This derived short path is non-empty as the vertex 2 serves if there is no other candidate. Now vertex 1 cannot be adjacent to any vertex v_j in this derived short path because it is not adjacent to any vertex $q_2, \ldots, q_m, 2, r_n, \ldots, r_2$ in the original path. Hence $r_1, 1, q_1, v_1, \ldots, v_p, r_1$ is a chordless cycle with at least four elements. This is impossible by supposition, hence q_1 and r_1 are adjacent.

Now construct the complete separating set, a, of vertices by including a vertex q in a if and only if q is a nearest neighbour of 1 on a short path from 1 to 2. This subgraph corresponding to a is complete because we have seen that each of these vertices must be adjacent. The set a separates vertices 1 and 2.

We have constructed a complete subset of vertices which separates 1 and 2. To show that G is reducible, we have to find an appropriate partition of K: take a as above; take b to consist of the vertex 1 and those vertices with paths to 1 that do not intersect a; and finally, take c to be the remainder, $c = K \backslash (a \cup b)$.

Clearly $\{a, b, c\}$ is a partition of the vertex set K, the set a is complete in the graph, and, b and c are not empty as b contains 1 and c contains 2. It only remains to show that b and c are separated by a. As the graph of G is connected all vertices in c have paths to 1, and from the way a and b are constructed every such path must intersect a. Hence no element in c can be adjacent to any element in b; because if so there would exist a path from c to 1 which did not pass through a; which is a contradiction. Hence the set a separates all elements of b from all those of c.

Now triangulation is inheritable under decomposition; so that if both components are complete then G is decomposable and the proof is finished; while if either component is incomplete, then a further application of this argument shows that it, too, must be reducible. Consequently, as the number of vertices in any component strictly decreases under decomposition, the process terminates with complete irreducible components, and G is decomposable. □

A corollary to the triangulation theorem, obvious from the construction of the complete separating set in the proof, is that two non-adjacent vertices of a decomposable model, which clearly can never occur together in the same irreducible component, can always be separated in just one step of the decomposition procedure. A second corollary, which we leave the reader to prove, is that triangulation is equivalent to the property: every subset of vertices that separates vertices i and j and is minimal is complete.

Full factorisations and absent edges

We return to the question of factorisation raised in Section 12.2. We have defined X to be decomposable if is reducible to a set of complete irreducible components, identifiable as the cliques of the graph, and in consequence, have seen that its density function is uniquely expressible as the product of marginal density functions on cliques. In this section we consider an alternative characterisation of decomposability phrased in terms of the edges absent from, rather than present in, the graph.

We know a little more about factorisation now: not any factorisation will do, we must only apply those independences with conditioning sets which are complete.

EXAMPLE 12.4.2 Such a restriction avoids the problems inherent in factorising the simple Markov chain

when starting with the independence $1 \perp\!\!\!\perp 5 | \{2, 3, 4\}$. The independence between vertices 2 and 4 then has to be used several times. □

There are two key ingredients needed to establish the equivalence with decomposability: the factors in the factorisation have to be marginal density functions, and the factorisation has to be full. Recall that a factorisation of the density function of X is *full* if and only if an independence statement corresponding to every pair of non-adjacent vertices in the independence graph of X is applied *exactly once* to factorise f into marginal density functions.

Proposition 12.4.3 Full factorisation. *The density function for a k- dimensional random vector X is decomposable if and only if it is fully factorisable with respect to the non-adjacent pairs of vertices in its independence graph.*

Proof: The essence of the argument is to note that any decomposition uses exactly the right number of independence statements: because (i) if X is reducible to X_{ab} and X_{ac} and if X_b and X_c are q- and r-dimensional respectively, exactly qr pairwise conditional independences are implicit in $X_b \perp\!\!\!\perp X_c | X_a$, and exactly this number is used to replace f_{abc} by $f_{ab} f_{ac} / f_a$. Furthermore, (ii) if X is reducible to a component X_{ab} then the independence graph of X_{ab} has no more edges than the induced subgraph of G on $a \cup b$. Hence no missing edges are 'lost' in a decomposition. Finally, (iii) once used in a decomposition no independence statement can be re-used, because the vertices are now in different components.

To finish the argument, if the graph is complete there is nothing to show. Otherwise, if X is decomposable, then X is reducible to complete irreducible components and at each step in the sequence of decompositions required to transform X into its irreducible components exactly the right number of independences is used, and the factorisation is full. Conversely if X is not decomposable, then it has at least one incomplete irreducible component. No two non-adjacent elements of an irreducible component have a complete separating set, so that application of one independence to factorise the density makes application of at least one other impossible. Hence the factorisation cannot be full. □

Our final remark is to make explicit the connection between direct estimates and decomposable models. In fitting graphical models to the Multinomial and Normal distributions the likelihood equations are characterised by the equality of the observed and fitted marginal densities on all margins corresponding to cliques in the graph. If the model is decomposable, so that the factorisation of the joint density function is full, the maximum likelihood estimate of the joint density is given by the product of the estimates of the marginal densities, which is just the product of the observed marginals.

12.5 Collapsibility

The independence graph of a k-dimensional vector is not usually a complete description of its independence structure. While the global Markov property makes some assertions it may not rule out other possibilities, and worrisome in applied work is the fear that the observed interaction in the X_a margin is entirely induced by marginalising the full vector $X_{ab} = (X_a, X_b)$ over X_b. For example, in the analysis of the clinic survival data, discussed Section 1.2 of the introduction, speculation about the strength of the observed marginal interaction between care and survival loses its point when it is realised that it entirely disappears after conditioning on clinic.

Collapsibility is concerned with the properties of the marginal density, f_a, when the joint distribution, f_{ab}, belongs to a graphical model. There are several related issues.

(i) *closure:* is the marginal distribution of X_a in the same parametric family of distributions as X_K?

(ii) *graphical collapsibility:* what may be deduced about the independence graph of X_a from knowledge of that of X_{ab}?

(iii) *parametric collapsibility:* are the interaction parameters in the marginal distribution equal to corresponding ones in the joint distribution, or more weakly, are they both zero together?

(iv) *commutativity of fitting and marginalisation:* can the predicted distribution obtained by marginalising the fitted model of the joint distribution be recovered by fitting a model to the marginal data?

(v) *test collapsibility:* can the values of test statistcs, such as the edge exclusion deviances, be recovered in the marginal distribution?

These versions of collapsibility are somewhat hierarchical, and for instance it makes no sense to check for commutativity if the distribution is not closed.

Collapsibility is important for two reasons. The first is that it breaks large problems down into small problems. In practice, a study may take observations on many variables but only use a few of these in any one analysis of the data. This may be because of the reduction in complexity (high dimensional interactions are usually difficult to interpret) or because of a paucity of data (requiring full measurements on all units is not always achieved in real life). The same is true for discrete and for continuous observations, for while a correlation matrix is not sparse, the efficiency with which parameters in the model can be estimated decreases as the number of dimensions increase, Altham (1984). If collapsing over extraneous variables does not lead to misleading inference then it should improve the efficiency of the analysis.

The second reason is because regression and recursive models are naturally formulated in terms of conditional and marginal distributions. Collapsibility explores the relationship to the joint model and thus checks if the regression model can be estimated by fitting the joint distribution. Rather than develop new algorithms for fitting $f_{b|a}$ directly we may fit indirectly using established methods for fitting f_{ab} and f_a.

There is a burgeoning literature on the subject which goes back to the 1951 paper of Simpson. The issue was brought to the fore with the publication of the text by Bishop *et al.* (1975); parametric collapsibility was treated by Whittemore (1978) and the relationship to graphical models is discussed by Asmussen and Edwards (1983). Much of the literature on collapsibility is of a highly technical nature; for instance, there are several similar definitions emphasising different aspects of related concepts, some of which lead to sufficient but not necessary conditions. We have no wish to get embroiled in an exhaustive summary, and aim just to give the reader a flavour of the ingredients involved. We discussed an aspect of parametric collapsibility in Section 4.6 where certain conditions are given for the invariance of mixed derivative measures of interaction and of divergence measures of edge strength.

Graphical collapsibility

Proposition 12.2.1, the last result of Section 12.2, showed that the independence graph of a component in a sequence of decompositions of a graph is identical to the induced subgraph obtained by deleting vertices (and corresponding edges) not in the component; that is, the full set of variables is graphically collapsible onto the component. The definition here generalises this idea to cope with marginalising with respect to an arbitrary partition. Suppose that X has the independence graph G.

Definition. The vector $X = X_{ab} = (X_a, X_b)$ is *graphically collapsible* over X_b if and only if the boundary (of each connected component) of X_b is complete in G. □

The next proposition shows which parts of the independence graph are invariant to collapsing.

Proposition 12.5.1 *If $X_{ab} = (X_a, X_b)$ is graphically collapsible over X_b then the conditional independences between the elements of X_a in the independence graph of (X_a, X_b) are preserved in the independence graph of X_a.*

Proof: First suppose that X_b consists of one connected component. Partition X_a into $(X_{a\backslash c}, X_c)$, where c is the boundary of b in a. Then the density function of $(X_{a\backslash c}, X_c, X_b)$, factorises

$$f_{ab} = f_{a\backslash c \cup c \cup b} = f_{a\backslash c|c \cup b} f_{c \cup b} = f_{a\backslash c|c} f_{b|c} f_c.$$

Integrating over b gives the marginal density

$$f_a = f_{a\setminus c|c} \int f_{b|c} f_c \, dx_b$$

exactly preserving the conditional density of $X_{a\setminus c}$ given the boundary X_c. By assumption the boundary is complete and so all independences are preserved. □

EXAMPLE 12.5.1 Suppose the independence graph of X is the graph G with diagram

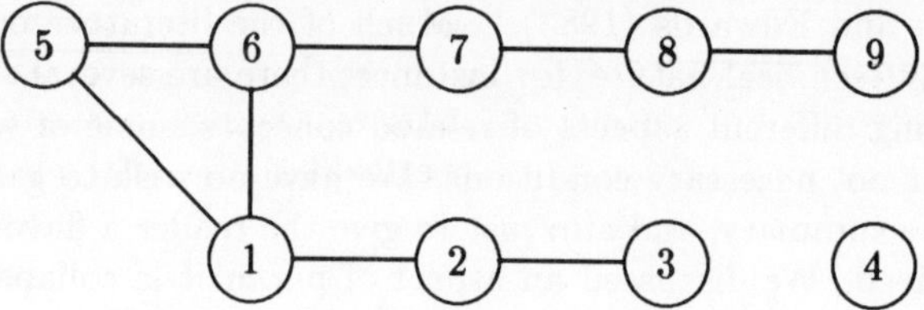

Then this graph

is collapsible over $b = \{5\}$ because the boundary $\text{bd}(5) = \{1,6\}$ which is complete in G;

is not collapsible over $b = \{8\}$ because the boundary is $\{7,9\}$ and these are not adjacent;

is collapsible over $b = \{5,8,9\}$ because the connected subgraphs of b are $\{5\}$ and $\{8,9\}$ and the boundary of each is complete; and

is not collapsible over $\{5,8\}$. □

As can be discerned from this example, a reformulation of the definition is that X is collapsible onto X_a if and only if there is a sequence of decompositions from which X_a is a component: if there exists a decomposition to components X_{ab} and X_{ac} then X is graphically collapsible onto either component.

Commutativity of fitting and marginalising is summarised in the diagram

$$\begin{array}{ccc} f_{ab} & \xrightarrow{\text{fit}} & \hat{f}_{ab} \\ \text{marginalise} \downarrow & & \downarrow \text{marginalise} \\ f_a & \xrightarrow{\text{fit}} & \hat{f}_a \end{array}$$

The model is collapsible in this sense if, for example, the fitted cell probabilities or the generalised likelihood ratio test statistic, are the same irrespective of the order of fitting and collapsing. The importance of graphical collapsibility is that together with closure, it is necessary and sufficient for the commutativity of the maximum likelihood estimates, see Asmussen and Edwards (1983) for further details. The commutativity of likelihood ratio tests requires collapsibility under both the null and the alternative models.

Parametric collapsibility

In general, there is no reason why a measure of interaction in a margin should be the same when it is conditioned on other variables, nor is there any reason to suggest that the conditional measure is more or less fundamental than the other; it is the context of the example that determines which interaction is of prime interest. We already know that the independence graph of a k-dimensional vector is not a complete description of the independence structure. For instance, while a decomposition 'preserves' the absence of edges in the graph, it may well not preserve an edge itself. We should like to know when parameters measuring marginal and conditional interaction are equal.

EXAMPLE 12.5.2 By constructing numerical examples of three dimensional contingency tables or correlation matrices it is easy to show that marginalising a three dimensional vector X over vertex 3 may send the independence graph 1 o o 2 or 1 o—o 2 .

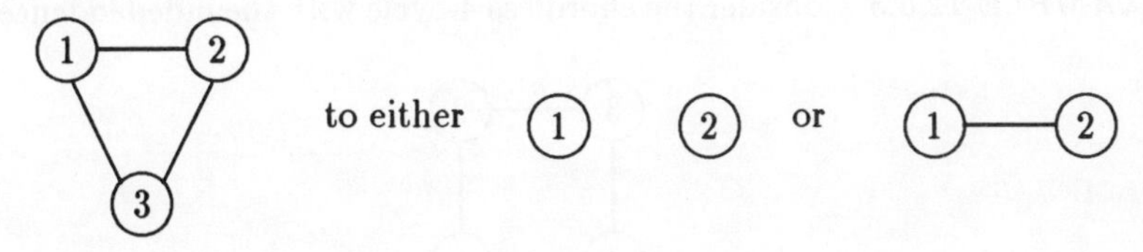

That is, in the class of three dimensional distributions that do not satisfy any conditional independence some exhibit marginal independence and some do not. Similarly

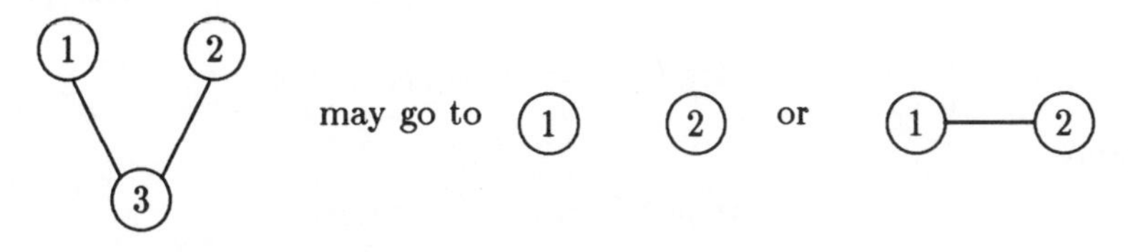

On the other hand note that

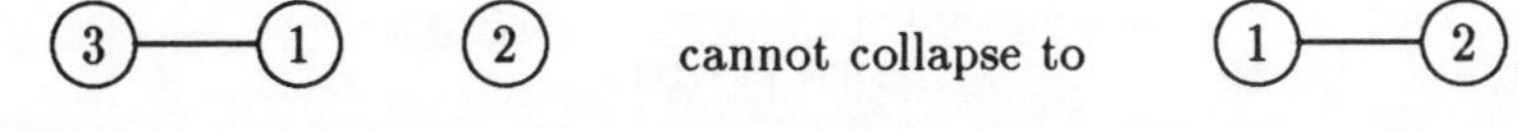

nor can

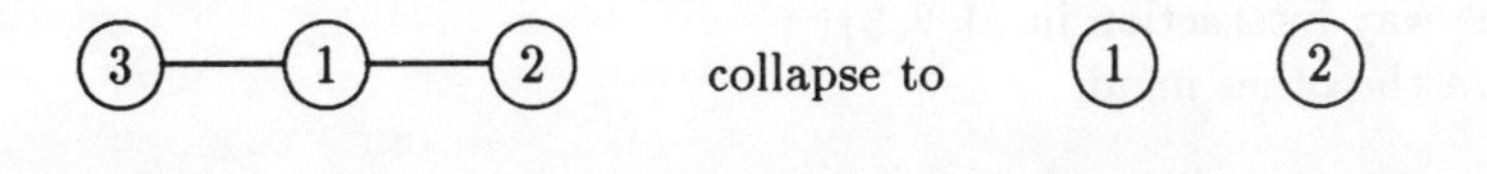

□

These examples, in which edges disappear under marginalisation, depend on precise numerical specification of the cell probabilities or the correlation matrices. Slight variation away from these values will induce the corresponding interaction and consequently preserve the edges. So requiring the collapsibility property to hold in a wide enough class of probability density functions will preserve edges, and Proposition 12.2.1 can then be strengthened to assert that component's independence graph and the corresponding induced subgraph are identical.

In relation to this last example, if the variables are Normal then the partial correlation between 1 and 2 given 3 is not the same as the marginal correlation between 1 and 2, because conditioning on 3 reduces the variance of 1. That is, correlation is not an invariant measure of interaction. Here we take up the connection between the parameters of the full and the marginalised distribution using the mixed derivative measure of interaction which is invariant. We saw (Chapter 2, Proposition 2.3.3) that in a three dimensional vector the independence of two variables given the third is a sufficient condition for collapsibility. Independence graphs allow us to explore the k-dimensional case.

EXAMPLE 12.5.3 Consider the chordless 4-cycle with the independence graph

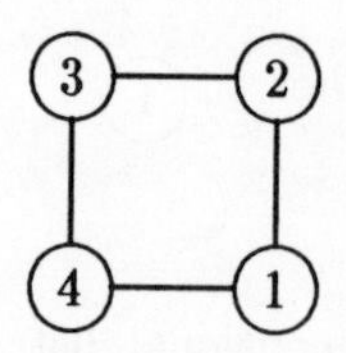

marginalised over X_4 onto (X_1, X_2, X_3). To begin, first consider the interaction in the margin $\{1,2\}$. Now $f_{1234} = f_{123} f_{4|123}$ and because $4 \perp\!\!\!\perp 2 | \{1,3\}$

$$D^2_{12}\log f_{4|123} = D^2_{12}\log f_{4|13} = 0.$$

Hence

$$D^2_{12}\log f_{1234} = D^2_{12}\log f_{123}.$$

The conditional independence $2 \perp\!\!\!\perp 4 | \{1,3\}$ implies the equality of the mixed derivative measures of interaction

$$i_{12|34} = i_{12|3}.$$

The same is true for the two-way interaction in the margin $\{2,3\}$ and the three-way interaction in $\{1,2,3\}$.

On the other hand,

$$i_{13|24} = D^2_{13}\log f_{1234} = 0$$

because of the independence $1 \perp\!\!\!\perp 3 | \{2,4\}$, but there is no reason to suppose that the corresponding derivative D^2_{13} on the marginal density $\log f_{123}$ is zero.

Thus the subgraph (a) of the independence graph, obtained by deleting vertex 4 and its edges to the boundary set, is not the independence graph of (X_1, X_2, X_3), which is given by (b):

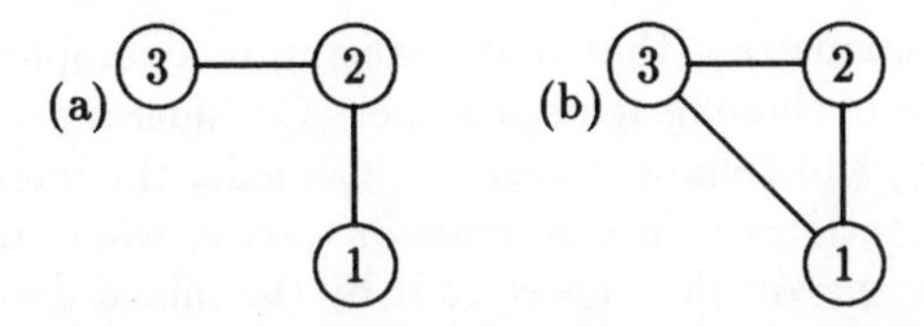

In this example the boundary of the set of collapsing variables is the set of vertices $\text{bd}(4) = \{1, 3\}$. The derivatives that are preserved under marginalisation have at least one vertex that is not in this boundary set. The trouble arises in the boundary set of the collapsed vertex, which, in this example, is not complete. □

EXAMPLE 12.5.4 A second related example is the graph consisting of the same 4-cycle but with an extra edge connecting vertices 1 and 3

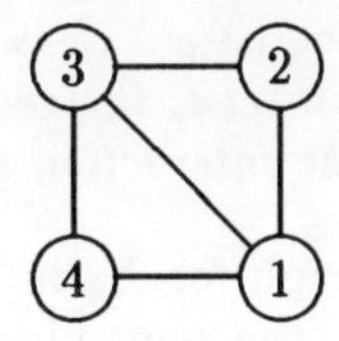

It is easily checked that the mixed derivative interactions on the margins $\{1, 2\}$, $\{2, 3\}$ and $\{1, 2, 3\}$ are invariant to marginalisation. There is no reason for the interaction on $\{1, 3\}$, the boundary of 4, to be the same in the marginal distribution; but while $i_{13|24}$ and $i_{13|2}$ may be different (recall Simpson's paradox) neither is necessarily zero. Consequently, and unlike the previous example, the subgraph obtained by deleting the vertex 4 and its edges to its boundary set gives the independence graph of the remaining variables, the complete graph on $\{1, 2, 3\}$ as in (b) above. □

EXAMPLE 12.5.5 Consider the independence graph

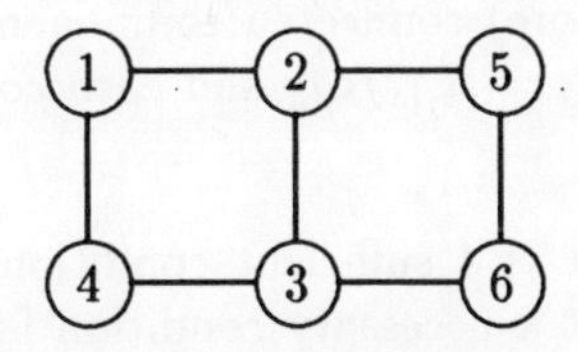

on 6 vertices collapsed over (X_4, X_5, X_6) onto the vector (X_1, X_2, X_3). The variables (X_4, X_5, X_6) in the collapsing set divide into (X_4) and (X_5, X_6) because they are separated by (X_1, X_2, X_3). These are the ***connected components*** of the collapsing set, because the subgraph of the collapsing set separates into two unconnected subgraphs. Each connected component has a boundary in X. We can examine each separately. □

These examples demonstrate that marginalising is a complicated issue. We formally bring some of these insights together. Consider a partition of the full vector into (X_a, X_b) and collapsed over X_b. Consider the interaction between the variables with vertices in the ***interactions set*** e, where of course $e \subseteq a$, and is of cardinality p. We shall measure it by the mixed derivative, denoted by D_e^p, obtained by taking the mixed partial derivative of $\log f$ with respect to each variable in e.

Definition. The vector X_K with density function f_K is ***parametrically collapsible*** over X_b with respect to the mixed derivative interaction if $D_e^p \log f_{ab}$ and $D_e^p \log f_a$ are identical. □

Another way of phrasing this condition is that K is parametrically collapsible if and only if $D_e^p \log f_{b|a} = 0$ for all x.

Proposition 12.5.2 *The vector (X_a, X_b) is parametrically collapsible over X_b with respect to an interaction set e, if, for each connected component of b, there is at least one vertex of the interaction set not in the boundary set.*

Proof: First suppose the subgraph for X_b in G is connected with boundary $c = \text{bd}(b) \subseteq a$. As c separates the remaining variables in X_a from X_b, the Markov property implies

$$f_{ab} = f_{b|a} f_a = f_{b|c} f_a$$

and the interaction measures are related by

$$D_e^p \log f_{ab} = D_e^p \log f_a + D_e^p \log f_{b|c}.$$

The distribution is parametrically collapsible if this second term $D_e^p \log f_{b|c}$ is zero. No vertex in e can be in b, so that if there exists a vertex $i \in a$ of e that is not in the boundary set c, then the second term is zero.

If there are two (or more) connected components of the marginalising set b, b_1 and b_2 say, then $f_{b|a} = f_{b_1|a} f_{b_2|a}$ and each component can be examined separately. □

That such a vertex exist is a sufficient condition that applies to the independence graph; it is not a necessary condition for parametric collapsibility because of the following generic example.

EXAMPLE 12.5.6 Suppose that $a = \{1, 2\}$ and the logarithm of the conditional density of X_b given X_a is additive in the interaction terms concerning

(X_b, X_1) and (X_b, X_2); that is, there exist functions g and h such that $f_{b|12}$ is of the form

$$\log f_{b|12} = g_{b\cup 1} + h_{b\cup 2}, \text{ for all } x.$$

Consider the interaction in the $\{1,2\}$ margin,

$$D^2_{12}\log f_{b|12} = D^2_{12}(g_{b\cup 1} + h_{b\cup 2}) = 0,$$

and hence any distribution satisfying this additivity condition is parametrically collapsible even though both vertices are in the boundary of b. Specific examples of tables of probabilities for which the conditional density $f_{a|b}$ is of this form with non-zero g and h are described by Whittemore (1978). □

12.6 Exercises

1: Find the irreducible components of $\log f_{1234} = h_{12} + h_{23} + h_{24}$ and give a factorisation.

2: Show that the model of no three-way interaction can be derived by marginalising the density function with an independence graph given by the chordless 4-cycle.

3: If the irreducible components of X on $K = \{1, 2, \ldots, 8\}$ have coordinates $\{1,2\}$, $\{2,3,4\}$, $\{5,6\}$, $\{2,6,7\}$, and $\{2,8\}$, find the factorisation of the density function of X.

4: The graph G has irreducible components $\{2,5\}$, $\{2,3,4\}$, $\{5,6\}$, $\{2,6,7\}$, and $\{2,8\}$. Find the diagram of the graph G^I that has a vertex corresponding to each component and an edge between any two vertices if the intersection of the corresponding components is not empty. Generally, show that G^I is triangulated.

5: The three dimensional vector of random variables $X = (X_1, X_2, X_3)$ has a joint density function of the form

$$f(x) = \exp(u + x_1x_2 + x_2x_3 + x_1x_3),$$

in the cube, $0 < x_i < 1, i = 1, 2, 3$. Find the marginal distribution of (X_1, X_2) and of X_1, and show that it is not possible to write f_{123} as the product

$$f_{12}f_{23}f_{13}/f_1f_2f_3.$$

6: Find a counter-example to the assertion that in any decomposition, a unique minimal complete separating set can be found.

7: Give a counter-example to the assertion that a model is necessarily irreducible if the nearest neighbours of every vertex are incomplete.

8: How many chordless k-cycles with $k \geq 4$ can be found in the graphs

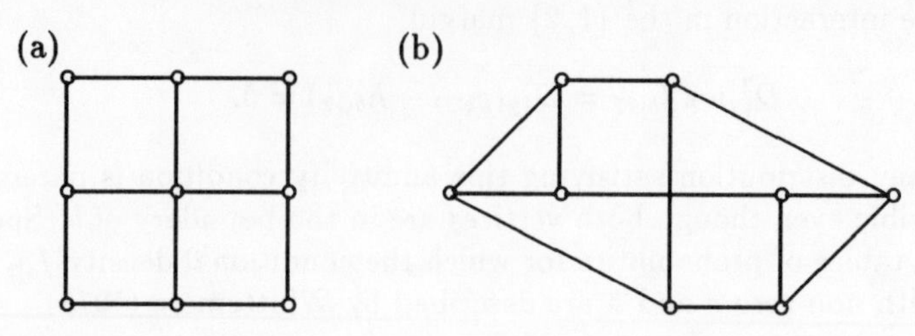

9: Consider the graph

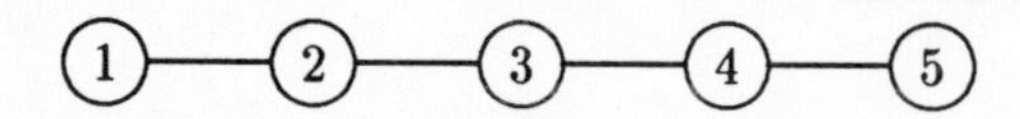

Discuss which margins it can be collapsed over.

10: In the context of a log-linear model of appropriate dimension, state which graphs are collapsible onto $\{1, 2, 3\}$ and why.

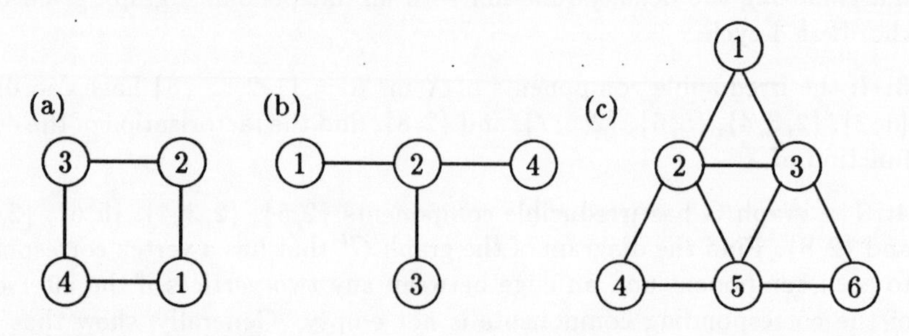

11: Write out the proof of Proposition 12.4.1 that p-cycles are inherited in any reduction of the graph.

12: Prove that triangulation is equivalent to the property that every subset of vertices which minimally separates two vertices is complete.

13: Show that a full factorisation of a density is unique.

14: Give an example of a directed graph with a moral graph that is not decomposable.

15: Prove that all graphical models with explanatory variables in which the response is one dimensional are decomposable.

Appendix A

A.1 Computing Packages

Because graphical modelling is intimately related to standard statistical methods such as regression analysis, analysis of variance and log-linear modelling, certain useful algorithms are readily available in standard statistical packages. For example, regression routines are available in GLIM, GENSTAT, MINITAB, BMDP, S, and SAS. However, there are two major deficiencies: firstly, graphical ideas are not implemented explicitly, so that the user interface is poor; and secondly, existing routines are generally limited to fitting graphical models with direct estimates which severely restricts the class of models, especially in the mixed case, that can be fitted. Though it is possible to develop macros in these packages to fit models with indirect estimates, it is a slow and tedious alternative.

In preparing this text, the author has employed a variety of packages for different parts of the analysis, but in particular, has been fortunate enough to have access to the experimental programs MIM and EXA, specifically developed as research tools for graphical modellers. MIM fits hierarchical mixed interaction models based on the joint CG family of distributions by maximum likelihood. EXA uses simulation to approximate exact conditional test procedures for graphical log-linear models with a recursive block structure. These cover almost all the models and techniques discussed in this text, with the exception of those CG regession models that cannot be fitted in the joint distribution; software is under development here, but is not yet available.

Some of the case studies used in this text, written in MIM, EXA or GLIM, used to prepare material for this book, can be obtained on disc (5.25in 1.2mb) by writing to

Joe Whittaker, Centre for Applied Statistics,
University of Lancaster, LA1 4YW, U.K.

and enclosing a blank formatted disc and self addressed envelope. Altenatively, you may email joe.whittaker@uk.ac.lancaster.central1 .

MIM: mixed interaction modelling

The program MIM is an interactive program that fits a class of statistical models, the hierarchical mixed interaction models, Edwards (1990), to data. Graphical models are those models entirely determined by their independence graph and form a subset of the hierarchical interaction models. Entirely continuous variables are assumed to have a multivariate Normal distribution, entirely categorical variables have a cross-classified Multinomial distribution, and mixed variables have the CG distribution. This framework includes graphical log-linear models, logit and logistic models, graphical Gaussian (covariance selection) models, multivariate ANOVA and some CG regression models.

The author is Edwards (1987, 1990), and more information about the program and the manual may be obtained by writing to

David Edwards, Bymarken 38, DK-4000 Roskilde, DENMARK.

The program is written in (Turbo) Pascal and runs on IBM-PC compatible computers in interactive and batch mode based on a simple command language. It provides an easy to use syntax for model formulae, which is used to issue fitting and plotting commands. The fitting algorithm is an adaptation of the iterative proportional fitting algorithm discussed in Chapter 4. The MIM package is not a complete tool for statistical modelling and it runs best in tandem with some other general purpose utility, such as MINITAB, to provide tools for data manipulation.

EXAMPLE A.1.1 The mathematics marks of Chapter 1. The file mm.mim contains the following commands

command and argument	comment
diaryon mim.log	% creates mim.log file
cont vwxyz	% declare 5 continuous variables
read vwxyz	% could read variance directly
77... 81	% first of 88 rows with 5 entries per row
......	
00...14!	% end of data
lab v "mech" ... z "stat"	% assign labels
print s v	% prints variance and partial correlations
mod // vwx,xyz	% declares 'butterfly' model
fit	% gives a deviance 0.8957, with 4 df
test	% generalised likelihood ratio test
print ih	% print fitted correlations
plot	% displays graph of fitted model
backsel	% computes edge exclusion deviances
help ; exit	% has a help facility

The program MIM is invoked and then the file is input using the MIM instruction 'input mm.mim'.

A graphical log-linear model is fitted to a four-way contingency table by using the commands in the file ll.mim

command and argument	comment
diaryon mim.log	% creates mim.log file
fact a2 b2 c2 d2	% declares four binary variables
statread a b c d	
15 23 ... 27 12 !	% four-way table of counts
print u v	% prints input table
mod acd,bcd//; fit	% model formulae for $A \perp\!\!\!\perp B \mid (C, D)$
exit	

□

EXA/DIGRAM

The program EXA tests conditional independence hypotheses between variables classifying a multi-way contingency table, using simulated approximations to exact conditional tests, described in Section 9.4. The author is Kreiner (1987), and upto date information on the program may be obtained by writing to,

Svend Kreiner, Buen 8, 2000 Frederiksburg, DENMARK.

The program EXA is written in (Turbo) Pascal and runs in a shell called DIGRAM, on IBM-PC compatible computers in either interactive or batch mode based on a simple command language.

EXAMPLE A.1.2 Diabetic data from Chapter 9. The example here is based on the diabetic data introduced in Exercise 15 of Chapter 9 concerning a three-way contingency table. Three files are pricipally involved: insulin.exa, insulin.cmd, insulin.rep; the first carries a description of the table and the table entries; the second contains a list of commands for batch execution; and the last contains the output of the analysis. Here is a listing of the first two files. insulin.exa:

command and argument	comment
diabetic data	name of table
3	3 variables classifying table
A 2 2	label, no. levels, type (nominal/ordinal)
I 2 2	
F 2 2	
1	number of recursive blocks
3	number of last variable in each block
T	table to follow, one row at a time
6 6	
1 36	
16 8	
2 48	
I = injections	comments (upto 10 lines)
A = age at onset	
F = family history	

insulin.cmd :

command and argument	comment
help exa	
exa	goto exa mode
read	read the table
new	initialize the graph
dia	open the diary file
show v	lists variables in diary file
exa 400 35	400 simulations and seed value
dev	use deviance
base	test pairwise cond. independences
plot	
sep -	check separation implications
test	test hypotheses
exit	

□

GLIM: generalised linear interactive modelling

The program GLIM is a statistical package developed under the auspices of the Royal Statistical Society that provides a framework for the fitting of generalised linear models to data. The essential purpose of GLIM is model fitting with one response variable, and it fits the classical linear models by specifying quantitative and qualitative covariates in a simple command language using the Wilkinson and Rogers (1973) syntax for model formulae. By exploiting the Poisson representation of the Multinomial distribution, the framework embraces graphical log-linear models for contingency tables, but it cannot fit graphical Gaussian models. The algorithm used in GLIM is iteratively re-weighted least squares which automatically calculates standard errors of parameters. It has useful high level calculating facilities, based on vector arithmetic, with program control structures for looping, branching and storing of macros. The theoretical basis for GLIM is described in Nelder and Wedderburn (1972) and McCullagh and Nelder (1983); a good introduction to the use of the package is Aitkin *et al.* (1989).

EXAMPLE A.1.3 Rochdale data. The GLIM code to fit the all two-way interaction model to the eight-way table concerning women's economic activity, Table 9.3.1, from the Rochdale data discussed in Chapter 9, is

directive and argument	comment
$unit 256	! declares length of vector to hold table
$data count $read	
5 0 2 1 ...	! reads the eight-way table
0 0 0 0	
$factor A 2 B 2 ... G 2 H 2	! declares factors
$calc A=%gl(2,128) : B=%gl(2,64)	! assigns values
... : G=%gl(2,2) : H=%gl(2,1)	
$yvar count $err p	! declares response variable
$fit A.B+A.C+···+G.H	! fits model
$disp e	! displays estimates
$stop	

□

A.2 Outline Answers

Outline answers to Exercises from Chapter 2

1: By total probability and independence $P(A \cap B^c) = P(A) - P(A \cap B) = P(A) - P(A)P(B) = P(A)P(B^c)$. Also $P(A \cap \Omega) = P(A) = P(A)P(\Omega)$, and similarly $A \perp\!\!\!\perp \phi$.

2: (a) $A \perp\!\!\!\perp B$ iff $P(A \cap B) = P(A)P(B)$ iff $P(A|B) = P(A)$. Now $A \perp\!\!\!\perp B$ iff $A \perp\!\!\!\perp B^c$ so that by same reasoning $P(A|B^c) = P(A)$. To go the other way use law of total probability $P(A) = P(A|B)P(B) + P(A|B^c)P(B^c)$. (b) Definitions and manipulation.

3: $P(A \cap (B \cup C)) = P(A \cap B) + P(A \cap C) - P(A \cap B \cap C) = P(A \cap B) + P(A \cap C) = P(A)(P(B) + P(C)) = P(A)P(B \cup C)$.

Counter-example:

P	$B \cap C$	$B \cap C^c$	$B^c \cap C$	$B^c \cap C^c$
A	0.25	0	0	0.25
A^c	0	0.25	0.25	0

and $P(A) = P(B) = P(C) = 0.5$, $P(A \cap B) = P(A \cap C) = 0.25$, $P(B \cup C) = 0.75$ but $P(A \cap (B \cup C)) = 0.25 + 0.25 + 0 = 0.5$.

4: (a) Use the result of question 3 above, and then the fact that any event can be obtained from a union of events in the partition of the sample space: $\{B \cap C, B \cap C^c, B^c \cap C, B^c \cap C^c\}$.
(b) The partition can be generated from these events.

5: $A \perp\!\!\!\perp [B, C]$ is equivalent to $A \perp\!\!\!\perp B$, $A \perp\!\!\!\perp C$ and $A \perp\!\!\!\perp B \cap C$. So $A \perp\!\!\!\perp [B, C]$ and $B \perp\!\!\!\perp C$ is equivalent to $A \perp\!\!\!\perp B$, $A \perp\!\!\!\perp C$, $B \perp\!\!\!\perp C$ and $P(A \cap B \cap C) = P(A)P(B)P(C)$. But this is symmetric. Definition of mutual independence.

6: (a) For the following table of probabilities:

P	$B \cap C$	$B \cap C^c$	$B^c \cap C$	$B^c \cap C^c$
A	1/8	0	3/8	0
A^c	0	3/8	0	1/8

$P(A) = P(B) = P(C) = 1/2$, none of $P(A \cap B), P(A \cap C), P(B \cap C)$ equal 1/4 but $P(A \cap B \cap C) = 1/8$. (b) Solution to 3 above.

7: (a)

C	B	B^c	C^c	B	B^c
A	0.1	0.1	A	0.3	0.1
A^c	0.1	0.1	A^c	0.1	0.1

(b) Use $A \perp\!\!\!\perp B|C \Rightarrow A \perp\!\!\!\perp B^c|C$ and $A \perp\!\!\!\perp B|C \Rightarrow B \perp\!\!\!\perp A|C$ twice.

8: Let C denote cancer and T a positive test. $P(T^c|C)$ is the probability of a false negative, $P(T|C^c)$ of a false positive. Given: $P(C) = 0.01$, $P(T|C) = 1 - P(T^c|C) = 1 - 0.2 = 0.8$ and $P(T|C^c) = 0.1$ Bayes theorem for one test gives

$$\begin{aligned} P(C|T) &= P(T|C)P(C)/P(T) \\ &= P(T|C)P(C)/(P(T|C)P(C) + P(T|C^c)P(C^c)) \\ &= (0.8)(0.01)/((0.8)(0.01) + (0.1)(0.99)) = 0.075. \end{aligned}$$

For two tests

$$P(C|T_1, T_2) = P(T_1, T_2|C)P(C)/(P(T_1, T_2|C)P(C) + P(T_1, T_2|C^c)P(C^c))$$

Conditional independence simplifies the denominator to

$$P(T_1|C)P(T_2|C)P(C) + P(T_1|C)P(T_2|C^c)P(C^c)$$

so that

$$P(C|T_1, T_2) = (0.8)(0.8)(0.01)/((0.8)(0.8)(0.01) + (0.1)(0.1)(0.99)) = 0.393$$

Note that it is not assumed that $T_1 \perp\!\!\!\perp T_2$ but that $T_1 \perp\!\!\!\perp T_2|C$ and $T_1 \perp\!\!\!\perp T_2|C^c$.

9: Let X and Y denote two discrete random variables, we have to show that $P(X = x, Y = y|X = x) = P(X = x|X = x)P(Y = y|X = x)$. But this is easy to verify. Similarly the second part.

11: $f_{X|Y}(x; y) = (1+y)^2 x \exp\{-(1+y)x\}$ on $x > 0$; a member of the Gamma family.
$i_{XY}(x, y) = -1$.

12: $X_1 \perp\!\!\!\perp X_4|(X_2, X_3)$ and $X_2 \perp\!\!\!\perp X_3|(X_1, X_4)$.

14: $X \sim B(1, 0.7)$, $X|Y = 0 \sim B(1, 0.75)$ and $X|Y = 1 \sim B(1, 0.666)$. Not independent.

15: $E(X_1) = p_1(1)$, $E_{1|2}(X_1) = p_{12}(1, x_2)/p_2(x_2)$, for $x_2 = 0, 1$ and $E(X_1X_2) = p_{12}(1, 1)$.

16: (a) 2/3, 3/8, 1/6

(c)

$p(x,y)$	y	0	1
x	0	$2a$	b
	1	a	b

has cpr $= 2$, and is a table of probabilities if $3a + 2b = 1$, and $a, b > 0$.

17: $\text{cpr}\,(X, Y) = E(XY)E(1 - X)(1 - Y)/EX(1 - Y)EY(1 - X)$.

18: (a) $u = -3.278$, $\text{cpr}\,(X, Y) = e$.

20: (a) cpr (outcome, treatment) = 6.20/6.20 = 1, independent. But cpr (., .|male) = 5/6 = cpr (., .|female).

(b)

	gender	male		female	
	treatment	treated	untreated	treated	untreated
outcome	success	1	1	1	2
	fail	2	2	1	2

21:

$$\begin{aligned} \text{cpr}\,(\text{care, surv}|\text{clinic 1}) &= (3)(293)/(4)(176) = 1.25 \\ \text{cpr}\,(\text{care, surv}|\text{clinic 2}) &= (17)(23)/(2)(197) = 0.99 \end{aligned}$$

$\log \text{cpr}\,(\text{care, surv}|x_1) = 0.22 - x_1(0.23)$, $x_1 = 0$ for clinic 1 and $x_1 = 1$ for clinic 2 so that cpr (care, surv|clinic 1) = exp(0.22) = 1.246 and so that cpr (care, surv|clinic 2) = exp(−0.01) = 0.99

22: Direct from the definition of the conditional cpr.

24: Grand total of table is 462; $\log 462\, p(x_1, x_2, x_3) = \log(2) + 3\log(2)x_1 + \log(2)x_2 + 2\log(2)x_3 + \log(2)x_1x_3$. Comment: $X_2 \perp\!\!\!\perp (X_1, X_3)$.

25: Full table: $\log \text{ const } p(x_1, x_2, x_3) = \theta x_1 + \theta x_2 x_3 - \theta x_1 x_2 - \theta x_1 x_3$. Margin: $\log \text{ const } p_{23}(x_2, x_3) = \alpha + (\log 2 - \alpha)x_2 + (\log 2 - \alpha)x_3 + 2(\alpha - \log 2)x_2x_3$ where $\alpha = \log(1 + e^{\theta})$.

26: Hard part is to note $E(XY) = E(X E_{Y|X}(Y)) = E\rho X^2 = \rho$. $E_{X|Y}(X) = \rho Y$ and $\text{var}_{X|Y}(X) = 1 - \rho^2$.

27: (a) $Y|X \sim N(\beta X, 1)$,
(b) $f_{XY}(x, y) = f_X(x) f_{Y|X}(y; x) = 1/2\pi \exp(-1/2(y^2 - 2\beta yx + (1 + \beta^2)x^2)$,
(c) $Y \sim N(0, 1 + \beta^2)$, (d) $i_{XY}(x, y) = \beta$.

28: Require $\rho \neq \pm 1$ for matrix inversion. Require $1 - \rho^2 > 0$ to conclude $f_{Y|X}$ has a finite integral over $-\infty, \infty$ and that it is Normal.

29: Variables 2 and 3.

30: $\rho_{21} = \rho_{23}\rho_{31}$ and $\rho_{23} = \rho_{21}\rho_{13}$, which imply $\rho_{21}(1 - \rho_{13}^2) = 0$. As no correlation can be unity, $\rho_{21} = 0$, similarly $\rho_{23} = 0$.

Outline answers to Exercises from Chapter 3

1:

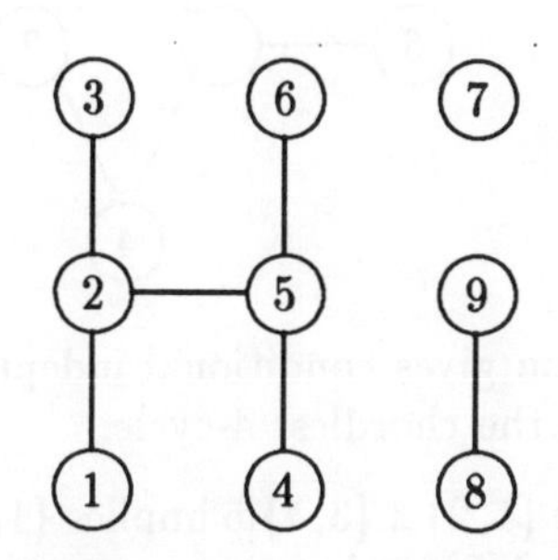

Neighbourhood set of 2 is $\{1, 3, 5\}$.

2: Paths A to F are A, B, D, E, F (short); A, B, D, C, F; A, B, C, F (short); and A, B, C, D, E, F. There are 11 distinct separating subsets: $\{B\}$, $\{C, D\}$, ..., $\{B, C, D, E\}$. Boundary $\{B, D, F\}$.

3:

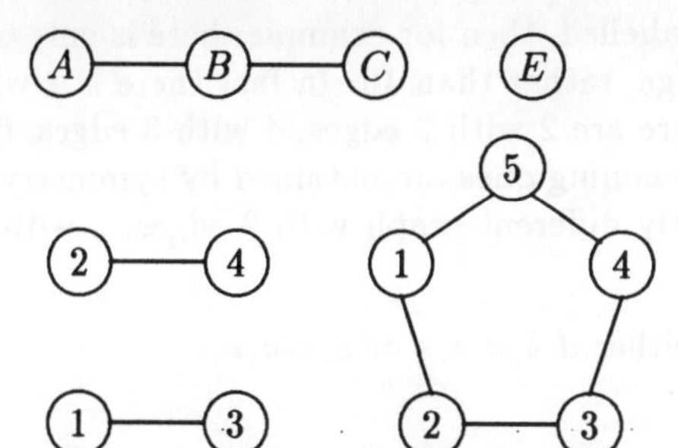

4: Cliques: $\{1,4\}$ $\{2,4\}$ $\{3,4\}$ $\{5\}$; all but $\{5\}$ separates; no chordless cycles.
Cliques: $\{1,2\}$ $\{2,3\}$ $\{3,4\}$ $\{4,5\}$ $\{5,6\}$ $\{1,6\}$ $\{2,5\}$; only $\{2,5\}$ separates; 2 chordless cycles.
Cliques: $\{1,2,5\}$ $\{2,3,5\}$ $\{3,4,5\}$ $\{1,4,5\}$; no separating cliques; one c-cycle $\{1,2,3,4\}$.

5:

$1 \perp\!\!\!\perp 3 | \{2,4,5,6\}$, $1 \perp\!\!\!\perp 4 | \{2,3,5,6\}$, $1 \perp\!\!\!\perp 5 | \{2,3,4,6\}$, $1 \perp\!\!\!\perp 6 | \{2,3,4,5\}$,
$2 \perp\!\!\!\perp 4 | \{1,3,5,6\}$, $2 \perp\!\!\!\perp 5 | \{1,3,4,6\}$, $2 \perp\!\!\!\perp 6 | \{1,3,4,5\}$,
$3 \perp\!\!\!\perp 5 | \{1,2,4,6\}$, $3 \perp\!\!\!\perp 6 | \{1,2,4,5\}$, $4 \perp\!\!\!\perp 6 | \{1,2,3,5\}$.

6: (a) $1 \perp\!\!\!\perp 3 | \{2\}$, $1 \perp\!\!\!\perp 4 | \{2\}$, $3 \perp\!\!\!\perp 4 | \{2\}$.
(b) $1 \perp\!\!\!\perp 3 | \{2\}$, $1 \perp\!\!\!\perp 4 | \{2\}$, $1 \perp\!\!\!\perp 4 | \{3\}$, $2 \perp\!\!\!\perp 4 | \{3\}$.
(c) $1 \perp\!\!\!\perp 3 | \{2,4\}$, $2 \perp\!\!\!\perp 4 | \{1,3\}$.
(d) $1 \perp\!\!\!\perp \{4,5,6\} | \{2,3\}$, $4 \perp\!\!\!\perp \{1,3,6\} | \{2,5\}$, $6 \perp\!\!\!\perp \{1,2,4\} | \{3,5\}$

7:

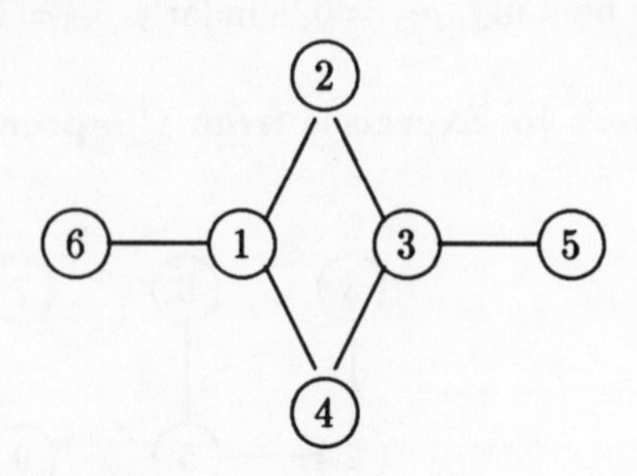

9: Factorisation criterion gives conditional independences between pairs 1,3 and 2,4; so the graph is the chordless 4-cycle.

10: Block independence $\{1,2\} \perp\!\!\!\perp \{3,4\}|5$ implies $\{1,2\} \perp\!\!\!\perp 3|\{4,5\}$ and $\{1,2\} \perp\!\!\!\perp 4|\{3,$ Continue to get $1 \perp\!\!\!\perp 3|\{2,4,5\}$ and so on.

11: $1 \perp\!\!\!\perp \{3,4,5\}|\{2\}$, $2 \perp\!\!\!\perp \{4,5\}|\{1,3\}$, $3 \perp\!\!\!\perp \{1,5\}|\{2,4\}$, $4 \perp\!\!\!\perp \{1,2\}|\{3,5\}$, $5 \perp\!\!\!\perp \{1,2,3\}|\{4\}$.

13: For a labelled graph there are $2^{10} = 1024$ different graphs. However if the graph is unlabelled, then for example there is only one distinctly different graph with 1 edge, rather than 10. In fact there is 1 with 0 edges, there is 1 with 1 edge, there are 2 with 2 edges, 4 with 3 edges, 6 with 4 edges, 6 with 5 edges. The remaining ones are obtained by symmetry, so for example there is just 1 distinctly different graph with 9 edges, 2 with 8edges, and so on: a total of 34.

17: Ordering: either d, b, a, c, e or d, b, a, e, c.

Outline answers to Exercises from Chapter 4

1: $I(\lambda; \nu) = \lambda \log(\lambda/\nu) - \lambda + \nu$

2: $2I((\mu, \sigma^2); (0,1)) = \mu^2 + \sigma^2 - \log \sigma^2 - 1 = 2I(\mu; 0) + 2I(\sigma^2; 1)$

3: $I(\alpha; \beta) = p(\alpha/\beta - \log \alpha/\beta - 1)$

4: $t \log t \to 0$ First and second derivatives are $1 + \log t$ and $1/t$. The latter is strictly positive for $t > 0$. Hence strictly convex.

6: Start with the identity

$$1 \leq 1/(1-z) = 1 + z + z^2 + \ldots + z^{n-1} + z^n/(1-z)$$

for $0 \leq z < 1$. Now integrate from 0 to t to give $\int dz \leq \int 1/(1-z) dz$.

15:

$$\text{Inf}(X \perp\!\!\!\perp (Y_1, Y_2, \ldots, Y_q)) = \sum_{i=1}^{q} \text{Inf}(X \perp\!\!\!\perp Y_i | (Y_1, Y_2, \ldots, Y_{i-1})).$$

Outline answers to Exercises from Chapter 5

1: (a)

$$\begin{aligned}
\text{cov}(X, X+Y) &= \text{cov}(X, X) + \text{cov}(X, Y) \\
&= v(1+r). \\
\text{cov}(X-Y, X+Y) &= \text{cov}(X, X+Y) - \text{cov}(Y, X+Y) \\
&= \text{cov}(X, X) + \text{cov}(X, Y) - \text{cov}(Y, X) - \text{cov}(Y, Y) \\
&= v(1+r-r-1) = 0. \\
\text{cov}(X, Y-rX) &= \text{cov}(X, Y) + \text{cov}(X, -rX) \\
&= \text{cov}(X, Y) - r\text{var}(X) \\
&= vr - rv = 0.
\end{aligned}$$

(b) As $\text{var}(X+Y) = 2v(1+r)$, and $\text{var}(X-Y) = 2v(1-r)$ the inequality follows as var is non-negative.
(c) $\text{var}(Y-bX) = \text{cov}(Y-bX, Y-bX) = \text{cov}(Y, Y) - 2b\text{cov}(Y, X) + b^2\text{var}(X)$, The derivative $d/db\text{var}(Y-bX) = -2\text{cov}(X, Y) + 2b\text{var}(X) = 0$, so that $b = \text{cov}(Y, X)/\text{var}(X) = rv/v = r$, and the min is $\text{var}(Y-bX) = v(1-r^2)$. (d) Now $\text{var}(aX+by) = a^2\text{var}(X) + 2ab\text{cov}(X, Y) + b^2\text{var}(Y) = (a^2+2abr+b^2)v$. Consider

$$\begin{aligned}
s &= a^2 + 2abr + b^2 - \theta(a^2 + b^2 - 1), \\
ds/da &= 2a + 2br - 2a\theta = 0 \\
ds/db &= 2ar + 2b - 2b\theta = 0
\end{aligned}$$

or $\begin{pmatrix} 1 & r \\ r & 1 \end{pmatrix}\begin{pmatrix} a \\ b \end{pmatrix} = \theta\begin{pmatrix} a \\ b \end{pmatrix}$, together with $ds/d\theta = a^2 + b^2 - 1 = 0$. These equations are solved by $a = b = 1/\sqrt{2}$ and $a = -b = 1/\sqrt{2}$. Maximum $v(1+r)$ at $(a, b) = (1, 1)/\sqrt{2}$, minimum $v(1-r)$ at $(a, b) = (1, -1)/\sqrt{2}$, if $r > 0$.

2: $\text{cov}(u^T X, v^T X) = u^T\text{var}(X)v = -7$, $\text{var}(u^T X) = 63$ and $\text{var}(v^T X) = 11$. Correlation is -0.266.

3: The variance matrices of Y are

(a) $\begin{pmatrix} 2 & 1 & 1 & 1 \\ 1 & 2 & 1 & 1 \\ 1 & 1 & 2 & 1 \\ 1 & 1 & 1 & 2 \end{pmatrix}$ (b) with $t = 1 + \alpha^2$ $\begin{pmatrix} t & \alpha & 0 & 0 \\ \alpha & t & \alpha & 0 \\ 0 & \alpha & t & \alpha \\ 0 & 0 & \alpha & t \end{pmatrix}$

(c) $\begin{pmatrix} 1 & 1 & 1 & 1 \\ 1 & 2 & 2 & 2 \\ 1 & 2 & 3 & 3 \\ 1 & 2 & 3 & 4 \end{pmatrix}$ (d) with $t = (1-\alpha^2)^{\frac{1}{2}}$ $\begin{pmatrix} 1 & t & t^2 & t^3 \\ t & 1 & t & t^2 \\ t^2 & t & 1 & t \\ t^3 & t^2 & t & 1 \end{pmatrix}$.

4: If $Y = AX$ then $\text{var}(Y) = A\text{var}(X)A^T = AA^T$. Now

$$A = \begin{pmatrix} 1 & 0 & 0 & 0 \\ 1 & 1 & 0 & 0 \\ 1 & 0 & 1 & 0 \\ 0 & 1 & 0 & 1 \end{pmatrix} \text{ so that } \text{var}(Y) = \begin{pmatrix} 1 & 1 & 1 & 0 \\ 1 & 2 & 1 & 1 \\ 1 & 1 & 2 & 0 \\ 0 & 1 & 0 & 2 \end{pmatrix}.$$

If $X = BY$ then $\text{var}(X) = B\text{var}(Y)B^T$ and as $\text{var}(X) = I$, $\text{var}(Y) = B^{-1}B^{-T}$. Consequently $\text{var}(Y)^{-1} = B^TB$. Now

$$B = \begin{pmatrix} 1 & 0 & 0 & 0 \\ -1 & 1 & 0 & 0 \\ -1 & 0 & 1 & 0 \\ 1 & -1 & 0 & 1 \end{pmatrix} \text{ so that } \text{var}(Y)^{-1} = \begin{pmatrix} 4 & -2 & -1 & 1 \\ -2 & 2 & 0 & -1 \\ -1 & 0 & 1 & 0 \\ 1 & -1 & 0 & 1 \end{pmatrix}.$$

5: (a) $aX_1 + bX_2$
(b) aX_2
(c) $\{(a-br)X_1 + (b-ar)X_2\}/(1-r^2) + cX_3$

8: Consider arbitrary predictor of the form b^TX and note that $\text{var}(AY - b^TX) = \text{var}(AY - A\hat{Y} + A\hat{Y} - b^TX)$. Then use $\text{cov}(Y-\hat{Y}, \hat{Y}) = 0$ and $\text{cov}(Y - \hat{Y}, X) = 0$.

9: Manipulation together with symmetries between X and Y :
(a) write $\hat{Y} = BX$, use $\text{cov}(\hat{X}, X) = \text{var}(\hat{X})$
(b) $\text{cov}(Y-\hat{Y}, X-\hat{X}) = 0 - B\text{cov}(X, X-\hat{X})$ and $\text{cov}(X, X-\hat{X}) = \text{var}(X - \hat{X})$
(c) use $\text{cov}(Y-\hat{Y}, X-\hat{X}) = 0 - \text{cov}(Y-\hat{Y}, \hat{X}) = -\text{cov}(Y, \hat{X}) + \text{cov}(\hat{Y}, \hat{X})$ and $\text{cov}(Y, \hat{X}) = \text{cov}(Y, X)$

10: The transform $\begin{pmatrix} 0 & I \\ I & 0 \end{pmatrix}$ is invertible.

11: Definition: $R(Y;X) = \text{corr}(Y, \hat{Y}(X)) = \text{cov}(Y,\hat{Y})/\text{var}(Y)\text{var}(\hat{Y})^{1/2} = \text{var}(\hat{Y})/\text{var}(Y)\text{var}(\hat{Y})^{1/2}$

13: Use $\hat{Y}(X) = E(Y) + \text{cov}(Y,X)\text{var}(X)^{-1}(X - EX)$. Then substitute $E(X) = P(X=1)$, $\text{cov}(X,Y) = P(X=1, Y=1) - P(X=1)P(Y=1)$ and $\text{var}(X) = P(X=0)P(X=1)$.

16: $\text{cov}(X, Y|Z) = 3/2$, $\text{corr}(X, Y|Z) = 0.51$, $R(X; (Y, Z)) = 0.38$.

17: $\text{var}(Y|X) = 1 - 2r^2/(1-r)$. This is zero if $r = 1/2$ ($r = -1$ is not admissible).

18: There are several ways to prove the first assertion. Because $p = 2$ it is possible to invert $\text{var}(X)$ and get the result explicitly. An alternative is to note that

$$\text{cov}(Y, X_1|X_2) = \text{cov}(Y, X_1) - \text{cov}(Y, X_2)\text{var}(X_2)^{-1}\text{cov}(X_2, X_1)$$

But because $\hat{Y}$ is the llsp $\text{cov}(Y, X_i) = \text{cov}(\hat{Y}, X_i)$ for $i = 1, 2$ Now use $\hat{Y} = bX_2$ and substitute to get the result. This generalises to arbitrary dimensions. No.

24: Any positive definite matrix with unit diagonals is interpretable as a correlation matrix. The diagonal elements of its inverse are of the form $1/(1-R^2)$ where R is a multiple correlation coefficient and so is greater than unity.

25: Replace $\text{var}(X, Y)$ by $\text{var}(X, Y - \hat{Y} + \hat{Y})$ and expand.

$$\begin{pmatrix} V_X & V_X B^T \\ BV_X & V + BV_X B^T \end{pmatrix}^{-1} = \begin{pmatrix} V_X^{-1} + B^T V^{-1} B & -B^T V^{-1} \\ -V^{-1} B & V^{-1} \end{pmatrix}$$

27: Partition $Z = (X, Y)$ where Y is one dimensional. Then $\det\text{var}(X, Y) = \det\text{var}(X)\det\text{var}(Y|X)$ and is maximised with respect to Y by maximising the scalar $\text{var}(Y|X) = 1 - \text{cov}(Y, X)\text{var}(X)^{-1}\text{cov}(X, Y)$. But $\text{var}(X)$ is positive definite and so the right hand side is maximised by setting $\text{cov}(Y, X) = 0$. Now attack X in the same manner. The process terminates as p is finite.

Outline answers to Exercises from Chapter 6

3:

$$(a)\ \begin{array}{c|cccc} 1 & * & & & \\ 2 & * & * & & \\ 3 & 0 & * & * & \\ 4 & * & 0 & * & * \end{array} \qquad (b)\ \begin{array}{c|cccc} 1 & * & & & \\ 2 & * & * & & \\ 3 & 0 & * & * & \\ 4 & 0 & * & 0 & * \end{array} \qquad (c)\ \begin{array}{c|cccccc} 1 & * & & & & & \\ 2 & * & * & & & & \\ 3 & * & * & * & & & \\ 4 & 0 & * & 0 & * & & \\ 5 & 0 & * & * & * & * & \\ 6 & 0 & 0 & * & 0 & * & * \end{array}$$

4: Find $\text{var}(Y)^{-1}$: Now $Y = AX$, where

$$A = \begin{pmatrix} 1 & 0 & 0 & 0 \\ 1 & 1 & 0 & 0 \\ 1 & 1 & 1 & 0 \\ 0 & 1 & 0 & 1 \end{pmatrix} \text{ and } A^{-1} = \begin{pmatrix} 1 & 0 & 0 & 0 \\ -1 & 1 & 0 & 0 \\ 0 & -1 & 1 & 0 \\ 1 & -1 & 0 & 1 \end{pmatrix}$$

and because $\text{var}(Y)^{-1} = A^{-T}A^{-1}$

$$\text{var}(Y)^{-1} = \begin{pmatrix} 3 & -2 & 0 & 1 \\ -2 & 3 & -1 & -1 \\ 0 & -1 & 1 & 0 \\ 1 & -1 & 0 & 1 \end{pmatrix}$$

From the zeros, the independence graph is

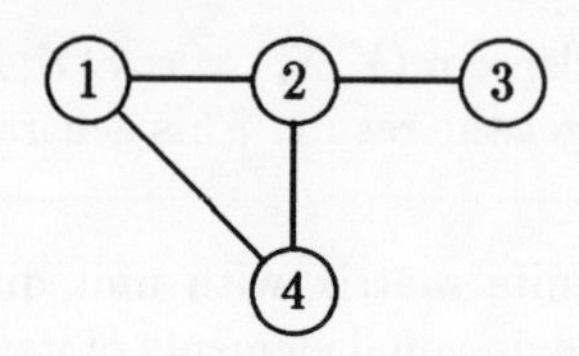

Conditionally on Y_2, Y_3 is independent of (Y_1, Y_4). Consequently the llsp

$$\begin{aligned} \hat{Y}_3(Y_1, Y_2, Y_4) &= \hat{Y}_3(Y_2) \\ &= \text{cov}(Y_2 + X_3, Y_2)\text{var}(Y_2)^{-1}Y_2 \\ &= Y_2. \end{aligned}$$

5: $Y|Z \sim N(BZ, I)$ and $Z \sim N(0, I)$ so that $(Y, Z) \sim N$ and $E(Y|Z) = BZ$ and $\text{var}(Y|Z) = I$. If $\hat{Y} = \hat{Y}(Z)$ is the llsp of Y then $\hat{Y} = E(Y|Z)$ because $(Y, Z) \sim N$. Now use

$$\text{var}(Y) = \text{var}(Y|Z) + \text{var}(\hat{Y}) = I + \text{var}(BZ) = I + BB^T$$

and

$$\text{cov}(Y, Z) = \text{cov}(Y - \hat{Y} + \hat{Y}, Z) = \text{cov}(\hat{Y}, Z) = B\text{var}(Z) = B$$

(Alternatively use $\text{var}(Y) = E\text{var}(Y|Z) + \text{var}(E(Y|Z))$.) Hence

$$\text{var}(Y, Z) = \begin{pmatrix} I + BB^T & B \\ B^T & I \end{pmatrix}$$

Check inverse by multiplication or by substitution in the inverse variance lemma. Check $\text{var}(Y)^{-1}$ by multiplication. The independence graph of (Y, Z) is

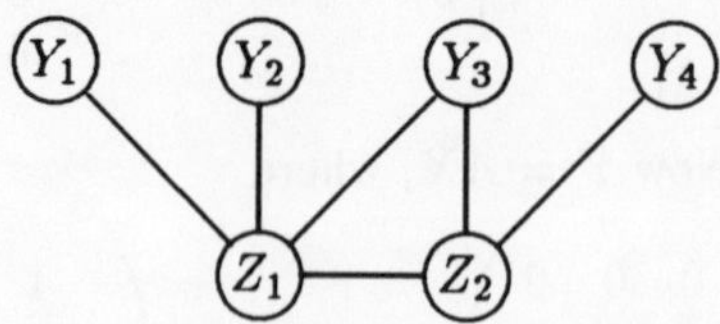

while the independence graph of Y is the complete graph on 4 vertices. The Z's serve to explain the Y's.

6: First, $|Z(v)| \leq c$ for all v, is equivalent to $max_v Z(v)^2 \leq c^2$. Now

$$Z(v)^2 = [v, X]^2/[v, Vv] = [v, VV^{-1}X]^2/[v, Vv]$$
$$\leq [v, Vv][V^{-1}X, VV^{-1}X]/[v, Vv] = [X, V^{-1}X]$$

irrespective of v, using the Cauchy-Schwarz inequality. Finally the right hand side has a chi-squared distribution.

9: $-1/2\log\{\det\mathrm{var}(Y, Z|X)/\det\mathrm{var}(Y|X)\det\mathrm{var}(Z|X)\}$ which is the (average) divergence against conditional independence.

10: From Section 6.4 $\mathrm{Inf}(X \perp\!\!\!\perp Y) = -1/2\log\det\mathrm{var}(X, Y)/\det\mathrm{var}(X)\det\mathrm{var}(Y) = -1/2\log\det\mathrm{var}(Y|X)/\det\mathrm{var}(Y)$, by Schur's identity. As Y is scalar, the determinant disappears and from Chapter 5 $R^2 = 1 - \mathrm{var}(Y|X)/\mathrm{var}(Y)$.

11: $I(f; g) = E\log f_{X,Y}/g_{X,Y} = E\log f_{Y|X}h(X)/g_{Y|X}h_X = E\log f_{Y|X}/g_{Y|X}$. Using the fact that the conditional densities are Normal $I(f; g) = E(-(Y - \beta X)^2/2\sigma^2 + Y^2/2\sigma^2)$ because the log det factor disappears, and finally $I = (-\mathrm{var}(Y|X) + \mathrm{var}(Y))/2\mathrm{var}(Y|X) = 1/2R^2/(1 - R^2)$.

13: Put $a = \{1, 2\}$, $b = \{3\}$, deviance is $-N\log\det S/\det S_{aa}\det S_{bb} = 52.68$.

14: The sample variance matrices of X_a and X_b are

mech	302.29		
vect	125.78	170.88	
alg	100.43	84.19	111.60

anal	217.88	
stat	153.77	294.37

The determinants are $\det S = 370213.53X10^5$, $\det S_{aa} = 2260218.50$ and $\det S_{bb} = 40491.94$. With $N = 88$, the deviance for the independence class of models is

$$-88\log(370213.53)10^5/(2260218.50)(40491.94) = 79.6461.$$

Compared to the percentage point of chi-square with $3.2 = 6$df the independence hypothesis is overwhelmingly rejected.

16: Apply the inverse variance lemma to the partitioned matrix $\mathrm{var}(X, Y)$ identified with

$$\begin{pmatrix} S & S_j^T \\ S_j & n \end{pmatrix}$$

Then $\mathrm{var}(X|Y) = S_j$, and $\mathrm{var}(Y|X) = N - S_j^T S^{-1} S_j$ and the result follows from $\det\mathrm{var}(X|Y)\det\mathrm{var}(Y) = \det\mathrm{var}(Y|X)\det\mathrm{var}(X)$.

Outline answers to Exercises from Chapter 7

1: Impossible, in contrast to a similar exercise for the Normal distribution.

2: Writing out the constraint gives the first assertion. Substituting for $x_1 = 0, 1$ and $x_2 = 0, 1$ gives

$$\begin{aligned}
\log p(0,0) &= u_\phi + u_2(0) + u_1(0) + u_{12}(0,0),\\
\log p(0,1) &= u_\phi + u_2(1) + u_1(0) + u_{12}(0,1),\\
\log p(1,0) &= u_\phi + u_2(0) + u_1(1) + u_{12}(1,0),\\
\log p(1,1) &= u_\phi + u_2(1) + u_1(1) + u_{12}(1,1).
\end{aligned}$$

Multiply the 2nd and 3rd equation by -1 and add.

3: Substituting for $x_1 = 0, 1$ and $x_2 = 0, 1, 2$ gives

$$\begin{aligned}
\log p(0,0) &= u_\phi\\
\log p(0,1) &= u_\phi + u_2(1)\\
\log p(0,2) &= u_\phi + u_2(2)\\
\log p(1,0) &= u_\phi + u_1(1)\\
\log p(1,1) &= u_\phi + u_2(1) + u_1(1) + u_{12}(1,1)\\
\log p(1,2) &= u_\phi + u_2(2) + u_1(1) + u_{12}(1,2)
\end{aligned}$$

4:

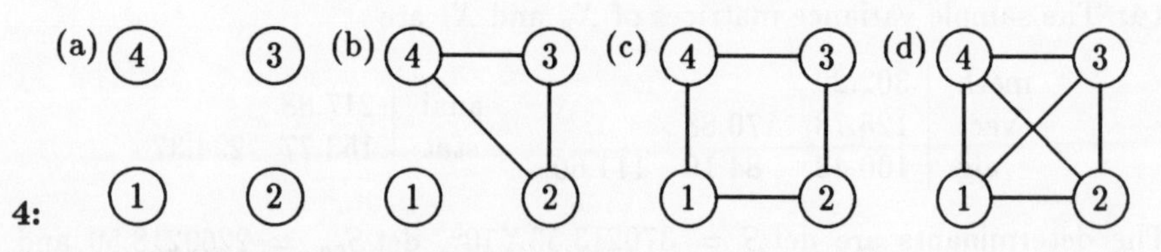

5: Grand total of table is 2000 and $\log p(x_1, x_2, x_3, x_4)$ is

$$\begin{aligned}
\log p = 5.375 - 1.792x_1 - 1.386x_2\\
- 2.197x_3 + 0.847x_4 + 2.773x_1x_2 + 3.584x_1x_3 - 3.045x_1x_4.
\end{aligned}$$

All the third and higher terms disappear, as do some of the second. The interaction graph is

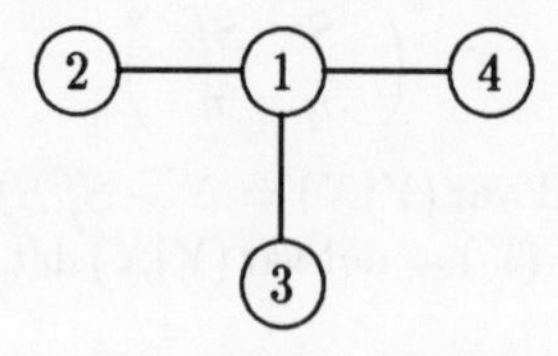

Given X_1, the variables X_2, X_3, X_4 are independent.

6: (a) $1.2 + 2.3 + 3.4 + 1.4$ (b) $1.2 + 2.3 + 2.4$ (c) $1.2.3 + 2.3.5 + 2.4.5 + 3.5.6$

7: Lattice diagram of equi-probability models in three dimensions:

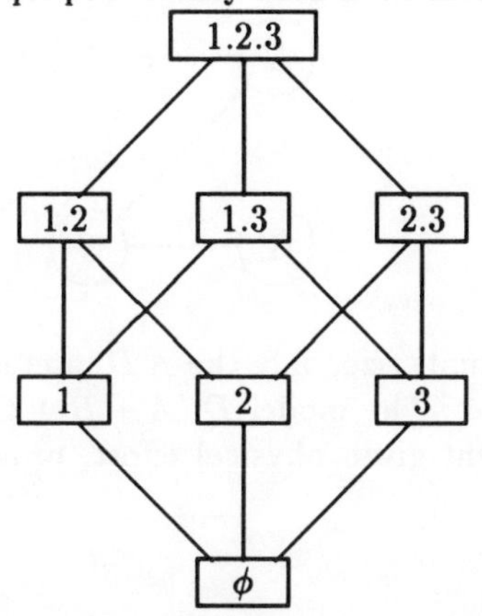

10: The log-likelihood function is $36u_\phi + 8u_{123}$. The constraint $\sum p = 1$ simplifies to $7\exp\{u_\phi\} + \exp\{u_\phi + u_{123}\} = 1$. Using Lagrange multipliers gives $\hat{p}_{123}(1,1,1) = 8/36 = 2/9$, so that $\hat{p}_{123}(x) = 1/9$, for $x \neq (1,1,1)$. The u-terms are $\hat{u}_\phi = -\log(9)$ and $\hat{u}_{123} = \log(2)$.

14: Note the zeros.

17: Each edge exclusion deviance has 4 df:

	1	2	3	4
1	*			
2	4.56	*		
3	5.51	27.18	*	
4	5.06	5.92	3.95	*

The model $1 + 2.3 + 4$ has a deviance of 18.19 on 10 df.

19: Variables 1 and 2 are see and buy at the first interview, 3 and 4 at the second interview. The exclusion deviances are

	1	2	3	4
1	*			
2	26.78	*		
3	253.9	31.63	*	
4	47.36	534.1	30.53	*

Outline answers to Exercises from Chapter 8

5: The model $E.N + N.S + S.W$ appears to fit as well as any other. It has a deviance of 3.87 on 3 df and the graph is

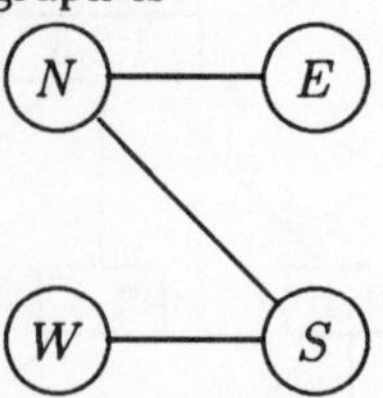

7: In relation to the sample size, only the $A.D$ interaction between noise and physical effort is sizable. The model $D.(A + B + C)$ in which the 3 other variables are independent given physical effort, provides a fair summary of the table.

Outline answers to Exercises from Chapter 9

2: $\log \text{cpr}\,(\text{smo,can}|\text{sex=male}) = 2.64$, $\log \text{cpr}\,(\text{smo,can}|\text{sex=female}) = 0.90$. The mle of a common value is $\log \text{cpr}\,(\text{smo,can}|\text{sex}) = 1.45$, but this is not an acceptable hypothesis, dev = 5.74 on 1 df. The estimate in the margin is $\log \text{cpr}\,(\text{smo,can}) = 1.09$.

6: Take 1 empty, implication is $N_2 \perp\!\!\!\perp N_3 | N_\phi$. Yes.

7: (a) dev=2.91, 1 df, asymp P-value=0.0880, exact P-value=0.250.
(b) dev=5.82, 1 df, asymp P-value=0.0158, exact P-value=0.020.

8: Prob $= \frac{3!6!5!\ 7!7!}{14!\ 1!4!2!2!2!3!} = 0.131$,
The tables are determined by the possible values for $n_{12}(1,1)$ and $n_{12}(1,1)$: in all 24 tables.

10: The deviance statistic is 15.34 on 4 df and Pearson's statistic is 14.76 on 4 df with asymptotic P-values of 0.0042 and 0.0053 respectively. A simulation experiment with 5000 trials estimated the value at 0.0039, while the exact value is 0.0038. The margins of the table are of reasonable size and the exact and asymptotic values are rather close.

11: Using 500 generated tables for the exact P-values, the P-values are

hypothesis	X^2	df	asymp	exact
$A \perp\!\!\!\perp I \| F$	49.12	2	0.000	0.000
$A \perp\!\!\!\perp F \| I$	1.04	2	0.596	0.623
$I \perp\!\!\!\perp F \| A$	0.05	2	0.977	1.000

Comment: exact and asymptotic are close, even with the extreme values exhibited. Either analysis leads to the model that family history is independent of age and insulin dependence, which themselves are closely related.

Outline answers to Exercises from Chapter 10

2: (a) (15.1, 25.0, 11.1) percent respectively,
(b)(i) complete independence: dev 224.57 on 12 df. (ii) dev($Y \perp\!\!\!\perp X$) 103.74 on 9 df. (iii) dev 85.69 on 3 df.
(d) dev of model with edges $X_1.X_2.X_3 + Y_1.Y_2.Y_3 + X_1.Y_2 + X_3.Y_2 + X_3.Y_3$ is 9.43 on 6 df.
(e) fitted partial correlations are

Y_1	.00	.09	−.00	1.00		
Y_2	.18	−.00	.20	.29	1.00	
Y_3	.00	−.00	.14	.12	.16	1.00
	X_1	X_2	X_3	Y_1	Y_2	Y_3

3: (b) The inclusion deviances are

moth	6.39	5.22	0.44	*
	min	max	rain	moth

(c) Analysis of variance:

	df	ss	ms	F
regression	2	0.767	0.38340	9?
residual	70	2.783	0.03976	
total	72	3.550	0.04931	

(e) The deviances for fitting the four regression models:

X	dev	df
(min,max,wind,cloud)	0.24	1
(max,wind,cloud)	2.78	1
(min, wind,cloud)	1.61	1
(wind,cloud)	7.80	2

5: Partition the variables into 3 blocks: $\{a, b\}$, $\{c, d\}$, $\{e, f\}$. The edge exclusion deviances from (i) the response model $a.b$, (ii) the treatment-response model $(a.b).c.d$, and (iii) the treatment-response model $(a.b.c.d).e.f$, are

b	16.16				
c	5.29	1.65			
d	0.33	1.09	2.85		
e	2.30	4.55	0.13	1.06	
f	5.74	0.32	0.61	2.12	67.91
	a	b	c	d	e

Inspection of these deviances leads to fitting

independence models			fitted model		
$a+b$	16.17	1	ab	0.00	0
$ab+c+d$	13.37	5	$ab+ca+d$	5.19	4
$abcd+e+f$	102.28	9	$abcd+eb+af+ef$	7.49	6
$a+b+c+d+e+f$	131.82	15	*total*	12.68	10

and the fitted model can be summarised by $/a.b/(a.b)+a.c+d/(a.b.c.d)+e.b+a.f+e.f/$ leading to a chain graph with 6 edges. There are some queries about the questionnaire design for the psychological variables, as the correlations are all rather small.

6: (b) The moderate size of standardised residuals, model reinforce the conclusion that the model fits well; however the distribution evinced by the histogram

interval		frequency
(-1.50,-1.0)	11	+++++++++++
<-0.5)	17	+++++++++++++++++
< 0.0)	6	++++++
< 0.5)	12	++++++++++++
< 1.0)	17	+++++++++++++++++
< 1.5)	7	+++++++
< 2.0)	1	+
< 2.5)	1	+

has too thicker tail to be well approximated by the Normal distribution.

7:

attributable to		dev	df	attr.	dev	df
dependence	(E,C) adj for D	80.40		(E,C)	86.30	
	(E,D)	38.78		(E,D) adj for C	32.88	
total dependence		119.2	5		119.2	5
random variation	E⫫ (A,B)|(C,D)	39.7	30			
total		158.9	35			

Outline answers to Exercises from Chapter 11

1: $X|Z=0 \sim N(0,1)$ and $X|Z=1 \sim N(0,1)$; hence $X \perp\!\!\!\perp Z$ and $X \sim N(0,1)$. Similarly $Y \sim N(0,1)$. If $Z=0, T \sim N(0,2)$ while if $Z=1, T \sim N(0,2+2r)$ and the marginal distribution of T is a mixture. The exponential function is not closed under addition.

2: 10

4: (i) $AB/BX/BX$, (ii) $AB/ABX, BY/ABX, BXY$, (iii) $B/BX, Y/BX, XY$, (iv) $AB/BX, Y/BX, XY$.

5: The density function satisfies $f(0,y) = f(1,y)$ for all y, so the density exhibits both conditional and marginal equiprobability.

6: $\log p(i) - k/2\log 2\pi - 1/2\log\det V(i) - 1/2[\mu(i), V(i)^{-1}\mu(i)]$.

7: $\text{var}\,(Y|I_1, I_2)^{-1} = \psi_\phi + \psi_1 + \psi_2$.

9: $l(\lambda_\phi, \lambda_1(1), \lambda_1(2), \eta_\phi, \eta_1(1), \eta_1(2), \psi_\phi, \psi_1(1), \psi_1(2)) = \lambda_\phi n_\phi + \lambda_1(1)n_1(1) + \lambda_1(2)n_1(2) + \eta_\phi t_\phi + \eta_1(1)t_1(1) + \eta_1(2)t_1(2) + \psi_\phi T_\phi + \psi_1(1)T_{11}(1) + \psi_1(2)T_{11}(2)$.

12: $I_1.I_2.(Y_1+Y_2)^2$ 0; $I_2.Y_2^2.(I_1+Y_1^2)$ 6; $I_1.Y_1^2.(I_2+Y_2^2)$ 6; $(I_1+Y_1^2).(I_2+Y_2^2)$ 11.

13: The log-linear model with formula $A + B.W.(L_A + Q_A)$ where L_A and Q_A denote the linear and quadratic components of A, has a deviance of 17.66 on 18 df and provides an excellent fit. It is possible to slightly simplify the model, by removing the quadratic interaction $-B.W.QA$ (change = +0.39 on 1 df) but not the linear interaction $-B.W.L_A$ (change = +19.61 on 1 df) and also to remove $-B.Q_A$ (change = +1.78 on 1 df). The parameters of interest in the resulting model are the interactions

	interaction	s.e.
$B.L_A$	−0.378	0.019
$W.L_A$	−0.265	0.056
$W.Q_A$	0.019	0.004
$B.W$	3.655	0.196
$B.W.L_A$	−0.127	0.029

The main feature of the data is how the strong interaction between breathlessness and wheeze, which starts from a very high estimated value of 3.66 for the log-cpr in the first age group, decreases steadily with age, taking 0.13 off the log-cpr for each age group. The sample log-cprs in each age group are 3.22 3.70 3.40 3.14 3.02 2.78 2.93 2.44 2.64 which confirms the analysis.

Outline answers to Exercises from Chapter 12

1:

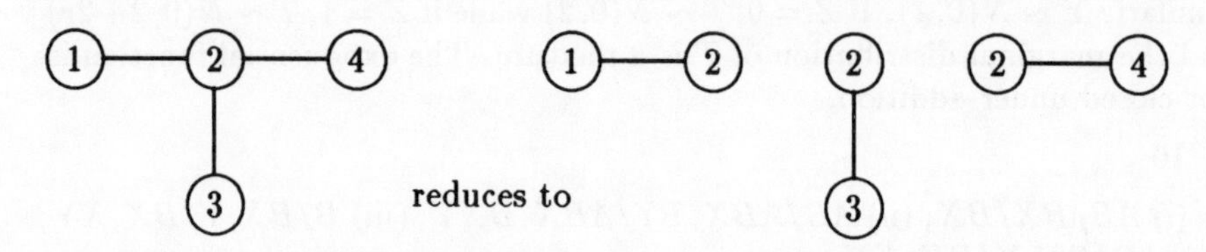

each of which is irreducible and complete. Hence it is decomposable. The joint density function factorises into $f_{1234} = f_{12} f_{23} f_{24} / f_2 f_2$.

2: If the density has the interaction expansion

$$\log f_{1234} = h_{12} + h_{23} + h_{34} + h_{14},$$

then the interaction expansion for the marginal distribution of (X_1, X_2, X_3) has no three-way interaction term:

$$\begin{aligned} f_{123} &= \exp(h_{12} + h_{23}) \int \exp(h_{34} + h_{14}) dx_4 \\ &= \exp(h_{12} + h_{23} + g_{13}). \end{aligned}$$

5: This relates to the existence of so-called perfect tables, see Darroch (1962), Whittemore (1978).

6:

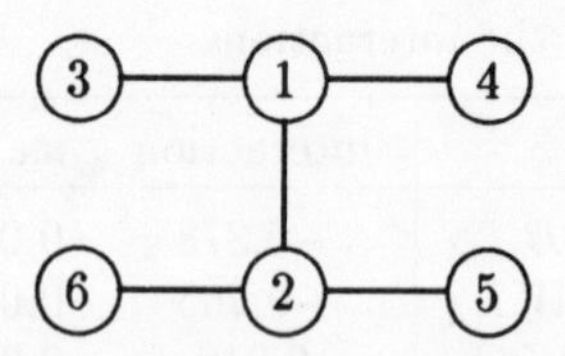

The separating set $\{1, 3\}$ contains a minimal subset: $\{1\}$. The separating set $\{1, 2\}$, contains two potentially minimal subsets: $\{1\}$ and $\{2\}$.

7:

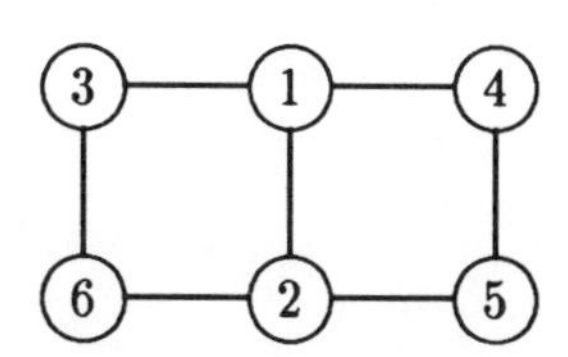

8: (a) $k = 4$, 4; $k = 8$, 1 (b) $k = 4$, 2; $k = 5$, 2; $k = 6$, 1.

9: $\{1\}$ because the boundary of $\{1\}$ is $\{2\}$;
not $\{2\}$ because the boundary of $\{2\}$ is $\{1,3\}$;
$\{1,2\}$ because the boundary of $\{1,2\}$ is $\{3\}$;
not $\{2,3\}$ because the boundary of $\{2,3\}$ is $\{1,4\}$;
$\{1,5\}$ because the boundary of $\{1\}$ is $\{2\}$ and the boundary of $\{5\}$ is $\{4\}$.

10: (a) No. The two-way interactions 12 and 23 are preserved, the zero three-way interaction 123 is preserved, the two-way interaction 13 is not preserved. It is not graphically collapsible as the boundary $\{1,3\}$ is not complete.
(b) All interactions are preserved. It is graphically collapsible.
(c) Yes. The three-way interaction 123 is preserved. It is graphically collapsible as the 3 connected components of $\{4,5,6\}$ have complete boundaries.

References

Agresti, A., Wacherly, D. and Boyett, J.M. (1979). Exact conditional tests for cross-classifications: approximation of attained significance levels. *Psychometrika*, **44**, 75-83.

Aitkin, M.A. (1979). A simultaneous test procedure for contingency table models. *Appl. Statist.*, **28**, 233-242.

Aitkin, M.A., Anderson, D.A., Francis, B.J. and Hinde, J. (1989). *Statistical Modelling in GLIM.* Oxford University Press: Oxford.

Akaike, H. (1973). Information theory and an extension of the maximum likelihood principle. *In* Petrov, B. N. and F. Csaki (Eds.) *2nd. Int. Symp. Information Theory*, pp. 267-281. Akademiai Kiado: Budapest.

Altham, P.M. (1984). Improving the precision of estimation by fitting a model. *J. Roy. Statist. Soc. B*, **46**, 118-119.

Andersen, A.H. (1974). Multi-dimensional contingency tables. *Scand. J. Statist.*, **1**, 115-127.

Andersen, E.B. (1980). *Discrete statistical models with social science applications.* North Holland: Amsterdam.

Anderson, T.W. (1954). Probability models for analysing time changes in attitudes. *In* P. F. Lazarfeld (Ed.) *Mathematical Thinking in the Social Sciences.* The Free Press: Glencoe, Illinois.

Asmussen, S. and Edwards, D. (1983). Collapsibility and response variables in contingency tables. *Biometrika*, **70**, 567-578.

Atkinson, A.C. (1985). *Plots, Transformations and Regression.* Oxford University Press: Oxford.

Barndorff-Nielsen, O.E. (1978). *Information and Exponential Families in Statistical Theory.* Wiley: New York.

Barnett, V. and Lewis, T. (1978). *Outliers in Statistical Data.* Wiley: Chichester.

Bartlett, M.S. (1935). Contingency table interactions. *J. Roy. Statist. Soc. Suppl.* **2**, 248-252.

Bartlett, M.S. (1954). A note on the multiplying factors for various chi-squared approximations. *J. Roy. Statist. Soc. B*, **16**, 296-298.

Bartowiak, A. (1987). Robust and ordinary confidence intervals in linear models: a case study for some epidemiological data. Working paper.

Benedetti, J.K. and Brown, M. (1978). Strategies for the selection of log-linear models. *Biometrics*, **34**, 680-686.

Berge, C. (1973). *Graphs and Hypergraphs.* Transl. from French by E. Miniecka. North Holland: Amsterdam.

Birch, M.W. (1963). Maximum likelihood in three-way contingency tables. *J. Roy. Statist. Soc.*, **25**, 220-223.

Birch, M.W. (1964). The detection of partial association I: the 2×2 case. *J. Roy. Statist. Soc.*, **26**, 313-324.

Birch, M.W. (1965). The detection of partial association II: the general case. *J. Roy Statist. Soc. B*, **27**, 114-124.

Bishop, Y.M. (1969). Full contingency tables, logits, and split contingency tables. *Biometrics,* **25**, 19-28.

Bishop, Y.M. (1971). Effects of collapsing multi-dimensional contingency tables. *Biometrics,* **27**, 119-128.

Bishop, Y.M., Fienberg, S. and Holland, P. (1975). *Discrete Multivariate Analysis.* M.I.T. Press: Cambridge, Mass.

Blalock, H.M. (Ed.) (1971). *Causal Models in the Social Sciences.* Aldine-Atheston: Chicago.

Brown, M. (1976). Screening effects in multi-dimensional contingency tables. *Appl. Statist.*, **25**, 1, 37-46.

Cochran, W.G. (1938). The omission or addition of an independent variate in multiple linear regresssion. *J. Roy. Statist. Soc. Suppl.,* **5**, 171-176.

Coleman, J.S. (1964). *Introduction to Mathematical Sociology.* The Free Press: Glencoe, Illinois.

Cook, R.D. and Weisberg, S. (1982). *Residuals and Influence in Regression.* Chapman and Hall: London.

Cox, D.R. (1970). *The Analysis of Binary Data.* Methuen: London.

Cox, D.R. (1984). Interaction. *Int. Statist. Rev.*, **52**, 1-31.

Cox, D.R. and Hinkley, D.V. (1974). *Theoretical Statistics.* Chapman and Hall: London.

Cox, D.R. and Snell, E.J. (1974). The choice of variables in observational studies. *Appl. Statistics,* **23**, 1, 51-59.

Cox, D.R. and Snell, E.J. (1981). *Applied Statistics, Principles and Examples.* Chapman and Hall: London.

Cramer, H. (1946). *Mathematical Methods of Statistics.* Princeton Univ. Press: Princeton.

Csiszar I. (1967). Information type measures of difference of probability distributions and indirect observations. *Studai Scientiarum Mathematicarum Hungarica,* **2**, 299-318.

Csiszar I. (1975). I-divergence geometry of probability distributions and minimisation problems. *Ann. Prob.*, **3**, 1, 146-158.

Darroch, J.N. (1962). Interactions in multifactor contingency tables. *J. Roy. Statist. Soc. B*, **24**, 251-263.

Darroch, J.N. (1976). No interaction in contingency tables. In *Proc. 9th Int. Biometric, Conf.*, **1**, 296-316. Biometric Society: Raleigh, North Carolina.

Darroch, J.N., Lauritzen, S.L. and Speed, T.P. (1980). Markov fields and log linear interaction models for contingency tables. *Ann. Stat.*, **8**, 522-539.

Darroch, J.N., and Ratcliff, D. (1972). Generalised iterative scaling for log-linear models. *Ann. Math. Statist.*, **43**, 1470-1480.

Darroch, J.N., and Speed, T.P. (1983). Additive and multiplicative models and interactions. *Ann. Statist.*, **11**, 3, 724-738.

Davies, R.B. (1987). The relationship between husband's unemployment and wife's economic activity: some new evidence. Lancaster University, Social Change and Economic Life Initiative, Working Paper 25.

Dawid, A.P. (1979). Conditional independence in statistical theory (with discussion). *J. Roy. Statist. Soc. B*, **41**, 1, 1-31.

Dawid, A.P. (1979b). Some misleading arguments involving conditional independence. *J. Roy. Statist. Soc. B*, **41**, 2, 249-252.

Dawid, A.P. (1980). Conditional independence for statistical operations. *Ann. Statist.*, **8**, 598-617.

Deming, W.E. and Stephan, F.F. (1940). On a least squares adjustment of a sampled frequency table when the expected marginal totals are known. *Ann. Math. Statist.*, **11**, 427-444.

Dempster, A.P. (1969). *Elements of Continuous Multivariate Analysis.* Addison-Wesley: Reading, Mass.

Dempster, A.P. (1972). Covariance Selection. *Biometrics*, **28**, 157-175.

Doll, R. and Hill, A.B. (1950). Smoking and carcinoma of the lung. *Brit. Med Journal*, ii, 739-748.

Draper, N.R. and Smith, H. (1981). *Applied Regression Analysis* (2nd Ed.) Wiley: New York.

Dudzinski, M.L. and Arnold, G.W. (1973). Comparison of diets of sheep and cattle grazing together on sown pastures on the southern tablelands of New South Wales by principal components analysis. *Austr. J. Agric. Res.*, **24**, 899-912.

Dyke, G.V. and Patterson, H.D. (1952). Analysis of factorial arrangements when the data are proportions. *Biometrics*, **8**, 1-12.

Edwards, D.E. (1987). A guide to MIM. Research Report 87/1. Statistical Research Unit, University of Copenhagen.

Edwards, D.E. (1990). Hierarchical interaction models, (with discussion). *J. Roy. Statist. Soc. B*, **52**, 1, 3-20.

Edwards, D.E. and Havranek, T. (1985). A fast procedure for model search in multi-dimensional contingency tables. *Biometrika*, **72**, 2, 339-351.

Edwards, D.E. and Havranek, T. (1987). A fast model selection procedure for large families of models. *J. Amer. Statist. Assoc.*, **82**, 205-211.

Edwards, D.E. and Kreiner, S. (1983). The analysis of contingency tables by graphical models. *Biometrika*, **70**, 3, 553-565.

Ekholm, A. (1985). A recursion formula for the log-linear parameters of a collapsed contingency table. Technical Report, 53, University of Helsinki.

Fienberg, S.E. (1977). *The analysis of cross-classified categorical data.* M.I.T. Press: Cambridge, Mass.

Fienberg, S.E. (1979). The use of chi-squared statistics for categorical data problems. *J. Roy. Statist. Soc. B*, **41**, 54-64.

Fisher, R.A. (1925). *Statistical Methods for Research Workers.* Oliver and Boyd: Edinburgh.

Fisher, R.A. (1935). The logic of inductive inference. *J. Roy. Statist. Soc.*, **98**, 39-54.

Frank, O. and Strauss, D. (1986). Markov graphs. *J. Amer. Statist. Assoc.*, **81**, 832-842.

Frets, G.P. (1921). Heredity of head form in man. *Genetica,* **3**, 193.

Frydenberg, M. (1986). Blandede interaktions modellar, kausale modeller, kollapsibilitet of estimation. *Statistiske Interna 42.* University of Aarhus.

Frydenberg, M. (1989). The chain graph Markov property. *Research Report, 186.* Department of Theoretical Statistics, University of Aarhus.

Frydenberg, M. and Edwards, D.E. (1989). A modified iterative proportional scaling algorithm for estimation in regular exponential families. *Comput. Statist. Data Anal.* to appear.

Frydenberg, M. and Lauritzen, S.L. , S.L. (1989). Decomposition of maximum likelihood in mixed graphical interaction models. *Biometrika,* **76**, 539-555.

Gabriel, K.R. (1969). Simultaneous test procedures: some theory of multiple comparisons. *Ann. Math. Statist.*, **40**, 1, 224-250.

Glonek, G.F., Darroch, J.N., and Speed, T.P. (1988). On the existence of maximum likelihood estimators for hierarchical log-linear models. *Scand. J. Statist.*, **15**, 187-193.

Goldberger, A.G. and Duncan, O.D. (1973). *Structural Equation Models in the Social Sciences.* Seminar Press: New York.

Golumbic, M.C. (1980). *Algorithmic Graph Theory and Perfect Graphs.* Academic Press: New York.

Goodman, L.A. (1968). The analysis of cross-classified data: independence, quasi-independence and interaction in contingency tables with or without missing cells. *J. Amer. Statist. Assoc.* **63**, 1091-1131.

Goodman, L.A. (1969). On partitioning and detecting partial association in three way contingency tables. *J. Roy. Statist. Soc. B*, **31**, 3, 486-498.

Goodman, L.A. (1970). The multivariate analysis of qualitative data: interaction among multiple classifications. *J. Amer. Statist. Assoc.* **65**, 226-256.

Goodman, L.A. (1971). Partitioning of chi-square, analysis of marginal contingency tables and estimation and expected frequencies in multi-dimensional contingency tables. *J. Amer. Statist. Assoc.* **66**, 339-344.

Goodman, L.A. (1973). The analysis of multi-dimensional contingency tables when some variables are posterior to others: a modified path analysis approach. *Biometrika*, **60**, 179-192.

Goodman, L.A. and Kruskal, W.H. (1979). *Measures of Association for Cross-classifications.* Springer-Verlag: New York.

Gower, J. (1966). Some distance properties of latent root and vector methods in multivariate analysis. *Biometrika*, **53**, 315-328.

Gratzer G. (1978). *General Lattice Theory.* Birkhauser-Verlag: Basel.

Haberman, S.J. (1974). *The Analysis of Frequency Data.* Univ. Chicago Press: Chicago.

Haberman, S.J. (1986). Adjustment by minimum discrimination information. *Ann. Statist.* **14**, 1.

Hald, A. (1952). *Statistical Theory with Engineering Applications.* Wiley: New York.

Harary, F. (1969). *Graph Theory.* Addison-Wesley: Reading, Mass.

Hartley, R.V. (1928). Transmission of information. *Bell System Tech. J.*, **7**, 535-563.

Havranek, T. (1984). A procedure for model search in multi-dimensional contingency tables. *Biometrics*, **40**, 95-100.

Hawkins, D.M. and Eplett, W.J. (1982). The Cholesky factorisation of the inverse correlation matrix in multiple regression. *Technometrics*, **24**, 191-198.

Hodge, R.W. and Treiman, D.J. (1968). Social participation and social status. *Amer. Sociol. Rev.*, **33**, 723-740.

Ireland, C.T. and Kullback, S. (1968). Contingency tables with given marginals. *Biometrika*, **55**, 179-188.

Isham, V. (1981). An introduction to spatial point processes and Markov random fields. *Int. Statist. Rev.*, **49**, 21-43.

Jeffers, J.N. R. (1967). Two case studies in the application of principal component analysis. *Appl. Statist.*, **16**, 225-36.

Joreskog, K.G. (1981). Analysis of covariance structures. *Scand. J. Statist.*, **8**, 65-92.

Joreskog, K.G. and Sorbom, D. (1981). *Lisrel User's Guide V*, Inter. Educ. Services: Chicago.

Jorgensen, B. (1987). Exponential dispersion models (with discussion). *J. Roy. Statist. Soc. B*, **49**, 2, 127-163.

Kemeny J.G., Snell, J.L., Knapp, A.W. and Griffeath, D. (1976). *Denumerable Markov Chains* (2nd Ed.) Springer-Verlag: Heidelberg.

Kemperman, J.H. (1967). On the optimum rate of transmitting information. In *Prob. and Inf. Th.*, 126-169. Lecture Notes in Mathematics. Springer-Verlag: Berlin.

Kendall, M.G. and Stuart, A. (1961). *The Advanced Theory of Statistics, II.*, Griffin: London.

Kent, J.T. (1983). Information gain and a general measure of correlation. *Biometrika*, **70**, 163-173.

Kerchoff, A.C. (1974). *Ambition and Attainment.* Rose Monograph Series.

Kiiveri, H. and Speed, T.P. (1982). Structural analysis of multivariate data: a review. *In* Leinhardt, S. (Ed.) *Sociological Methodology.* Jossey Bass: San Francisco.

Kiiveri, H., Speed, T.P. and Carlin, J.B. (1984) Recursive causal models. *J. Aust. Math. Soc.*, **36**, 30-52.

Knuiman, A. (1978). Covariance selection. *Suppl. Adv. Appl. Prob.*, **10**, 123-130.

Kreiner, S. (1987). Analysis of multi-dimensional contingency tables by exact conditional tests: techniques and strategies. *Scand. J. Stat.*, **14**.

Krippendorf, K. (1986). *Information Theory: Structural Models for Qualitative Data.* Quantitative Applications in the Social Sciences. **62.** Sage Publications: London.

Kullback, S. (1959, 1968). *Information Theory and Statistics.* Wiley: New York.

Kullback, S. (1967). A lower bound for discrimination information in terms of variation. *IEEE Trans. Information Theory*, **13**, 126-127.

Kullback, S. (1968). Probability densities with given marginal. *Ann. Math. Statist.*, **39**, 1236-1243.

Kullback S. and Leibler, R.A. (1951). On information and sufficiency. *Ann. Math. Statist.*, **22**, 79-86.

Lang, S. (1970). *Linear Algebra* (2nd Ed.) Addison-Wesley: Mass.

Lauritzen, S.L. (1979, 1982). *Lectures on Contingency Tables.* (2nd Ed.) University of Aalborg Press: Aalborg.

Lauritzen, S.L. (1988). section 1.3 Mixed graphical association models. Technical Report. R 88-21. Institut fur Electroniske Systemer. Aalborg University. to be published by *Scand. J. Statist.*

Lauritzen, S.L., Dawid, A.P., Larsen, B.N. and Leimer, H.G. (1988). Independence properties of directed Markov fields. Research Report. R-88-32. Institute of Electronic Systems. Aalborg University.

Lauritzen, S.L., Speed, T.P. and Vijayan, K. (1984). Decomposable graphs and hypergraphs. *J. Austral. Math. Soc. A*, **36**, 12-29.

Lauritzen, S.L. and Spiegelhalter, D.J. (1988). Local computations with probabilities on graphical structures and their application to expert systems (with discussion). *J. Roy. Statist. Soc. B*, **50**, 2, 157-224.

Lauritzen, S.L. and Wermuth, N. (1984). Mixed interaction models. Research

Report. R-84-8. Institute of Electronic Systems. Aalborg University.

Lauritzen, S.L. and Wermuth, N. (1989). Graphical models for associations between variables, some of which are qualitative and some quantitative. *Ann. Statist.*, **17**, 31-57.

Lawley, D.N. (1956). A general method for approximating to the distribution of the likelihood ratio criteria. *Biometrika*, **43**, 295-303.

Lawley, D.N. and Maxwell, A.E. (1971). *Factor Analysis as a Statistical Method* (2nd Ed.) Butterworths: London.

Lee, S.K. (1977). On the asymptotic variances of u-terms in log-linear models of multi-dimensional contingency tables. *J. Amer. Statist. Assoc.*, **72**, 412-419.

Leimer, H. (1989). Triangulated graphs with marked vertices. *Ann. Discrete Math.*, **41**, 311-324.

Lindley, D.V. and Smith, A.F.M. (1972). Bayes estimate for the linear model. *J. Roy. Statist. Soc. B*, **34**, 1-18.

Mardia, K.V., Kent, J.T. and Bibby, J.M. (1979). *Multivariate Analysis.* Academic Press: London.

Markov, A.A. (1906). *Izvestia Phys.-Math. Society, Kazan Univ.*, **15**, 7.

McCullagh, P. and Nelder, J.A. (1983). *Generalised Linear Models.* Chapman and Hall: London.

Miller, R.G. Jnr (1981). *Simultaneous Statistical Inference* (2nd Ed.) McGraw-Hill: New York.

Morrison, D.F. (1976). *Multivariate Statistical Methods.* McGraw-Hill: New York.

Morrison, A.S., Black, M.M., Lowe, C.R., Macmahon, B. and Yuasa, S. (1973). Some international differences in histology and survival in breast cancer. *Int. J. Cancer*, **11**, 261-267.

Muirhead, R.J. (1982). *Aspects of Multivariate Statistical Theory.* Wiley: New York.

Nelder, J.A. (1977). A reformulation of linear models. *J. Roy. Statist. Soc. A*, **140**, 48-77.

Nelder, J.A. and Wedderburn, R.W.M., (1972). Generalised linear models. *J. Roy. Statist. Soc. A*, **135**, 370-384.

Newton, R.G. and Spurrell, D.J. (1967). A development of multiple regression for the analysis of routine data. *Appl. Statist.*, **16**, 51-64.

Oliver, R.N. and Smith, J.Q. (1990). *Influence Diagrams, Belief Nets and Decision Analysis.* Wiley: Chichester.

Ouellette, D.V. (1981). Schur complements and statistics. *Linear Algebra Applic.*, **36**, 187-295.

Patefield, W.M. (1981). An efficient method of generating random RxC tables with given row and column totals. *Appl. Statist.*, **30**, 91-97.

Pearl, J. (1986). Bayes and Markov networks: a comparison of two graphical representations of probabilistic knowledge. Technical Report R-46. Cognitive Systems Laboratory, UCLA.

Pearl, J. (1988). *Probabilistic Reasoning in Intelligent Systems.* Morgan and Kaufman: San Mateo.

Penn, R. (1988). Ten hypotheses about male unemployment and female activity. Lancaster University, Social Change and Economic Life Intitiative, Working Paper 31.

Pinsker, M.S. (1964). *Information and Information Stability of Random Variables and Processes.* Holden-Day: San Francisco.

Plackett, R.L. (1974). *The Analysis of Categorical Data.* Griffin: London.

Porteous, B.T. (1985a). Properties of log-linear and covariance selection models. Ph.D. thesis. Univ. of Cambridge.

Porteous, B.T. (1985b). Improved likelihood ratio statistics for covariance selection models. *Biometrika*, **72**, 97-101.

Rao, C.R. (1948). Tests of significance in multivariate analysis. *Biometrika*, **35**, 58-79.

Rao, C.R. (1973). *Linear Statistical Inference and its Applications* (2nd Ed.) Wiley: New York.

Reinis, Z. *et al.* (1981). Prognostic significance of the risk profile in the prevention of coronary heart disease. (In Czech). *Bratis. Lek. Listy.*, **76**, 2, 137-150.

Renyi, A. (1970). *Probability Theory.* North Holland: Amsterdam.

Ries, P.N. and Smith, H. (1963). The use of chi-square for preference testing in multi-dimensional problems. *Chem. Eng. Progress*, **59**, 39-43.

Rose, D.J. (1970). Triangulated graphs and the elimination process. *J. Math. Anal. Appl.*, **32**, 597-609.

Sakamoto, Y. (1982). Efficient use of Akaikes information criterion for model selection in high dimensional contingency table analysis. *Metron*, **60**, 259-275.

Sakamoto, Y. and Akaike, H. (1978). Analysis of cross classified data by AIC. *Ann. Inst. Statist. Math.*, **30**, B, 185-197.

Schur, J. (1917). Uber Potenzreihen, die im Innern des Einheitskreises beshrankt sind. *J. Reine Angew. Math.*, **147**, 205-232.

Seber, G.A. F. (1977). *Linear Regression Analysis.* Wiley: New York.

Seber, G.A. F. (1984). *Multivariate Observations.* Wiley: New York.

Sewell, W.H. and Shah, V.P. (1968). Social class, parental encouragemnt and educational aspirations. *Amer. J. Sociol.*, **73**, 559-572.

Shannon, C.E. (1948). A mathematical theory of communication. *Bell System Tech. J.*, **27**, 379-423, 623-653.

Simpson, C.H. (1951). The interpretation of interaction in contingency tables. *J. Roy. Statist. Soc. B*, **13**, 238-241.

Smith, J.Q. (1989) Influence diagrams for statistical modelling. *Ann. Statist.*,

17, 654-672.

Speed, T.P. (1978a). *Graph-theoretic methods in the analysis of interaction.* Lecture Notes, Institute of Mathematical Statistics, University of Copenhagen.

Speed, T.P. (1978b). Relations between models for spatial data, contingency tables and Markov fields on graphs. *Suppl. Adv. Appl. Prob.*, **10**, 111-122.

Speed, T.P. (1979). A note on nearest neighbour Gibbs and Markov probabilities. *Sankhya A*, **41**, 184-197.

Speed, T.P. and Kiiveri, H. (1986). Gaussian Markov distributions over finite graphs. *Ann. Statist*, **14**, 1, 138-150.

Spiegelhalter, D.J. (1986). Probabilistic reasoning in predictive expert systems. *In* Kanal, L. N. and J. Lemner (Eds.) *Uncertainty in Artificial Intelligence* pp. 47-68. North Holland: Amsterdam.

Spitzer, F. (1971). Markov random fields and Gibbs ensembles. *American Math. Monthly*, **78**, 142-154.

Spjotvoll, E. (1972). Multiple comparison of regression functions. *Amer. Math. Statist.*, **43**, 1076-1088.

Stouffer, S.A. and Toby, J. (1951). Role conflict and personality. *Amer. J. Sociol.*, **56**, 395-406.

Styan, G.P. (1985). Schur complements and linear statistical models. *In* Pukkila, T. and S. Puntanen (Eds.) *Proc. First Inter. Tampere Seminar on Linear Statistical Models.* pp. 37-75. University of Tampere, Finland.

Sundberg, R. (1975). Some results about decomposable (or Markov-type) models for multi-dimensional contingency tables: distribution of marginals and partitioning of tests. *Scand. J. Statist.*, **2**, 71-79.

Tarjan, R.E. (1985). Decomposition by clique separators. *Discrete Mathematics*, **55**, 221-232.

Tarjan, R.E. and Yannakakis, M. (1984). Simple linear time algortithms to test chordality of graphs, test acyclicity of hypergraphs and selectively reduce acyclic hypergraphs. *SIAM J. Comput.*, **13**, 566-579.

Wermuth, N. (1976a). Analogies between multiplicative models in contingency tables and covariance selection. *Biometrics*, **32**, 95-108.

Wermuth, N. (1976b). Model search among multiplicative models. *Biometrics*, **32**, 253-264.

Wermuth, N. (1980). Linear recursive equations, covariance selection and path analysis. *J. Amer. Stat. Ass.*, **75**, 963-972.

Wermuth, N. (1986a). Collapsibility of different measures of association in contingency tables with a dichotomous response variable. Technical report, University of Mainz. 86-1.

Wermuth, N. (1986b). Implications of the Yule-Simpson paradox for simple multicentric studies. Technical report, University of Mainz. 86-1.

Wermuth, N. (1988). Introduction to the use of graphical chain models. Unpublished tutorial notes to Compstat 88.

Wermuth, N. and Lauritzen, S.L. (1983). Graphical and recursive models for contingency tables. *Biometrika*, **70**, 3, 537-552.

Wermuth, N. and Lauritzen, S.L. (1990). On substantive research hypotheses, conditional independence graphs and graphical chain models. *J. Roy. Statist. Soc. B*, **52**, 1, 21-50.

Whittaker, J. (1982). Glim syntax and simultaneous tests for graphical log-linear models. *In* R. Gilchrist (Ed.) *GLIM 82. Lecture notes in Statistics*, **14**, 98-108. Springer Verlag.

Whittaker, J. (1984a). Model interpretation from the additive elements of the likelihood function. *Appl. Statist.*, **33**, 1, 52-64.

Whittaker, J. (1984b). Fitting all possible decomposable and graphical models to multiway contingency tables. *In* Havranek, T. *et al.* (Eds.) *Compstat*, Physica-Verlag: Vienna. 401-406.

Whittaker, J. (1986). A mixed derivative measure of interaction. Technical Report, Department of Mathematics, University of Lancaster.

Whittaker, J. and Aitkin, M. (1978). A flexible strategy for fitting complex log-linear models. *Biometrics*, **34**, 487-495.

Whittaker, J., Iliakopoulis., T. and Smith, P. (1988). Graphical modelling with large numbers of variables: a comparison with principal components. *In* Edwards, D.G. and N. E. Raun (Eds.) *Compstat 88.* Physica Verlag: Heidelberg. 73-80.

Whittemore, A.S. (1978). Collapsibility of multi-dimensional contingency tables. *J. Roy. Statist. Soc. B*, **40**, 328-340.

Wilkinson, G.N. and Rogers, C.E. (1973). Symbolic description of factorial models for the analysis of variance. *Appl. Statist.*, **22**, 392-399.

Wilks, S.S. (1938). The large sample distribution of the likelihood ratio for testing composite hypotheses. *Ann. Math. Statist.*, **9**, 60-62.

Williams, D.A. (1976). Improved likelihood ratio statistics for complete contingency tables. *Biometrika*, **63**, 33-37.

Wright, S. (1921). Correlation and causation. *J. Agric. Res.*, **20**, 557-585.

Wright, S. (1923). The theory of path coefficients: a reply to Niles' criticism. *Genetics,* **8**, 239-255.

Wright, S. (1934). The method of path coefficients. *Ann. Math. Statist.*, **5**, 161-215.

Wright, S. (1954). The interpretation of multivariate systems. *In* Kempthorne, O. *et al.* (Eds.) *Statistics and Mathematics in Biology.* pp. 11-23. Iowa State University Press: Ames.

Author Index

Subject Index

Applied Probability and Statistics (Continued)

GREENBERG and WEBSTER • Advanced Econometrics: A Bridge to the Literature
GROSS and HARRIS • Fundamentals of Queueing Theory, *Second Edition*
GROVES • Survey Errors and Survey Costs
GROVES, BIEMER, LYBERG, MASSEY, NICHOLLS, and WAKSBERG • Telephone Survey Methodology
GUPTA and PANCHAPAKESAN • Multiple Decision Procedures: Theory and Methodology of Selecting and Ranking Populations
GUTTMAN, WILKS, and HUNTER • Introductory Engineering Statistics, *Third Edition*
HAHN and SHAPIRO • Statistical Models in Engineering
HALD • Statistical Tables and Formulas
HALD • Statistical Theory with Engineering Applications
HAND • Discrimination and Classification
HEIBERGER • Computation for the Analysis of Designed Experiments
HOAGLIN, MOSTELLER and TUKEY • Exploring Data Tables, Trends and Shapes
HOAGLIN, MOSTELLER, and TUKEY • Understanding Robust and Exploratory Data Analysis
HOCHBERG and TAMHANE • Multiple Comparison Procedures
HOEL • Elementary Statistics, *Fourth Edition*
HOEL and JESSEN • Basic Statistics for Business and Economics, *Third Edition*
HOGG and KLUGMAN • Loss Distributions
HOLLANDER and WOLFE • Nonparametric Statistical Methods
HOSMER and LEMESHOW • Applied Logistic Regression
IMAN and CONOVER • Modern Business Statistics
JESSEN • Statistical Survey Techniques
JOHNSON • Multivariate Statistical Simulation
JOHNSON and KOTZ • Distributions in Statistics
Discrete Distributions
Continuous Univariate Distributions—1
Continuous Univariate Distributions—2
Continuous Multivariate Distributions
JUDGE, GRIFFITHS, HILL, LÜTKEPOHL and LEE • The Theory and Practice of Econometrics, *Second Edition*
JUDGE, HILL, GRIFFITHS, LÜTKEPOHL and LEE • Introduction to the Theory and Practice of Econometrics, *Second Edition*
KALBFLEISCH and PRENTICE • The Statistical Analysis of Failure Time Data
KASPRZYK, DUNCAN, KALTON, and SINGH • Panel Surveys
KAUFMAN and ROUSSEEUW • Finding Groups in Data: An Introduction to Cluster Analysis
KEENEY and RAIFFA • Decisions with Multiple Objectives
KISH • Statistical Design for Research
KISH • Survey Sampling
KUH, NEESE, and HOLLINGER • Structural Sensitivity in Econometric Models
LAWLESS • Statistical Models and Methods for Lifetime Data
LEAMER • Specification Searches: Ad Hoc Inference with Nonexperimental Data
LEBART, MORINEAU, and WARWICK • Multivariate Descriptive Statistical Analysis: Correspondence Analysis and Related Techniques for Large Matrices
LINHART and ZUCCHINI • Model Selection
LITTLE and RUBIN • Statistical Analysis with Missing Data
McNEIL • Interactive Data Analysis
MAGNUS and NEUDECKER • Matrix Differential Calculus with Applications in Statistics and Econometrics
MAINDONALD • Statistical Computation

Applied Probability and Statistics (Continued)

MALLOWS • Design, Data, and Analysis by Some Friends of Cuthbert Daniel
MANN, SCHAFER and SINGPURWALLA • Methods for Statistical Analysis of Reliability and Life Data
MARTZ and WALLER • Bayesian Reliability Analysis
MASON, GUNST, and HESS • Statistical Design and Analysis of Experiments with Applications to Engineering and Science
MIKÉ and STANLEY • Statistics in Medical Research: Methods and Issues with Applications in Cancer Research
MILLER • Beyond ANOVA, Basics of Applied Statistics
MILLER • Survival Analysis
MILLER, EFRON, BROWN, and MOSES • Biostatistics Casebook
MONTGOMERY and PECK • Introduction to Linear Regression Analysis
NELSON • Applied Life Data Analysis
NELSON • Accelerated Testing: Statistical Models, Test Plans, and Data Analyses
OCHI • Applied Probability and Stochastic Processes in Engineering and Physical Sciences
OSBORNE • Finite Algorithms in Optimization and Data Analysis
OTNES and ENOCHSON • Applied Time Series Analysis: Volume I, Basic Techniques
OTNES and ENOCHSON • Digital Time Series Analysis
PANKRATZ • Forecasting with Univariate Box-Jenkins Models: Concepts and Cases
PLATEK, RAO, SARNDAL and SINGH • Small Area Statistics: An International Symposium
POLLOCK • The Algebra of Econometrics
RAO and MITRA • Generalized Inverse of Matrices and Its Applications
RÉNYI • A Diary on Information Theory
RIPLEY • Spatial Statistics
RIPLEY • Stochastic Simulation
ROSS • Introduction to Probability and Statistics for Engineers and Scientists
ROUSSEEUW and LEROY • Robust Regression and Outlier Detection
RUBIN • Multiple Imputation for Nonresponse in Surveys
RUBINSTEIN • Monte Carlo Optimization, Simulation, and Sensitivity of Queueing Networks
RYAN • Statistical Methods for Quality Improvement
SCHUSS • Theory and Applications of Stochastic Differential Equations
SEARLE • Linear Models
SEARLE • Linear Models for Unbalanced Data
SEARLE • Matrix Algebra Useful for Statistics
SKINNER, HOLT, and SMITH • Analysis of Complex Surveys
SPRINGER • The Algebra of Random Variables
STAUDTE and SHEATHER • Robust Estimation and Testing
STEUER • Multiple Criteria Optimization
STOYAN • Comparison Methods for Queues and Other Stochastic Models
STOYAN, KENDALL and MECKE • Stochastic Geometry and Its Applications
THOMPSON • Empirical Model Building
TIJMS • Stochastic Modeling and Analysis: A Computational Approach
TITTERINGTON, SMITH, and MAKOV • Statistical Analysis of Finite Mixture Distributions
UPTON and FINGLETON • Spatial Data Analysis by Example, Volume I: Point Pattern and Quantitative Data
UPTON and FINGLETON • Spatial Data Analysis by Example, Volume II: Categorical and Directional Data
VAN RIJCKEVORSEL and DE LEEUW • Component and Correspondence Analysis
WEISBERG • Applied Linear Regression, *Second Edition*